COMPUTER AIDED DESIGN AND MANUFACTURING

Computer
Aided
Design and Manufacturing

K. LALIT NARAYAN
Associate Professor
Department of Mechanical Engineering
Sir C.R.R. College of Engineering
Eluru

K. MALLIKARJUNA RAO, Ph.D.
Head
Department of Mechanical Engineering
JNTU College of Engineering
Kakinada

M.M.M. SARCAR, Ph.D.
Professor
Department of Mechanical Engineering
A.U. College of Engineering
Andhra University
Visakhapatnam

PHI Learning Private Limited
Delhi-110092
2015

₹ 475.00

COMPUTER AIDED DESIGN AND MANUFACTURING
K. Lalit Narayan, K. Mallikarjuna Rao and M.M.M. Sarcar

ISBN-978-81-203-3342-0

The export rights of this book are vested solely with the publisher.

Third Printing ... ... **August, 2015**

Published by Asoke K. Ghosh, PHI Learning Private Limited, Rimjhim House, 111, Patparganj Industrial Estate, Delhi-110092 and Printed by Mohan Makhijani at Rekha Printers Private Limited, New Delhi-110020.

CONTENTS

Part 1   CAD—Fundamentals of Design, Computers and Controllers

Part II   CAD—Hardware and Software Components

6. PRINCIPLES OF INTERACTIVE COMPUTER GRAPHICS   117–140

7. TRANSFORMATION SYSTEMS   141–156

10. SOLID MODELLING

Part III   CAD—Design Aspects of Industrial Components

11. FINITE ELEMENT MODELLING AND ANALYSIS

Part IV   CAM—Numerical Control Production Systems

Part V   Information Systems in Manufacturing

Part VI   Quality Control and Automated Inspection

17. COMPUTER-AIDED INSPECTION AND QUALITY CONTROL   387–400

18. MACHINE VISION   401–412

Part VII Integration of Manufacturing Systems

19. COMPUTER INTEGRATED PRODUCTION PLANNING SYSTEMS   415–443

Part VIII Intelligent CAD and Manufacturing Systems

FOREWORD

Computer Aided Design and Computer Aided Manufacturing (CAD/CAM) emerged as a single field in 1950s. Today it has evolved into several areas like Geometric modelling, Machine vision, and Flexible manufacturing systems. The central focus of this book is to compile the available literature in the subject in a reader-friendly format.

CAD/CAM is a part of the core courses of Mechanical Engineering in various universities. The textbooks currently available in this field do not conform to the syllabus of any particular university. This book will make an excellent text/reference for all undergraduate and postgraduate courses of various universities, especially JNTU.

This book, authored by Prof. K. Mallikarjuna Rao of JNTU College of Engineering, Kakinada, Prof. M.M.M. Sarcar of Andhra University, Visakhapatnam and Mr. K. Lalit Narayan, Associate Professor of C.R. Reddy College of Engineering, Eluru provides a comprehensive coverage of the subject. It discusses all the state-of-the-art concepts at length. It caters to the large fraternity of CAD/CAM learners, primarily teachers, students, researchers besides industrial practitioners.

The authors with their decades of teaching experience have carefully designed the contents and translated the essentials into their work.

I congratulate the authors and hope the readers will reap the best of benefits out of the book.

Dr. K. RAJAGOPAL
Vice-Chancellor
Jawaharlal Nehru Technological University
Hyderabad

PREFACE

Computer aided design and manufacturing form the core of the engineering subjects. The engineering curriculum and the engineering academic process attempt to provide students and engineers, with a sufficient number of tools to perform, among other things, design and manufacturing. Mathematics, computers and computational techniques, communication methods and drafting skills are the essential tools a designer needs. In the past fifteen years the interactive computer graphics and CAD/CAM technology have been impacting the drafting, design and manufacturing tools significantly.

The purpose of writing book on CAD/CAM is to present CAD/CAM principles and tools in generic and basic forms with enough depth. These principles are supplemented with engineering and design applications. The presentation of these principles and tools maintains a balance between both theory and practice. The book is concerned with developing the proper attitude and approach to utilize the existing CAD/CAM technology and software packages in engineering. It will attempt to expand the reader's domain of imagination beyond just creating interactive graphics.

The book is targeted for the need of students, engineers and professionals who are interested in the CAD/CAM technology and its application to design. A coherent realization of the current CAD/CAM tools and their relationships to one another form an essential core to the learning process. Thus, learning the basics of existing tools enhances both the utilization of current systems and the development of new design and manufacturing applications. It provides a description of both the hardware and software of CAD/CAM systems.

The technology of Computer Aided Design and Drafting (CADD) has immense potential for application in several industries. The impact of this technology in automobile engineering, marine engineering and aerospace engineering has been tremendous. In these industries, it calls for training of design engineers in the various aspects of CADD. Manufacturing and the subject of using computers in manufacturing are receiving particular prominence as industries seek to improve product quality, increase productivity and flexibility and to reduce inventory costs. Therefore, emphasis has been attributed to the subject of CAD and its integration with CAM.

ORGANIZATION OF THE BOOK

Part I   Fundamentals of Design, Computers and Controllers

Chapter 1 provides an in-depth discussion of the CAD system definition, reasons for implementing CAD, design process, application of computers in design, benefits of computer aided design and creating the manufacturing database. Chapter 2 is devoted entirely to computer systems. Students will learn the basic hardware components of a digital computer, mass storage devices, input/output devices, and system configuration. Microcomputers, minicomputers, main frame computers, supercomputers and programmable logic controllers are covered in Chapter 3.

Part II   Hardware and Software Components

Chapter 4 presents the basic theoretical concepts of CAD system hardware. It also includes the design workstation, graphics terminal, graphics input devices, graphic display devices, graphics output devices, modes of operation and CAD system configuration. Chapter 5 discusses CAD system software. It covers graphics software functions of a graphics standards for graphics programming, product data based format, standard for the exchange of product model data drawing exchange format, dimensional measurement interface specification, parasolid and ACIS.

Principles of interactive computer graphics are explored in Chapter 6.

Part III   Design Aspects of Industrial Components

Chapter 7 includes transformation principles, two-dimensional geometric transformations, three-dimensional geometric transformations, linear transformations, display, windowing and clipping, display files for 3D data and visualization of 3D data. Wire frame modelling, surface modelling and solid modelling are described in Chapters 8, 9 and 10.

Chapter 11 provides introduction to FE analysis and in-depth discussion about general procedure for finite element analysis, mesh generation techniques, automatic mesh generation, and CAD application to finite element modelling.

Part IV   Numerical Control Production Systems

Chapter 12 provides introduction to what is numerical control. It also discusses about the NC procedure, NC coordinate systems, elements of NC system, classification of NC systems, absolute programming, incremental programming, open-loop and closed-loop system, advantages of NC systems and disadvantages of NC systems.

Computer numerical control, components of CNC, functions of CNC, features of CNC, advantages of CNC, industrial applications of CNC, DNC system, types of DNC, advantages of DNC are covered in Chapter 13.

Chapter 14 provides introduction to NC part programming.

Part V Information Systems in Manufacturing

Chapter 15 provides introduction to group technology and discusses about part families, parts classification and coding systems, coding method, codes and coding system structure.

Chapter 16 provides introduction to what is process planning and discusses about computer aided process planning, retrieval type CAPP system, generative CAPP system, hybrid CAPP system, process planning systems, machinability data systems, benefits of CAPP and computer programming languages for CAPP.

Part VI Quality Control and Automated Inspection

Chapter 17 introduces computer-aided inspection and quality control and discusses about quality assurance and quality control, inspection and testing.

Chapter 18 speaks about machine vision and discusses about imaging devices, vidicon tube camera, analog to digital conversion, image processing and analysis and application of machine vision.

Part VII Integration of Manufacturing Systems

Chapter 19 reviews computer integrated production planning systems and discusses about production planning and control, master production schedule, material requirement planning, capacity requirement planning, inventory control, computerized inventory management system, manufacturing resource planning (MRP II), Just in time, shop floor control, functions of shop floor control and computer process monitoring.

Chapter 20 provides introduction to robot and discusses about structure and operation of robot, robot anatomy, robot specifications, types of robots, robot technology levels, common robot configurations, classification of robots, sensory devices, programming robots, programming methods, robot programming functions, basic types of robot programming languages, robot applications, industrial applications, non-industrial applications, definition of A.G.V, types of A.G.V's, applications and function of A.G.V's.

Chapter 21 and 22 cover flexible manufacturing systems and computer integrated manufacturing.

Part VIII Intelligent CAD and Manufacturing Systems

Chapter 23 provides introduction to artificial intelligence and discusses about expert systems, natural language processing, machine vision, artificial neural network, fuzzy logic, artificial intelligence in CAD, applications of AI in design, expert systems, structure of an expert system, development of an expert system, expert system shells, knowledge representation, inference engine, role of AI in manufacturing, characteristics of an expert system, languages for expert systems and benefits of expert system.

Chapter 24 provides an indepth discussion about communication networks in manufacturing, hierarchy of computers in manufacturing, communications standards, manufacturing communication systems, internet, intranet, hardware elements of a network, networking in a manufacturing company, issues in inter-system communication, fundamentals of computer communications and wire communication path.

K. Lalit Narayan

Dr. K. Mallikarjuna Rao

Dr. M.M.M. Sarcar

ACKNOWLEDGEMENTS

We M.M.M. Sarcar and K. Mallikarjuna Rao owe special debt of thanks to K. Lalit Narayan, one of the co-authors for sharing the major labour of compiling the concepts of the subject into the present form.

We must express our continued gratitude to our wives and families without whose patience and forbearance over many months the work would never have been completed. This book is dedicated to them.

We are grateful, in particular, to the management of Sir C.R.R. College of Engineering especially Sri Kommareddi Ram Babu, President Sir C.R.R. Institutions and Sri K. Rajendra Vara Prasada Rao, Secretary, Sir C.R.R. Institutions for providing a conducive work atmosphere.

We are indebted to Prof. Alam Appa Rao, Principal, A.U. College of Engineering, Dr. E.V. Prasad, Principal, J.N.T.U. College of Engineering, Kakinada and Dr. K. Mohan Rao, Principal, and Dr. Y.V.S.S.S.V. Prasada Rao, Director, Sir C.R.R. College of Engineering, Eluru for their support throughout the work. We also thank our colleagues of our respective departments who were involved directly or indirectly for helping in their own individual ways to shape the book into final form. We are thankful to G. Kumar and P. Siva Kumar for the error-free typing of the manuscript. We are grateful to the respective Librarians of our institute for giving continuous encouragement and moral support during the work. We wish to express our appreciation to the various members of Prentice-Hall of India who handled the book at various stages, particularly to Mr. K.C. Devasia who gave encouragement and cooperation for writing this book.

K. Lalit Narayan
Dr. K. Mallikarjuna Rao
Dr. M.M.M. Sarcar

Part I

CAD

Fundamentals of Design, Computers and Controllers

CAD

Chapter

FUNDAMENTALS OF CAD

1

1.1 INTRODUCTION

Computer aided design (CAD) can be defined as the use of computer systems to assist in the creation, modification, analysis or optimization of a design.

The computer systems consist of the hardware and software to perform the specialized design functions required by the particular user firm. The CAD hardware typically includes the computer, one or more graphic display terminals, keyboards and other peripheral equipment. The CAD software consists of the computer programs to implement computer graphics on the system plus application programs to facilitate the engineering functions of the user company. Examples of these application programs include stress-strain analysis of components, dynamic response of mechanisms, heat transfer calculations and numerical control part programming. The collection of application programs will vary from one user firm to the other because their product lines, manufacturing processes and customer markets are different. These factors give rise to differences in CAD system requirements.

1.2 THE CAD SYSTEM—DEFINITION

Computer aided design involves any type of design activity, which makes use of the computer to develop, analyze or modify an engineering design. Modern CAD systems are based on interactive computer graphics (ICG). Interactive computer graphics denotes a user-oriented system in which the computer is employed to create, transform and display data in the form of pictures or symbols. The user in the computer graphics design system is the designer who communicates data and commands to the computer through any of several input devices. The computer communicates with the user via a cathode ray tube (CRT). The designer creates an image on the CRT screen by entering commands to call the desired software subroutines stored in the computer. In most systems, the image is constructed out of basic geometric elements points, lines, circles and so on. It can be modified according to the commands of the designer, enlarged, reduced in size, moved to another location on the screen, rotated and other transformations. Through these various manipulations the required details of the image are formulated.

The typical ICG system is a combination of hardware and software. The hardware includes a central processing unit, one or more workstations (including the graphics display terminals) and peripheral devices such as printers, plotters and drafting equipment. The software consists of the computer programs needed to implement graphics processing on the system. The software would also typically include additional specialized application programs to accomplish the particular engineering functions required by the user company.

The ICG system is one component of a computer aided design system. The other major component is the human designer. Interactive computer graphics is a tool used by the designer to solve a design problem. In effect, the ICG system magnifies the powers of the designer. This has been referred to as the synergistic effect. The designer performs the portion of the design process that is most suitable to human intellectual skills, the computer performs the task best suited to its capabilities and the resulting system exceeds the sum of its components.

1.3 REASONS FOR IMPLEMENTING CAD

1. *To increase the productivity of the designer:* CAD helps the designer to visualize the product and its component sub-assemblies and parts. This reduces the time required to synthesize, analyze and document the design. This productivity improvement results not only into lower design cost but also into shorter design project completion times.

2. *To improve the quality of design:* A CAD system permits a thorough engineering analysis within a short time using various software and a larger number of design alternatives can be investigated. Design errors are also reduced by the accuracy built into the system by means of calculations and checks available with the system. These factors lead to improvement in the quality and accuracy in the design.

3. *To improve communications through documentation:* The use of CAD system provides better engineering drawings, more standardization in the drawings, better documentation of the design, fewer drawing errors and greater legibility for the drawing.

4. *To create a database for manufacturing:* In the process of creating the documentation for the product design (geometry and dimension of components, bill of materials, etc.) much of the required database to manufacture is also created which can be applied for several computer integrated manufacturing (CIM) applications like CNC programming, programming of robots, process planning and so on.

1.4 DESIGN PROCESS

Design is the act of devising an original solution to a problem by a combination of principles, resources and products in design. Design process is the pattern of activities that is followed by the designer in arriving at the solution of a technological problem.

The design process is an iterative procedure as shown in Fig. 1.1. A preliminary design is made based on the available information and is improved upon as more and more information is generated. There have been several attempts to provide a formal description of the stages or elements of the design process. The design progresses in a step-by-step manner from some

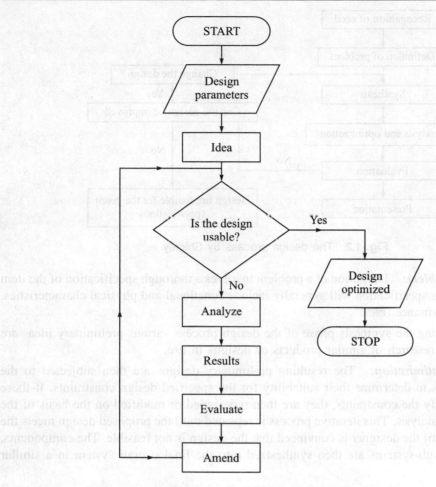

Fig. 1.1 Iterative design procedure.

statement of need through identification of the problem, a search for solutions and development of the chosen solution to trial production and use. These descriptions of design are known as **models of the design process**.

There are four models of the design process, which are iterative in nature. These models are defined by Shigley, Pahl and Beitz, Ohsuga and Earle.

1.4.1 Shigley Model

The Shigley model is shown in Fig. 1.2. It involves six basic steps which are explained as follows:

1. *Recognition of need:* Recognition of need involves the realization by someone that a problem exists for which some feasible solution is to be found. This might be the identification of some defect in a current machine design activity by an engineer or the perception of a new product marketing opportunity by a salesman.

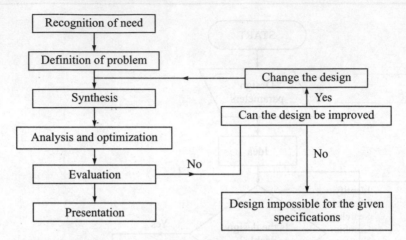

Fig. 1.2 The design process by Shigley.

2. *Definition of problem:* Definition of a problem involves a thorough specification of the item to be designed. This specification will generally include functional and physical characteristics, cost, quality, performance, etc.

3. *Synthesis:* During the synthesis phase of the design process various preliminary ideas are developed through research of similar products or designs in use.

4. *Analysis and optimization:* The resulting preliminary designs are then subjected to the appropriate analysis to determine their suitability for the specified design constraints. If these designs fail to satisfy the constraints, they are then redesigned or modified on the basis of the feedback from the analysis. This iterative process is repeated until the proposed design meets the specifications or until the designer is convinced that the design is not feasible. The components, sub-assemblies or sub-systems are then synthesized into the final overall system in a similar iterative manner.

5. *Evaluation:* The assessment or evaluation of the design against the specifications established during the problem definition phase is then carried out. This often requires the fabrication and testing of a prototype model to evaluate operating performance quality, reliability, etc.

6. *Presentation:* The final phase in the design process is the presentation of the design. This includes documentation of the design through drawings, material specifications, assembly lists and so on.

1.4.2 Pahl and Beitz Model

In this model, the design process is described by a flow diagram, as shown in Fig. 1.3 comprising four main phases, which may be summarized as follows:

1. *Clarification of the task:* This phase involves the collection of information about the requirements to be embodied in the solution, and also about the constraints on the design and describing these in a specification.

2. *Conceptual design:* The conceptual design phase involves the establishment of the functions to be included in the design, and identification and development of suitable solutions.

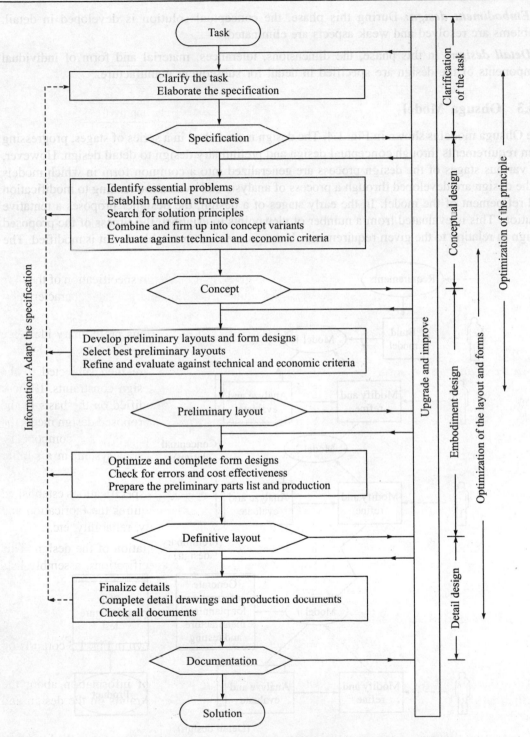

Fig. 1.3 Steps of the design process according to Pahl and Beitz.

3. *Embodiment design:* During this phase, the conceptual solution is developed in detail. Problems are resolved and weak aspects are eliminated.

4. *Detail design:* In this phase, the dimensions, tolerances, material and form of individual components of the design are specified in detail for subsequent manufacture.

1.4.3 Ohsuga Model

The Ohsuga model is shown in Fig. 1.4. The design is described in a series of stages, progressing from requirements through conceptual design and preliminary design to detail design. However, the various stages of the design process are generalized into a common form in which models of the design are developed through a process of analysis and evaluation leading to modification and refinement of the model. In the early stages of a design, the designer proposes a tentative solution. This is evaluated from a number of viewpoints to establish the fitness of the proposed design in relation to the given requirements. If the proposal is unsuitable then it is modified. The

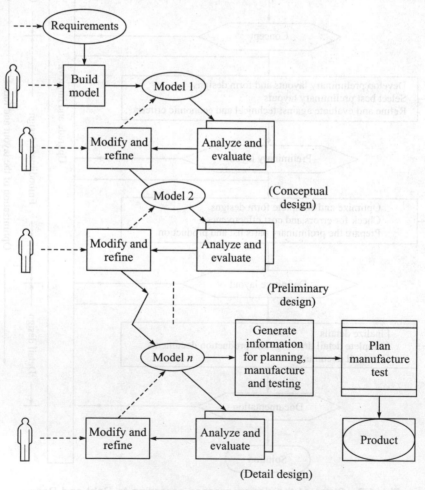

Fig. 1.4 The design process according to Ohsuga.

process is repeated until the design is at a point where it can be developed in more depth and the preliminary design stage will start. In this stage, the design is refined, and the evaluation and modification are repeated at a greater level of detail.

1.4.4 Earle Model

The steps in the design process, proposed by Earle (Fig. 1.5), are as follows:

1. Problem identification
2. Preliminary ideas
3. Design refinement
4. Analysis
5. Decision
6. Implementation

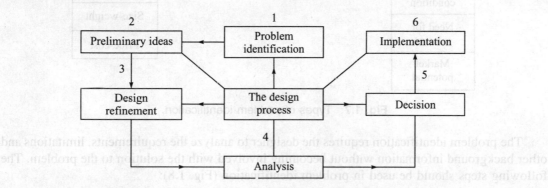

Fig. 1.5 The design process by Earle.

1. *Problem identification:* Problem identification need to gather data of several types: Fixed data, opinion surveys, historical records, personal observations, experimental data and physical measurements and characteristics as shown in Fig. 1.6. Problem identification can be of two general types: (i) Identification of a need or (ii) Identification of design criteria.

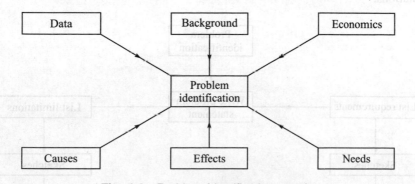

Fig. 1.6 Problem identification needs.

Identification of a need is the beginning point of the process. It may be a defect or a shortcoming in an existing product or a new product.

Identification of design criteria is that part of the problem where the designer conducts an in-depth investigation of the specifications that must be met by a new design. Types of problem identification are shown in Fig. 1.7.

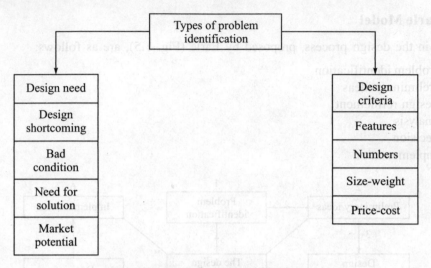

Fig. 1.7   Types of problem identification.

The problem identification requires the designer to analyze the requirements, limitations and other background information without becoming involved with the solution to the problem. The following steps should be used in problem identification (Fig. 1.8):

(a) **Problem statement**   Write down the problem statement to begin the thinking process. The statement should be complete and comprehensive, but concise.

(b) **Problem requirements**   List the positive requirements that must be achieved in the design.

(c) **Problem limitations**   List negative factors that confine the problem specified limitations.

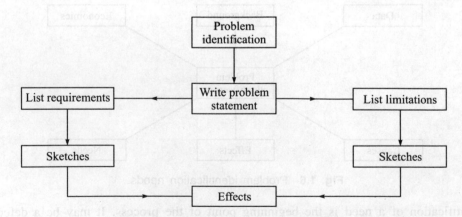

Fig. 1.8   Steps in problem identification.

(d) *Sketches* Make sketches of physical characteristics of the problem. Add notes and dimensions that would make these sketches more understandable.

(e) *Gather data* The data might be population trends of related designs, physical characteristics, sales records and market studies. This data should be graphed for easy interpretation.

2. *Preliminary ideas:* Preliminary ideas is the generation of as many ideas for solution as possible (Fig. 1.9). These ideas should be sufficiently broad to allow for unique solutions that could revolutionize present methods. All ideas should be recorded in written form with sketches. A systematic approach should be used to gather preliminary ideas for the design problem.

The following sequence of steps is suggested

(a) Hold brainstorming session
(b) Prepare sketches and notes
(c) Research existing design
(d) Conduct surveys

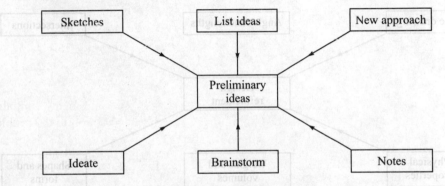

Fig. 1.9 Generation of ideas.

Brainstorm is defined as the practice of a conference technique by which a group attempts to find a solution for a specific problem by amassing all the ideas spontaneously contributed by its members. The rules for brainstorming are:

(i) No criticism
(ii) Think wild
(iii) Quantify ideas
(iv) Seek combination and improvement

The organization steps of brainstorming session are:

(i) Select a panel of about 12 members with and without knowledge of the subject.
(ii) Prepare a one page note pertaining to the session and should be distributed to the members about two days in advance.
(iii) The problem should be concisely defined.
(iv) A moderator and recorder should be appointed for the session.
(v) Hold the session by initiating the problem.
(vi) The recorder should reproduce the list of ideas gathered during the session for distribution to the participants.

Sketching is the designers most important medium for developing preliminary ideas. Computer graphics can be used for modifying and developing a number of ideas for consideration. Preliminary ideas can be obtained through research of similar products and designs from technical magazines, general magazines, manufacturers brochures, patents and consultants.

Survey methods are used to gather opinions and reactions to a preliminary design or completed design, especially when a product is being designed for the general market. This could be accomplished by personal interview, telephone interview and mail questionnaire.

3. *Design refinement:* Several of the better preliminary ideas are selected for further refinement to determine their true merits. Rough sketches are converted to scale drawings that will permit space analysis, critical measurements and the calculation of ideas and volumes affecting the design. Consideration is given to spatial relationships, angles between planes, lengths of structural members, intersection of surfaces and planes. The sequence for refinement of ideas is shown in Fig. 1.10. The determination of physical properties of the proposed solutions is the important concern of the designer. Descriptive geometry can be applied for this purpose. Computer graphics is a powerful tool that can be used to refine the preliminary idea.

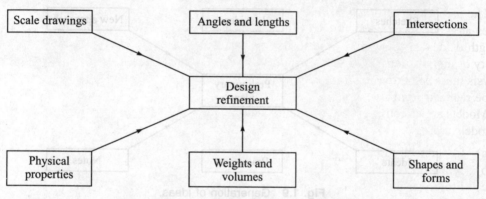

Fig. 1.10 Refinement of ideas.

4. *Analysis:* It involves the evaluation of the best designs to determine the comparative merits of each with respect to cost, strength, function and market-appeal. The analysis phase of design is shown in Fig. 1.11. The general areas of analysis are:

    (a) Functional analysis
    (b) Human engineering
    (c) Market and product analysis
    (d) Specification analysis
    (e) Strength analysis
    (f) Economic analysis
    (g) Model analysis

Engineering graphics and descriptive geometry are valuable tools for analysis. The analysis of a design's functional characteristics can be performed using mathematics, graphics and engineering disciplines. A product must be analyzed to determine its acceptance by the market before it is released for production. Areas of product analysis are:

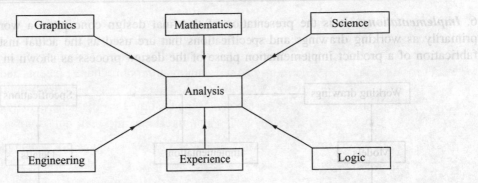

Fig. 1.11 Analysis phase of design.

(a) Potential market evaluation
(b) Market outlets
(c) Advertising methods
(d) Sales features

The physical specification of a product must be analyzed to finalize the design, e.g. sizes, ranges and shipping specifications. Much of the engineering is devoted to the analysis of the strength of design to support dead loads, withstand shocks and to endure repetitive usages of variety of motions ranging from slow to fast. Before a product is released for production, a cost analysis must be performed to determine the items production cost and the margin of profit that can be realized from it.

Models are effective aids in analyzing a design in final stages of its development. The types of models are:

(a) Conceptual models
(b) Mock-up models
(c) Prototype models
(d) System layout models

5. *Decision:* At this stage, a single design is accepted as the solution of the design problem. Graphics is a primary means of presenting the proposed design for a decision. The graphs must compare costs of manufacturing, weights, operational characteristics and other data that would be considered in arriving at the final decision. Figure 1.12 shows the decision phase of the design process.

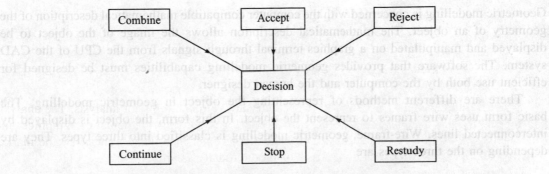

Fig. 1.12 Decision phase of design process.

6. *Implementation:*   It is the presentation of the final design concept in a workable form, primarily as working drawings and specifications that are used as the actual instruments for fabrication of a product implementation phase of the design process as shown in Fig. 1.13.

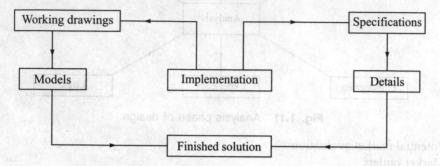

Fig. 1.13 Implementation phase of design process.

1.5 APPLICATION OF COMPUTERS IN DESIGN

Engineering design has traditionally been accomplished on drawing boards with the design being documented in the form of a detailed engineering drawing. This process is iterative in nature and is time consuming. The computer can beneficially be used in the design process in CAD. The design task is performed by a CAD system rather than a single designer working over a drawing board. The various design related tasks, which are performed by the CAD system, can be grouped into four functional areas:

1. Geometric modelling
2. Engineering analysis
3. Design review and evaluation
4. Automated drafting

These functional areas are discussed in the following sections. The CAD modified design process is shown in Fig. 1.14.

1.5.1 Geometric Modelling

Geometric modelling is concerned with the computer compatible mathematical description of the geometry of an object. The mathematical description allows the image of the object to be displayed and manipulated on a graphics terminal through signals from the CPU of the CAD system. The software that provides geometric modelling capabilities must be designed for efficient use both by the computer and the human designer.

There are different methods of representing the object in geometric modelling. The basic form uses wire frames to represent the object. In this form, the object is displayed by interconnected lines. Wire-frame, geometric modelling is classified into three types. They are depending on the three types are

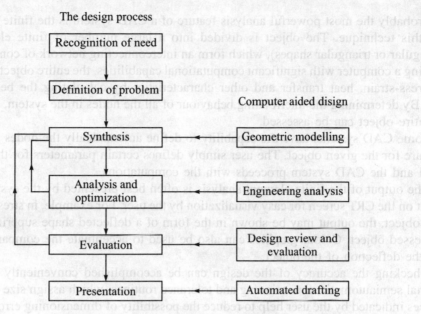

Fig. 1.14 CAD modified design process.

1. 2D   Two-dimensional representation is used for a flat object.
2. 2½D   This goes somewhat beyond the 2D capability by permitting a three-dimensional object to be represented as long as it has no side–wall details.
3. 3D   This allows for full three-dimensional modelling of a more complex geometry.

The most advanced method of geometric modelling is solid modelling in three dimensions. This method typically uses solid geometry shapes called **primitives** to construct the object as a matter of fact with the use of solid modelling software. A design can be directly conceived at the graphic terminal itself using feature based approach.

1.5.2   Engineering Analysis

In the formulation of nearly any engineering design project, some type of analysis is required. The analysis may involve stress-strain calculations, heat transfer calculations or computation of natural frequencies, acceleration, velocity and displacement response or other parameters. Analysis can be carried out using standard CAD software like finite element modelling and analysis or by software developed internally in the design offices. Most CAD software suits incorporate engineering analysis software to carry out finite element analysis and to determine mass properties of the component.

The analysis of mass properties is the analysis feature of a CAD system that has probably the widest application. It provides properties of a solid object being analyzed such as the surface area, weight, volume, centre of gravity and moment of inertia for a plane surface (or a cross-sectional of a solid object). The corresponding computations include the parameter area and inertia properties.

Probably the most powerful analysis feature of a CAD system is the finite element method with this technique. The object is divided into a large number of finite elements (usually rectangular or triangular shapes), which form an interconnecting network of concentrated nodes. By using a computer with significant computational capabilities, the entire object can be analyzed for stress-strain, heat transfer and other characteristics by calculating the behaviour of each node. By determining the interrelating behaviour of all the nodes in the system, the behaviour of the entire object can be assessed.

Some CAD systems have the capability to define automatically the nodes and the network structure for the given object. The user simply defines certain parameters for the finite-element model and the CAD system proceeds with the computations.

The output of the finite-element analysis is often best presented by the system in graphical format on the CRT screen for easy visualization by the user. For example, in stress-strain analysis of an object, the output may be shown in the form of a deflected shape superimposed over the un-stressed object. Colour graphics can also be used to accentuate the comparison before and after the deflection of the object.

Checking the accuracy of the design can be accomplished conveniently on the graphics terminal semiautomatic dimensioning and tolerance routines, which assign size specifications to surfaces indicated by the user help to reduce the possibility of dimensioning errors. The designer can zoom in on part design details and magnify the image on the graphics screen for close scrutiny.

1.5.3 Design Review and Evaluation

A procedure called **layering** is often helpful in design review. For example, a good application of layering involves over-layering the geometric image of the final shape of the machined part on top of the image of the rough casting. This ensures that sufficient material is available on the casting to accomplish the final machined dimensions. This procedure can be performed in stages to check each successive step in the processing of the part.

Another related procedure for design review is interference checking. This involves the analysis of an assembled structure in which the risk that the components of the assembly may occupy the same space. This risk occurs in the design of large chemical plants, air-separation cold boxes and other complicated piping structure.

One of the most interesting evaluation features available on some computer aided design system is kinematics; the available kinematics packages provide the capability to animate the motion of simple designed mechanisms such as hinged components and linkages. This capability enhances the designers visualization of the operation of the mechanism and helps to ensure against interference with other components without graphical kinematics on a CAD system. A designer must often restart to the use of pin and cardboard models to represent the mechanism.

1.5.4 Automated Drafting

Automated drafting involves the creation of hardcopy engineering drawings directly from the CAD database. In some early computer aided design departments, automation of the drafting represented the principal justification for investing in the CAD system. Indeed, CAD systems can increase productivity in the drafting function by roughly five times over manual drafting.

Some of the graphics features of computer aided design systems lend themselves especially well to the drafting process. These features include automatic dimensioning generation of crosshatched areas, scaling of the drawing and capability to develop sectional views and enlarged views of particular part details, the ability to rotate the part or to perform other transformations of the image (e.g., oblique isometric or perspective views).

1.6 BENEFITS OF COMPUTER AIDED DESIGN

There are many benefits of computer aided design, only some of which can be easily measured. Some of the benefits are intangible which are reflected in improved work quality and more pertinent and usable information. Some of the benefits are tangible which are discussed hereinafter.

1. *Productivity improvement in design:* CAD helps in increased design productivity by reducing the time for developing conceptual design, analysis and drafting. It is also possible to reduce the manpower requirements for a given project. Productivity improvement in computer aided design process is dependent on factors such as

   Complexity of the drawing

   Degree of repetitiveness of features in the designed parts

   Degree of symmetry in the parts

   Extensive use of library of user defined shapes and commonly used entities

2. *Shorter lead times:* Interactive CAD is inherently faster than traditional manual design process. CAD tools reduce the number of iterations. It speeds up the task of preparing reports and bill of materials using a CAD system. A finished set of component drawings and documentation can be prepared in a relatively short time. Shorter lead times in design result in reduction of the elapsed time between receipt of customer order and delivery of the finished product. The enhanced productivity of the designers working in CAD environment will reduce the importance of design, engineering analysis and drafting as critical time elements in the over-all manufacturing lead time.

3. *Design analysis:* The design analysis routines available in a CAD system help to optimize the design into an appropriate logical work pattern. The use of design analysis softwares such as finite element analysis and kinematics analysis reduces the time and improves the design accuracy. Instead of having feedback sessions between design and analysis groups, the designer can perform the analysis while working on a CAD workstation. This enhances the concentration of designers, since the process is interactive in nature. Calculation of mass properties can be made almost instantaneously.

4. *Fewer design errors:* Interactive CAD systems have inherent capability for avoiding errors in design, drafting and documentation. These errors occur during manual handling. Errors are avoided because interactive CAD systems perform time consuming and repetitive functions such as multiple symbol placements.

5. *Flexibility in design:* Interactive CAD systems apart from generating designs with repetitive accuracy offers the advantage of easy modification of design to satisfy customer's specific requirements.

6. *Standardization of design, drafting and documentation:* The single database and operating system used in CAD provide a common basis for design, analysis and drafting process with interactive CAD systems, drawings are "standardized" as they are drawn. It is also possible to reuse previous modules in developing a range of products.

7. *Drawings are more understandable:* With the increase in the use of 3D views and solid modelling, it has become easier to comprehend the features of the component readily. One does not have to reconstruct mentally the solid shape from 2D objects. Many software packages allow 3D view generation from a 2D model. This has several advantages from the manufacturing point of view.

8. *Improved procedures for engineering changes:* Control and implementation of engineering changes can be significantly improved with computer-aided design. Original drawings and reports are stored in the database of the CAD system and are easily accessible. Revision information can be retained and new drawings with changes can be created without destroying previous features.

9. *Benefits in manufacturing:* The benefits of computer aided design can be used as a basis for a number of downstream manufacturing operations. Some of the manufacturing benefits are:

    (a) Tool and fixture design for manufacturing
    (b) Computer aided process planning
    (c) Computer aided inspection
    (d) Preparation of numerical control programs for manufacturing of components on computer numerical control machines
    (e) Preparation of assembly lists and bill of materials for production
    (f) Coding and classification of components
    (g) Production planning and control
    (h) Assembly sequence planning

1.7 CREATING THE MANUFACTURING DATABASE

The important reason for using a CAD system is that it offers the opportunity to develop the database needed to manufacture the product. In the conventional manufacturing cycle, engineering drawings were prepared by design draftsmen and then used by manufacturing engineers to develop the process plan (i.e., route sheets). The activities involved in designing the product were separated from the activities associated with process planning. Basically a two-step procedure was employed. This was both time consuming and involved duplication of effort by design and manufacturing engineers. In an integrated CAD/CAM system, a direct link is established between product design and manufacturing. It is the goal of CAD/CAM not only to automate certain phases of design and certain phases of manufacturing but also to automate the transition from design to manufacturing. Computer based systems have been developed which create much of the data and documentation required to plan and manage the manufacturing operations for the product.

    The manufacturing database is an integrated CAD/CAM database. It includes all the data on the product generated during design, i.e., geometry data, bill of material and assembly lists, material specifications, etc. as well as additional data required for manufacturing, much of which is based on the product design. Figure 1.15 shows how the CAD/CAM database is related to design and manufacturing.

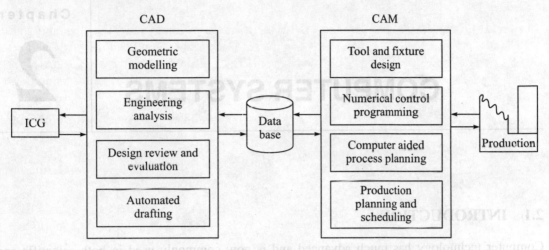

Fig. 1.15 Relationship of CAD/CAM database to CAD and CAM.

REVIEW QUESTIONS

1.1 What is computer aided design? Discuss the various design related tasks performed by CAD.

1.2 Discuss the reasons for implementing CAD.

1.3 What are the advantages to be gained by implementation of CAD?

1.4 What is the design process? Explain briefly the four models for the design process.

1.5 What are the functions that get benefited by the use of computers in design?

1.6 Differentiate between the Pahl and Beitz model for the design process.

1.7 Explain the various steps for the design process as outlined by Shigley.

1.8 Describe the six steps proposed by Earle for the design process.

1.9 How computer aided design differs from conventional design?

1.10 What are the applications of computers for design?

COMPUTER SYSTEMS

2.1   INTRODUCTION

Computer technology has much advanced and is now commonly used in both scientific and commercial fields. Computers have significant impact on design and manufacturing information functions. Recent advances in microelectronics has led to the development of computers that have considerable computing power and which are physically small, reliable and economically affordable to make their use acceptable for many CAD/CAM applications. They have made it possible to perform functions that are just too complex and time consuming to do manually.

The digital computer is a major and essential ingredient of CAD/CAM system. To understand CAD/CAM, it is important to be familiar with the technology of the digital computer.

The digital computer is an electronic computing machine that can perform mathematical and logical computations and data processing functions according to a predetermined series of instructions known as a **program**. The essential components making up a computer system are referred to as **hardware**, whereas the various programs are commonly called the **software**.

2.2   HARDWARE COMPONENTS

The three basic hardware components of a digital computer are as follows:

1.   Central processing unit (CPU)
2.   Memory
3.   Input/Output devices

The relationship of these three components is illustrated in Fig. 2.1.

2.2.1   Central Processing Unit

The central processing unit is the centre of a digital computer, since it coordinates and controls the activities of all the other units. The CPU consists of three separate subsections:

(a)   Control unit

20

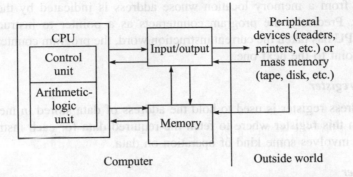

Fig. 2.1 Basic hardware structure of a digital computer.

(b) Arithmetic Logic Unit (ALU)

(c) Memory

The control unit basically acts as an administrator in a computer. It coordinates the operations of all the other components. It controls the input and output of information between the computer and the outside world through I/O devices, synchronizes the transfer of signals between the various sections of the computer and regulates the other sections to perform their individual functions.

The control unit coordinates the various operations specified by the program instructions. These operations include receiving data, which enter the computer, and deciding how and when the data should be processed. The control unit directs the operation of the ALU. It sends data to the ALU and tells the ALU what functions to perform on the data and where to store the results. The capability of the control unit to accomplish these operations is provided by a set of instructions called an **executive program**, which is stored in memory.

The arithmetic logic unit as its name implies, carries out the arithmetic and logic manipulations of data. It performs the calculations such as addition, subtraction, multiplication, division and comparison according to the programmed instructions.

These operations are carried out on data in binary form. The logic section can also be used to alter the sequence in which instructions are executed and to perform other functions such as editing and masking data for arithmetic operations.

Both the control unit and the ALU make use of registers to perform their functions. Computer registers are small memory units with in the CPU that can receive, hold and transfer data. Each register contains binary cells to represent a binary digit called **bit**. The number of bits in the register constitutes the computer's word length, which can indicate to a certain degree the processing capability of the computer. Different computers have a different number of registers depending on the complexity of the architecture of their CPU. These registers are arranged in such a way that they provide support for the CPU to carryout many functions. To accomplish a given sequence of programmed instructions, the functions of these registers can be classified as follows:

Program counter

The program counter is referred to as the control register, contains the address of the next instruction to be performed by the CPU. During program execution, the CPU fetches each instruction word from a memory location whose address is indicated by the contents of the program counter. Precisely, the program counteracts as a pointer to instruction words of a program. After CPU has taken the current instruction word, the program counter is automatically incremented to point to the next one.

Memory address register

The memory address register is used to hold the address of data stored in memory so that the CPU knows from this register where to fetch the required data for each instruction as almost every instruction involves some kind of operation on data.

Instruction register

The instruction register is used to hold the instruction for decoding. Decoding is translation into the corresponding machine code so that the CPU carries out the desired operation.

Accumulator

An accumulator is a temporary storage register used during an arithmetic or logic operation. For example, in the addition of two numbers, the accumulator first stores one number while the second number is being fetched. The second number is added to the first and stores the sum. The sum is stored in the accumulator and would then be operated on or transferred to temporary storage according to the next instruction in the program.

Status register

A status register is used to indicate the internal condition of the CPU. It is used to hold a status word, which contains information about the conditions of the CPU such as the result of a logical decision, overflow during an arithmetic operation, and interrupt state (used in process control).

Arithmetic logic unit (ALU)

The ALU provides the circuitry required to perform the various calculations and manipulations of data. Most ALUs can add and subtract, but there are now some ALUs that are capable of performing multiplication and division and even other complex mathematical functions. ALUs with simplex circuits are capable of being programmed to perform these more complicated operations, but more computing time is required. The more complex arithmetic logic units are faster, but these units are more costly.

2.2.2 Memory

The memory section consists of binary storage units, which are organized into bytes. The memory section stores all the instructions and data of a program. Therefore the CPU must transfer these instructions and data to and from the memory throughout the execution of the program.

The type of memory is a very important consideration in the design of the entire computer system. There are two basic categories of computer memory:

(a) Main memory (Primary storage)
(b) Auxiliary memory (Secondary storage)

Main memory (*Primary storage*)

The purpose of the main memory is essentially to provide a working area for the current program and it only retains information on a temporary basis until the termination of the program. The main memory is used to contain instructions and data of a program during its execution. It includes the working registers and memory devices, which are closely configured to the CPU. Primary storage can be divided into three main categories:

(a) **Main data storage.** This storage is characterized by its close proximity to the CPU fast access rate, low storage capacity and costly when compared to other forms of memory such as magnetic core or solid state memory.

(b) **Control storage.** It commonly contains the microprograms that assist the CPU circuitry in performing its functions.

(c) **Local storage.** The high-speed working registers used in the arithmetic and logical operations.

Auxiliary memory (*Secondary Storage*)

Programs and data files are not generally kept in main memory but are stored on auxiliary memory devices and loaded into main memory when required. Primary storage is very expensive and has a limited capacity. Computers may often operate on an enormous amount of data that exceeds what it can hold at one time. Backing storage is used to supplement the main memory and to save the data on a permanent basis. Information saved in backing storage can be retrieved and quickly transferred to the CPU when it is needed.

There are several types of devices commonly referred to as auxiliary devices, which constitute the secondary storage that can provide this kind of backing storage. There are two forms of secondary storage:

(a) **Sequential access storage.** It is also known as **serial access storage.** A sequential access storage unit can be distinguished by the fact that to read one particular record in the file, all the records preceding it must also be read. If additional data is to be added into the file, it is only possible to do at the end of the file, otherwise the rest of the file will have to be rearranged.

(b) **Direct access storage.** With direct access storage, individual records can be located and read immediately without reading any other records. As the information is readily available at random, direct access storage is also known as **random access storage**.

Because of its method of operation, the random access storage is much more efficient than the sequential access storage and consequently the random access storage reads or writes data more quickly. On the other hand, random access storage is more expensive in terms of cost per bit because its technology is more sophisticated.

For these reasons, sequential access storage is used mainly to store files as permanent records in an archive where they are very rarely accessed, whereas direct access storage is most suitable for storing files that require frequent access.

2.2.3 Memory—ROM and RAM

For a computer to work, it must contain memory where it can store data and programs until they are needed. In the computer, storage is on memory chips of many kinds which basically, however, fall into two categories: read-only memory (ROM) and random-access memory (RAM). Both are very fast because data is read to and from the chips electronically; they have no moving parts.

Read-only memory (ROM)

Read-only memory (ROM) is static and unchanging; as a result it is called **non-volatile memory**. The data in ROM is permanently recorded on memory chips by the computer's manufacturer prior to their being sold or shipped and you can't change it. Turning the computer off will not affect it; the data will still be there when you turn it back on. ROM is generally used to store programs and instructions that the computer frequently needs, sometimes called **firmware**. For example, it contains the instructions the computer follows to start up when you first turn it on and the system's set-up program that you use to change settings such as the time and date.

Random-access memory (RAM)

Random-access memory (RAM) is used to store an ever-changing parade of programs and data. When you load a program into the computer or create a word processing document, the data you enter using the keyboard is stored in random-access memory (RAM) also called **main**, **primary** or **user memory**. Usually, if you turn off the computer, any programs or data stored in this memory are lost; thus RAM is said to be volatile memory.

The term random comes from the way the data in memory can be located or accessed by the computer. One way to understand random is to think of the differences between a tape cassette and an audio CD disc. To play the third song on a tape cassette, you must first advance the tape and past the first two songs. This is called **sequential access** because you access each song in sequence. With an audio CD disc you can go directly to the track where the third song begins. This is called **random access** because you can randomly access songs without first advancing through the songs that precede them.

Types of memory chips

Memory comes in a variety of forms. The computer designer's decision which to use where depends on what their function is to be, and on their speed and cost. Dynamic RAM (DRAM) is the most common type of main memory. It is dynamic because each memory cell quickly loses its charge so it must be refreshed hundreds of times each second to prevent data from being lost. Here are some of the types of DRAM that have been or will be popular in most desktop systems (listed from oldest to newest):

- Fast page mode (FPM) DRAM was used in most computers until EDO RAM came along.
- Extended data out (EDO) DRAM is slightly faster than FPM. One variation called burst EDO (BEDO) DRAM assumes that the next data-address to be requested by the CPU follows the current one so it sends that also.

- Synchronous DRAM (SDRAM) can synchronize itself with the clock that controls the CPU. This makes data transfers more reliable and faster because timing delays are eliminated. It is anticipated that this form of memory will replace EDO as the most common form of memory.
- Rambus DRAM (RDRAM) is the latest design and Intel has announced that all of their future systems will require it. RDRAM is very fast, but the system must be slightly redesigned to use it. RDRAM sends data down a high-bandwidth "channel" 10 times faster than standard DRAM.

Static RAM (SRAM) is like DRAM but it is a lot faster, larger, and more expensive, and consumes more power. It is static because it doesn't need to be continually refreshed. Because of its speed, SRAM is used mainly in a special area of memory called a **cache**. Flash memory chips are one of the latest storage devices. These chips, a form of static RAM (SRAM) chips, store data much like those used in the computer's primary storage. However, the data stays recorded even when the power is turned off. Flash memory is non-volatile. Since flash memory devices have no moving parts, and are therefore very fast, they may eventually replace slower, mechanical hard disk drives.

Video RAM (VRAM) is used to accelerate the display of graphics on the screen. It does this by using two "ports," one connected to the CPU and the other to the screen. Data flows in one port and out the other very smoothly. A variation of this is Window RAM (WRAM) that supports memory.

Memory modules—SIMMs and DIMMs

In most systems you don't add memory by adding individual chips. Instead, memory is plugged into the computer's system board using memory modules. These are circuit boards into which memory chips have been inserted. They in turn are plugged into a socket on the computer's system board. One type is the Single In-line Memory Module (SIMM). Another is the Dual In-line Memory Module (DIMM). These names refer to the package, not to the type of chips they contain. DIMMs are preferred in newer systems because they can be used singly instead of in-pairs like the older SIMMs.

In addition to the RAM in your system's memory chips there are other types of memory and other ways to use it. These include ways to store more data, store it so it can be accessed faster, or store it so it can be moved within an application or between applications.

Virtual memory

Very large programs and very large files of data can strain the capacity of even a large computer's memory. To solve this problem, modern operating systems and many application programs store parts of their programs or data outside RAM until needed. This type of storage is called **virtual memory**. On a system that uses virtual memory, only the parts of the program or data file currently needed are stored in the computer's memory; the rest are stored on a hard disk drive. When the computer needs a part of the file stored on the disk, it moves something in memory to the disk to make memory available. It then moves the program or data from the disk into memory. For example, as you scroll through a long document, only part of it may be in RAM at any one time. The rest remains on the disk until you scroll toward it. Virtual memory reduces the overall cost of the system because it is cheaper to store data on a hard disk drive than it is to add additional memory chips to the computer. Prices fluctuate a lot, but in 1997 hard disk space cost 9 cents a megabyte. A megabyte of chip memory cost $9 over 100 times as expensive.

Caches

When a computer is operating, it is frequently moving data between the CPU, disk and memory. To process data, the data must first be moved to the CPU from wherever it is stored. Since this takes time, special techniques have been developed to speed up this operation. The basic technique is to store data in a special area of high-speed memory called a **cache** that is located between the CPU and main memory. When the processor needs data, it will check in high-speed cache first to see if the data is there. If not, then the processor will retrieve the data from slower main memory.

When you ask for something that the computer needs to go to RAM or the disk for, it also gets the data that follows it. The assumption is that if you look at the first name in a database, the next one you want to look at is the second name. It then moves all of this data into the cache so the CPU can get it from there rather than going to the disk for it. Unfortunately, the computer can't always anticipate correctly what should be in the cache.

When the CPU finds the right data there, it is called a **cache hit**. When it doesn't, it is called a **cache miss**.

Cache memory uses special chips, often SRAM (Static RAM) chips. On some systems, these chips are four times as fast as regular memory. However the chips cost six times as much. It is this cost that keeps them from being used for all of the system's memory. There are two levels of cache memory:

- Level 1 cache (also called L1, internal, or primary cache) is physically a part of the microprocessor chip. This cache typically ranges up to 256 kilobytes on desktop systems.
- Level 2 cache (also called L2, external, or secondary cache) is in the form of memory chips mounted in special sockets on the system board. This cache is larger than L1, ranging up to 1 MB or more.

Buffers

Some programs allocate a small portion of the computer's random-access memory as a buffer. Buffers have several applications. For example, a buffer stores data that you enter from the keyboard until the CPU can accept it. (If the buffer becomes full, the computer beeps, and any keys you press are not stored, so you must reenter them when the CPU empties the buffer.) Another buffer temporarily saves a record of the most recently used commands and deleted text so you can "undo" any mistakes. Still another buffer (sometimes called a clipboard) is used as a storage area when you cut or copy data so that you can later paste it elsewhere. For example, you can create a graph in a spreadsheet program and copy it to the clipboard. You can then open a word processing program and paste the graph into a report.

2.3 Mass Storage Devices

The most common hardware devices used for computer storage technology are:

(a) Magnetic tape storage

(b) Magnetic disk storage

(c) Magnetic drum storage

Magnetic tape storage

Magnetic tape storage is a good example of sequential access storage technology. Data are stored on magnetic tape, similar to that used in audio systems. The major advantages of magnetic tape are that it is relatively cheap when compared with other types of storage medium and that it can easily hold a large amount of data for its size. Magnetic tape, unlike punched paper tapes or cards, can be used again by simply overwriting previously stored data.

Since data are stored sequentially access time is relatively slow. However, the low cost per bit and high capacity of magnetic tape make it ideal for system backup. It is most suitable for applications, which may be required in payroll, personnel management, inventory control and customer invoicing where a large amount of data is to be processed sequentially.

Magnetic tape may be used not only as a backing storage medium but also as an input/output medium. Information can be read into the computer from the tape for processing and information can be written to tape where it is stored until it is required at a later time or until it becomes obsolete and redundant. When the tape can be reused for some other purpose and its contents be overwritten magnetic tape is used to backup the data in a computer system, so that in the event of a crash, a recent copy of the system on tape can be reloaded, if necessary. Magnetic tape is also widely used for transferring data between computer installations that are not linked together.

Magnetic disk storage

Magnetic disk storage is also a random access storage device. The storage medium is a magnetically coated disk. There are several types and sizes of disks each best suited to a particular set of applications.

Floppy disk

Floppy disks come in two standard sizes: the larger one is 8 inches in diameter and the smaller is 5¼ inches and is referred to as mini floppy. The storage capacity of an 8-inch floppy can be between 250 Kbytes and 1.5 Mbytes whereas that of a 5¼-inch mini floppy lies in the region between 125 Kbytes and 500 Kbytes, depending on recording density and whether single-sided or double-sided.

The hard disk is a thin metal disk, which is coated on both sides with magnetic ferric oxide. Data are recorded in the form of magnetized spots on tracks on the disk surface. Several disks are combined into a single disk pack and these are separated by a fixed distance and joined by a vertical shaft. The disk pack is rotated at several thousand revolutions per minute by a disk drive unit. Data are transferred by moving a set of read/write heads (one per recording surface) to the appropriate track. Only one of the heads is used to transfer data at a time, although all heads are moved together. The particular track is read until the appropriate data are found. This means that the access time is dependent on the rotational speed of the disks and the capacity of the head to read from the disk surface. This rate is usually several thousand bytes per second.

In general, floppy disk cannot hold, as much data as conventional hard disk and its data transfer rate is less than desirable for use as fast access backing storage. However because of the fact that it is relatively inexpensive compared to its storage capacity and access speed. It is particularly suitable for use with small and low cost microcomputer systems.

Magnetic drum storage

The magnetic drum is a direct access storage device with high capacity and high access rates. The magnetic drum consists of a magnetically coated cylinder during operation. The drum is rotated at a constant speed and data are recorded in the form of magnetized spots. The drum can be read repeatedly without causing data loss. Read/write heads are used to read data to and from the drum as it rotates. The drum surface is divided into tracks each with its own read/write head.

Since there is one read/write head for each track of the drum, the time required to store or retrieve one particular item of data on the drum is usually less than for magnetic disk where read/write heads must normally be moved across the disk surface to reach the appropriate track. As a result of the way in which the heads are arranged every item of data will pass under a head during one revolution of the drum. So data transfer rates are on average higher than disk storage and range from 3,00,000 to 1.5 million characters per second. Access time is also faster than disk, whereas storage capacity is comparable and it is possible to have a multiple drum system, which can considerably increase the storage capacity. However, a drum storage unit generally costs more than disk storage and the drum is not removable from the drive unit.

Compact disk (CD)

CD-ROM disks are similar in concept to the audio disks (CDs) so popular in the music recording industry. These disks can store up to 660 mega bytes of data. Like other media, they are rated by their access times and transfer rates. Since many of their characteristics are fixed by specifications that make them compatible with the millions of drives, performance is boosted by spinning them faster in the drive. The faster the spin, the faster are the access times and the higher the transfer rates. The spin rate is designed by the 2X, 4X, 6X, 16X and 52X designations each drive carries. A 16X drive spins 16 times faster than the original drives.

Most CD-ROM disks can only be read but there are other versions that you can write to.

- D-Recordable (CD-R) disks can be written on once. These disks have a thin layer of gold with a layer of green dye below. To record data the laser forms bumps in the dye layer. When plate back, the computer reads a bump as 1 and the absence of a bump as 0. The fact that you can only write this disk once isn't always a drawback. Many firms need archive copies of data and the fact that the data can't be modified once it has been copied is actually a benefit.
- D-ReWritable (CD-RW) can be recorded, erased and reused just like a hard disk. These disks record data by changing a material from a well structured crystalline state to a less ordered amorphous state.

DVD disks

This latest generation of optical disks went through so many name changes (Digital Video Disk and Digital Versatile Disk) that only the DVD acronym remains in use. In one massive shift, this new medium is about to replace music CDs, video, tapes, 12 inch video laser disks and CD-ROMS. These new DVD disks and the computer cousin DVD-ROM disks will initially store 4.7 gigabytes of digital information on a single sided, single layer disk the same size as today's CD-ROMs. With compression that is enough to contain a full length, full screen, full motion feature film, including 3 channels of CD quality audio and 4 channels of subtitles. Future plans include 9.4 gigabyte double-sided and double-layer disks that will store about 17 gigabytes. That nearly 30 times the capacity of today's CD-ROM discs.

Currently there are three different types of DVD drives that have been defined:

DVD-ROM.   These are the drives with only the reading capability. They are used basically for removable mass storage for large volumes of data such as encyclopaedia and are currently available.

DVD-R.   These are drives with write once capability. DVD-R drives are also called **write once, read many (WORM)** drives and are currently available. These are similar to the CD-R drives and WORM capability.

DVD-RAM.   These are drives with both read and write capability. DVD-RAM drives are also called **write many, read many (WMRM)** drives. Unfortunately there is no agreed format in this category. As a result there are a number of different formats that are being pushed by the various groups in the DVD forum. What was approved by the forum is phase change design that can hold 2.6 GB of data per side on single or double sided disks. The single-sided disks will come in removable cartridges, but to protect the sensitive recording layer, double-sided disks will be permanently mounted in cartridges.

DVD+RW.   This is supported by Hewlett-Packard, Philips and Sony. DVD + RW's single layer phase change disks have more capacity than DVD-RAM disks—4.7 GB per side, and use a higher density recording process. The DVD+RW format does not rely on cartridges to hold the disks.

DVD-R/W.   This is put forward by pioneer and the first one to be available commercially. It will use random-access media that can hold up to 4.7 GB. One of this technologies key characteristics is that its phase-change media have a higher reflectivity and as a result, can be read in existing DVD-ROM drives and DVD players without modification.

2.4   INPUT/OUTPUT DEVICES

Card readers

A card reader is an input device that transfers data from the punched card to the computer system. There are two types of punched cards: the 80-column Hollerith card and the 96-column IBM card. There are two types of card readers currently in use:

(a)   **The brush reader:**   In this device, the punched cards are moved past a set of electrically conductive brushes. Holes in the cards allow circuits to be closed, producing electrical pulses, which correspond to binary data.

(b)   **The photoelectric reader:**   This device utilizes a set of photocells and a light source, which shines through the holes in the card to produce electrical pulses at the photocell junctions. This type of reader has certain advantages as a result of having fewer moving and contact parts.

Cardpunches

A cardpunch is an output device that transfer data from the computer system onto punched card. The punch outputs the data as a pattern of rectangular holes in a card, and then reads it to verify that the card was punched correctly. Cards are transferred from a hopper, punched and then reread to assure correct punching. Card punching speeds typically range from 100 to 300 cards per minute. Card readers and punches are often integrated into one single unit.

The major disadvantage of punched cards as a storage medium is that they are cumbersome to handle and too bulky to keep because they occupy a relatively large space when a bulk volume is involved.

Paper tape reader

A paper tape reader operates in much the same way as a card reader. It translates the data punched in code on paper tape and transmits it into a central processing unit. The tape passes through a reading unit where the presence or absence of holes is detected and converted to electrical pulses.

Paper tape data entry is usually slower than magnetic tape. A common application of punched tape readers is in numerical control programming. The major disadvantage of paper tape is that errors are difficult to correct and that readers are relatively slow operating at approximately 500 and 300 characters per second respectively.

Paper tape punches

A paper tape punch operates on data in the opposite direction to a paper taper reader. That is it outputs the information in a central processing unit in the form of punched holes on paper tape. Data from main storage are converted into the appropriate code and punched on the tape as it is fed through the punching unit paper tape readers and punches are often built into a single unit.

Keyboard input devices

Many input devices employ a typewriter like keyboard, which can be used by a typist with little additional training. Some of these devices input data and programs directly to the computer. Others produce data on a special medium for subsequent input to the computer system.

The keypunch

The keypunch contains an electromechanical keyboard device, which converts operator keystrokes into machine-readable holes on card.

Blank cards are entered into the keypunch machine, and a corresponding combination of holes is punched when a key is depressed.

The cards are then read through a card reader to the computer.

Key-to-tape unit

The key-to-tape unit is an electronic typewriter device that converts operator keystrokes into machine-readable codes on magnetic tape. Two types of units are available. The first produces computer compatible tape, but is now obsolete. The other unit produces magnetic cartridges or tape cassettes.

Alphanumeric display

An alphanumeric display consists of a typewriter-like keyboard and a display screen, usually a cathode ray tube (CRT) that can be used to display data. Stand-alone CRT terminals include the following components: screen, keyboard, communications interface, buffer memory and a local microprocessor used for editing. The CRT terminal can be connected directly to a computer for on-line operation or it can be used with independent devices for off-line operation.

Transmission speeds are usually selectable, from 110 to 9600 baud. A baud is a unit representing the number of discrete signal changes per second. For a binary system it is equal to the number of bits per second. Communication-line quality limits the speed of data transmission programming and data input on a CRT terminal are faster than for other keyboard entry devices because of enhancements in keyboard design, screen formatting, prompting capabilities and local editing.

Teleprinters

A teleprinter consists of an electromechanical or electronic typewriter keyboard and a hardcopy-printing device. It can function both as a remote data entry terminal and as an on-line output terminal. During input, data are usually transmitted character by character as keys are depressed, although some units have buffer memory available to permit batched continuous transmission. Transmission speed depends on the device design. Older electromechanical units print data at a rate between 110 and 300 baud. The newer electronic units operate at speeds of upto 9600 baud.

Visual display units (VDUs)

A VDU contains a keyboard and a display device, which is usually a cathode ray tube (CRT). The display device provides a screen to print information and essentially performs the same functions as the carriage and paper of the teleprinter. As each character is typed for input it is displayed on the screen. At the end of the input line, carriage return key is pressed and the position indicator or cursor is returned to the beginning of the next line for further input. As lines are displayed, preceding lines are successively scrolled up until the top line disappears at the top of the screen for output lines arc displayed in a similar manner. A VDU is formatted to display 24 lines of 80 characters each.

The major advantages of the VDU over the teletypewriter are that its data communication rate is much faster between 110 and 19,200 baud but 9600 baud is the most common and that its printing speed is far superior.

Printer

A printer is an output device that converts data from computer and prints it in a readable form on paper. There are a number of different categories of printers used to produce printed output, such as line printers, serial printers and laser printers.

Line printer. A line printer prints a whole line at a time and operates on the same basic principle as a typewriter. The speed of line printers varies from 60 to 5000 lines per minute and there are normally 80 to 132 characters per line, so it is very suitable for high volume output, to ensure sufficient and uninterrupted supply of paper for these high printing speeds, continuous stationery is used. Continuous stationery is a long piece of paper with sprocket holes on the edges and with perforations to separate the paper into convenient sizes. One disadvantage of line printer is that the quality of printing is not very good in comparison with that of typewriters.

Serial printer. A serial printer, also known as **character printer,** is another category of printer, which outputs one character at a time, as opposed to one line at a time in the case of the line printer. A serial printing device can normally be operated using continuous stationery or separate A4 size sheets. A serial printer is much slower, but cheaper than a line printer. The most common type of serial printer is the dot matrix printer. The print head consists of a matrix of tiny needles typically seven rows of nine needles each, which hammers out characters in the form of patterns of small dots. The shape of each characte, i.e. the dot pattern is obtained from information held electronically in the printer. The printing speed of dot matrix printers varies between 45 and 220 characters per second.

Dot-matrix printer. Dot-matrix printers use a print head containing pins or wires, arranged in a column to print characters. As the print head passes across the paper, the computer tells it which pins in the print head are to be fired to form a particular character. As the pins are fired they strike an inked ribbon against the paper. The printed dots are arranged in an invisible matrix. The number of pins or wires and dots determine the characters resolution. The latest printers have 18 or 24 wires in their print heads. Dot-matrix printers have been killed off by newer technologies.

Inkjet printer. Colour inkjet printers use three separate inkjets, one for each of the primary colours (Red, Green and Blue). Inkjet printers are very quiet and provide laser-like quality at a much lower cost although supplies are relatively expensive. Although you can print on regular paper, better results are obtained when using special paper that doesn't allow the ink to soak in.

Liquid inkjet printers use ink cartridges from which they spray ink into the paper. A cartridge of ink attached to a print head with 50 or so nozzles, each thinner than a human hair, moves across the paper. The number of nozzles determine the printer's resolution. A digital signal from the computer tells each nozzle when to propel a drop of ink onto the paper. On some printers, this is done with mechanical vibrations.

Solid inkjet printers use solid ink sticks that are melted into a reservoir and sprayed in precisely controlled drops onto the page where the ink immediately hardens. High pressure rollers flatten and fuse the ink to the paper to prevent smearing. This produces an exceptionally high-quality image with sharp edges and good colour reproduction. Solid inkjet printers are also the best for producing low-cost but high quality transparences. These printers are sometimes referred to as phase change printers because the ink moves from a solid to a liquid phase to be printed then back to a solid phase on the page. As a final step, the paper moves between two rollers to cold-fuse the image and improve the surface texture.

Laser printer. A laser printer is a relatively new technological innovation in which a combination of electronics lasers and copier techniques are used. An electronic page of different size can be printed at one time by a laser printer. It is capable of producing very high quality print very quickly and a wide selection of character fonts is also available. Most laser printers operate at speeds between 30 and 250 pages per minute.

Other input/output devices

In previous sections, the common input/output devices that are most commonly found in general purpose computer systems have been described. There are many other types of input/output devices which are used specifically for some applications.

The process of preparing data for input from a form acceptable to humans to a form acceptable to computer is time-consuming and tedious. The ideal solution would be a method of input that could do away with these data preparation procedures. That is the use of documents that could serve both as source documents and as input medium. There are number of devices that have been developed to achieve this objective.

Magnetic ink character recognition (MICR). In MICR readers, the recording of data is based on specially designed fonts in which the characters are printed with ink containing ferromagnetic materials. Magnetic ink characters are used primarily on bank cheques to indicate the code number of the bank, the customer account number and the cheque sequence number. These devices are capable of reading up to 1600 documents per minute. When a magnetic ink character is being scanned it induces a current, which will be proportional to its area. The patterns of the varying currents can then be compared with and identified as bit patterns of the selected character, which is translated into the internal machine code of the computer and transferred to the main memory.

Optical character recognition (OCR). In the operation of the OCR reader, a mechanical drum is used to rotate documents past an optical scanning station. A light source and lens system can distinguish the patterns of the character. These patterns are converted into electrical pulses, which are interpreted as each individual character. An OCR can be programmed to read a variety of character sets, and scan a document for marks made by hand or printed by line printers.

Optical barcode reader (OBR). An optical barcode reader senses the configuration of shaded bars of different widths and correlates them to previously defined characters. Barcode readers use a scanner or light pen to enter information to a computer through a terminal device. The scanner or light pen is stroked across the pattern of bars; a sequence of bits corresponding to that pattern is generated and stored in the computer. An OBR can read 400 characters per second, and is limited mainly by the operators ability to manipulate the items to be scanned.

Optical mark readers. Optical mark readers sense the physical position of marks on a document. The position is correlated to a previously defined character. The marks can be read from 80 column cards or full page documents; data transmission speed depends on the speed of the feeding device. A rate of 1500 forms per minute is not unusual.

2.5 PROGRAMMING LANGUAGES

In general, there are three basic levels of computer programming language:

    (a) Machine language

    (b) Assembly language

    (c) High-level language

Machine language

Data representations in a computer are all in binary numbers, therefore, programs and data have to be written in binary code. The binary coded instructions that computers can understand are called **machine language** or **machine code**. The language used by the computer is called **machine language**. A machine language instruction consists of two parts, namely the operation code and the operand. The operand might be a memory address, a device address or data in machine language programming, storage locations are designated for the program and data and these are used throughout the program to refer to specific data or program steps, in addition the programmer must be familiar with the specific computer system since machine language instructions are different for each computer. A machine language is machine dependent in that it can be recognized and understood only by the computer for which it was designed. Programming in machine language is tedious, complicated and time consuming. To overcome the difficulties in writing programs in binary, symbolic languages were developed which substitute an English-like mnemonics for each binary instruction. Mnemonics are easier to remember than binary so they help speed up the programming process.

Assembly language

A language consisting of mnemonic instructions is called an **assembly language**. Assembly languages are considered to be low-level languages. The programmer must be very knowledgeable about the computer and equipment being programmed. Low-level languages are the most efficient in terms of fast execution on the computer. Although there exist various versions of assembly language for different computers, they include normally the same kind of instruction sets that perform the four fundamental operations within a computer such as input/ output operations, arithmetic operations, movement of data within the CPU and logical or comparison operations.

Assembly language programs are not in a form directly executable by the computer. Assembly language programs must be converted into machine language before the computer can execute them. A program called an **assembler** carries out the conversion. The assembler takes the assembly language program, performs the necessary conversions and produces two new programs: the machine language version and an assembly listing. The assembly listing shows the mnemonics instructions and their associated machine language equivalents and any errors the original assembly language program may have contained. Assemble languages are hardware dependent. Hence they differ from one system to the other.

High-level language

Both machine language and assembly language are machine-oriented. High-level languages by contrast are problem-oriented. They are very much independent of the computer on which they are used. A high-level language program written on one computer can be executed on a different computer without major modifications to the program itself.

High-level languages consist of English-like statements and traditional mathematical symbols. Each high-level statement is equivalent to many instructions in machine language. The significant advantage of high-level language is that it is not necessary for the programmer to be familiar with machine language. The program is written as an English-like algorithm to solve a problem. Like an assembly language high-level languages must also be converted into machine code. A special program called a **compiler** achieves this. The compiler takes the high-level program and converts it into a lower-level code, such as the machine language. A compiler is able to detect statement errors in the program and corresponding error messages are printed in a special program listing to indicate the error conditions. There are different high-level languages used for various applications. The most common high-level languages used for business and engineering applications are FORTRAN, Basic, COBOL, Pascal, APLRPG, PL/1.

2.6 OPERATING SYSTEM

An operating system may generally be defined as a program which organizes a collection of hardware into a consistent and coherent entity so that it can be used for developing and running programs. It can assist users and their programs to make the most efficient use of system resources such as memory capacity and processing power with minimal programming effort by providing software to supplement and enhance the system facilities available. Effectively an operating system acts as an interface between the user and the computer. There are a number of different processing techniques widely used for implementing this human computer interface and each technique has its own characteristics:

    (a) Batch processing
    (b) On-line processing
    (c) Multiple programming
    (d) Time-sharing
    (e) Real-time

Batch processing

In this mode of processing, programs to be run are collected several at a time and are entered into the computer as a batch of programs, hence the name **batch processing**. The programs are read in and placed on disk storage to form an execution queue. When one program is completed, the next one in the queue is loaded for processing. In other words, programs are processed sequentially one at a time.

Batch processing provides no or very little interaction between the user and the computer. The user usually prepares programs and their associated data on punched cards, submits them to the computer, and after some time receives a printout of results which can be examined at leisure. The results may also be output as a graph or drawing produced by the computers graph plotter.

On-line processing

In an on-line mode of operation, the user communicates directly with the computer through its terminal devices. Information and instructions entered via a terminal are executed virtually immediately and a response is received as soon as possible, often within seconds. The user can also use all the available devices to their best advantage. On-line processing is normally used where a high degree of user-computer interaction is necessary.

Multiple programming

There can be only one program at a time in main memory for execution in batch processing. As the CPU is normally the fastest component in the computer system, its expensive memory and the full capabilities of the ALU may be underutilized because not every program will be large enough to fill main memory and not every program will need to use the ALU all the time. Multiple programming attempts to balance the CPU's speed with the slower devices by enabling more than one program to reside in main memory at the same time.

The objective is to make more efficient use of the CPU by keeping more parts of it busy most of the time. For example, when program A is receiving data or output results, program B can then be processed. If both program A and program B are performing I/O functions, a third program C can be executed.

Time-sharing

When a computer system operates in time-sharing mode its resources are shared by several users through terminals which are often connected to the computer by land or telephone lines and are commonly referred to as remote terminals. The basic idea behind time-sharing is to allow all users to have a brief share of the CPU in turn. Although the computer actually services each user in sequence, the high speed of the computer makes it appear that the users are all handled simultaneously. Each user is allocated a short interval of time on the computer, often known as a **time slice,** during which it has sole control of the CPU. Once the interval has expired, the control of the CPU is passed over to another user, thus one single user beyond a fixed time limit cannot monopolize the CPU. For example, a computer system can support upto five terminals and user 1's program is allowed to run when its time slice is over, or if it requests an input/output function and therefore temporarily does not require the CPU. The CPU is promptly scheduled for user 2's program. User 1's program is then swapped out of memory until its next turn to run has come, and user 3's program is swapped in to be ready for execution. During this time interval, user 2's program is being processed. A program is in main memory only when it is being executed and it is on backing storage when it is not being executed. As the swapping process, sometimes called 'roll in–roll out', happens many times within a few seconds, the only feasible backing storage devices are disks which transfer data at a much faster rate.

Real-time

The term **real-time** is used to refer to any system, which produces a response almost immediately as a result of input data. The processing of these input data must occur virtually at the same time as the event generating them, so that the response is fast enough to affect subsequent actions to be taken promptly within time limits in order for it to be useful to the real world. Real-time systems are often used in process control and airline ticket reservation.

2.7 SYSTEM CONFIGURATION

Apart from the central processing unit, a computer system consists of other devices such as backing storage devices and input/output devices which are collectively called **peripheral devices**. A peripheral unit is defined as any hardware device that is connected to the CPU and performs a specific task such as input/output for the system. There are normally many peripherals in a computer system and the way in which they are arranged and what they are determine the configuration of the system. There are in general a number of established types of system configuration, which are as follows:

(a) Simple configuration
(b) Network configuration

Simple configuration

A simple configuration of computer system is shown in Fig. 2.2. The system comprises essentially one CPU, various backing storage devices and several input/output devices. Such a configuration is rather small and its processing power is undoubtedly limited. When the number of users increases and their demands for system resources grow, this kind of system configuration cannot really cope and there is very little room to expand by way of introducing additional hardware. Because of the lack of flexibility of this type of configuration, a new concept, known as **network configuration,** has been developed and has now become widely used in many computer applications.

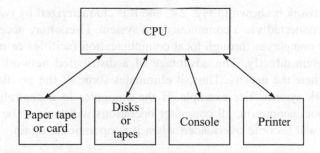

Fig. 2.2 A simple configuration of a computer system.

Network configuration

A computer network system is formed by linking distinct computer installations together through a communication system. The purpose of setting up a network is to share computing power and to improve the flow and to allow the exchange of information within organizations and beyond. Network systems may be considered as a step towards establishing a means by which people can communicate with each other through a computer. A local area network (LAN) allows organizations operating on one site to link their computers, user terminals and workstations and often peripheral devices in an efficient and cost effective way. A wide area network (WAN), which links various sites, computer installations and user terminals and may even permit LANs to communicate with each other.

There are three main classes of configuration for computer network systems:

(a) Centralized network
(b) Distributed network
(c) Ring network

A centralized network is shown in Fig. 2.3, and it is characterized by a centralized computer complex, a communication system and a number of users who can access and interact with the computer system via local terminal devices. A centralized network is particularly useful for organizations that require a centralized database or a centralized processing facility.

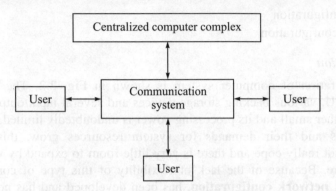

Fig. 2.3 Centralized computer network.

A distributed network is shown in Fig. 2.4, and it is characterized by two or more computer complexes that are connected via a communication system. Users may access and interact with one of the computer complexes through local communication facilities or may be linked to the communications system directly. The advantage of a distributed network system is to place computing power where the user is. Thus, it eliminates some of the problems associated with a centralized network system. For example, if the computer in a centralized system fails to function, which is not uncommon, all computer operations will have to be suspended. Also as a result, the system will become overloaded when it is operational again.

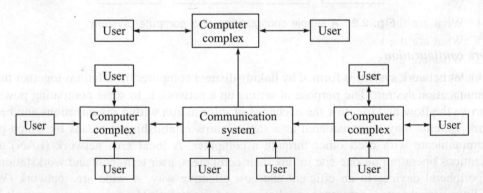

Fig. 2.4 Distributed computer network.

Figure 2.5 shows a typical ring network system, which is a special case of a distributed network. Every computer system is connected to exactly two other systems in the ring network, which is employed by the large organizations to coordinate its various branches in different locations.

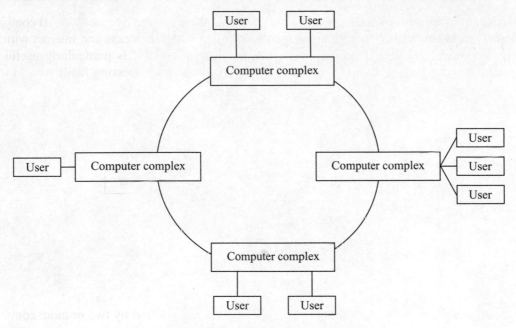

Fig. 2.5 Ring computer network.

REVIEW QUESTIONS

2.1 What is the basic hardware structure of a digital computer?

2.2 What do you understand by the term CPU?

2.3 What is the importance of a CPU in a computing system?

2.4 What are the functions of registers in both the control unit and ALU?

2.5 What are the various categories of computer memory available in CPU?

2.6 What are the various mass storage devices used for computer storage technology?

2.7 What are the types of printers that would be useful for printing graphic information?

2.8 Discuss the following terms in relation with computer systems:

    (a) Programming languages

    (b) Operating system

    (c) System configuration

2.9 What is the difference between a low-level and high-level computer language?

2.10 What is the difference between direct access and sequential access in the case of secondary storage devices?

2.11 What are the functions of an operating system?

2.12 What is the initial capacity of DVD ROM?

2.13 How do you differentiate between ROM and RAM?

2.14 What is virtual memory?

2.15 What is the difference between co-processor and parallel processor?

2.16 What is buffer?

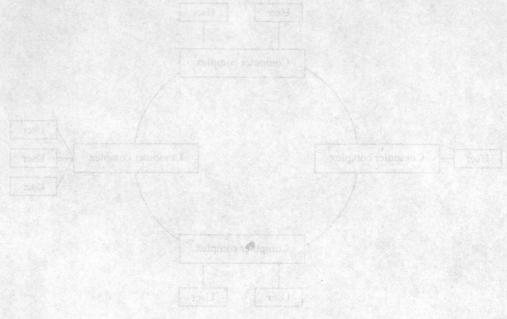

Part II

CAD

Hardware and Software Components

(d) Supercomputer.

Chapter

USE OF COMPUTERS IN CAD/CAM SYSTEM

3

3.1 INTRODUCTION

The microcomputer is making a great impact on everyday activity of mankind, and plays an important role in day-to-day working of industrialized societies. In the early years of computer development, the emphasis was on larger and more powerful computers. Large computers were designed to solve complex scientific and engineering problems and handle records of large corporations and government agencies. Only large institutions and organizations were able to purchase large digital computers. A trend started in the middle of 1960's to design smaller computers for smaller organizations. In the late 1960's, minicomputers were available for use in office, college, medium-sized business organization, small factory and laboratory. The smaller organizations which were not able to afford large computer could now purchase a minicomputer.

The microcomputer is the outcome of the trend towards smaller computers which started in the middle of 1960's. With rapid advances in semiconductor technology it became possible to fabricate the whole CPU of a digital computer on a single chip using LSI or VLSI technology. The term LSI refers to ICs containing components usually transistors in the range of 1000 to 10000. A VLSI chip contains more than 10000 transistors. A CPU built in to a single LSI or VLSI chip is called a **microprocessor**. It is the latest development in the field of computer technology. A digital computer whose CPU is a microprocessor is called a **microcomputer**. A microprocessor combined with memory and input/output devices forms a microcomputer. The prefix micro indicates its physical size, not its computing power. As far as the computing power of a microcomputer is concerned the latest 32 bit microcomputers are as powerful as earlier mainframe computers. At present it is possible to construct a microcomputer possessing most of the features of second and third generation mainframe computers using just a handful of ICs.

Classification of computers

Computers are generally considered to fall into four size categories, all of which are based on roughly the same architecture. The four size categories are:

(a) Microcomputer
(b) Minicomputer

(c) Large mainframes

(d) Supercomputer

A microcomputer uses a microprocessor as the basic central processing unit. The microprocessor consists of integrated circuits contained on LSI chips. The LSI chips can be manufactured in large quantities very inexpensively. The microprocessor is capable of performing virtually all the functions of the conventional CPU (e.g. arithmeticlogic operations or fetching data from memory). Accordingly the microprocessor can be connected to a memory unit and the appropriate input/output devices to form a microcomputer.

Minicomputers are smaller versions of the large mainframe computers. The trend towards miniaturization in computer technology provides two alternative approaches in the design of a computer. The first is to package greater computational power into the same physical size with each new computer generation. The second approach is to package the same computational power into a smaller size. The minicomputer manufacturers elected the second approach in developing their product lines. The cost of a minicomputer ranges from less than a thousand dollars upto about $50,000. Minicomputers can be utilized for the same two general functions as a large mainframe computer. The size of the jobs to be processed must be smaller to be within the capacity of the minicomputer.

The large mainframe computer is distinguished by its cost, capacity and function. The price of a new corporate general purpose computer can run into millions of dollars. The main memory capacity is several orders of magnitude larger than the minicomputer and the speed with which computations can be made is several times the speed of a minicomputer or microcomputer. The functions which are accomplished on a large mainframe computer can typically be classified into two categories:

(a) Complex engineering and scientific problems: examples include iterative calculation procedures often required is heat transfer analysis, fluid dynamics analysis or structural design analysis.

(b) Large scale data processing: examples include corporate accounting and pay roll operations, production scheduling, compiling production costs and maintenance of large information files.

3.2 MICROCOMPUTER BASED SYSTEMS

The microcomputer is a complete computing system built around a microprocessor. A complete computing system has a microprocessor based CPU and it has memory and input/output functions. Some microcomputer systems are not personal computers but rather dedicated computational systems based on a microprocessor. The system has a card cage containing a CPU card, a random-access memory card, a disk controller card and a video (I/O) card. The microcomputer also has a video display and a keyboard. The video display given an alphanumeric display (output) on a cathode ray tube (CRT). That is, it displays both letters and numbers on a screen just like a television's picture tube. The keyboard provides alphanumeric input. The floppy disks are connected to the disk controller card. They provide a way to store programs and data and to interchange programs and data between different microcomputers. They are a form of memory called **mass storage**. This system also has a power supply and packaging.

Hence this system is self-contained. All the cards in the microcomputer are connected by a bus. The bus consists of many signal lines that allow the different cards 'talk' to one another. All the cards communicate through the bus signals because each uses the same set of signals usually any microcomputer card will work if plugged into any slot on the bus.

The microprocessor is on the CPU card. The CPU card also has a clock generator, which generates timing signals for the microprocessor. A read-only memory (ROM) stores a few program instructions which are used to input other programs from the floppy disk. Like all the other cards the CPU card has a number of ICs at the bottom. These ICs electrically interface the CPU card circuits to the bus, which usually requires high current levels than those produced normally.

Besides the ICs that interface with the bus the random-access memory (RAM) card has RAM ICs. This card has all the microcomputers working data and program storage space.

The memory card also has the input/output function which is implemented by a universal synchronous receiver-transmitter (UART) and a clock IC. The UART converts parallel data to serial so the microcomputer can talk to a printer, modem or other remote devices. A fourth card used in this microcomputer is for video display. Besides the ICs that interface with the bus, the front panel card has circuits to drive a video display. The video display actually displays the contents of a special section of memory (RAM), therefore, this card includes the video RAM.

The power supply of this microcomputer system is another separate component. It is connected to all the boards by the bus.

3.2.1 Overview of Microcomputer Systems

A microcomputer system, just as another computer system, includes two principal components, hardware and software.

The hardware is of course the circuitry, cabinetry, etc. and the software is the collection of programs which direct the computer while it performs its tasks.

Hardware

The architecture of a computer is the general layout of its major components, their principal features and how they are connected together. Figure 3.1 depicts the overall architecture of a typical microcomputer system. The components shown in the figure are the central processing unit (CPU), timing circuitry, memory, input/output (I/O) subsystem, bus control logic and system bus. In a microcomputer, the CPU is a microprocessor and is often referred to as the microprocessor unit (MPU). Its purpose is to decode the instructions and use them to control the activity within the system. It also performs all arithmetic and logical computations. The timing circuitry or clock generates one or more trains of evenly spaced pulses and is needed to synchronize the activity within the microprocessor and the bus control logic. The pulse trains output by the timing circuitry normally have the same frequency but are offset in time, i.e., are different phases. For many of the recent microprocessors the timing circuitry, except for the oscillator, is included in the same integrated circuit (IC) as the processing circuitry. The memory is used to store both the data and instructions that are currently being used. It is normally broken into several modules, with each module containing several thousand locations. Each location may contain part or all of a datum or instruction and is associated with an identifier called a **memory**

address. The CPU does its work by successively inputting, or fetching instructions from memory and carrying out the tasks dictated by them.

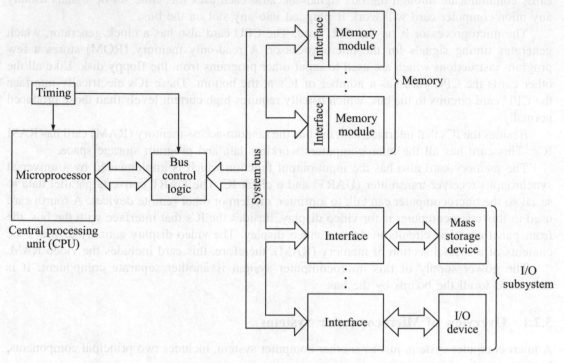

Fig. 3.1 Architecture of a typical microprocessor.

The I/O subsystem may consist of a variety of devices for communicating with the external world and for storing large quantities of information. Card readers, paper tape readers and analog-to-digital converters are examples of input equipment and line printers, plotters, card punches, paper tape punches and digital-to-analog converters are output devices. Some devices such as terminals, provide both input and output capabilities. Computer components for permanently storing programs and data, referred to as **mass storage equipment,** are magnetic tape and disk units, but recent technology has made available magnetic bubble memories (MBMs) and charge-coupled devices (CCDs). Although a mass storage unit may be used to store programs as well as data. Programs must be transferred to memory before they are executed.

The system bus is a set of conductors that connects the CPU to its memory and I/O devices. It is over these conductors, which may be wires in a cable or lines on a printed circuit board (PCB), that all information must travel. Exactly how information is transmitted over the bus is determined by the bus's specifications. Normally, the bus conductors are separated into three groups:

1. The data lines for transmitting the information
2. The address lines, which indicate where the information is to come from or is to be placed
3. The control lines which regulate the activity on the bus

The signals on the bus must be coordinated with the signals created by the various components connected to the bus. The circuitry needed to connect the bus to a device is called **interface** and the bus control logic is the interface to the CPU. Most manufacturers provide a variety of IC devices for facilitating the design of interfaces and the bus control logic depending on the complexity of the system. The bus control logic may be partially or totally contained in the CPU IC.

Memory interfaces consist primarily of the logic needed to decode the address of the memory location being accessed and to buffer the data onto or off of the bus, and of the circuitry to perform memory reads or writes.

All I/O interfaces must be capable of buffering data onto and/or off of the system bus, receiving commands from the CPU and transmitting status information from their associated devices to the CPU. In addition those connected to mass storage units must be capable of communicating directly with memory and this requires them to have the ability to control the system bus. The communication between an I/O interface and the data bus is accomplished through registers which are referred to as I/O ports.

Software

Computer software is generally divided into two broad categories. System software is the collection of programs which are needed in the creation, preparation and execution of other programs. User software consists of those programs generated by the various users of the system in their attempts to apply the computer to solving their problems.

How much system software accompanies a particular computer system depends on the application. A computer whose function is static throughout its life time may execute only one program. On the other hand, a computer that is used for general purpose data processing may require an extensive array of system programs. The general purpose setting programs normally fit into a well defined hierarchy similar to the one shown in Fig. 3.2.

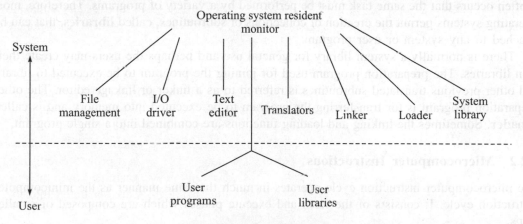

Fig. 3.2 Software hierarchy.

The operating system is a collection of system programs that provide an interface between the user and the machine and enable the machine to be used efficiently. The term **operating system** is not precisely defined and exactly which system programs are included in the operating system depends on the manual or books being referenced. The most important part of the operating system is the resident monitor. It is the part of the operating system that is in the computer's memory at all times while the computer is turned on. The resident monitor must be capable of receiving commands from the users and initiating the corresponding actions to be performed by the operating system.

The operating system also includes programs, **called I/O drivers,** and file management routines for performing I/O operations and for handling large collections of data that are stored on mass storage devices. Whenever a user program or other system program needs to use an I/O device, it does not normally carryout the operation itself, but instead it requests the operating system to use an I/O driver to perform the task. This gives the operating system better control of the computer and alleviates the need to include I/O subroutines within user programs. The file management routines are used in conjunction with the mass storage I/O drivers and for accessing copying and otherwise manipulating files.

In addition the system software may encompass a variety of high-level language translators, an assembler, a text editor and programs for aiding in the preparation of other programs. There are three levels of programming. They are:

(a) Machine language
(b) Assembly language
(c) High-level language

These are already discussed in Chapter 2.

A text editor is a program for inputting or modifying text (letters, numbers, punctuation marks, etc.) that is to be stored or is stored on a mass storage device. The text may constitute an assembler or high-level language program, a set of data or a report. Text editors may be used for several purposes. Our interest in them is due to their being used to create programs.

There are two other system programs that are required in preparing programs for execution. It often occurs that the same task must be performed by a variety of programs. Therefore, most operating systems permit the creation of collections of subroutines, called **libraries**, that can be attached to any system or user program.

There is normally a system library for general use and perhaps the users may create their own libraries. The preparation program used for joining the program to be executed to library and other previous translated subroutines is referred to as a linker or linkage editor. The other preparation program is for transferring the program to be executed into memory and is called a **loader**. Sometimes the linking and loading functions are combined into a single program.

3.2.2 Microcomputer Instructions

The microcomputer instruction cycle operates in much the same manner as the minicomputer instruction cycle. It consists of the fetch and execute phases which are composed of smaller steps.

The instruction set for a microcomputer consists of the instructions for operations which can be accomplished by the CPU. Different microcomputers possess varying capabilities and accordingly there are differences in their instruction sets. The types of instructions generally provided include

(a) Data transfer and input/output instructions
(b) Arithmetic operations
(c) Logical operations
(d) Branching instructions

Data transfer and input/output instructions

This type of instruction is concerned with the transfer of data between a microprocessor register and another component, such as memory, an I/O device, or another register. Each instruction specifies a source register and a destination register. If the destination register is part of the CPU, this is usually implied in the instruction. This type of register is called an **internal register**. External registers, on the other hand, must be explicitly named in the instruction.

Arithmetic operations

Simple arithmetic operations (such as addition and subtraction) as well as more complex operations (multiplication, division, etc.) can be executed by means of the appropriate instructions in this category. Instructions for binary addition are always provided in the microprocessor's instructions set. Subtraction is usually provided also. Multiplication and division can be accomplished either directly in hardware by means of special hardware circuits or by software in the form of subroutines. These more complex operations are not typically included within the instruction sets for 8 bit microprocessors. Microprocessors with 16 bit and greater capacity would usually include multiplication and division instructions.

Logical operations

Most microprocessors include instructions for carrying out the logical operations AND, OR, NOT and EXCLUSIVE-OR.

Branching operations

A branch instruction causes program control to be transferred from the current value of the program counter to another nonconsecutive instruction. In the Intel 8085A microprocessor branching is implemented through the "jump" instruction. Branching operations make possible the use of program loops in which a series of program instructions can be executed repeatedly.

Branching can be made conditional on a set of circumstances or it can be made unconditional in the unconditional jump instruction the address specified in the instruction is loaded into the program counter, and control of execution is passed there. Conditional branches can be performed such that if the given condition is true, the operation is carried out otherwise it is ignored and program execution proceeds sequentially.

3.3 MINICOMPUTER BASED SYSTEMS

The development of LSI (large scale integrated) circuits, and now VLSI has changed the basic principles of computer architecture and has directly led to the proliferation of minicomputers during the 1970's. Minicomputers began to take over from mainframe computers in the 1970's as it was possible to configure them at costs less than the basic cost of communication networks with a mainframe. Early versions of minis were 16 bit word slow and limited storage computers. The Digital Equipment Corporation (DEC) PDP series offers a typical example. In the late 1970's the arrival of super minicomputers such as the VAX II/1780, with 32 bit word and virtual memory operating systems, boosted CAD/CAM applications and helped decentralize them from mainframes. Minicomputers have enable the rapid growth of the CAD/CAM industry. Their low-cost end-user control, generally trouble-free programming and small size have all been significant factors in getting vendors and customers interested. The super minicomputers of today are available with speeds, accuracy, and storage that are more adequate for sophisticated CAD/CAM including computations.

A minicomputer is typically distinguished as a low cost, general purpose word addressable computer with 4 K to 32 K of 8 to 16 bit words of main storage. The 32 bit minicomputers are becoming more and more common. Factors such as speed, ease of programming and other operating characteristics are determined by the particular architecture of the minicomputer. The architecture of a minicomputer follows the general computer structure. The fundamental subunits are the memory, arithmetic and logic processor, control unit and input/output unit.

The memory stores the instructions and data of both user and control programs. The arithmetic logic processor receives data from memory and performs operations on it as directed by the program instructions. The control unit regulates the flow of data in the system, fetches program instruction from memory and decodes the instructions in instruction registers. Together the arithmetic logic processor and the control unit are known as the **CPU of the system**. The input/output unit provides the interface to peripheral devices, transferring data to and from the external world.

A microprogrammed minicomputer differs from the conventional machine in the sense that the control unit, instead of being hard wired logic is a stored program device. The control unit consists of two functional parts, the control decode and the control store. The control decode operates all elements of the computer system including main storage and high speed control store. The control store contains microprograms called **firmware**. The firmware corresponds to conventional computer sequences (software).

Microprogramming increases the speed with which the minicomputer can operate. For certain operations, a microprogrammed minicomputer can perform at higher speeds than a conventional minicomputer. When a large microinstruction set is available the machine can be optimized to perform a broad variety of high speed tasks.

3.3.1 Minicomputer Instructions

The basic minicomputer instruction consists of two parts, an operation code and an operand. The operation code specifies the function to be performed (Arithmetic operation, data transfer, etc.). This function is executed when its operation code is properly decoded by the CPU control logic. The operation is performed on data that are contained in memory. The operand typically consists of an address and an addressing mode. The address specifies the memory location of the data. The addressing mode refers to the method by which the data can be accessed. The data might be accessed directly by going to the address specified in the instruction.

The instruction is carried out during the minicomputer instruction cycle. This consists of a series of events which can be grouped into two phases, fetch and execute. During the fetch phase, the control logic fetches the address of the instruction from the program counter. This address is decoded by the decoder circuitry. The instruction is brought from this address in memory and loaded into the instruction register, where it is decoded. During the execution phase the operations such as addition, subtraction, comparison defined by the contents of the instruction register are carried out. Each instruction has a unique sequence of steps associated with its execution.

Most of the processing that occurs in the minicomputer during program execution takes place in registers. The number and type of registers in the computer architecture determine its performance. Generally as the number of registers in a minicomputer is increased, the required program size and the execution time decrease. Therefore, an indication of computing power in a minicomputer is its number of registers.

A minicomputer is provided with the capability to perform a variety of operations and functions such as data transfer, arithmetic operations and logic functions. These operations and functions are known as the **instruction** set.

The instructions can be classified as memory or non-memory. Memory instructions are used to transfer data to and from memory locations. Non-memory instructions carry out the other functions of the computer. The non-memory instructions can be divided into two categories:

(a) Arithmetic and logic functions
(b) Input/output instructions

The arithmetic instructions include functions such as addition, subtraction, multiplication and comparisons. Logical functions include Boolean operations such as AND, OR and EXCLUSIVE OR.

Input functions are concerned with the transfer of data between main storage and peripheral devices. The factors that determine the I/O capability of a minicomputer are its I/O transfer speed the number and types of I/O software schemes available and the priority control structure which permits external devices to interrupt the execution of the CPU.

Input/output functions are either controlled or interrupt driven controlled. I/O functions are usually achieved by means of the system programming. In this case the data are transferred to and from peripheral devices through an I/O bus.

The I/O bus is equipped with circuitry that can decode the address of a specific peripheral in order to effect the transfer of data to it. In some computers, the device address is given directly and the circuitry transfers the data to that device during a clocked time period associated with that specific device. This method of transfer is called **time multiplexing**.

Interrupt driven input/output occurs when the peripheral device is capable of stopping the execution of the current programs in the CPU and transferring program control to special subroutines. These subroutines when called must save the contents of the working registers, the status register and the program counter. So that when control is returned to the main program, no changes will have taken place and execution of the previous program can continue.

3.4   MAINFRAME-BASED SYSTEMS

At one point all CAD/CAM systems had been mainframe-based since that was the only type of computer available. A typical mainframe-based CAD/CAM system consists of one or more design/drafting station. Each of these include at a minimum a graphics display, an alphanumeric control display (integrated with the graphics screen) and a keyboard. The designer/drafting station is linked to the mainframe computer. To the mainframe are also linked plotters, printers, storage devices, digitizing boards, alphanumeric terminals and other output devices. Design/drafting stations may be equipped with input devices such as light pens, joysticks or other cursor control components. They may also have the programmed function keyboard (PFK) instead of a conventional keyboard or a digitizing tablet to enter commands.

Figure 3.3(a) shows a schematic of the over all system components as well as details of a workstation. The computer environment is typically divided into two: (i) The user environment and (ii) the system environment. The first signifies the components and the area to which the user can have access. These include primarily workstation and peripherals. There is always a maximum number of workstations that a host can support to avoid degradation of response time between users and the system. Users spend most of their time on workstations to perform their work. They normally know how to use and operate the various components of the stations and peripherals. A typical workstation consists mainly of two major segments, input and output devices. The former includes cursor control devices, for graphic input and text input devices. The cursor can be controlled via a lightpen, joystick, mouse, electronic pen (stylus) with a digitizing tablet, thumb wheel or track ball. Text can be input through a keyboard which may have programmed function keys. Output devices consist of a graphic display with a hard copy printer (a dot matrix printer) to provide convenient raster plots of full screen contents. Figure 3.3(b) shows the details of a workstation with specific input and output devices. In this figure, the cursor control device is an electronic pen and a digitizing tablet and the keyboard is the text input device. The graphics display is shown as a display monitor and processor.

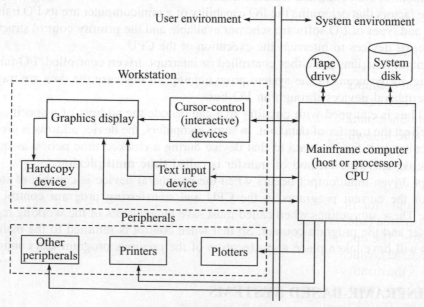

(a) Overall view of the system

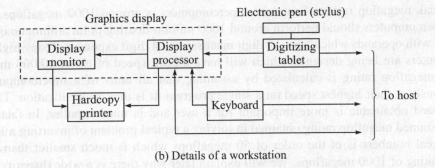

(b) Details of a workstation

Fig. 3.3 Schematic diagram of a mainframe-based CAD/CAM system.

3.5 SUPERCOMPUTERS

In the last three years a lot of interest has been generated in our country on supercomputers. Supercomputers have become part of the vocabulary of all educated persons as press reports and cartoons regularly appear in the newspapers on India's order for a supercomputer and the delays in procuring it due to export controls on high technology by the countries manufacturing these machines. There is thus a widespread curiosity to know what is a supercomputer and how it is different from other computers.

3.5.1 Defining a Supercomputer

The fastest and the most expensive computers available at any given time are generally called **supercomputers**. This is not a satisfactory definition but we are forced to use such a definition as computer technology has been evolving rapidly. In the past 20 years there has been a thousand-fold increase in the speed of arithmetic operations of computers and it is thus difficult to give a time invariant definition of the speed of a supercomputer. Although the cost of computers for a specified speed and size has been going down, typically the cost of supercomputers has remained constant around US$ 10 million. Currently for a computer to be called a supercomputer it must have the following characteristics:

High computing speed

The computing speed of a supercomputer is measured in megaflops. A mega is a million and flops is an abbreviation for floating point operations per second. A floating point operation is an arithmetic operation (add, subtract, multiply or divide) on operands which are real numbers with fractional parts. The operands are expressed as a pair (mantissa, exponent). For example, the number 185.67827 which equals 0.18567827×10^3 is expressed as (0.18567827, 3) where 0.18567827 is the mantissa and 3 is the exponent. In supercomputers the number of digits in the mantissa is 15 whereas it is 8 in other computers. By other computers we mean personal, mini and mainframe computers normally used in computing laboratories. The range of the exponent in supercomputers is also high, around ±5000 compared to ±99 in other computers. Thus, the arithmetic precision and range of numbers used in supercomputers is significantly higher than those used in other computers.

The peak megaflop rating of modern supercomputers is around 1000 megaflops. In other words, supercomputers should perform around 1000 million floating point arithmetic operations per second with operands which have 15 digit mantissa and 4 digit exponent. Today higher speed supercomputers are being designed which will have a peak speed of about 10,000 megaflops. The peak megaflop rating is calculated by assuming that all units of a supercomputer work simultaneously at their highest speed on a single program. It is thus an idealization. The steady average speed obtainable is more important for a user and is much smaller. In fact the best average sustained megaflop rating obtained in solving a typical problem of inverting a 100 ´ 100 matrix of real numbers is of the order of 50 megaflops which is much smaller than the peak megaflop rating of 1000 megaflops. We will explain later why there is a wide disparity between the peak megaflop rating and the average sustained megaflop rating. In contrast with supercomputers, the average megaflop rating of, for instance, VAX 8810 a mainframe computer made by the Digital Equipment Corporation (DEC) is around 1 and that of an IBM PC is 0.01.

High precision of stored numbers

We mentioned that supercomputers use 15 digits for mantissa which is double the number used normally in computers. One may wonder why such a high precision for representing numbers is needed. The reason is the possibility of accumulation of the small rounding errors made by the computer while performing arithmetic operations. Larger the number of arithmetic operations, larger will be the accumulation of errors. A computer carrying out 100 million operations per second will also be making 100 million rounding errors per second! Special care should thus be taken to prevent accumulation of rounding errors. More significant digits are used to represent real numbers in supercomputers to provide better factor of safety in arithmetic computation.

All numbers are stored as binary numbers in computers. Each real number is stored in a supercomputer's memory as a 64 bit unit called a **word**. Out of these 64 bits 49 bits are used to represent the mantissa (giving around 15 digit significance) and 15 bits are used to store the exponent and its sign. This provides an exponent range of $10^{\pm 5000}$.

Large, fast main memory

The size of the main memory and its speed are important parameters of a supercomputer. As pointed out in the previous paragraph a word of a supercomputer is 64 bits long. The width of one data path from the main memory to the processing unit of a supercomputer is 64 bits. The size of the main memory is at least 8 million words and recent machines provide 256 million words of main memory. This is to be contrasted with mainframe computers whose main memories are around 8 million 32 bit words. Besides the size, the time to access data from the main memory should be comparable to the arithmetic speed. Thus, the memory is organized as an interleaved set of memory banks. Typically around 64 banks are interleaved allowing 64 words to be read in almost the same time as it takes to read one word. This time is around 50 ns giving an average access of one word from the main memory to the processor every 0.78 ns (1 ns = 10^{-9} s). This is to be contrasted with mainframe computers whose average access time of data from main memory to the processor is at least 100 times slower.

Large, fast secondary memory

As the computation speed of supercomputers is high, the data to be processed must be readily available in the main memory. For a processing speed of 150 megaflops a maximum of 150 million pairs of operands would be needed per second and all of them cannot be stored in the main memory. As the data is retrieved from the main memory and processed new data should be moved to the main memory from the secondary memory. The size of the secondary memory should be large and the speed with which data is to be transferred to main memory should be compatible with the speed of the main memory. Normally atleast 40 Gigabytes (Giga = 10^9, Gigabytes usually abbreviated as GB) storage is provided in supercomputers. The rate at which data is transferred from the secondary memory (usually magnetic disk memory) is around 40 Megabytes per second (Mega = 10^6, Megabytes abbreviated as MB). Disks in mainframe computers provide a data transfer rate of around 5 MB per second.

To summarize, a computer may be classified as a supercomputer if its average arithmetic speed is of the order of 50 Megaflops with 64 bit operands; it has a large main memory capacity of the order of 256 million 64 bit words with an access time compatible with the arithmetic speed and it has around 40 GB of secondary memory with a data transfer rate to main memory of around 100 MB per second. The main point to note is that mere megaflop rating is not sufficient to call a computer a supercomputer; there must be matching high speed high capacity main memory and high capacity disk drives with very high speed data transfer capability.

3.5.2 Why do We Need Supercomputers?

Of late simulation has emerged as an important method in science, complementing theoretical analysis and experimental observations. Numerical simulation has many advantages:

- It is cheaper than setting up big experiments or building prototypes of physical systems.
- It is possible in a numerical model to change many parameters and observe their effect.

Experiments do not allow easy change of many parameters.

- Numerical modelling is versatile. A wide range of problems can be simulated on a computer.
- Observations and interactions allow models to be refined. Such refinements provide a better understanding of physical problems which cannot be obtained from experiments.

Numerical simulation is also known as "numerical experimentation" as the philosophy is similar to conducting experiments. In some instances it may in fact be the only feasible substitute for experiments, for example, nuclear fusion experiments and finding out the damage to an aircraft and type of injuries to passengers when an aircraft crashlands.

The role of experiments, theoretical models and numerical simulation models is shown in Fig. 3.4. It is seen that experiments, theory and numerical simulation now form three interacting methods in science and engineering.

With the advance in science, the models used incorporate more detail. This has increased the demand for computing speed and storage capacity. For example, in order to design supersonic aircraft it is necessary to realistically simulate turbulent aerodynamic flows round its wings and body. This is a general non-linear problem modelled by using partial differential equations and

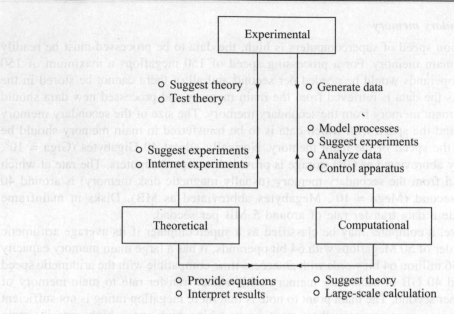

Fig. 3.4 Interaction between theory, experiments and simulation.

cannot be solved analytically. Such a problem is solved by numerical simulation which requires the surface of the body and wings to be modelled by 10^7 tiny squares-bounded by parallel lines (or a grid) as shown in Fig. 3.5.

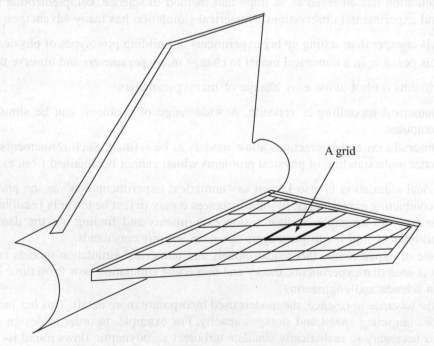

A grid

Fig. 3.5 Grids on the surface of an aircraft wing used for simulation.

The partial differential equations are discretized to difference equations which are in turn solved as a set of simultaneous algebraic equations. For each grid point 5 to 30 real numbers are stored to represent vector quantities such as velocity, acceleration and pressure. Sets of these equations are normally solved numerically using iterative methods (i.e., trial and error methods). In such iterative methods several, trials (100 to 1000) are needed for each grid point before the results converge. The calculation of each trial value normally requires around 100 to 500 floating point arithmetic operations. Thus, the total number of floating point operations required for each simulation run is approximately given by

Number of floating point operations per simulation =
Number of grid points $\times$
Number of values per grid point $\times$
Number of trials $\times$
Number of operations per trial $= 10^7 \times 20 \times 100 \times 500 = 10^{13}$

If each floating point operation takes 1 microsecond (corresponding to 1 megaflop speed) then the time taken for each simulation run is 10^7 seconds which is 115 days.

A computer with 1 megaflop speed will thus take 115 days to complete one simulation run if it is operated 24 hours a day. In other words, this problem cannot be solved using such a computer. If the sustained megaflop rating of a supercomputer is 200 megaflops then each simulation run will take 13.8 hours, which is still quite high, but manageable.

There are many other problems which also require large computational time. A common application is to model global weather. The behaviour of the earth's atmosphere affects global weather. The behaviour is modelled by partial differential equations in which the most important variables are the wind speed, air temperature, humidity and atmospheric pressure. The objective of a numerical weather model is to predict the status of the atmosphere at a particular area at a future time based on the current and past observations of the atmosphere. This is done by approximating the partial differential equations by a system of difference equations in which physical quantities are specified at points on a three-dimensional grid over the surface of the earth as shown in Fig.3.6. In the horizontal plane the grid points are defined by 87 parallels of latitude between the north and south poles and 144 meridian circles equally spaced around the globe. In the vertical direction nine layers describe the atmosphere and five physical variables are used for the description. In this example also around 10^{10} floating point calculations are needed for each solution.

In general, the problem complexity is described by the formula: $PC = G \times V \times T \times A$ where

PC = Problem complexity
G = Geometry of the grid system
V = Variables per grid point
T = Number of steps per simulation for solving problem
A = Number of floating point operations per variable

For the supersonic aircraft design example:

$$G = 10^7, \ V = 20, \ T = 100 \quad \text{and} \quad A = 500$$
$$PC = G \times V \times T \times A = 10^{13} \text{ operations}$$

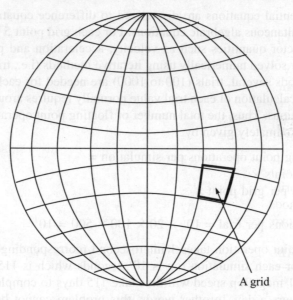

A grid

Fig. 3.6 Grid for numerical weather model for the Earth.

For the weather modelling problem,

$$G = 144 \times 87 \times 9 = 112752, V = 5, T = 200, A = 400 \text{ giving } PC \sim 10^{11} \text{ operations.}$$

Currently a large number of realistic applications require 10^{12} to 10^{14} operations per solution. If each solution has to be done in about an hour then the sustained speed of a computer should be $10^{14}/60 \times 60$ operations per second which is equal to 27,700 megaflops!

If computing time per solution exceeds 100 hours, we call it an intractable problem. Thus with a machine giving 100 megaflop sustained speed a problem requiring 10^{14} floating point operations for a solution is intractable. From the earlier examples it is clear why we need supercomputers with speeds in the thousands of megaflop range to solve problems of current interest to scientists and engineers.

3.5.3 How do Supercomputers Achieve Their Speed?

One method of increasing the speed of computers is to use faster semiconductor components to build units of a computer. Higher speed components cost more and normally dissipate more heat. For example, low speed personal computers use integrated circuits based on silicon semiconductors. Currently Gallium Arsenide semiconductor devices, which are faster, are used in some supercomputers. The rate of growth of speed by using better devices and technology is relatively slow. For instance, the time to add two floating point numbers in a high performance computer in 1980 was 20 nanoseconds and in 1990 it was 6 nanoseconds.

Another method to increase the speed of computation is to design the computer so that the different units of the computer work simultaneously. For instance, while the processor is computing, data which may be needed later could be fetched from memory and simultaneously an I/O operation can be going on. Such an overlap of operations is achieved by using both hardware and software features. This method is called **architectural method**.

Besides overlapping operations of various units of a computer the arithmetic unit itself may be designed to exploit parallelism inherent in the problem being solved. For example, if two vectors are to be added, all pairs of components of the vector may be added simultaneously by a set of adders thereby reducing the time taken to add the vectors. This type of parallelism is called **data parallelism**. Time taken to add two vectors may also be reduced by designing an adder known as a **pipeline adder** which uses temporal parallelism. Still another method of increasing speed of computation is to organize a set of computers to work simultaneously and cooperate to carry out tasks in a program. These methods are also classified as architectural methods.

As the increase in speed of electronic components is limited by physical constraints, supercomputers primarily use architectural methods to exploit parallelism. Table 3.1 summarizes the architectural methods used to increase the speed of computers. In the subsequent chapters of this book we will explain the architectural methods.

TABLE 3.1 Methods used to Increase the Speed of Computers

- Use faster devices such as GAs (Genetic algorithms) to build computers
- Use architectural methods to exploit parallelism

The architectural methods are:

- Overlap operation of different units of a computer
- Execute many instructions simultaneously with multiple functional units
- Increase speed of arithmetic logic unit by exploiting data and/or temporal parallelism.

3.5.4 Applications of Supercomputers

Supercomputers are indispensable tools for scientists and engineers in their research and development. As was pointed out in the first chapter, numerical experimentation using computers is now as important as hypothesis or theory formulation and experimental verification. Numerical experimentation, namely, simulation on a computer of the mathematical model enhances a scientist's ability to reason about complex phenomenon in many ways. Simulation allows the study of phenomenon which are difficult to study experimentally. Examples of such situations are simulating accidents in nuclear reactors, simulating crashes of a motor car or aeroplane or spread of fire in oil wells. Simulation allows study of complex, non-linear models which are difficult to solve analytically. Examples are the study of fusion reactors, formulating atmospheric models and modelling spread of oil spills in oceans. Simulation is also useful to test whether a proposed theory is correct. Examples are numerical study of drug reactions and synthesis of drugs. We illustrate these with examples in this chapter. Besides their use in science and engineering which may be thought of as traditional applications, many other applications have now emerged which were not foreseen by supercomputer designers. The interesting ones are the use of supercomputers in Hollywood to make cartoons and by advertising agencies to make innovative graphics based video presentations. Economists now use it for large economic models, security agencies use them to design secret codes and cryptologists to break these codes. The enormous speed of supercomputers, wider availability, reduction in cost and the emergence of excellent high resolution graphics have all led to an increase in the variety and novelty of supercomputer applications.

Motor car crash simulation: An interesting application of supercomputers is to simulate the effect of various types of accidents involving a motor car, its driver and passengers. The primary aim is to design a car body in such a way that if it collides with another vehicle on the side, or head on, or crashes against a barrier the driver and the passengers in the car will be safe. In western countries standards have been specified for motor car bodies to ensure safety which should be adhered to by manufacturers. Thus, manufacturers design their vehicles with care so that it is a safe vehicle to drive.

In earlier days (1970s and 80s) manufacturers used to make model cars and actually crash them against a barrier at different speeds and study the effect. This has many disadvantages. It is very expensive to make a car and crash it. Thus, the number of experiments carried out would be limited by this cost. If it is found after the crash that major modifications are needed in the body design, there will be a long delay before the new model car is ready to be marketed. This delay will reduce the competitiveness of the manufacturer and hence his sales. Thus, the extensive crash tests were not feasible.

With the advent of computers a natural question to ask is "Is it possible to model the motor car body mathematically in enough detail and simulate crashes using this mathematical model on a computer?". To answer this question let us examine how to mathematically model the body of the car. This is done by taking the surface of the body and dividing it into small squares called **finite elements**, as shown in Fig. 3.7.

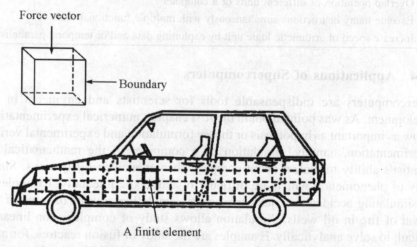

Force vector

Boundary

A finite element

Fig. 3.7 Division of a motor car body as finite elements.

The force due to collision is modelled as force vectors on each finite element. With the known geometry of the element and the property of the material of the motor car body the stresses and resultant deformation is computed. If the body is divided using a 100×100 grid, each corner has three translational and three rotational degrees of freedom so that there are roughly 6^4 unknown quantities to be computed. The basic matrix equation to compute these is given by:

$$[K] \{U\} = \{P\}$$

where $[K]$ is the stiffness matrix generated from the known characteristics of the material used for the body $\{U\}$ is the unknown displacements to be computed and $\{P\}$ is the vector of forces applied to the grid points and generated from the type of crash to be simulated.

The computations required to generate $[A*]$ $\{/*\}$ and subsequently solve for $\{(/\}$ are substantial. They are around 10^9 floating point operations for each simulation run. The stiffness matrix is sparse with over 90% of the element values being zeros. This problem takes too long a time on mainframe computers. It is found that the time to run one simulation on a 200 megaflop (peak speed) supercomputer is 6 hours. In spite of this, manufacturers find this technique more cost-effective than testing with physical car models. The main advantage is that simulation can be performed very early in the design cycle of a motor car. Many body designs can be simulated and only the good ones kept for further detailed design and analysis. Fewer real crash tests on physical prototypes are needed. The body design can be optimized to meet very strict safety specifications. This is one of the applications for which car manufacturers abroad use supercomputers extensively.

Similar ideas are used to design bodies of aircrafts and space vehicles. The problems are a lot more complex and need very large supercomputers.

3.6 PROGRAMMABLE LOGIC CONTROLLER

The programmable controller (PC) was introduced in the late 1960s as a substitute for electromechanical relay logic systems. At that time relay control systems were typically used to regulate the operation of production equipment. The problems with electromechanical relays are their physical size and programming inflexibility. The programmable controller could perform the same functions as a relay logic system with greater flexibility and space requirements. The PC can be defined as a sequential logic device which generates output signals according to logic operations performed on input signals. The sequence of instructions that determines the inputs, outputs and logical operations constitute the program. The initial success of the programmable controller was due to its apparent similarity to relay control systems. Wiring charts, called **ladder diagrams,** have been used for many years to set up relay panels shop personnel are familiar with these diagrams for wiring and maintenance of the panels. Most PCs are programmed using the same type of ladder diagrams. PCs can be interfaced with computers in production operations. The programmable controller is used to direct certain aspects of the operations while computers are used for other monitoring, control and data processing functions.

A programmable logic controller is a solid-state system designed to perform the logic functions previously accomplished by components such as electromechanical relays, drum switches, mechanical timer/counters etc. for the control and operation of manufacturing process equipment and machinery. A programmable controller is an industrial computer in which control devices such as limit switches, push buttons, proximity or photoelectric sensors, float switches or pressure switches to name a few provide incoming control signals into the unit. An incoming control signal is called **input**.

Incoming control signals or inputs, interact with instructions specified in the user program, which tells the PLC how to react to the incoming signals. The user program also directs the PLC on how to control field devices like motor starters, pilot lights and solenoids. A signal going out of the programmable logic controller to control a field device is called **output**.

The National Electrical Manufacturing Association (NEMA) defines a programmable controller (PC) in the following terms.

A programmable controller is a digitally operating electronic apparatus which uses a programmable memory for storage of instructions to implement specific functions such as logic, sequencing, timing, counting and arithmetic to control through digital or analog input/output modules, various types of machines or processes.

3.6.1 Components of a PLC

A typical PLC can be divided into five components:

    (a) Input/output section
    (b) Processor
    (c) Memory
    (d) Programming device
    (e) Power supply

These components will be contained in suitable housings and cabinets to permit them to withstand the shop environment. The configuration of the PC system is presented in Fig. 3.8.

The programmable controller is designed to be connected to industrial equipment. This connection is accomplished by means of the input/output section. The input/output section consists of input modules and output modules. The number of input and output modules necessary is dictated by the requirements of the equipment that is to be controlled by a PLC. The input section is designed to receive process and machine signals and convert them into an acceptable form for the PC. The output section converts PC control signals into a form which can be used by the process equipment. The input section is separate from the output section and both types are designed to be modular for flexibility.

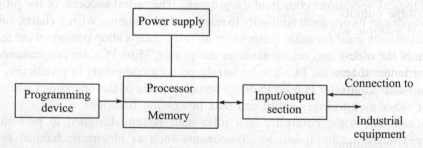

Fig. 3.8 Typical programmable controller configuration.

The processor is the central component of the PC and is referred to as the central processing unit (CPU). It executes the various logic functions, performs operations on inputs and determines the appropriate outputs. The processor unit houses the processor which is the decision-maker or "brain" of the system. The brain is a microprocessor based system that replaces control relays, counters, timers, sequencers and so forth and is designed so that the user can enter the desired program in RELAY LADDER logic. The processor then makes all the decisions necessary to carry out the user program, based on the status of the inputs and outputs for control of a machine or process. It can also perform arithmetic functions, data manipulation and communications between the local/input/output section, remotely located I/O sections, and/or other networked PLC systems.

The PC memory is used to store the program which specifies the logic of the input/output processing. Memory for a programmable logic controller is specified the same way as for a digital computer (e.g. 1 K equals 1024 bits of storage). Memory capacities of commercial PCs range from less than 1 K upto more than 48 K.

The program is loaded into the PC memory by means of a programming device. Either of two types of programming devices can be used for this purpose. The first is the CRT terminal. The CRT permits the programmer to use either a relay ladder diagram or other programming language to input the control logic into memory. The second type of programming device is a small manual keyboard device. With this device, the control logic and other data are entered by means of special function buttons and thumbwheels. Manual programming devices are less expensive and more portable, but the CRT is more convenient for programming.

The power supply drives the PC and serves as a source of power for the output signals. It is also used to help protect the PC against noise in the electrical power lines. The power supply is necessary to convert 120 or 240 volts AC voltages to the low voltage DC required for the logic circuits of the processor and for the internal power required for I/O modules. The power supply does not supply power for the actual input or output devices themselves; it only provides the power needed for the internal circuitry of the input ad output modules. DC power for the input and output devices, if required must be provided from a separate source.

3.6.2 Open PLC Systems

Programming the PLC: The PLC can do nothing until someone develops a user program and loads it into the CPU's memory. Once the CPU has the program in memory and has been put into run mode, it can look at inputs and as a result of solving the user program ladder logic instructions, it can control the outputs and their associated field devices.

There are multiple ways to program a PLC:

1. One of the oldest methods of programming a PLC is by pressing buttons on a handheld programming terminal to enter a user program.

2. The most popular method of PLC programming is using a personal desktop computer and either a dos or windows operating system to run the manufacturers software for that specific PLC.

3. PLC programming can be accomplished using a laptop personal computer running PLC programming software and a personal computer memory card international association (PCMCIA) interface card or in some cases a direct connection between the personal computer serial port and the PLC CPU.

4. Using a industrial computer and the PLC manufacturers program software.

5. Using third party "open software" and running a personal computer as the PLC's CPU.

Even though PLCs from different manufacturers function similarly, supporting multiple brands can be a real maintenance pain in the neck as well as a programming nightmare. Each PLC has its own propriety software packages and programming techniques. In some cases, each

PLC brand has its own proprietary network. Many of today's PLCs are called **closed systems** because you must use the software that was manufactured specifically for that manufacturer's PLC.

PLC software has evolved at a rapid pace over the last 10 to 15 years. Today the costs for PLC hardware are declining with the emergence of smaller, smarter and faster "micro-PLCs". Software is becoming more sophisticated as PLC functionality increases. As a result, software is becoming a proportionally larger cost item in a PLC system. Few would disagree that software development, debugging, installation, control system commissioning, troubleshooting and future system modification would be greatly simplified if there were some standards governing the development, usability and reusability of software. With this in mind, a group of technical experts was commissioned in 1979 to develop the first draft of comprehensive programmable controller standard. This standard would attempt to standardize PLC programming and consequently make a program developed on one system usable on other PLC platforms with minimum modification. The first draft from this committee was issued in 1982. This standardization is known as the **IEC 1131 standard**.

The IEC 1131 standard for programmable controllers comprised five parts:

Part 1: General information

Part 2: Hardware requirements

Part 3: Programming languages

Part 4: User guidelines

Part 5: Communication

Part three of the standard; IEC 1131-1 (Programming languages) has attracted the most attention from the international community.

3.6.3 The IEC 1131-3 Programming Standard

Standard IEC 1131-3 defines a consistent set of programming languages for programmable controllers. The specification consists of four traditional languages and one higher level programming language. The languages are broken down into two graphical languages, ladder diagram and function block diagram and two text based languages. Instruction list and structured text along with flowchart type programming is called **sequential function chart programming**.

Ladder diagram (LD)

Ladder diagram programming is similar to relay ladder logic. When most people in North America think of PLC programming, they associated ladder rungs and ladder programming with PLCs under the IEC 1131-3 standard. Individuals familiar with relay ladder diagrams or PLC ladder programming can continue to program with relay ladder logic.

Advanced features supported by ladder diagram programming include function blocks, which simplify process-orientated applications. ASAP Inc. of Chagrin Falls, Ohio, is one of the many software developers. ASAP has an open-control software solution called **ASIC-100 software**. ASIC relay ladder logic programming software offers a comprehensive set of control elements and function blocks. These elements are organized into a palette for easy access. Power flow, I/O forcing and on-line changes are provided during run time.

Function block diagram (FBD)

Function block diagram programming is based on a graphical language widely used in Europe. Process flow applications can be depicted graphically as function blocks that are wired together like circuit diagrams. Function blocks are standard blocks that execute algorithms. The function blocks are controlled by external parameters.

Instruction list (IL)

Instruction list is a low-level, assembly-type language. Instructions are organized into a list-like format. Instruction list programming allows only one operation to be performed per line. As an example, storing a value in a memory location would be a single operation. Instruction list programming is usually used in smaller applications.

Structured text (ST)

Structured text is an English-like programming language that resembles BASIC programming. Structured text programming can be used to perform most of the tasks currently done with ladder logic. Figure 3.9 illustrates a timer ladder rung. The structured text program equivalent is as follows.

IF (I:1/0) AND (!T4:0.DN) THEN TON (T4:0, 1.0,4,0);

The first line includes our two input instructions. The exclamation mark before the T4:O.DN identifies the input instruction as an XIO bit instruction. Our timer instruction is on the second line. The TON instruction is identified before the parentheses, while the timer address, time base, and preset and accumulated values are listed inside the parentheses.

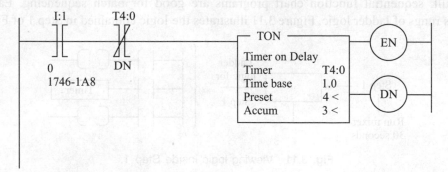

Fig. 3.9 Ladder logic to be represented as a structured text program.

Sequential function chart (SFC)

Sequential function chart (SFC) programming is similar to flow chart programming. SFC programs consist of steps and transitions. Each step is represented by a box that contains one or more major actions. When all actions in the box are satisfied, the box is exited. A transition step must be true before moving onto the next step. Once leaving a particular step, the processor executes the next step. Previous steps are no longer executed. ASAP Inc.'s ASIC 100 is a sophisticated flowchart-like programming language that offers conditional branching, parallel branching, control loops and jumps. ASIC 100s patented extensions to SFC add motion control capability, icon-based programming, new control functionality and enhanced diagnostics.

Programming with a sequential function chart lets you develop segmented programming. Rather than developing one long ladder program, the program can be divided into several sequences or steps (see Fig. 3.10). There is a transition between each step. The advantage to sequential function charts is that only the logic in the active step is scanned until it is time-to-transition to the next step. Compare this to scanning the entire ladder program when only a few rungs are active. The major disadvantage of sequential function chart programs is that alarms cannot be monitored.

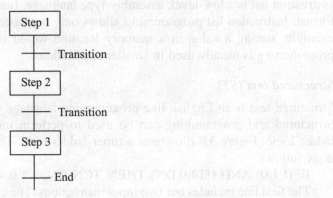

Fig. 3.10 Three-step sequential function chart with transition points identified.

Each step corresponds to a control task, for example, turning on a mixer for a specified time. As a result, sequential function chart programs are good for batch sequencing. Each step comprises rungs of ladder logic. Figure 3.11 illustrates the logic contained in Step 1 of Fig. 3.10.

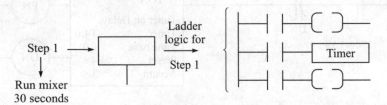

Fig. 3.11 Viewing logic inside Step 1.

The logic in the current step runs continuously from rung one to the last rung. When the last rung of a step is scanned, the processor will look at the transition to see if the processor is to go onto the next step. The transition is the logic condition that directs the processor to progress to the next step. The transition can be as simple as one rung of logic. When the transition logic is true, the processor will scan Step 2 repeatedly until its transition is true.

The PLC processor checks the current steps transition at the end of scanning each rung of the step. If the transition is true, the processor goes onto scan the logic in the next step. If the transition is false, the current steps logic is rescanned (see Fig. 3.12).

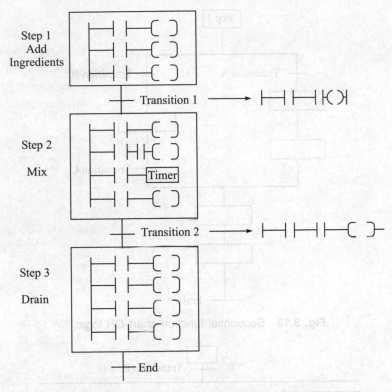

Fig. 3.12 Entire sequential function chart program with logic rungs visible.

Step 1, which comprises rungs one, two and three, is continuously scanned until all ingredients are added. The transition rung will be true when all ingredients have been added. With Transition 1 true, the processor leaves Step 1 and begins scanning Step 2. Step 2 is the mixing sequence. This step will be continuously scanned until the mixing time has elapsed. The timer will make the rung in Transition 2 true when the mixing time expires. With Transition 2 true, the processor will stop scanning Step 2 and move on to Step 3. Step 3 will execute until the transition is true, at which time the sequence will stop. The PLC processor will now look at Step 1 for a start signal to begin the sequence over again.

Selection branch. A selection branch is equivalent to OR logic. When Step 1 in Fig. 3.13 is completed, the processor will look at the transition at the beginning of each of the branches. The transition that becomes true first will determine the direction of program flow.

If both transitions A and B become true at the same time, priority will be given to the left-most one, Transition A.

Simultaneous branch. A simultaneous branch executes parallel paths simultaneously. In Fig. 3.14, when Step A is scanned and Transition B is true, the parallel path of steps C and F will be run at the same time as the path with Step D.

The transitions B and G are both outside the branch. Each time the processor finishes executing logic in the simultaneous branch, the common Transition G in Fig. 3.14 will be tested to see whether it is true. If it is true and both branches have been executed at least once, the control will pass to Step H.

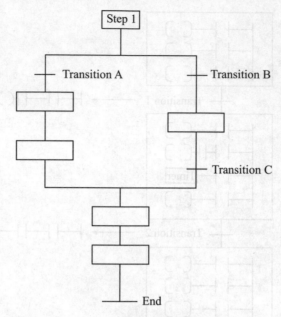

Fig. 3.13 Sequential function chart OR logic.

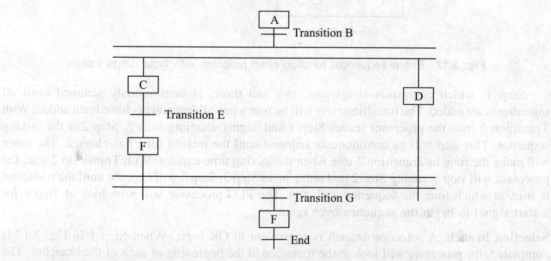

Fig. 3.14 Sequential function chart illustrating a simultaneous branch.

There are many answers to the question of what hardware to use to program a PLC. We have looked at the traditional methods of PLC programming in this chapter. However, there is a relatively new solution to PLC programming and control. What if we combine the separate personal computer as a programming terminal and the PLC's CPU? In this way, we can eliminate the PLC's CPU and directly control our system from our personal computer. In the past, the standard personal computer was a fragile device with questions regarding reliability and high failure rates. The unpredictability of the personal computer made it a risky choice as a factory floor controller. As a result, the electricians and maintenance individuals who installed and maintained PLC control systems—and the engineers who designed them—had little confidence that a personal computer could do the job of an industrially hardened, easily maintainable, expandable, reliable PLC and its CPU.

Today's personal computer hardware has been standardized and made more reliable, and computational ability has been increased. With computer processing speed and memory size increased, along with the increased ability to connect to networks like Ethernet, why not consider using a personal computer as the control processor? When a personal computer is used to replace the traditional programmable logic controller, this is called a soft PLC or an **open-architecture control system**.

3.7 MANUFACTURING AND ASSEMBLY PROCESSES

In an industrial setting, PLCs are used to automate manufacturing and assembly processes. By process, we mean a step-by-step procedure whereby a product is manufactured and assembled. It is the responsibility of the product engineering (PE) department to plan for the manufacture of new or modified products. According to Rehg and Kraebber, in their book, **Computer-Integrated Manufacturing,** 2nd ed., the PE plan has seven elements:

1. Process planning
2. Production machine programming
3. Tool and fixture engineering
4. Work and production standards
5. Plant engineering
6. Analysis for manufacturability and assembly
7. Manufacturing cost estimating

Other processes might involve the filling and capping of bottles, the printing of newspapers or the assembly of automobiles. The list is endless. In many such manufacturing situations, PLCs play an important role in carrying out the various processes.

3.8 ADVANTAGES AND DISADVANTAGES OF PLC

Following are the major advantages of using a programmable controller:

Flexibility

In the past, each different electronically controlled production machine required its own controller; 15 machines might require 15 different controllers. Now, it is possible to use just one model of a PLC to run any one of the 15 machines. Furthermore, you would probably need fewer than 15 controllers because one PLC can easily run many machines. Each of the 15 machines under PLC control would have its own distinct program.

Implementing changes and correcting errors

With a wired relay-type panel, any program alterations require time for rewiring of panels and devices. When a PLC program circuit or sequence design change is made, the PLC program can be changed from a keyboard sequence in a matter of minutes. No rewiring is required for a PLC-controlled system. Also, if a programming error has to be corrected in a PLC control ladder diagram, a change can be typed in quickly.

Large quantities of contacts

The PLC has a large number of contacts for each coil available in its programming. Suppose that a panel-wired relay has four contacts and all are in use when a design change requiring three more contacts is made. Time would have to be taken to procure and install a new relay or relay contact block. Using a PLC, however, only three more contacts would be typed in. The three contacts would be automatically available in the PLC. Indeed, a hundred contacts can be used from one relay—if sufficient computer memory is available.

Lower cost

Increased technology makes it possible to condense more functions into smaller and less expensive packages. Now you can purchase a PLC with numerous relays, timers and counters, a sequencer and other functions for under hundred dollars.

Pilot running

A PLC programmed circuit can be prerun and evaluated in the office or lab. The program can be typed in, tested, observed and modified if needed, saving valuable factory time. In contrast, conventional relay systems have been best tested on the factory floor, which can be very time consuming.

Visual observation

A PLC circuit's operation can be seen during operation directly on a CRT screen. The operation or misoperation of a circuit can be observed as it happens. Logic paths light up on the screen as they are energized. Troubleshooting can be done more quickly during visual observation.

In advanced PLC systems, an operator message can be programmed for each possible malfunction. The malfunction description appears on the screen when the malfunction is detected by the PLC logic (for example, "MOTOR #7 IS OVERLOADED"). Advanced PLC systems also may have descriptions of the function of each circuit component.

Speed of operation

Relays can take an unacceptable amount of time to actuate. The operational speed for the PLC program is very fast. The speed for the PLC logic operation is determined by scan time, which is a matter of milliseconds.

Ladder or Boolean programming method

The PLC programming can be accomplished in the ladder mode by an electrician or technician. Alternatively, a PLC programmer who works in digital or Boolean control systems can also easily perform PLC programming.

Reliability and maintainability

Solid-state devices are more reliable, in general, than mechanical systems or relays and timers. The PLC is made of solid-state components with very high reliability rates. Consequently, the control system maintenance costs are low and downtime is minimal.

Simplicity of ordering control system components

A PLC is one device with one delivery date. When the PLC arrives, all the counters, relays and other components also arrive. In designing a relay panel, however, you may have 20 different relays and timers from 32 different suppliers. Obtaining the parts on time involves various delivery dates and availabilities. With a PLC you have one product and one lead time for delivery. In a relay system, forgetting to buy one component would mean delaying the startup of the control system until that component arrives. With the PLC, one more relay is always available— provided that you ordered a PLC with enough extra computing power.

Documentation

An immediate printout of the true PLC circuit is available in minutes, if required. There is no need to look for the blueprint of the circuit in remote files. The PLC prints out the actual circuit in operation at a given moment. Often, the file prints for relay panels are not properly kept up to date. A PLC printout is the circuit at the present time; no wire-tracing is needed for verification.

Security

A PLC program change cannot be made unless the PLC is properly unlocked and programmed. Relay panels tend to undergo undocumented changes. People on late shifts do not always record panel alterations made when the office area is locked up for the night.

Ease of changes by reprogramming

Since the PLC can be reprogrammed quickly, mixed production processing can be accomplished. For example, if Part B comes down the assembly line while Part A is still being processed, a program for Part B's processing can be reprogrammed into the production machinery in a matter of seconds.

These are some of the advantages of using a PLC. There will, of course, be other advantages in individual applications and industries.

Following are some of the disadvantages of, or perhaps precautions involved in, using PLCs:

Newer technology

It is difficult to change the thinking of some personnel from ladders and relays to the PLC computer concept. Although today, with the pervasive use of computers not only at home and in the office but on the factory floor, acceptance of the computer as a powerful and reliable productivity-enhancing tool is, if not universal, almost so. Electricians and technicians are lining up to take courses on PLCs because they know that doing so contributes to job security and advancement.

Fixed program applications

Some applications are single-function applications. It does not pay to use a PLC that includes multiple programming capabilities if they are not needed. One example is in the use of drum controller/sequencers. Some equipment manufacturers still use a mechanical drum with pegs at an overall cost advantage. Their operational sequence is seldom or never changed, so the reprogramming available with the PLC would not be necessary.

Environmental considerations

Certain process environments, such as high heat and vibration, interfere with the electronic devices in PLCs, which limit their use.

Fail-safe operation

In relay systems, the stop button electrically disconnects the circuit; if the power fails, the system stops. Furthermore, the relay system does not automatically restart when power is restored. This, of course, can be programmed into the PLC; however, in some PLC programs, you may have to apply an input voltage to cause a device to stop. These systems are not fail-safe. This disadvantage can be overcome by adding safety relays to a PLC system, as shown later in this text.

Fixed-circuit operation

If the circuit in operation is never altered, a fixed control system (such as a mechanical drum) might be less costly than a PLC. The PLC is most effective when periodic changes in operation are made.

REVIEW QUESTIONS

3.1   What are the various computer generations? Discuss main features of each generation.

3.2   Describe the important applications of microcomputers.

3.3   Classify computers from their computing power consideration. Discuss the important features of each category.

3.4   Describe the important applications of minicomputers.

3.5   What do you understand by word length of a computer? Explain with an example.

3.6   Describe the important applications of mainframe computers.

3.7   What do you understand by processing speed of a computer?

3.8   Differentiate between the terms MIPS and MFLOPS.

3.9   Describe the important applications of supercomputers.

3.10   What is Programmable Logic Controller (PLC)?

3.11   What are the different components of a PLC?

3.12   What are the functions that can be performed by PLC?

3.13   How PLC is different from minicomputer?

3.14   What are the different ways to program a PLC?

Fig. 4.1 Typical ... system.

CAD SYSTEM HARDWARE

Chapter

4

4.1 INTRODUCTION

In the present chapter we examine the various hardware components that make up a modern CAD system.

Hardware components for computer aided design are available in a variety of sizes, configurations and capabilities. Hence it is possible to select a CAD system that meets the particular computational and graphics requirements of the user firm. Engineering firms that are not involved in production would choose a system exclusively for drafting and design related functions. Manufacturing firms would choose a system to be part of a company-wide CAD/CAM system. Of course, the CAD hardware is of little value without the supporting software for the system and we shall discuss the CAD system software in the following chapter.

A modern CAD system is based on interactive computer graphics (ICG). However the scope of CAD includes other computer system as well. For example, computerized design has also been accomplished in a batch mode, rather than in an interactive mode. With interactive graphics the system provides an immediate response to inputs by the user.

The user and the system are in direct communication with each other. The user enters commands and responds to the questions generated by the system. The hardware that we discuss in this chapter is restricted to CAD systems which utilize interactive computer graphics. Typically a stand-alone CAD system would include the following hardware components:

Graphics terminal

Operator input devices

Central processing unit (CPU)

One or more plotters and other output devices

Secondary storage

These hardware components would be arranged in a configuration as shown in Fig. 4.1.

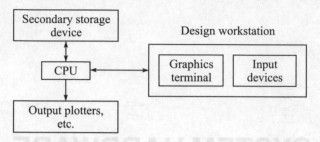

Fig. 4.1 Typical configuration of hardware components in a stand-alone CAD system.

4.2 THE DESIGN WORKSTATION

The CAD workstation is the system interface with the outside world. It represents an important factor in determining how convenient and efficient it is for a designer to use the CAD system. The workstation must accomplish five functions:

1. It must interface with the central processing unit.
2. It must generate a steady graphic image for the user.
3. It must provide digital descriptions of the graphic image.
4. It must translate computer commands into operating functions.
5. It must facilitate communication between the user and the system.

The use of interactive graphics has been found to be the best approach to accomplish these functions. A typical interactive graphics workstation would consist of the following hardware components:

Graphics terminal
Operator input devices

4.3 GRAPHICS TERMINAL

There are various different approaches which have been applied to the development of graphics terminals. The technology continues to improve their products and reduce their costs.

4.3.1 Image Generation in Computer Graphics

All computer graphics terminals which are in use today use the cathode ray tube (CRT) as the display device. Television sets use a form of the same device as the picture tube. The operation of the CRT is shown in Fig. 4.2. A heated cathode emits a high speed electron beam onto a phosphor-coated glass screen. The electrons energize the phosphor coating causing it to glow at the points where the beam makes contact. By focusing the electron beam, changing its intensity and controlling its point of contact against the phosphor coating through the use of a deflector system, the beam can be made to generate a picture on the CRT screen.

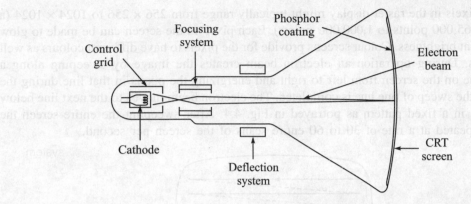

Fig. 4.2 Diagram of cathode ray tube (CRT)

There are two basic techniques used in current computer graphics terminals for generating the image on the CRT screen. They are:

1. Stroke writing
2. Raster scan

Other names for the stroke writing technique include line drawing, random position, vector writing, and directed beam. Other names for the raster scan technique include digital TV and scan graphics.

The stroke writing system uses an electron beam which operates like a pencil to create a line image on the CRT screen. The image is constructed out of a sequence of straight line segments. Each line segment is drawn on the screen by directing the beam to move from one point on the screen to the next where each point is defined by its x and y coordinates. The process is illustrated in Fig. 4.3.

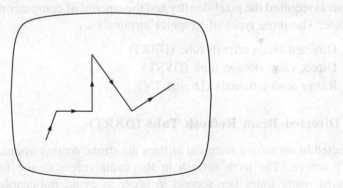

Fig. 4.3 Stroke writing for generating images in computer graphics.

Although the procedure results in images composed of only straight lines, smooth curves can be approximated by making the connecting line segments short enough.

In the raster scan approach, the viewing screen is divided into a large number of discrete phosphor picture elements called **pixels**. The matrix of pixels constitutes the raster. The number of separate pixels in the raster display might typically range from 256×256 to 1024×1024 (a total of over 65,000 points to 1,000,000 points). Each pixel on the screen can be made to glow with a different brightness. Colour screens provide for the pixels to have different colours as well as brightness. During operation an electron beam creates the image by sweeping along a horizontal line on the screen from left to right and energizing the pixels in that line during the sweep when the sweep of one line is completed. The electron beam moves to the next line below and proceeds in a fixed pattern as potrayed in Fig. 4.4. After sweeping the entire screen the process is repeated at a rate of 30 to 60 entire scans of the screen per second.

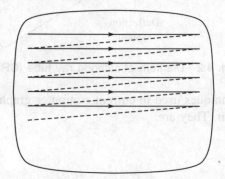

Fig. 4.4 Raster scan approach for generating images in computer graphics.

4.4 GRAPHICS TERMINALS FOR CAD

The two approaches described earlier are used in the majority of current day CAD graphics terminals. There are also a variety of other technical factors which result in different types of graphics terminals. These factors include the type of phosphor coating on the screen, whether the colour is required the pixel density and the amount of computer memory available to generate the picture. The three types of graphics terminals are:

1.   Directed beam refresh tube (DBRT)
2.   Direct view storage tube (DVST)
3.   Raster scan terminals (Digital TV)

4.4.1 Directed Beam Refresh Tube (DBRT)

The directed beam refresh terminal utilizes the stroke writing approach to generate the image on the CRT screen. The term refresh in the name refers to the fact that the image must be regenerated many times per second in order to avoid noticeable flicker of the image. The phosphor elements on the screen are capable of maintaining their brightness for only a short time. In order for the image to be continued, these picture tubes must be refreshed by causing the directed beam to retrace the image repeatedly on densely filled screens (very detailed line images or many characters of text). It is difficult to avoid flickering of the image with this process.

There are several advantages associated with the directed beam refresh systems. Because the image is being continually refreshed, selective erasure and alteration of the image is readily accomplished. It is also possible to provide animation of the image with a refresh tube.

The directed beam refresh system is the oldest of the modern graphics display technologies. Other names sometimes used to identity this system include vector refresh and stroke writing refresh. Early refresh tubes were very expensive, but the steadily decreasing cost of solid state circuitry has brought the price of these graphics systems down to a level which is competitive with other types.

4.4.2 Direct View Storage Tube (DVST)

DVST terminals also use the stroke writing approach to generate the image on the CRT screen. The term **storage tube** refers to the ability of the screen to retain the image which has been projected against it, thus avoiding the need to rewrite the image constantly.

What makes this possible is the use of an electron flood gun directed at the phosphor-coated screen which keeps the phosphor elements illuminated once they have been energized by the stroke writing electron beam. The resulting image on the CRT screen is flicker-free. Lines may be readily added to the image without concern over their effect on image density or refresh rates. However the penalty associated with the storage tube is that individual lines cannot be selectively removed from the image.

Storage tubes have been the low-cost terminals and are capable of displaying the large amounts of data either graphical or textual. Because of these features there are probably more storage tube terminals in service industry. The principal disadvantage of a storage CRT is that selective erasure is not possible instead if the user wants to change the picture. The change will not be manifested on the screen until the entire picture is regenerated. Other disadvantages include its lack of colour capability, the inability to use a light pen as a data entry device and its lack of animation capability.

4.4.3 Raster Scan Terminals (Digital TV)

Raster scan terminals operate by causing an electron beam to trace a zigzag pattern across the viewing screen. The operation is similar to that of a commercial television set. The difference is that a TV set uses analog signals originally generated by a video camera to construct the image on the CRT screen, while the raster scan ICG terminal uses digital signals generated by a computer. For this reason the raster scan terminals used in computer graphics are sometimes called **digital TVs**.

The introduction of the raster scan graphics terminal using a refresh tube had been limited by the cost of computer memory. For example, the simplest and lowest cost terminal in this category uses only two beam intensity levels: ON or OFF. This means that each pixel in the viewing screen is either illuminated or dark. A picture tube with 256 lines of resolution and 256 addressable points per line to form the image would require 256×256 or over 65,000 bits of storage. Each bit of memory contains the ON/OFF status of the corresponding pixel on the CRT screen. This memory is called the **frame buffer** or **refresh buffer**. The picture quality can be improved in two ways by increasing the pixel density or adding a gray scale (or colour). Increasing pixel density for the same size screen means adding more lines of resolution and more addressable points per line. A 1024×1024 raster screen would require more than 1 million bits of storage in the frame buffer. A gray scale is accomplished by expanding the number of intensity levels, which can be displayed on each pixel to store the intensity level. Two bits are required for four levels, three bits for eight levels and so on. Five or six bits would be needed to achieve an approximation of a continuous gray scale. For a colour display three times as many bits are required to get various intensity levels for each of the three primary colours—red, blue and green. A raster scan graphics terminal with high resolution and gray scale can require a very large capacity refresh buffer. Until recent developments in memory technology, the cost of this storage capacity was prohibitive for a terminal with good picture quality.

The capability to achieve colour and animation was not possible except for very low-resolution levels. It is now possible to manufacture digital TV systems for interactive computer graphics at prices, which are competitive with the other two types. The advantages of the present raster scan terminals include the feasibility to use low cost TV monitors, colour capability and the capability for animation of the image. Many of the important characteristics of the three types of graphics terminals are summarized in Table 4.1.

TABLE 4.1 Comparison of Graphics Terminal Features

		Directed beam refresh	DVST	Raster scan
1	Image generation	Stroke writing	Stroke writing	Raster scan
2	Picture quality	Excellent	Excellent	Moderate to good
3	Data content	Limited	High	High
4	Selective erase	Yes	No	Yes
5	Gray scale	Yes	No	Yes
6	Colour capability	Moderate	No	Yes
7	Animation capability	Yes	No	Moderate

4.5 GRAPHICS INPUT DEVICES

Graphics input devices are provided at the graphics workstation to facilitate convenient communication between the user and the system. Workstations generally have several types of input devices to allow the operator to select the various preprogrammed input functions. These functions permit the operator to create or modify an image on the CRT screen or to enter alphanumeric data into the system. This results in a complete part on the CRT screen as well as a complete geometric description of the part in the CAD database.

Different CAG system vendors offer different types of (graphics) operator input devices. These devices can be divided into three general categories:

- Cursor control devices
- Digitizers
- Alphanumeric and other keyboard terminals

Cursor control devices and digitizers are both used for graphical interaction with the system. Keyboard terminals are used as input devices for commands and numerical data.

There are two basic types of graphical interaction accomplished by means of cursor control and digitizing creating and positioning new items on the CRT screen. Pointing at or otherwise identifying locations on the screen, usually associated with existing images ideally, a graphical input device should lend itself to both of these functions. However this is difficult to accomplish with a single unit and that is why most workstations have several different input devices.

4.5.1 Cursor Control Devices

The cursor manually takes the form of a bright spot on the CRT screen that indicates where lettering or drawing will occur. The computer is capable of reading the current position of the cursor. Hence the user's capability to control the cursor position allows locational data to be entered into the CAD system database. A typical example would be for the user to locate the cursor to identify the starting point of a line. Another, more sophisticated case, would be for the user to position the cursor to select an item from a menu of functions displayed on the screen. For instance, the screen might be divided into two sections one of which is an array of blocks which correspond to operator input functions. The user simply moves the cursor to the desired block to execute the particular function.

There are a variety of cursor control devices which have been employed in CAD systems These include

Thumb wheels
Direction keys on a keyboard
Joysticks
Tracker ball
Light pen
Electronic tablet/pen
Mouse

The first four items in the list provide control over the cursor without any direct physical contact of the screen by the user. The last two devices in the list require the user to control the cursor by touching the screen with a pen-type device.

The thumb wheel device uses two thumbwheels: one to control the horizontal position of the cursor, the other to control the vertical position. This type of device is often mounted as an integral part of the CRT terminal. The cursor in this arrangement is often represented by the intersection of a vertical line and a horizontal line displayed on the CRT screen. The two lines are like cross hairs in a gun sight which span the height and width of the screen.

Direction keys on the keyboard are another basic form of cursor control used not only for graphics terminals but also for CRT terminals without graphics capabilities. Four keys are used for each of the four directions in which the cursor can be moved (right or left and up or down).

The joystick apparatus is pictured in Fig. 4.5. It consists of a box with a vertical toggle stick that can be pushed in any direction to cause the cursor to be moved in that direction. The joystick gets its name from the control stick that was used in old air planes.

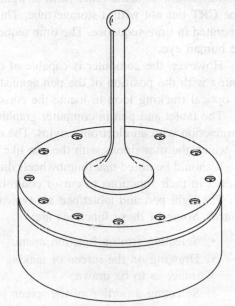

Fig. 4.5 Joystick input device for interactive computer graphics.

The tracker ball is pictured in Fig. 4.6. Its operation is similar to that of the joystick except that an operator controlled ball is rotated to move the cursor in the desired direction on the screen.

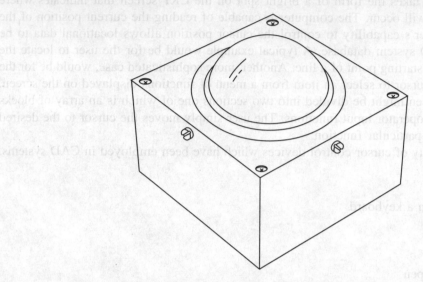

Fig. 4.6 Tracker ball input device for interactive computer graphics.

The light pen is a pointing device in which the computer seeks to identify the position where the light pen is in contact with the screen. Contrary to what its name suggests the light pen does not project light. Instead it is a detector of light on the CRT screen and uses a photodiode, phototransistor or some other form of light sensor. The light pen can be utilized with a refresh type CRT but not with a storage tube. This is because the image on the refresh tube is being generated in time sequence. The time sequence is so short that the image appears continuous to the human eye.

However, the computer is capable of discerning the time sequence and it coordinates this timing with the position of the pen against the screen. In essence the system is performing as an optical tracking loop to locate the cursor or to execute some other input function.

The tablet and pen in computer graphics describe an electronically sensitive tablet used in conjunction with an electronic stylus. The tablet is a flat surface separate from the CRT screen, on which the user draws with the pen like stylus to input instructions or to control the cursor.

It should be noted that thumbwheels direction keys, joysticks and tracker balls are generally limited in their functions to cursor control.

The light pen and tablet/pen are typically used for other input functions as well as cursor control. Some of these functions are:

- Selecting from a function menu.
- Drawing on the screen or making strokes on the screen or tablet which indicate what image is to be drawn.
- Selecting a portion of the screen for enlargement of an existing image.

A light pen resembles a fountain pen in the method of holding, but it works on the principle of light rather than ink, hence the name. Light pens are not used for writing on the screen as is erroneously believed by many, but actually only to detect the presence of light on the screen as shown in Fig. 4.7, with the help of a light detecting resistor. Their normal use in graphic applications is to identify the objects or locations on the display screen for possible graphics handling. These are to be used only with refresh type display devices. The resolution of the light pen is poor, as the field of view of the photosensitive element is conical. Since the light pen points to the graphic display directly, it is a natural graphic interactive tool. However, as the operator has to hold the light pen against the gravity along with its cable connecting the graphics adapter card for making any selection, ergonomically it is inconvenient to use it over long periods.

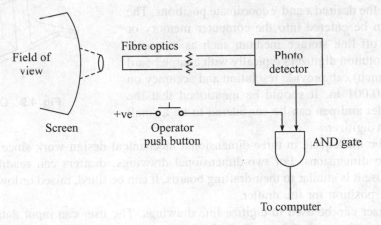

Fig. 4.7 Light pen

The mouse shown in Fig. 4.8 is the pointing device, which has been gaining importance with the advent of the microprocessors, and the pull down menus associated with the application software. The mouse operates on three basic principles—Mechanical, Optical and Opto-mechanical. The mechanical mouse contains a free floating ball with rubber coating on the underside which, when moved on a firm plane surface, would be able to follow the movement of the hand. The motion of the ball is resolved into x and y motions by means of the two rollers pressed against the ball. They in turn control the cursor on the screen, which can then be utilized for any desired applications by means of the clicking of the buttons on the mouse.

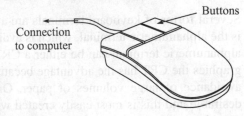

Fig. 4.8 Mouse

This can only suffice to point, on the screen but not for giving positional data. In case of the optical mouse, a special reflective plain surface with etched fine grids is required. The LEDs present inside the mouse (in place of the rubber ball) would reflect the number of grid lines crossed in the x and y directions, thereby showing the distance moved. The life of the optical mouse is high since it has no moving parts, but it has not gained as much acceptance as the mechanical mouse because of the special surface needed for its operation. The operation of the opto-mechanical mouse is similar to that of the mechanical mouse, but the position resolvers used are based on the optical principle.

4.5.2 Digitizers

The digitizer shown in Fig. 4.9 is an operator input device which consists of a large, smooth board (the appearance is similar to a mechanical drawing board) and an electronic tracking device which can be moved over the surface to follow existing lines. It is a common technique in CAD systems for taking x, y coordinates from a paper drawing. The electronic tracking device contains a switch for the user to record the desired x and y coordinate positions. The coordinate can be entered into the computer memory or stored on an off-line storage medium such as magnetic tape. High resolution digitizers typically with a large board (e.g. 42×60 inch) can provide resolution and accuracy on the order of 0.001 in. It should be mentioned that the electronic tablet and pen can be considered to be a small low-resolution digitizer.

Fig. 4.9 Digitizer

It would be inadequate in three-dimensional mechanical design work since the digitizer is limited to two dimensions. For two-dimensional drawings, drafters can readily adapt to the digitizer because it is similar to their drafting boards. It can be tilted, raised or lowered to assume a comfortable position for the drafter.

The digitizer can be used to digitize line drawings. The user can input data from a rough schematic or large layout drawing and edit the drawing to the desired level of accuracy and detail. The digitizer can also be used to free hand a new design, with subsequent editing to finalize the drawing.

4.5.3 Keyboard Terminals

Several forms of keyboard terminals are available as CAD input device. The most familiar type is the alphanumeric terminal, which is available with nearly all interactive graphics systems. The alphanumeric terminal can be either a CRT or a hard-copy terminal, which prints on paper. For graphics the CRT has the advantage because of its faster speed, the ability to easily edit, and the avoidance of large volumes of paper. On the other hand, a permanent record is sometimes desirable and this is most easily created with a hard-copy terminal. Many CAD systems use the graphics screen to display the alphanumeric data, but there is an advantage in having a separate CRT terminal so that the alphanumeric messages can be created without disturbing or over-writing the image on the graphics screen.

The alphanumeric terminal is used to enter commands, functions and supplemental data to the CAD system. This information is displayed for verification on the CRT or typed on paper. The system also communicates back to the user in a similar manner. The computer, as part of the interactive procedure, can display menu listings, program listings, error messages and so forth.

Some CAD systems make use of special function keyboards, as pictured in Fig. 4.10. These function keyboards are provided to eliminate extensive typing of commands or calculate coordinate positions and other functions. The number of function keys varies from about 8 to

80. The particular function corresponding with each button is generally under computer control so that the button function can be changed as the user proceeds from one phase of the design to the next. In this way the number of alternative functions can easily exceed the number of buttons on the keyboard. Also lighted buttons are used on the keyboards to indicate which functions are possible in the current phase of design activity. A menu of the various function alternatives is typically displayed on the CRT screen for the user to select the desired function.

Fig. 4.10 Special function keyboard for interactive computer graphics.

4.6 GRAPHICS OUTPUT DEVICES

There are various types of output devices used in conjunction with a computer aided design system. These output devices include

    Pen plotters

    Hardcopy units

    Electrostatic plotters

    Computer-output-to-micro-film (COM) units

    We discuss these devices in the following sections.

4.6.1 Pen Plotters

The accuracy and quality of the hardcopy plot produced by a pen plotter are considerably greater than the apparent accuracy and quality of the corresponding image on the CRT screen. In case of the CRT image, the quality of the picture is degraded because of lack of resolution and because of losses in the digital-to-analog conversion through the display generators. On the other hand, a high precision pen plotter is capable of achieving a hard-copy drawing whose accuracy is nearly consistent with the digital definitions in the CAD database.

    The pen plotter uses a mechanical ink pen (either wet ink or ball point) to write on paper through relative movement of the pen and paper. There are two basic types of pen plotters currently in use:

Drum plotters
Flat-bed plotters

The drum plotter, shown in Fig. 4.11, is generally the least expensive. It uses a round drum, usually mounted horizontally and a slide, which can be moved along a track mounted axially with respect to the drum. The paper is attached to the drum and the pen is mounted on the slide. The relative motion between pen and paper is achieved by coordinating the rotation of the drum with the motion of the slide. The drum plotter is fast and it can make drawings of virtually unlimited length. The length of the drum however limits the width. These lengths typically range between 8½ inch (216 mm) and 42 inch (1067 mm).

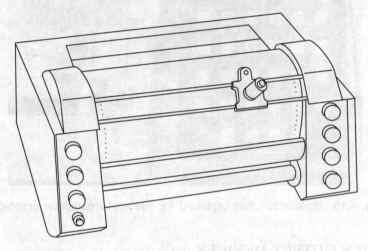

Fig. 4.11 Drum plotter.

The flat-bed plotter, shown in Fig. 4.12, is more expensive. It uses a flat drawing surface to which the paper is attached on some models. The surface is horizontal while other models use a drawing surface which is mounted in a nearly vertical orientation to conserve floor space. This type is shown in Fig. 4.13. Parallel tracks are located on two sides of the flat surface. A bridge is driven along these tracks to provide the x-coordinate motion. Attached to the bridge is another

Fig. 4.12 Flat-bed x-y plotter in near vertical position.

Fig. 4.13 Belt-bed plotter in near vertical position.

track on which rides a writing head movement of the writing head relative to the bridge produces the y-coordinate motion. The writing head carries the pen or pencil, which can be raised or lowered to provide contact with the paper as desired. The size of these automated drafting tables can range up to roughly 5 ft (1.5 m) by 20 ft (6.1 m) with plotting accuracies approaching ± 0.001 in (± 0.025 mm).

The pen plotter accepts digitized data either on-line from the computer or off-line in the form of magnetic tape or punched tape on modern pen plotters, a microprocessor is often used as the control unit. This allows certain shapes such as circles and ellipses to be programmed in the form of simple instructions to the plotter. In this way the digital data for a complicated shape can be made more compact and efficient.

Many plotters work with several pens of different colours to achieve multi-colour plots. Also in some models the pen may be replaced by a highly focused, high intensity light and the conventional drafting paper by a photosensitive paper. This arrangement would be used for certain artwork applications.

Another option available on a flat-bed plotter is to combine the plotter function with the operation of a digitizer. Such a device is called a **digitizer plotter**.

4.6.2 Hardcopy Unit

A hard copy unit is a machine that can make copies from the same image data displayed on the CRT screen. The image on the screen can be duplicated in a matter of seconds. The copies can be used as records of intermediate steps in the design process or when rough hard copies of the screen are needed quickly. The hardcopies produced from these units are not suitable as final drawings because the accuracy and quality of the reproduction is not nearly as good as the output of a pen plotter.

Most hardcopy units are dry silver copiers that use light sensitive paper exposed through a narrow CRT window inside the copier. The window is typically 8½ inch (216 mm) corresponding to the width of the paper, by about ½ inch (12 mm) wide. The paper is exposed by moving it past the window and coordinating the CRT beam to gradually transfer the image.

A heated roller inside the copier is used to develop the exposed paper. The size of the paper is usually limited on these hardcopy units to 8½ by 11 in. Another drawback is that the dry silver copies will darken with time when they are left exposed to normal light.

4.6.3 Electrostatic Plotters

Hardcopy units are relatively fast but their accuracy and resolution are poor. Pen plotters are highly accurate but plotting time can take many minutes. The electrostatic plotter offers a compromise between these two types in terms of speed and accuracy. It is almost as fast as the hardcopy unit and almost as accurate as the pen plotter. The electrostatic copier consists of a series of wire styli mounted on a bar which spans the width of the charge sensitive paper. The styli have a density of upto 200 per linear inch. The paper is gradually moved past the bar and certain styli are activated to place dots on the paper by coordinating the generation of the dots with the paper travel. The image is progressively transferred from the database into hardcopy form. The dots overlap each other slightly to achieve continuity. For example, a series of adjacent dots gives the appearance of a continuous line.

A limitation of the electrostatic plotter is that the data must be in the raster format (i.e., in the same format used to drive the raster-type CRT) in order to be readily converted into hardcopy using the electrostatic method. If the data are not in raster format, some type of conversion is required to change them into the required format. The conversion mechanism is usually based on a combination of software and hardware.

An advantage of the electrostatic plotter, which is shared, with the drum type pen plotter is that the length of the paper is virtually unlimited. Typical plotting widths might be upto 6 ft (1.83 m). Another advantage is that the electrostatic plotter can be utilized as a high-speed line printer, capable of upto 1200 lines of text per minute.

4.6.4 Computer-Output-to-Microfilm (COM) Units

COM units reproduce the drawings on microfilm rather than as full size engineering drawings. It is an expensive piece of equipment. However, for the large corporation able to afford a COM unit, there are several important advantages. One advantage is storage capability. A large engineering department may have tens of thousands of engineering drawings to be stored. Reducing the size of each drawing to microfilm achieves a significant storage benefit. If a full size hard copy drawing is ever required, the microfilm can be easily retrieved to be photographically enlarged to full size. Another advantage is speed COM units produce a microfilm copy much faster than a pen plotter, perhaps several hundred times faster for a complicated line drawing. Computer-output-to-microfilm is also faster than electrostatic plotters.

Disadvantages of the COM process are that the user cannot write notes on the microfilm as is possible with a paper copy. Also enlargements of the microfilm onto paper, although adequate, are not of as high quality as the output from a pen plotter.

4.7 MODES OF OPERATION

A suitable choice of the graphics output device for a CAD system is obviously crucial to the overall performance of the system, but the mode of graphics output operation is just as important to meet the needs of a particular application. Modes of operation in this case means the methods used to supply data to the graphics output device. In plotter, there are four modes of graphics output operation: hardcopy, on-line, off-line and remote.

4.7.1 Hard Copy Mode

Figure 4.14 illustrates the operation in hardcopy mode. If a permanent copy on paper is required of an image on a CRT display screen, it can be produced using the plotter in this mode of operation. The term hardcopy distinguishes the permanent physical presence of the image on paper from the 'soft' temporary visual image on the screen.

In fact a hard copy may be simply defined as a plot produced by any output operation mode. Hardcopy mode provides a quick way to reproduce on paper the same image on the screen. It is basically a straight copy, or screen dump of the image. A hardcopy is sometimes very useful in some applications where it is necessary to check rough preliminary or intermediate output before the final version is produced on some other plotter with the required quality. An electrostatic plotter and a dot-matrix impact plotter are most suitable for working in this mode because the plot data from the CRT are in a raster format. Pen plotters are not really appropriate for hard copy mode.

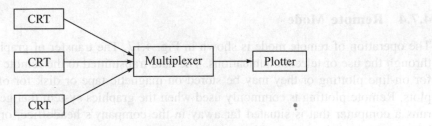

Fig. 4.14   Hard copy.

4.7.2 On-line Mode

Figure 4.15 shows the on-line mode of plotter operation. This is the simplest mode in which the plotter is connected directly to the same computer that is running the CAD software. On-line mode has the advantage that intermediate data transfer steps between the plotter and the computer are avoided, therefore related hardware for this operation is minimized, hence the cost as well. One possible drawback of on-line mode is that additional processing demand is placed on the computer.

Fig. 4.15   On-line.

4.7.3 Off-line Mode

Although on-line mode is simple and involves less hardware investment, it does have its disadvantage. Plotting is usually slow relative to the speed of the computer. In situations where continuous plotting to the extent that other essential functions of the system cannot really be performed to any acceptable level off-line mode is frequently used as a solution to this problem. Figure 4.16 illustrates how off-line mode works. In this mode, the graphics data are transferred from the computer to the plotter through some removable storage volume such as a magnetic tape or a disk. The computer outputs the data to a storage volume which is then removed and physically taken by a human operator to the plotter system which reads the storage volume and sends the stored data to the plotter to produce the drawing with this off-line procedure. It is more efficient and flexible. There is much freedom to choose whether to produce the plot immediately or keep them for later plotting or use them repeatedly. If the multiple copies are required and the computer is relieved of the task of plotting, it can perform other functions much more quickly.

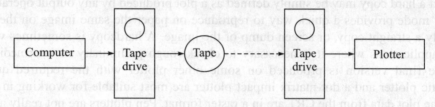

Fig. 4.16 Off-line.

4.7.4 Remote Mode

The operation of remote mode is shown in Fig. 4.17. The transfer of graphics data is achieved through the use of telecommunications. The data transmitted to the remote system may be used for on-line plotting or they may be stored on magnetic tape or disk for off-line production of plots. Remote plotting is commonly used when the graphics system that generates the plot data runs a computer that is situated far away in the company's headoffice, or on a computer that belongs to a service bureau from whom computer time is purchased. The cost of the modems required for sending and receiving the data is relatively low but the major disadvantage is the high cost of telecommunications time.

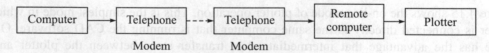

Fig. 4.17 Remote.

4.8 CAD SYSTEM CONFIGURATION

In the early days of computer aided design, graphics peripherals were connected to the computer system to form a typical configuration of a CAD system, as shown in Fig. 4.18. Each graphics device communicates directly with the computer which is very often one of the commercially available general purpose minicomputers. Traditionally most CAD minicomputers are 16 bit word machines but the new minicomputers tend to use 32 bit processors and are also known as **super minicomputers**. The advantages of using a 32 bit minicomputer include higher processing speed, greater accuracy and more main memory. In some powerful CAD systems, a mainframe computer may even be used. For secondary storage, disks and magnetic tapes are commonly employed so that the fast access and efficient storage of data are achieved. CAD systems software and its associated database for drawings are generally stored on disks because they provide rapid random access of data. As a result of this feature the computer can quickly load and swap programs and files between main and secondary memory as and when needed. Since magnetic tape is a sequential access device, it is only used to store programs and files that are not frequently required by the system. In a CAD system, magnetic tape is mainly used for disk backup permanent archival drawing files and data transfer to output devices such as plotter or other computer installations.

Clearly, it can be seen that the CAD system, illustrated in Fig. 4.18, has a centralized configuration which has the advantage of being simple and straightforward because each graphics peripheral is connected directly to the computer. However, this configuration does not bring out the best possible performance from the system. As there is only one computer in the system for processing, it cannot support too many peripherals. In addition graphics functions are usually input/output intensive, so the computer may be unnecessarily held up.

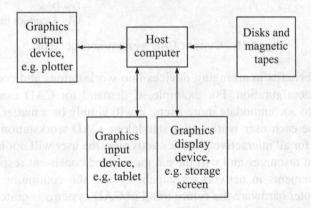

Fig. 4.18 Conventional CAD system configuration.

The current trend in CAD systems is to use a workstation in a distributed configuration, as shown in Fig. 4.19. A CAD workstation would consist of a graphics display and graphics input devices, which are linked to a local computer with its own secondary storage. The concept of workstation is that a collection of hardware is combined together to form a single unit capable of acting as a stand-alone system to perform computations and essential functions required for a particular application. As a workstation has its own computer and secondary storage, it is sometimes referred to as an intelligent workstation. The local computer in a CAD workstation is usually one of the microcomputers based on a 16 bit microprocessor. So it is sufficiently powerful to handle most of the processing demands in CAD. Each CAD workstation is treated basically as a functional entity, which is connected to the host computer through one of the data communication lines. The host computer takes on the role of central database for storing drawings and other data created at the different workstations by various users. In addition the host computer will run those complex analysis programs with which the local computer does not have the resources to cope. Sometimes, the host computer acts as a intermediary for communications between separate workstations.

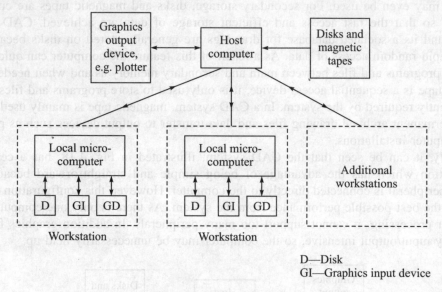

D—Disk
GI—Graphics input device

Fig. 4.19 Current CAD system configuration.

There are many benefits in arranging devices into workstations and connecting them into a distributed network configuration. For example, if demand for CAD usage requires that the system be expanded to accommodate more users, it will simply be a matter of introducing more workstations. Because each user works on a standalone CAD workstation that is dedicated to provide total support for all interactive graphics activities, the user will not have to compete with other users for system resources and will have a guaranteed consistent response. With the rapid technological improvements in networking capability and the continuing miniaturization and falling costs of computer hardware the future trend in CAD systems is quite clear. Gradually the local microcomputer in a workstation will become more and more powerful to the extent that all CAD processing is performed on the microcomputer and the host computer is relegated to the role of just providing a central database for storage of drawings and other information. It may also be possible for one workstation to communicate directly with the other workstations in the system configuration without the need to go through the central host computer.

REVIEW QUESTIONS

4.1 Describe the important functions of a design workstation.

4.2 What are the two basic techniques used in current computer graphics terminals for generating the image on the CRT screen?

4.3 Discuss and compare the working principles of the following types of graphic terminals:

 (a) DBRT (b) DVST

 (c) Raster scan (d) Flat screens

4.4 What is meant be raster scanning? Why is it preferred to the storage tube in the display of graphics information?

4.5 Differentiate between stroke writing and raster scan techniques.

4.6 What are the various cursor control devices that can be employed in CAD systems?

4.7 Discuss the working principles of the following graphics input devices:

 (a) Tracker ball (b) Light pen

 (c) Mouse (d) Digitizer

4.8 How does a workstation differ from PC? Give the minimum configuration of a CAD workstation.

4.9 Describe the various hardware components in a stand-alone CAD system.

4.10 What are the various types of plotters? Differentiate between a flat bed plotter and a drum plotter.

4.11 What are the various methods employed to supply data to the graphics output device?

4.12 Discuss the conventional CAD system configuration and the current CAD system configuration.

4.13 Can a light pen be used with a storage tube display? Why?

4.14 How do you choose a plotter for graphics application? Justify your choice.

4.15 What are the various display devices that are used for displaying graphic information? Outline their capabilities, merits and demerits of the various interactive input devices.

4.16 What are the merits and demerits of the various interactive input devices?

4.17 How do you specify a digitizer for the CAD application? Justify it.

CAD SYSTEM SOFTWARE

INTRODUCTION

CAD has been acknowledged as the key to improving manufacturing productivity and the best approach for meeting the recent critical design requirements. CAD software provides engineers with the tools needed to perform their technical jobs efficiently and free them from the tedious and time consuming tasks that require little or no technical expertise. Experience has shown that CAD software speeds the design process, therefore increasing productivity, innovation and creativity of designers. In some design cases such as VLSI, CAD software has provided the only means to meet the new technological design and production requirements of increased accuracy and uniformity. The need for the software in the future will be even greater due to the expected intricate design and manufacturing requirements.

The CAD hardware would be useless without the software to support it. An investigation of existing software in general reveals that it has common characteristics regardless of the hardware it runs on. It is an interactive program typically written in a standard programming language. It is hardware-dependent and seems different to the user from conventional software due to the user interface. The database structure and database management system of the software determines its quality, speed and ease of information retrieval.

The most important characteristic of CAD software is its fully three-dimensional, associative, centralized and integrated database. Such a database is always rich in information needed for both the design and manufacturing processes. The centralized concept implies that any change in or addition to a geometric model in one of its views is automatically reflected in the existing views or any views that may be defined later. The integrated concept implies that a geometric model of an object can be utilized in all various phases of a product cycle.

The associativity concept implies that input information can be retrieved in various forms. For example, if the two end points of a line are input, the line length and its dimension can be output.

CAD software is typically a large, complex program that has been developed over the years. Users of the software are usually faced with learning its related semantics and syntax of its user interface. Semantics specifies how the software functions and what information is needed for each operation on an object. For example, a block requires three lengths and an orientation to create.

Syntax defines the formats of inputs and outputs. It is considered the grammar of the software. It specifies the rules that users must follow to achieve the desired semantics.

Performance is another common characteristic of software. The larger the number of interactive users, the longer the interactive response time. The software occasionally "locks" and ceases to respond to or accept user commands. This is typically referred to as "system crash". When this happens, the user looses the work performed after the last filing or save command is issued and rebooting the system is required. This is why users are always advised to file or save their work frequently.

In this chapter, the software aspects of a CAD system will be discussed. Generally the software of a CAD system can roughly be divided into three main levels, namely the operating system, the graphics system and the application system. At the lowest level of the CAD system software is the operating system which provides software facilities for the user to develop the graphics system and the application system. The graphics system offers some basic graphics capabilities with which the application system can be built.

5.1 GRAPHICS SOFTWARE

The graphics software is the collection of programs written to make it convenient for a user to operate the computer graphics system. It includes programs to generate images on the CRT screen, to manipulate the images and to accomplish various types of interaction between the user and the system. In addition to the graphics software there may be additional programs for implementing certain specialized functions related to CAD/CAM. This includes design analysis programs (e.g. finite element analysis and kinematic simulation) and manufacturing planning programs (e.g. automated process planning and numerical control part programming).

The graphics software for a particular computer graphics system is very much a function of the type of hardware used in the system. The software must be written specifically for the type of CRT and the types of input devices used in the system. The details of the software for a stroke writing CRT would be different than for a raster scan CRT. The differences between storage tube and a refresh tube would also influence the graphics software. Although these differences in software may be invisible to the user to some extent, they are important considerations in the design of an interactive computer graphics system.

Newman and Sproull listed six "ground rules" that should be considered in designing graphics software.

- **Simplicity.** The graphics software should be easy to use.
- **Consistency.** The package should operate in a consistent and predictable way to the user.
- **Completeness.** There should be no inconvenient omissions in the set of graphics functions.
- **Robustness.** The graphics system should be tolerant of minor instances of misuse by the operator.

- **Performance.** Within limitations imposed by the system hardware, the performance should be exploited as much as possible by software. Graphics programs should be efficient and speed of response should be fast and consistent.
- **Economy:** Graphics programs should not be so large or expensive as to make their use prohibitive.

5.1.1 The Software Configuration of a Graphics System

In the operation of the graphics system by the user, a variety of activities takes place which can be divided into three categories:

- Interact with the graphics terminal to create and alter images on the screen.
- Construct a model of something physical out of the images on the screen. The models are sometimes called **application models**.
- Enter the model into computer memory and/or secondary storage.

In working with the graphics system, the user performs these various activities in combination rather than sequentially. The user constructs a physical model and inputs it to memory by interactively describing images to the system. The reason for separating these activities in this fashion is that they correspond to the general configuration of the software package used with the interactive computer graphics system. The graphics software can be divided into three modules according to a conceptual model suggested by Foley and Van Dam.

- The graphics package also called the graphics system
- The application program
- The application database

This software configuration is illustrated in Fig. 5.1. The central module is the application program. It controls the storage of data into and retrieves data out of the application database. The application program is driven by the user through the graphics package which is also known as the **graphics system**.

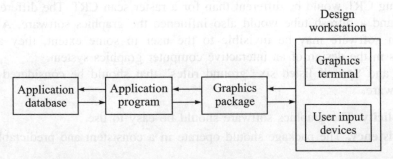

Fig. 5.1 Model of graphics software configuration.

The application program is implemented by the user to construct the model of a physical entity whose image is to be viewed on the graphics screen. Application programs are written for particular problem areas. Problem areas in engineering design would include architecture, construction, mechanical components, electronics, chemical engineering and aerospace engineering. Problem areas other than design would include flight simulators, graphical display of data, mathematical analysis and even art work. In each case, the application software is developed to deal with images and conventions which are appropriate for that field.

The graphics package is the software support between the user and the graphics terminal. It manages the graphical interaction between the user and the system. It also serves as the interface between the user and the application software.

The graphics package consists of input subroutines and output subroutines. The input routines accept input commands and data from the user and forward them to the application program. The output subroutines control the display terminal and converts the application models into two-dimensional or three-dimensional graphical pictures.

The third module in the ICG software is the database. The database contains mathematical, numerical and logical definitions of the application models such as electronic circuits, mechanical components, automobile bodies and so on. It also includes alphanumeric information associated with the models, such as bills of materials, mass properties and other data. The contents of the database can be readily displayed on the CRT or plotted out in the hard copy form.

5.2 FUNCTIONS OF A GRAPHICS PACKAGE

The function of the graphics software for a CAD system is to provide graphics capabilities so that the various applications can make use of them to help solve design problems. As a result of this objective, the graphics software has to be written and organized into a structure that is sufficient to meet the requirements of many different and diverse applications of CAD. A graphics package should essentially provide a system for handling user actions, a set of basic graphic functions and utilities and a system for the operation of application programs. It is of paramount importance that a graphics package be designed in such a manner as to allow applications systems to be incorporated into the CAD system without the application programmer having to be concerned with low-level data, detail system programming or peripheral handling.

To fulfil its role in the software configuration, the graphics package must perform a variety of different functions. These functions can be grouped into function sets. Each set accomplishes a certain kind of interaction between the user and the system. Some of the common function sets are:

- Generation of graphic elements
- Transformations
- Display control and windowing functions
- Segmenting functions
- User input functions

5.2.1 Generation of Graphic Elements

A graphic element in computer graphics is a basic image entity such as a dot (or point), line segment, circle and so on. The collection of elements in the system could also include alphanumeric characters and special symbols. There is often a special hardware component in the graphic system associated with the display of many of the elements. This speeds up the process of generating the element. The user can construct the application model out of a collection of elements available in the system.

The term *primitive* is often used in reference to graphic elements. Accordingly, a primitive is a three-dimensional graphic element such as a sphere, cube or cylinder. In three-dimensional wire frame models and solid modelling, primitives are used as building blocks to construct the 3D model of the particular object of interest to the user.

5.2.2 Transformations

Transformations are used to change the image on the display screen and to reposition the item in the database. Transformations are applied to the graphic elements in order to aid the user in constructing an application model. These transformations include enlargement and reduction of the image by a process called **scaling**, repositioning the image or translation and rotation.

5.2.3 Display Control and Windowing Functions

This function set provides the user with the ability to view the image from the desired angle and at the desired magnification. In effect it makes use of various transformations to display the application model the way the user wants it shown. This is sometimes referred to as windowing because the graphic screen is like a window being used to observe the graphics model. The notion is that the window can be placed wherever desired in order to look at the object being modelled.

Another aspect of display control is hidden line removal. In most graphic systems, the image is made up of lines used to represent a particular object. Hidden line removal is the procedure by which the image is divided into its visible and invisible (or hidden) lines. In some systems, the user must identify which lines are invisible so that they can be removed from the image to make it more understandable. In other systems the graphics package is sufficiently sophisticated to remove the hidden lines from the picture automatically.

5.2.4 Segmenting Functions

Segmenting functions provide users with the capability to selectively replace, delete or otherwise modify portions of the image. The term *segment* refers to a particular portion of the image which has been identified for purposes of modifying it. The segment may define a single element or logical grouping of elements that can be modified as a unit.

Storage type CRT tubes are unsuited to segmenting functions. To delete or modify a portion of the image on a storage tube requires erasing the entire picture and redrawing it with the changes incorporated. Raster scan refresh tubes are ideally suited to segmenting functions because the screen is automatically redrawn 30 or more times per second. The image is regenerated each cycle from a display file, a file used for storage that is part of the hardware in the raster scan CRT. The segment can readily be defined as a portion of that display file by giving it a name. The contents of that portion of the file would then be deleted or altered to execute the particular segmenting function.

5.2.5 User Input Functions

User input functions constitute a critical set of functions in the graphics package because they permit the operator to enter commands or data to the system. The entry is accomplished by means of operator input devices. The user input functions must of course be written specifically for the particular component of input devices used on the system. The extent to which the user input functions are well designed has a significant effect on how "friendly" the system is to the user, that is how easy it is to work on the system.

The input functions should be written to maximize the benefits of the interactive feature of ICG. The software design compromise is to find the optimum balance between providing enough functions to conveniently cover all data entry situations without flooding the user with so many commands that they cannot remember. One of the goals that are sought after by software designers in computer graphics is to simplify the user interface so that a designer with little or no programming experience can function effectively on the system.

5.3 CONSTRUCTING THE GEOMETRY

5.3.1 The Use of Graphics Elements

The graphics system accomplishes the definition of the model by constructing it out of graphic elements. These elements are called by the user during the construction process and added one by one to create the model. There are several aspects about this construction process which will be discussed.

First, as each new element is being called but before it is added to the model, the user can specify its size, its position and its orientation. These specifications are necessary to form the model to the proper shape and scale. For this purpose the various transformations are utilized.

A second aspect of the geometric construction process is that graphics elements can be subtracted as well as added. Another way of saying this is that the model can be formed out of negative elements as well as positive elements.

Figure 5.2 illustrates this construction feature for a two-dimensional object, C. The object is drawn by subtracting circle B from rectangle A.

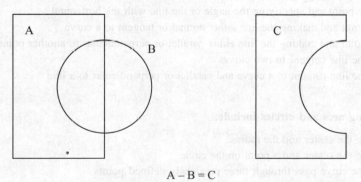

$$A - B = C$$

Fig. 5.2 Example of two-dimensional model construction by subtraction of circle B from rectangle A.

A third feature available during model building is the capability to group several elements together into units which are sometimes called **cells**. A cell, in this context, refers to a combination of elements which can be called to use anywhere in the model. For example, if a bolt is to be used several places in the construction of a mechanical assembly model, the bolt can be formed as a cell and added anywhere to the model. The use of graphic cells is a convenient and powerful feature in geometric model construction.

5.3.2 Defining the Graphic Elements

The user has a variety of different ways to call a particular graphic element and position it on the geometric model. Table 5.1 lists several ways of defining points, lines, arcs, circles and other components of geometry through interaction with the ICG system. These components are maintained in the database in the mathematical form and referenced to a three-dimensional coordinate system. For example, a point would be defined simply by its x, y and z coordinates. A polygon would be defined as an ordered set of points representing the corners of the polygon. A circle would be defined by its centre and radius. Mathematically, a circle can be defined in the x, y plane by the equation:

$$(x - m)^2 + (y - n)^2 = r^2 \qquad (5.1)$$

TABLE 5.1 Methods of Defining Elements in Interactive Computer Graphics

Points

Methods of defining points in computer graphics include:

1. Pointing to the location on the screen by means of cursor control
2. Entering the coordinates via the alphanumeric keyboard
3. Entering the offset (distance in x, y, and z) from a previously defined point
4. The intersection of two points
5. Locating points at fixed intervals along an element

Lines

Methods of defining lines include:

1. Using two previously defined points
2. Using one point and specifying the angle of the line with the horizontal
3. Using a point and making the line either normal or tangent to a curve
4. Using a point and making the line either parallel or perpendicular to another point
5. Making the line tangent to two curves
6. Making the line tangent to a curve and parallel or perpendicular to a line

Arcs and Circles

Methods of defining arcs and circles include:

1. Specifying the center and the radius.
2. Specifying the center and a point on the circle
3. Making the curve pass through three previously defined points
4. Making the curve tangent to three lines
5. Specifying the radius and making the curve tangent to two lines or curves

(Contd.)

Conics

Conics, including ellipses, parabolas, and hyperbolas, can be defined in any plane by methods which include:

1. Specifying five points on the element
2. Specifying three points and a tangency condition

Curves

Mathematical splines are used to fit a curve through given data. For example, in a cubic spline, third-order polynomial segments are fined between each pair of adjacent data points. Other curve generating techniques used in computer graphics include Bezier curves and B-spline methods. Both of these methods use a blending procedure which smoothens the effect of the data points. The resulting curve does not pass through all the points. In these cases the data points would be entered to the graphics system and the type of curve-fining technique would be specified for determining the curve.

Surfaces

The methods described for generating curves can also be used for determining the mathematical definition of a surface. Automobile manufacturers use these methods to represent the sculptured surfaces of the sheet metal car body. Some of the methods for generating surfaces include:

1. Using a surface of revolution formed by rotating any lines and/or curves around a specific axis.
2. Using the intersection line or surface of two intersecting surfaces. For example, this could be used to generate cross sections of parts, by slicing a plane through the part at the desired orientation.

This specifies that the radius of the circle is r and the x and y of the centre are m and n. In each case, the mathematical definition can be converted into its corresponding edges and surfaces for filing in the database and display on the CRT screen.

5.3.3 Editing the Geometry

A computer-aided design system provides editing capabilities to make corrections and adjustments in the geometric model. When developing the model the user must be able to delete, move, copy and rotate components of the model. The editing procedure involves selecting the desired portion of the model (usually by means of one of the segmenting functions) and executing the appropriate command (often involving one of the transformation functions).

The method of selecting the segment of the model to be modified varies from system to system with cursor control, a common method is for a rectangle to be formed on the CRT screen around the model segment. The rectangle is defined by entering the upper left and lower right corners of the rectangle. Another method involving a light pen is to place the pen over the component to be selected with the electronic pen and tablet, the method might be to stroke a line across the portion of the model which is to be altered.

The computer must somehow indicate to the user which portion of the model has been selected. The reason for this is verification that the portion selected by the computer is what the user intended various techniques are used by different ICG systems to identify the segment. These include: placing a mark on the segment, making the segment brighter than the rest of the image and making the segment blink.

Some common editing capabilities available in commercial CAD systems are as follows:

1. *Move an item to another location.* This involves the translation of the item from one location to another.

2. *Duplicate an item at another location*. The copy function is similar to the move function except that it preserves a copy of the item at its original location.

3. *Rotate an item*. This is the rotation transformation, in which the item is rotated through a specified angle from its original orientation.

4. *Mirror an item*. This creates a mirror image of the item about a specified plane.

5. *Delete an item*. This function causes the selected segment of the model to be removed from the screen and from the database.

6. *Remove an item from the display* (without deleting it from the database). This removes the particular segment from the current image on the screen. However, it is not removed from the database. Therefore, repainting the screen from the database will cause the segment to reappear.

7. *Trim a line or other component*. This function would remove the portion of the line that extends beyond a certain point.

8. *Create a cell out of graphic elements*. This feature provides the capability to construct a cell out of selected elements. The cell can then be added to the model in any orientation as needed.

9. *Scale an item*. A selected component can be scaled by a specified factor in x, y, and z directions. The entire size of the model can be scaled, or it can be scaled in only one or two directions.

5.4 OPERATING SYSTEM

An operating system may be defined as a program which organizes a collection of hardware into a consistent and coherent entity so that it can be used for developing and running programs. It can assist users and their programs to make the most efficient use of system resources such as memory capacity and processing power with minimal programming effort for providing software to supplement and enhance the system facilities available. In general an operating system is usually provided with the computer by the vendor as a complete package, so there is no need for the user to write it. There are numerous different types of operating systems. Most of them are interactive, which means that a command entered by a user is executed immediately and the results are returned as soon as possible. Some sophisticated operating systems provide multi-user and multi-tasking capabilities which allow more than one user to access the system and more than one program to run in the system. Typically no matter which type of operating system is used, it will invariably have utility programs such as editor, an assembler and a compiler for some high-level programming languages. These are the most basic facilities needed to develop a program. An editor enables a user to create and modify source programs which are just ASCII characteristics representing instructions in some programming languages. An assembler is used to assemble programs written in the systems assembly language into binary code before they can be executed. A complier serves essentially the same function as an assembler, but for programs written in a high-level language such as C & C++. Very often an operating system provides a system library which contains subroutines to perform basic functions such as input/output and file handling.

5.5 SOFTWARE MODULES

There are considerable numbers of software packages for the various types of CAD/CAM systems discussed in Chapter 4. Each package has its own strengths and uniqueness and is usually targeted toward a specific market and group of users. For example, there are mechanical, electrical and architectural CAD and CAM software for the respective users. Investigating an existing software on various systems reveals that it has a generic structure and common modules. Awareness of such structure and modules enables users to understand better system functions for both evaluation and training purposes. The major available modules are discussed hereinafter.

5.5.1 Operating System Module

This module provides users with utility and system commands that deal with their accounts and files. Typical functions such as file manipulations (delete, copy, rename, etc.) managing directories and sub-directories using text editors, programming and accounts set-ups are supported by the OS module. Files that are generated in a users CAD/CAM account by the OS can be classified into two groups: The first group includes all conventional files (text files). The second group includes graphics-related files. A model geometry and its shaded images are stored in these files.

Due to the distinction between the OS and graphics functions on a CAD/CAM system, two working levels are available to the user. These are the OS and graphics levels. The user can easily invoke one level from the current one. The software usually provides its users with a command or procedure to go back and forth between the two levels to achieve maximum flexibility and increase user's productivity.

5.5.2 Graphics Module

This module provides users with various functions to perform geometric modelling and construction, editing and manipulation of existing geometry, drafting and documentation. The typical graphic operations that users can engage in are model creation, clean-up, documentation and plotting.

5.5.3 Applications Module

The creation of a geometric model of an object represents a means and not a goal to engineers. Their ultimate goal is to be able to utilize the model for design and manufacturing purposes. This module varies from one software to the other. However, there are common applications shared by most packages. Mechanical applications include mass property calculations, assembly analysis, tolerance analysis, sheet metal design, finite element modelling and analysis, mechanisms analysis, animation techniques and simulation and analysis of plastic injection moulding. Manufacturing applications include process planning, NC, CIM, robot simulation and group technology.

5.5.4 Programming Module

Typically this module provides users with system dependent and standard programming languages. The former is provided for graphic purposes while the latter is used for analysis and calculations.

5.5.5 Communications Module

This module is crucial if integration is to be achieved between the CAD/CAM system, other computer systems and manufacturing facilities. It is common to network the system to transfer the CAD database of a model for analysis purposes or to transfer its CAM database to the shop floor for production. This module also serves the purpose of translating databases between CAD/CAM systems using graphics standards such as IGES and ISO STEP.

5.6 GRAPHICS STANDARDS

Graphics software needs standards, without them users find that having purchased new computers graphics hardware, the further cost of modifying their existing software to run on the new equipment is prohibitive and the procedure is time-consuming. There are three major benefits of introducing standards for basic computer graphics. The first and obvious benefit is 'program portability' which permits application programs involving graphics to be easily transferred to and executed on almost any graphics installation. Graphics standards can help applications programmers in the understanding and use of graphics techniques. This gives rise to the second benefit of 'programmer portability' which means a programmer can leave one graphics application project for another without the need to learn a new set of graphics commands again. The third benefit is the graphics standards serve as a guideline for manufacturers of graphics equipment in providing useful combinations of graphics capabilities in a device, thus save programmers from having to create their own set of graphics commands and to write the interpreter for them. This advantage greatly helps to speed up the development of projects.

5.7 STANDARDS FOR GRAPHICS PROGRAMMING

Computer graphics became a very versatile tool for many applications that employ graphics. However, most of these graphics applications ran on machines of different sizes and types, ranging from small personal computers to the top level CAD/CAM systems. A graphics standard planning committee (GSPC) was formed in 1974 by ACM-SIGGRAPH. Its objective was to produce a draft proposal that could be used as a basis for a computer graphics application interface standard. In 1977, the GSPC published the so-called CORE specification which many assumed was going to become the ground work for this international graphics standard. In 1979 after over two years of implementation and review, the GSPC released another CORE specification which included raster graphics as well as the machine/device interface.

A committee for the development of computer graphics standard was formed by Deutsches Institute for Normung (DIN) in 1975. Since this adoption of GKS by DIN, the International

Standards Organization (ISO) and American National Standards Institute (ANSI) have cooperated to refine GKS in an attempt to make it a truly acceptable graphics standard. A significant development in CAD standards is the publication of GKS in 1982.

5.7.1 Features of GKS

GKS were designed in accordance with the following six requirements:

- First, GKS has to provide all the capabilities that are important for the whole range of graphics, from simple passive output to highly interactive applications.
- Second, different types of graphics devices such as vector and raster devices, micro-film recorders, storage tube displays, refresh displays and colour displays must be controlled by GKS in a consistent manner
- Third, GKS must include all the capabilities required by a majority of applications without becoming excessively large.
- Fourth, a complete suite of display management functions, cursor control and other features are provided.
- Fifth, graphic functions are defined in 2D or 3D.
- Sixth, all text or annotations are in a natural language like English.

5.7.2 GKS-3D

The objective of GKS-3D is to enhance GKS to 3D by introducing.

- The definition and the display of 3D graphical primitives
- Mechanisms to control viewing transformations and associated parameters
- Mechanisms to control the appearance of primitives including optional support for hidden line and/or hidden surface elimination but excluding light source, shading and shadow computation
- Mechanisms to obtain 3D input

The condition is that GKS-3D retains the general style of capabilities provided in GKS-2D. So that existing GKS application programs could execute without any modifications.

5.7.3 Programmers Hierarchical Interactive Graphics Standard (PHIGS)

PHIGS includes in its functionality three-dimensional output primitives and transformations. It has dynamic control over the visual appearance of attributes of primitives in a segment. The objective of PHIGS is to provide capabilities that support such functions as:

- Definition, display and modification of either 2D or 3D graphical data
- Definition, display and manipulation of geometrically related objects
- Rapid dynamic articulation of graphical entities

The PHIGS standard defines a set of device-independent logical concepts. Application programmers can use these concepts within a set of PHIGS rules.

Difference between GKS-3D and PHIGS

The various capabilities of GKS and PHIGS are compared in Table 5.2.

TABLE 5.2 Comparison of GKS and PHIGS

S.No.	Features	GKS 3D	PHIGS
1.	Picture composition	In non-normalized coordinate system	In normalized coordinate system
2.	Transformation	Normalization transformation	Modelling transformation
3.	Primitives	Primitives grouped together in a segment	No grouping of primitives
4.	Graphical output elements	Can be created inside and outside workstation	Output created from structure elements
5.	Display	Primitives displayed on all active workstations	Primitives displayed on workstation to which structures producing primitives are ported.
6.	Transformation, visibility, highlighting detectability	Workstation independent	Determined from a NAME SET attribute and workstation dependent filters
7.	Priority of primitives	Workstation independent	Based on sequence of posting

5.7.4 Initial Graphics Exchange Specification (IGES) Graphics Standard

The IGES committee was established in the year 1979. The CAD/CAM integrated information network (CIIN) of Boeing served as the preliminary basis of IGES. IGES version 1.0 was released in 1980, version 2.0 in 1983 and version 3.0 in 1986. IGES continues to undergo revisions and a brief revision history is given in Table 5.3.

TABLE 5.3 Revision History of IGES

Version	Year	Features
1.0	1980	Mechanical 2D and 3D drawings
2.0	1983	Sculpture surfaces, rational B-splines, finite elements and electrical drawings
3.0	1986	AEC, piping, etc.
4.0	1988	Constructive solid geometry
5.0	1990	Rationalization of existing formats
6.0	1991	B-REP solids

A number of vendors have implemented IGES. They include Applicon, Autotrol, CADAM, Calcomp, Calma, Computer Vision, Control Data, Gerber, IBM, Intergraph, Mc Auto, Prime, Summa Graphics, Tektronix and T & W systems.

IGES enables data transfer from one CAD system to another. This is shown in Fig. 5.3.

Fig. 5.3 CAD data transfer using IGES.

The software which translates data from CAD system to IGES is called a **pre-processor**. The software which translates IGES data to a CAD system is called **post-processor**.

Like most CAD systems IGES is based on the concept of entities, which range from simple geometric objects such as points, lines and circles to more sophisticated items like dimensions and sculptured surfaces.

Entities in IGES are divided into three categories:

- *Geometry*. Lines, circles, surfaces, etc. that define an object
- *Annotation*. Dimension, notes, title block
- *Structure*. Ways in which CAD systems combine other entities to make description of object easier (Block, cell or DITTO in CAD systems)

IGES files are divided into six sections:

Flag section

This is the optional section. It is not output by the post-processor and is found only in compressed ASCII files, never in fixed length ASCII files. It is identified by a "C" in column 73 indicating that the file is in compressed ASCII format. This section contains information that will be required by a post-processor. A post-processor is a program such as IGESIN. It translates a file from IGES format to the database form of a specific CAD.

Start section

An "S" in column 73 identifies this section. It contains that you can use to describe the drawing, identify its source, comment on its format and so forth.

Global section

A "G" in column 73 identifies this section. It contains a series of values that describe global characteristics of the IGES file, such as the name of the file, the system that created the file, the parameter and record delimiters, units of measure and precision.

Directory entry section

A "D" in column 73 identifies this section. It describes all the entities in the drawing. There is one entry for each entity in the file. Each entry consists of two lines organized into 20 eight-character fixed-format fields. This section contains attributes such as colour, view, line style, pointers to transformation matrices, pointers to parameter data for the entities and so on.

This section also provides an index to the entities in the file. IGES entities are identified by their type number (fields 1 and 11l) and a form number (filed 15).

Parameter data section

A "P" in column 73 identifies this section. It contains the data to describe each entity, such as point coordinates, coefficients of curve and surface equations, pointers to other entities, text characters and other attributes. The kind of data found in the PD section is different for each entity type. This section contains at least one record for each entity in the file. It contains a combination of real numbers, integers and text.

Terminate section

A "T" in column 73 identifies this section. This section is a single record organized into 10 eight-character fields. The first four fields contain the sequence number of the last line from each of the four preceding sections. The next five fields are unused and last field is the sequence number of the terminate section itself, this record must always have a sequence number of 1.

Major limitations of IGES are:

- IGES is complex and wordy.
- IGES files are about five times larger than an equivalent picture file.
- Several entities required for specialized CAD applications are yet not available.
- Since IGES is undergoing frequent revisions, many of the deficiencies are likely to be eliminated in future.

5.8 PRODUCT DATA BASED FORMAT

The experience gained from standards based on exchanging shape and non-shape data coupled with the need to automate CAD/CAM functions led to the development of exchange standards which include design and manufacturing data. The earliest standard developed is product data definition interface (PDDI) developed by us AIRFORCE ISO technical committee TCI 84 (industrial automation systems) evolved in 1984 STEP (standard for transfer and exchange of product model data) ESPRIT is another effort in this direction.

The need to have an exchange format incorporating the functionalities of design, analysis, manufacturing quality assurance, testing and support led to the initiation of PDES (product data exchange using STEP) in 1985.

This is designed to support any industrial application (mechanical, electrical, plant design, architecture, engineering and construction, etc.) in order to support industrial automation PDES codes information in a form directly interpreted by the computer. PDES methodology involves a three-layer architecture, reference models, formal languages and coordination with other standard efforts. The three layers are:

- Application layer
- Logical layer
- Physical layer

Figure 5.3 shows the three layers. The application layer is the interface between the user and PDES. It contains all descriptions and informations of various application areas. These descriptions and informations are expressed formally with PDES via information modelling techniques. The purpose of a logical layer is to provide a consistent and computer independent description of the data constructs that contain the information to be exchanged. The physical layer deals with the data structures and data format for the exchange file itself.

The main goal is here to establish and maintain efficiency in file size and processing time.

The input to the PDES methodology is a set of reference or discipline models. A discipline model represents an application expert's view of a discipline like kinematic analysis or robot design. The discipline model is created by the expert. This discipline model is used to develop a logical layer model. This is the summation of geometry, topology presentation and geometry topology associatives. In order to convert the logical layer model back to its discipline model the set of correspondence (mappings) from the discipline specific entities to the generic entities must be maintained. A global model is introduced for this purpose. This is a composite model, containing discipline specific and generic entities.

From the global model, information model is generated using a data specification language (DSL). This is in the form of a text file. The exchange of information using PDES is shown in Fig. 5.4.

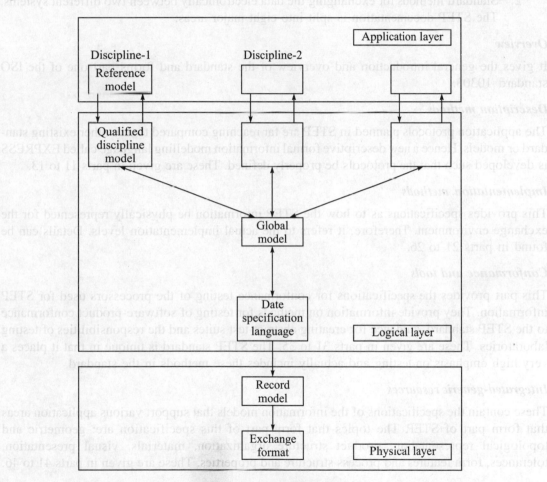

Fig. 5.4 PDES methodology.

5.9 STANDARD FOR THE EXCHANGE OF PRODUCT MODEL DATA (STEP)

Standard for the exchange of product model data (STEP), officially the ISO standard 10303, product data representation and exchange, is a series of international standards with the goal of defining data across the full engineering and manufacturing life cycle. The ability to share data across applications, across vendor platforms and between contractors, suppliers and customers, is the main goal of this standard.

The broad scope of STEP is as follows:

1. The standard method of representing the information necessary for completely defining a product throughout its entire life, i.e., from the product conception to the end of useful life.
2. Standard methods for exchanging the data electronically between two different systems.

The STEP documentation is split into eight major areas:

Overview

It gives the general introduction and overview of the standard and forms part one of the ISO standard 10303.

Description methods

The application protocols planned in STEP are far reaching compared to any other existing standard or models. Hence a new descriptive formal information modelling language called EXPRESS is developed such that the protocols be properly defined. These are given in parts 11 to 13.

Implementation methods

This provides specifications as to how the STEP information be physically represented for the exchange environment. Therefore, it refers to the actual implementation levels. Details can be found in parts 21 to 26.

Conformance and tools

This part provides the specifications for conformance testing of the processors used for STEP information. They provide information on methods for testing of software-product conformance to the STEP standard, guidance for creating abstract-test suites and the responsibilities of testing laboratories. These are given in parts 31 to 35. The STEP standard is unique in that it places a very high emphasis on testing and actually includes these methods in the standard.

Integrated-generic resources

These contain the specifications of the information models that support various application areas that form part of STEP. The topics that form part of this specification are: geometric and topological representation, product structure organization, materials, visual presentation, tolerances, form features and process structure and properties. These are given in parts 41 to 46.

Application information models

These specify the information models to be used for specific application areas such as draughting, finite element analysis, kinematics, building core model and engineering analysis core. These are given in part numbers 101 onwards.

Application interpreted constructs

These relate to the specific resources useful for defining the generic structures useful for applications. These are reusable groups of information resource entities that make it easier to express identical semantics in more than one application protocol. Examples include edge-based wireframe, draughting elements, constructive solid geometry, etc. These are given in part numbers 501 onwards. Examples include edge-based wireframe, shell-based wireframe, geometry-bounded 2D wireframe, draughting annotation, drawing structure and administration, draughting elements, geometry-bounded surface, non-manifold surface, manifold surface, geometry-bounded wireframe, etc.

Application protocols

These define the context for the use of product data for a specific industrial need. These are more complex data models used to describe specific product-data applications. These parts are known as **application protocols** and describe not only what data is to be used in describing a product, but how the data is to be used in the model. The application protocols use the integrated information resources in well-defined combinations and configurations to represent a particular data model of some phase of product life.

These are the main protocols to be used as subsets of STEP information model for exchange of data between specific application systems (such as between two finite element systems or between a CAD and process planning system). These are given in part numbers 201 onwards.

Application protocols currently in use are the Explicit Draughting AP 201 and the Configuration Controlled Design AP 203. Some details of these are given in Figs. 5.5 to 5.7. Other examples are associative draughting, mechanical design using boundary rep, mechanical design using surface rep, sheet metal die planning and design, electro-technical design and installation, numerical control (NC) process plans for machined parts, core data for automotive mechanical design processes, ship arrangements and mechanical parts definition for process planning using machining features.

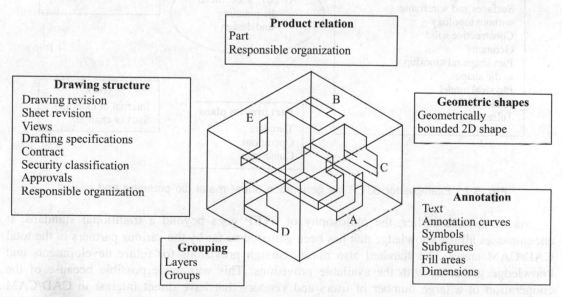

Fig. 5.5 STEP application protocol AP 203 explicit draughting.

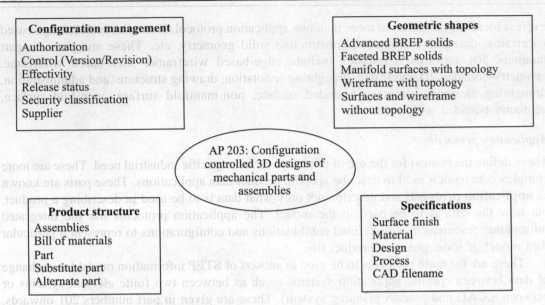

Configuration management

Authorization
Control (Version/Revision)
Effectivity
Release status
Security classification
Supplier

Geometric shapes

Advanced BREP solids
Faceted BREP solids
Manifold surfaces with topology
Wireframe with topology
Surfaces and wireframe
without topology

AP 203: Configuration controlled 3D designs of mechanical parts and assemblies

Product structure

Assemblies
Bill of materials
Part
Substitute part
Alternate part

Specifications

Surface finish
Material
Design
Process
CAD filename

Fig. 5.6 STEP application protocol AP 207 configuration controlled design.

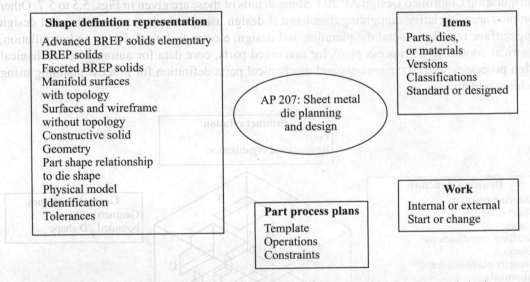

Shape definition representation

Advanced BREP solids elementary
BREP solids
Faceted BREP solids
Manifold surfaces
with topology
Surfaces and wireframe
without topology
Constructive solid
Geometry
Part shape relationship
to die shape
Physical model
Identification
Tolerances

AP 207: Sheet metal die planning and design

Items

Parts, dies,
or materials
Versions
Classifications
Standard or designed

Part process plans

Template
Operations
Constraints

Work

Internal or external
Start or change

Fig. 5.7 Example for STEP file generation sheet metal die planning and design.

As mentioned earlier, the philosophy of STEP goes beyond a traditional standard. It encompasses all the knowledge that has been gathered so far by the various partners in the total CAD/CAM arena. The standard also makes enough provisions for future developments and knowledge gathering with the available provisions. This was made possible because of the cooperation of a large number of users and vendors that have direct interest in CAD/CAM systems. This is also the reason why it took so long (ISO initiated the work on STEP in 1985) for coming to this form. Even today (early 2001) the standard is still evolving and only some parts are released for use by the industry as ISO 10303. The other parts are in various stages of proposal to standard stages and would take some time before completion.

5.10 DRAWING EXCHANGE FORMAT (DXF)

The DXF format has been developed and supported by Autodesk for use with the AutoCAD drawing files. It is not an industry standard developed by any standards organization, but in view of the widespread use of AutoCAD made it a default standard for use of a variety of CAD/CAM vendors.

A drawing interchange file is simply an ASCII text file with a file extension of .DXF and specially formatted text. The overall organization of a DXF file is as follows:

HEADER section

This section contains general information about the drawing similar to the global section of IGES. It consists of the AutoCAD database version number and a number of system variables. Each parameter contains a variable name and associated value. This information is used for database conversion purpose.

CLASSES section

It holds the information for application-defined classes, whose instances appear in the BLOCKS, ENTITIES and OBJECTS sections of the database. A class definition is permanently fixed in the class hierarchy.

TABLES section

This contains definitions for the following symbol tables which directly relates to the object types available in AutoCAD:

- Linetype table
- Layer table
- Text style table
- View table
- User coordinate system table
- Viewport configuration table
- Dimension style table
- Application identification table
- Block reference table

BLOCKS section

This contains block (symbol) definition and drawing entities that make up each block reference in the drawing.

ENTITIES section

This contains the graphical objects (entities) in the drawing, including block references (insert entities).

OBJECTS section

This contains the non-graphical objects in the drawing. All objects that are not entities or symbol table records or symbol tables are stored in this section. Examples of entries in the OBJECTS section are dictionaries that contain mline (multiple lines) styles and groups. A DXF file is composed of many groups, each of which occupies two lines in the DXF file. The first line is a group code. The second line is the group value, in a format that depends on the type of group specified by the group code. DXF files are either standard ASCII text or special binary form files which are more compact.

5.11   DMIS

Dimensional Measurement Interface Specification (DMIS) is a new standard in communication being established by CAM-I for manufacturing. Most of the standards that are existing or discussed in this book pertain to the translation of data when the data is generated in design form. However, this standard tries to establish a means of knowing what has been made by the CAM process. The objective of DMIS is therefore to provide a bi-directional communication of inspection data between computer systems and inspection equipment so as to see what has to be made and has been made.

The database in the form of geometric instructions and manufacturing information is already present, which is being used by some of the part programming systems for automatically converting into CNC part programs. From the same database, it is also possible to generate the inspection programs for the coordinate measuring machines (CMM).

The type of instructions needed for CMM are inspection probe selection, speed for positioning the probe, the path to be followed by the probe, speed and angle at which the probe approaches the workpiece, tolerance based information, etc. After a part has been produced on the CNC machine tool, the part would be checked on a CMM with the inspection program down loaded from the computer directly into the CMM for checking the part. After CMM checks the part, data about the part is sent back to the computer, where the original part geometry is stored as shown in Fig. 5.8. Thus, part geometry as designed is compared with the part produced and the resultant deviations existing could be identified, which would help in identifying problems in manufacturing and be suitably rectified.

5.12   PARASOLID

Most high-end CAD/CAM systems have their proprietary geometric modeller embedded within the modelling application. Parasolid is a portable "kernel" that can be used in multiple systems —both high-end and mid-range. In the early 1990s a number of software developers decided to adopt Parasolid. By adopting Parasolid, start-up software companies have eliminated a major barrier to application development—a high initial investment. This enabled them to effectively market softwares with strong solid modelling functionality at lower cost.

Today Parasolid is used as component software in many of the world's leading CAD/CAM systems. Unigraphics and Solid Edge, Parametric Technology, Solid Works Corporation (a

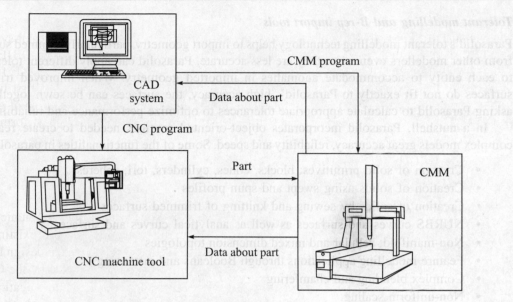

Fig. 5.8 Introduction of measurement in product development cycle.

subsidiary of Dassault Systems SA), ANSYS, Mechanical Dynamics, MacNeal-Schwendler and Bentley Systems are today using Parasolid. Parasolid is used for internal development by companies such as Boeing, Israel Aircraft Industries and Mitsubishi Motors.

5.12.1   Modelling Functionalities of Parasolid

Parasolid's modelling abilities include:

Complex blending

Parasolid provides excellent functionality of blending edges and faces. The functions include, rolling ball, variable, face-face, cliff-edge, overlapping, overflowing, continuous, tangent hold line, disc and conic-section and conic hold line.

Hollowing, shelling, offsetting and thickening surfaces

These are methods for creating thin-walled parts which are very common in many industries. To the CAD/CAM user, these are conceptually simple, but the topological changes required are a rigorous test of a modeller reliability.

Tapering and draft angles

These operations are used primarily for mould and die design. Parasolid can apply taper to complex designs following complex parting lines. Non-uniform scaling is another useful feature.

Complex modelling using B-surfaces

These are methods for creating thin-walled parts which are very common in many industries. To the CAD/CAM user these are conceptually simple, but the topological changes required are a rigorous test of a modeller reliability.

Tolerant modelling and B-rep import tools

Parasolid's tolerant modelling technology helps to import geometry, particularly trimmed surfaces from other modellers even when they are less accurate. Parasolid can apply different tolerances to each entity to accommodate anomalies in imported geometry. When improved trimmed surfaces do not fit exactly to Parasolid's high accuracy, the surfaces can be sewn together by asking Parasolid to calculate appropriate tolerances to optimize performance and reliability.

In a nutshell, Parasolid incorporates object-oriented routines needed to create real life, complex models great accuracy, reliability and speed. Some of the functionalities in parasolid are:

- Creation of solid primitives: blocks, cones, cylinders, tori, spheres
- Creation of solids using swept and spun profiles
- Creation of solids by sewing and knitting of trimmed surfaces
- NURBS curves and surfaces as well as analytical curves and surfaces
- Non-manifold, cellular and mixed dimension topologies
- Feature modelling applications through Booleans and instancing
- Complex blending and chamfering
- Non-uniform scaling
- Tolerant modelling enables reliable modelling with imported data and B-rep imp
- Geometry and topology inquiries
- Interference detection and clearance calculations
- Calculation of mass and inertia properties
- Foreign geometry interface to allow import from proprietary geometric systems
- Full range of geometric output formats including trimmed surface
- Graphical data output for wire-frame, silhouettes, hidden line and faceting

5.12.2 Applications

Computer aided design

CAD systems create mathematical models based on user input. Individual models can then be combined as components of an assembly to create a product. Parasolid adds value by being able to model real life objects more accurately and more reliably.

Computer aided engineering

CAE systems analyze designs created by CAD systems by calculating stress, applying the principles of thermodynamics, vibration, mechanisms, acoustics, aerodynamics and electromagnetism. A design is optimized by analyzing, making design changes and re-analyzing.

Computer aided manufacturing

CAM systems generate programs for CNC machine tools to manufacture a component. CAM covers every type of machining process from simple turning on a lathe to 5 axis milling of complex surfaces. Parasolid adds value by permitting the import of models from many Parasolid-based CAD systems and the interrogation of the geometry and topology.

Translators

Parasolid is capable of accepting data from other modeller formats. Its unique tolerant modelling functionality can accommodate and compensate for less accurate data.

5.13 ACIS

ACIS is another object oriented three-dimensional geometric modelling engine. It is designed for use as the geometry foundation in a 3D modelling application. ACIS is written in C++ and consists of a set of C++ classes and functions. A developer uses these classes and functions to create an end user 3D application. ACIS complements existing applications by offering a unified environment for the modelling of curves, surfaces and solids. ACIS provides a foundation of common modelling functionality and flexibility to be adapted and extended for particular application requirements. ACIS integrates wireframe, surface and solid modelling by allowing these alternatives representations to coexist naturally in a unified data structure, which is implemented in a hierarchy of C++ classes. ACIS bodies can have any of these forms or combinations of them. Linear and quadratic geometry is represented analytically and non-uniform rational B-splines represent free-form geometry. ACIS is a boundary-representation (B-rep) modeller, which means that it defines the boundary between solid material and empty space. This boundary is made from a closed set of surfaces.

5.13.1 ACIS Architecture

The ACIS has been designed using software component technology. A 'software component' is a functionally specialized unit of software—a collection of software items (functions, classes, etc.) grouped together to serve some distinct purpose. It serves as a constituent part of a whole software system or product. A 'product' is one or more software components that are assembled together and sold as a package. Components can be arranged in different combinations to form different products.

Some of the ACIS components are:

Advanced blending	–	Optional husk for blending beyond standard ACIS blending
AG spline	–	Spline library and interface
ACISMFC	–	Support for MFC based applications
Boolean	–	Unite, intersect and subtract operations
Clearance	–	Determine minimum distance between bodies or faces
Constructors	–	Basic topology construction; wireframe construction and editing; analysis (area, length, mass properties)
Cellular topology	–	Divide lumps into sets of cells
Euler Operations	–	Expand, separate and combine lumps
Faceter	–	Generate faceted (polygonal) representation
Generic attributes	–	Attributes that allow applications to exchange data
Graphic interaction	–	Commonly needed graphic display functionality

Kernel	–	Basic entity and attribute support, topology and geometry ENTITY classes; construction geometry classes
Local OPS	–	Optional husk for locally manipulating models
Meshing	–	Optional husk for representing a surface as a network of polygonal elements
Sweeping	–	Sweep a profile along a path

REVIEW QUESTIONS

5.1 What are the ground rules that should be considered in designing graphics software?

5.2 What are the functions of a graphics package?

5.3 What are the functions of an operating system?

5.4 What is meant by interactive software?

5.5 Why are graphic standards necessary?

5.6 Explain the various standards which work at various levels of graphics systems.

5.7 List and discuss the major available modules in CAD software packages.

5.8 Differentiate between GKS-3D and PHIGS.

5.9 What are the features of GKS?

5.10 What are the limitations of IGES?

5.11 Describe the structure of an IGES file.

5.12 What is the need for product data exchange?

5.13 Compare the IGES and PDES methodologies.

5.14 Explain the importance of IGES in the Indian manufacturing scene.

PRINCIPLES OF INTERACTIVE COMPUTER GRAPHICS

Chapter

6

INTRODUCTION

Computer graphics is a topic of rapidly growing importance in the computer field. It has always been one of the most visually spectacular branches of computer technology, producing images whose appearance and motion make them quite unlike any other form of computer output. Computer graphics is also an extremely effective medium for communication between man and computer, the human eye can absorb the information context of a displayed diagram or perspective view much faster than it can scan a table of numbers. We have all come across applications of computer graphics in everyday life, for example, in television titles and weather forecast presentations or in video games. Images like these are examples of non-interactive or passive computer graphics, the observer has no control over the image. We can give the observer some control over the image by providing him with an input device, such as the lever of the ping-pong game, so that he can signal his requests to the computer. We then have an example of interactive computer graphics.

There are a number of variations in the characteristics of computer graphics that may be classified into three categories as follows:

- The first category defines the control the user has over the image. In passive computer graphics the user has no control, in interactive graphics the user may interact with the graphics and with the programs generating them.

- The second category concerns the way, the image is generated. In vector graphics, the image comprises a number of lines, whereas raster graphics involve the manipulation of the colour and/or intensity of points, known as **picture elements** or **pixels**, in a matrix making up the image.

- The third category distinguishes between image space graphics, in which image itself is directly manipulated to create a picture and object – space graphics in which the image is a representation of a separate model.

CAD may be categorized as an application of interactive object-space graphics, in that the objective is to develop interactively a model of a design. It also involves both vector graphics and raster graphics. Line drawings and diagrams make extensive use of vector graphics and they are also used in the production of wire frame images for three-dimensional models of all types, although raster hardware is practically universally used for the display of the line images. Raster graphics are used extensively for the display of surface and solid models and highly realistic displays can be achieved with the use of appropriate rendering software to generate the image of the model.

6.1 GRAPHIC PRIMITIVES

A picture can be described in several ways. Assuming we have a raster display, a picture is completely specified by the set of intensities for the pixel position in the display. At the other extreme, we can describe a picture as a set of complex objects, such as trees and terrain or furniture and walls, positioned as specific coordinate locations within the scene shapes and colours of the objects can be described internally with pixel arrays or with sets of basic geometric structures, such as straight line segments and polygon colour areas. The scene is then displayed either by loading the pixel arrays into the frame buffer or by scan converting the basic geometric structure specifications into pixel patterns. Typically graphics programming packages provide functions to describe a scene in terms of these basic geometric structures referred to as output primitives. Each output primitive is specified with input coordinate data and other information about the way that object is to be displayed points and straight line segments are the simplest geometric components of pictures. Additional output primitives that can be used to construct a picture include circles and other conic sections, quadric surfaces, spline curves and surfaces, polygon colour areas and character strings.

A drawing is created by an assembly of points, lines, arcs and circles. Point, line, circle, arc, ellipse polygon and polyline are called **entities**. These entities are found in a typical CAD package by any user. The fundamentals of generation of some of these entities are discussed hereinafter.

6.1.1 Point Plotting

Point plotting is accomplished by converting a single coordinate position furnished by an application program into appropriate operations for the output device in use. With a CRT monitor, for example, the electron beam is turned on to illuminate the screen phosphor at the selected location. How the election beam is positioned depends on the display technology.

A random-scan (vector) system stores the position of the electron beam at the screen locations to be plotted during each refresh cycle. For a black and white raster system, on the other hand, a point is plotted by setting the bit value corresponding to a specified screen position within the frame buffer to 1. Then, as the electron beam sweeps across each horizontal scan line, it emits a burst of electrons whenever a value of 1 is encountered in the frame buffer. It is loaded with the colour codes for the intensities that are to be displayed at the screen pixel positions.

6.1.2 Drawing of Lines

Line drawing is accomplished by calculating intermediate positions along the line path between two specified end point positions. An output device is then directed to fill in these positions between the end points. For analog devices, such as a vector pen plotter or a random-scan display, a straight line can be drawn smoothly from one end point to the other linearly varying horizontal and vertical deflection voltages are generated that are proportional to the required changes in the x and y directions to produce the smooth line.

Straight line segments are used a great deal in computer generated pictures. The following salient points have been formulated for line drawing displays:

(a) Lines should appear straight.
(b) Lines should have constant density.
(c) Line density should be independent of length and angle.
(d) Lines should terminate accurately.
(e) Lines should be drawn rapidly.

The process of turning on the pixels for a line segment is called **vector generation**. If the end points of the line segment are known, there are several schemes for selecting the pixels between the end pixels. One method of generating a line segment is the symmetrical digital differential analyzer (DDA).

Digital devices display a straight line segment by plotting discrete points between the two end points. Discrete coordinate positions along the line path are calculated from the equation of the line. For a raster video display, the line colour (intensity) is then loaded into the frame buffer at the corresponding pixel points. Reading from the frame buffer, the video controller then "plots" the screen pixels. Screen locations are referenced with integer values, so plotted positions may only approximate actual line positions between two specified end points.

6.2 LINE DRAWING ALGORITHMS

The Cartesian slope-intercept equation for a straight line is

$$y = mx + b \tag{6.1}$$

where m representing the slope of the line and b as the y intercept. Given that the two end points of a line segment are specified at positions (x_1, y_1) and (x_2, y_2), as shown in Fig. 6.1, we can determine values for the slope m and y intercept b with the following calculations:

$$m = \frac{y_2 - y_1}{x_2 - x_1} \tag{6.2}$$

$$b = y_1 - mx_1 \tag{6.3}$$

Algorithms for displaying straight lines are based on the line equation [Eq. (6.1)] and the calculations given in Eqs. (6.2) and (6.3).

For any given x interval Δx along a line, we can compute the corresponding y interval Δy from Eq. (6.2) as:

$$\Delta y = m \, \Delta x \tag{6.4}$$

Similarly, we can obtain the x interval Δx corresponding to a specified Δy as:

$$\Delta x = \frac{\Delta y}{m} \tag{6.5}$$

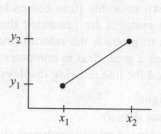

Fig. 6.1 Line path between end point positions (x_1, y_1) and (x_2, y_2).

Equations (6.4) and (6.5) form the basis for determining deflection voltages in analog devices. For line with slope magnitude $|m| < \Delta x$ can be set proportional to a small horizontal deflection voltage and the corresponding vertical deflection is then set proportional to Δy as calculated from Eq. (6.4). For lines whose slopes have magnitudes $|m| > \Delta y$ can be set proportional to a small vertical deflection voltage and the corresponding horizontal deflection is then set proportional to Δx as calculated from Eq. (6.5). For lines with $m = 1$, $\Delta x = \Delta y$ and the horizontal and vertical deflections voltages are equal. In each case, a smooth line with slope m is generated between the specified end points.

6.2.1 DDA Algorithm

The digital differential analyzer generates lines from their differential equations. The DDA works on the principle that x and y are simultaneously incremented by small steps proportional to the first derivatives of x and y. In the case of a straight line, the first derivatives are constant and are proportional to Δx and Δy, where Δ is a small quantity.

In the real world of limited precision displays, addressable pixels only must be generated. This can be done by rounding to the next integer after each incremental step. After rounding a pixel is displayed at the resultant X and Y locations. An alternative to rounding is the use of arithmetic overflow. X and Y are kept in registers that have integer and fractional parts. The incrementing values which are less than unity are repeatedly added to the fractional part and whenever the result overflows the corresponding integer part is incremented. The integer parts of X and Y are used to plot the line. This would normally have the effect of truncating. The DDA is therefore initialized by adding 0.5 in each of the fractional parts to achieve true rounding.

The symmetrical DDA generates reasonably accurate lines since a displayed pixel is never away from a true line by half the pixel unit. A Pascal procedure for a simple DDA is given as follows.

```
PROCEDURE LINE DDA (Xa, Ya, Xb, Yb: integer);
VAR
dx, dy, STEPS, k: INTEGER;
X INCREMENT, Y INCREMENT X, Y: REAL;
BEGIN
dx: = xb - xa;
dy: = yb - ya;
```

```
IF abs (dx) > abs (dy) THEN STEPS: = abs (dx)
ELSE STEPS: = abs (DY);
x INCREMENT : = dx / STEPS;
y INCREMENT : = dy / STEPS;
x: = xa;
y: = ya;
SET PIXEL (ROUND(X), ROUND(Y), 1);
FOR K:=1 to STEPS DO
BEGIN
X: = X + X INCREMENT;
Y:= Y + Y INCREMENT;
SET PIXEL (ROUND(X), ROUND(Y), 1);
END
END: (LINE DDA)
```

6.2.3 Bresenham's Line Algorithm

An accurate and efficient raster line—generating algorithm developed by Bresenham scan converts lines using only incremental integer calculations that can be adopted to display circles and other curves. Bresenham's algorithm is an algorithm which enables the selection of optimum raster locations to represent a straight line. In this algorithm pixels along X or Y directions are incremented by one unit depending upon the slope of the line. The increment in the other direction is determined by examining the error or distance between actual line location and the nearest grid locations. X like the simple DDA, it is designed so that each iteration changes one of the coordinate values by ±1. The other coordinate may or may not change, depending on the value of an error term maintained by the algorithm. This error term records the distance, measured perpendicular to the axis of greatest movement, between the exact path of the line and the actual dots generated. In the example of Fig. 6.2 where the X-axis is the axis of greatest movement, the error term e is shown measured parallel to the Y-axis. The following description of the algorithm assumes this particular orientation of the line.

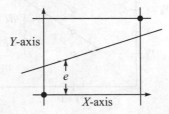

Fig. 6.2 Bresenham's algorithm error term e measures distance between path of line and dots generated.

At each iteration of the algorithm, the slope of the line, $\Delta y/\Delta x$, is added to the error term e. Before this is done, the sign of e is used to determine whether to increment the y coordinate of the current point. A positive e value indicates that the exact path of the line lies above the current point; therefore, the y coordinate is incremented and 1 is subtracted from e. If e is negative, the y coordinate value is left unchanged X.

To illustrate Bresenham's approach, we first consider the scan-conversion process for lines with positive slope less than 1. Pixel positions along a line path are then determined by sampling at unit x intervals. Starting from the left end point (x_0, y_0) of a given line, we step to each successive column (x position) and plat the pixel whose scan-line y value is closest to the line path.

Figure 6.3 demonstrates the kth step in this process. Assuming we have determined that the pixel at (x_k, y_k) is to be displayed, we next need to decide which pixel to plot in column x_{k+1}. Our choices are the pixels at positions (x_{k+1}, y_k) and (x_{k+1}, y_{k+1}).

$$
\begin{array}{|c|c|c|c|}
\hline
y_{k+3} & Y = mx + b & & \\
\hline
y_{k+2} & & & \\
\hline
y_{k+1} & & & \\
\hline
y_k & & & \\
\hline
\end{array}
$$

Fig. 6.3 Section of the screen grid showing a pixel in column x_k on scan line y_k that is to be plotted along the path of a line segment with slope $0 < m < 1$.

At sampling position x_{k+1}, we label vertical pixel separations from the mathematical line path as d_1 and d_2 (Fig. 6.4). The y coordinate on the mathematical line at pixel column position x_{k+1} is calculated as:

$$Y = m(x_k + 1) + b \qquad (6.6)$$

Then

$$
\begin{aligned}
d_1 &= y - y_k \\
&= m(x_k + 1) + b - y_k
\end{aligned} \qquad (6.7)
$$

And

$$
\begin{aligned}
d_2 &= (y_k + 1) - y \\
&= y_{k+1} - m(x_{k+1}) - b
\end{aligned} \qquad (6.8)
$$

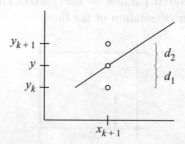

Fig. 6.4 Distances between pixel positions and the line y coordinate at sampling position x_{k+1}.

The difference between these two separations is:

$$d_1 - d_2 = 2m(x_k + 1) - 2y_k + 2b - 1 \qquad (6.9)$$

A decision parameter p_k for the kth step in the line algorithm can be obtained by rearranging Eq. (6.9) so that it involves only integer calculations. We accomplish this by substituting

$m = \Delta y / \Delta x$, where Δy and Δx are the vertical and horizontal separations of the end point positions and defining

$$p_k = \Delta x \, (d_1 - d_2)$$
$$= 2\Delta y x k - 2\Delta x \, y k + c \qquad (6.10)$$

The sign of p_k is the same as the sign of $d_1 - d_2$, since $\Delta x > 0$ for our example. Parameter c is constant and has the value $2\Delta y + \Delta x \, (2b - 1)$, which is independent pixel at y_k. If it is closer to the line path than the pixel at y_{k+1} (i.e. $d_1 < d_2$) then decision parameter p_k is negative. In that case, we plot the lower pixel, otherwise we plot the upper pixel.

Coordinate changes along the line occur in unit steps in either the x or y directions. Therefore, we can obtain the values of successive decision parameters using incremental integer calculations.

At step $k + 1$, the decision parameter is evaluated from Eq. (6.10) as:

$$p_k + 1 = 2\Delta y x k + 1 - 2\Delta x y k + 1 + c \qquad (6.11)$$

Subtracting Eq. (6.10) from Eq. (6.11), we have

$$p k + 1 - p_k = 2\Delta y \, (x k + 1 - x k) - 2\Delta x \, (y k + 1 - y k) \qquad (6.12)$$

where the term $y_{k+1} - y_k$ is either 0 or 1, depending on the parameter p_k.

This recursive calculation of decision parameters is performed at each integer x position, starting at the left coordinate end point of the line.

The first parameter, p_0 is evaluated from Eq. (6.10) at the starting pixel position (x_0, y_0) and within evaluated at $\Delta y / \Delta x$

$$p_0 = 2\Delta y - \Delta x \qquad (6.13)$$

We can summarize Bresenham line drawing for a line with a positive slope less than in the following listed steps. The constants $2\Delta y$ and $2\Delta y - 2\Delta x$ are calculated once for each line to be scan converted, so the arithmetic involves only integer addition and subtraction of those two constants.

6.2.4 Bresenham's Line-Drawing Algorithm For $|m| < 1$

1. Input the two line end points and store the left end point in (x_0, y_0)
2. Load (x_0, y_0) into the frame buffer, that is plot the first point.
3. Calculate constants Δx, Δy, $2\Delta y$ and $2\Delta y - 2\Delta x$, and obtain the starting value for the decision parameter as $p_0 = 2\Delta y - \Delta x$.
4. At each x_k along the line starting at $k = 0$, perform the following test:
   If $p_k < 0$, the next point to plot is (x_{k+1}, y_k) and $p_{k+1} = p_k + 2\Delta y$. Otherwise, the next point to plot is $(x_k + 1, y_k + 1)$ and $p_k + 1 = p k + 2\Delta y - 2\Delta x$.
5. Repeat Step 4 Δx times.

6.3 BRESENHAM'S CIRCLE ALGORITHM

Since the circle is a frequently used component in pictures and graphs, a procedure for generating either full circles or circular is included in most graphics packages. More generally, a single procedure can be provided to display either circular or elliptical curves.

6.3.1 Properties of Circles

A circle is defined as the set of points that are all at a given distance r from a centre position (X_c, Y_c). This distance relationship is expressed by the Pythagorean theorem in cartesian coordinates as:

$$(x - x_c)^2 + (y - y_c)^2 = r^2 \tag{6.14}$$

We could use this equation to calculate the position of points on a circle circumference by stepping along the x-axis in unit steps from $(x_c - r)$ to $(x_c + r)$ and calculating the corresponding y values at each position as

$$y = y_c \pm \sqrt{r^2 - (x_c - x)^2} \tag{6.15}$$

But this is not the best method for generating a circle. One problem with this approach is that it involves considerable computation at each step. Moreover the spacing between plotted pixel positions is not uniform as shown in Fig. 6.5. We could adjust the spacing by interchanging x and y (stepping through y values and calculating x values) whenever the absolute value of the slope of the circle is greater than 1. But this simply increases the computation and processing required by the algorithm.

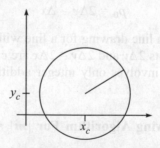

Fig. 6.5 Circle with centre coordinates (x_c, y_c) and radius r.

Another way to eliminate the unequal spacing shown in Fig. 6.5 is to calculate points along the circulars boundary using polar coordinates r and θ (Fig. 6.6). Expressing the circle equation in parameter polar form yields the pair of equations:

$$x = x_c + r \cos \theta$$
$$y = y_c + r \cos \theta \tag{6.16}$$

When a display is generated with these equations using a fixed angular step size, a circle is plotted with equally spaced points along the circumference. The step size chosen for θ depends on the application and the display device. Larger angular separations along the circumference can

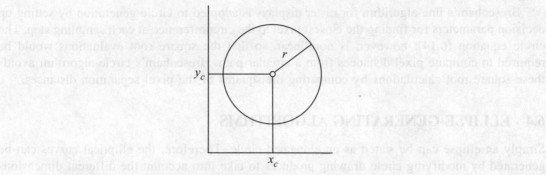

Fig. 6.6 Positive half of a circle with Eq. (6.15) and with $(x_c, y_c) = (0, 0)$.

he connected with straight line segments to approximate the circular path. For a more continuous boundary on a raster display, we can set the step size at $1/r$. This plots pixel positions that are approximately one unit apart.

Computation can be reduced by considering the symmetry of circles. The shape of the circle is similar in each quadrant. We can generate the circle section in the second quadrant of the xy plane by noting that the two circle sections are symmetric with respect to the y-axis and circle sections in the third and fourth quadrants can be obtained from sections in the first and second quadrants by considering symmetry between octants. Circle sections in adjacent octants within one quadrant are symmetric with respect to the 45° line dividing the two octants. These symmetry conditions are shown in Fig. 6.7, where a point at position (x, y) on a one-eighth circle sector is mapped into the seven circle points in the other octants of the xy plane. Taking advantage of the circle symmetry in this way, we can generate all pixel positions around a circle by calculating only the points within the sector from $x = 0$ to $x = y$.

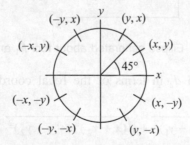

Fig. 6.7 Symmetry of a circle calculation of a circle point (x, y) in one octant yields the circle points shown for the other seven octants.

Determining pixel positions along a circle circumference using either Eq. (6.14) or Eq. (6.15) still requires a good deal of computation time. The cartesian equation (6.14) involves multiplication and square root calculations, while the parametric equations contain multiplications and trigonometric calculations. More efficient circle algorithms are based on incremental calculation of decision parameters, as in the Bresenham's line algorithm, which involves only simple integer operations.

Bresenham's line algorithm for raster displays is adopted to circle generation by setting up decision parameters for finding the closest pixel to the circumference at each sampling step. The circle equation (6.14), however, is non-linear, so that the square root evaluations would be required to compute pixel distances from a circular path. Bresenham's circle algorithm avoids these square root calculations by comparing the squares of the pixel separation distances.

6.4 ELLIPSE-GENERATING ALGORITHMS

Simply an ellipse can be stated as an elongated circle. Therefore, the elliptical curves can be generated by modifying circle drawing produces to take into account the different dimensions of an ellipse along the major and minor axes.

Properties of ellipse

An ellipse is defined as the set of points such that the sum of the distances from two fixed positions (foci) is the same for all points (Fig. 6.8). If the distances to the two foci from any point $p = (x, y)$ on the ellipse are labelled d_1 and d_2, then the general equation of an ellipse can be stated as:

$$d_1 + d_2 = \text{Constant} \tag{6.17}$$

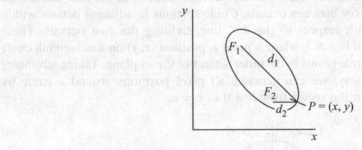

Fig. 6.8 Ellipse generated about foci F_1 and F_2.

Expressing distances d_1 and d_2 in terms of the focal coordinates $f_1 = (x_1, y_1)$ and $f_2 = (x_2, y_2)$, we have

$$\sqrt{(x + x_1)^2 + (y - y_1)^2} + \sqrt{(x + x_2)^2 + (y - y_2)^2} = \text{Constant} \tag{6.18}$$

By squaring this equation, isolating the remaining radical and then squaring again, we can rewrite the general ellipse equation in the form:

$$Ax_2 + By^2 + Cxy + Dx + Ey + F = 0 \tag{6.19}$$

where the coefficients A, B, C, D, E, and F are evaluated in terms of the focal coordinates and the dimensions of the major and minor axes of the ellipse. The major axis is the straight line segment extending from one side of the ellipse to the other through the foci. The minor axis spans the shorter dimension of the ellipse, bisecting the major axis at the halfway position (ellipse centre) between the two foci.

An interactive method for specifying an ellipse is an arbitrary orientation is to input the two foci and a point on the ellipse boundary. With these three coordinate positions, we can evaluate the constant in Eq. (6.18). Then the coefficient in Eq. (6.19) can be evaluated and used to generate pixels along the elliptical path.

Ellipse equations are greatly simplified if the major and minor axes are oriented to align with the coordinate axes. In Fig. 6.9, we show an ellipse in "standard position" with major and minor axes oriented parallel to the x and y axes.

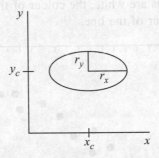

Fig. 6.9 Ellipse centred at (x_c, y_c) with semi-major axis r_x and semi-minor axis r_y.

Parameter r_x for this example labels the semi-major axis and parameter r_y labels the semi-minor axis. The equation of the ellipse shown in Fig. 6.9 can be written in terms of the ellipse centre coordinates and parameters r_x and r_y as:

$$\left(\frac{x - x_c}{r_x}\right)^2 + \left(\frac{y - y_c}{r_y}\right)^2 = 1 \tag{6.20}$$

Using polar coordinates r and θ, we can also describe the ellipse in standard position with the parametric equations:

$$x = x_c + r_x \cos\theta$$
$$y = y_c + r_y \sin\theta \tag{6.21}$$

Symmetry considerations can be used to further reduce computations. An ellipse in standard position is symmetric between quadrants, but unlike a circle, it is not symmetric between the two octants of a quadrant. Thus, we must calculate pixel positions along the elliptical arc throughout one quadrant, then we obtain positions in the remaining three quadrants by symmetry (Fig. 6.10).

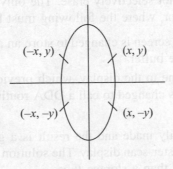

Fig. 6.10 Symmetry of an ellipse calculation of a point (x, y) in one quadrant yields the ellipse points shown for the other three quadrants.

6.5 SCAN CONVERSION

In order to display vectors on a frame buffer display, each vector must be converted from a conventional geometric representation into a raster representation using a process called **scan conversion**. Figure 6.11 shows the result of scan converting a straight line segment on a small 16×16 raster. Most of the pixels are white, the colour of the background, those in the path of line segment are black, the colour of the line.

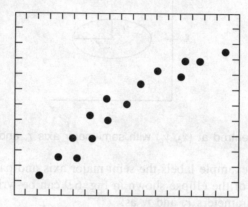

Fig. 6.11 A line drawn on a raster using an incremental point-plotting algorithm.

Figure 6.11 will remind us immediately of the methods described for generating lines on a point-plotting display. This is no coincide, the raster shown in Fig. 6.11 is in fact generated with a DDA algorithm.

A small modification must be made to the DDA algorithm to adapt it to scan conversion. After computing each dot coordinate position (x, y) the algorithm plots the point, not by adding a dot to the display file but by calling set pixel (frame buffer, x, y, intensity), where intensity is the shade of the line. Any of the incremental point-plotting techniques can be similarly adopted to scan conversion. These algorithms can be applied repeatedly to build up images containing many lines.

A graphics package can easily be adopted to drive frame buffer display. The simplest approach is to treat the display as the logical equivalent of a storage tube terminal. Thus, we can add to its contents, but we cannot selectively erase. The only changes to the graphics package are in the display code generator, where the following must be done:

1. The routine to clear the screen is changed to store an appropriate background value into every byte in the frame buffer.

2. The routine to add a line to the display, which previously generated commands to the storage tube terminal, is changed to call a DDA routine that sets the appropriate pixels in the frame buffer.

These two changes are easily made and the result is a graphics package that allows line drawings to be displayed on a raster-scan display. The solution is far from optimal, however, and provides no better performance than a storage tube.

Somewhat better performance can be achieved by allowing selective modifications to the frame buffer. To erase an individual line, for example, the same DDA algorithm can be used, but each call to set pixel will specify the intensity of the background. Thus all the pixels altered when the line was originally drawn with the DDA are now reset to the background value. This selective erasure can be used to delete all the lines in a display file segment whenever the segment is unposted or deleted, a strategy that is often faster than clearing the buffer and scan converting lines from all posted segments. Although line erasures may leave holes in remaining lines, the holes can be filled in by a small amount of additional scan conversion. This strategy first makes the important changes to the display (lines are erased) and later touches up the image to remove defects introduced by the erasing method.

Even if a graphics system uses selective erasure, it has not fully used the facilities of a raster display. Further advantages derive from the ability to place arbitrary patterns in the frame buffer and from the capability of any raster display to show solid areas of tone. These capabilities require further extensions to the scan-conversion process.

There are several techniques of scan conversion:

- Real time scan conversion
- Run length encoding
- Cell encoding
- Frame buffer memory

Real time scan conversion

In this technique, the picture is randomly represented in terms of visual attributes like colour, shade and intensity and geometrical properties like x, y coordinates, slopes and text, which are ordered in y. The scan conversion program scans through this information and calculates the intensity of each pixel on the screen.

Run length encoding

In run length encoding scheme, the number of pixels of same intensity and colour in a given scan line specifies. In the simplest case, the encoded data will show the intensity and run length.

For example, suppose we have a pixel arrangement in a scan line as shown in Fig. 6.12. The encoding is done as per intensity run length fashion. For the scan line of Fig. 6.12 the encoding is shown in Fig. 6.13. For the first four pixels in the scan line, the line intensity is zero. Intensity is one for the next pixel and so on. Run length encoding has the advantage of large data compression. Its disadvantage is that since the run lengths are stored sequentially, addition or deletion of lines is difficult.

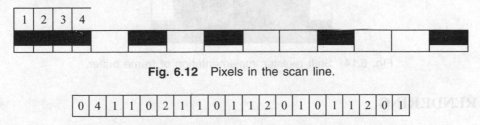

Fig. 6.12   Pixels in the scan line.

| 0 | 4 | 1 | 1 | 0 | 2 | 1 | 1 | 0 | 1 | 1 | 2 | 0 | 1 | 0 | 1 | 1 | 2 | 0 | 1 |

Fig. 6.13   Run length encoding of scan line shown in Fig. 6.12.

Cell encoding

Run length encoding stores a picture sequentially line by line. In contrast, cell encoding considers representation of picture information by dividing the display area into cells of suitable sizes. For example, a display area of 512×512 can be divided into 4096 cells of 64 pixels. In the case of drawings combinations of these adjacent cells can be used to construct complete lines.

Frame buffers

A frame buffer is a common method of implementing a raster CRT graphics device. It can be considered to be a large (Contiguous) piece of computer memory. In the simplest case there can be one memory bit for each pixel in the raster. This amount of memory is called a **bit plane**. A 320×320 raster requires 64 K memory bits in a single plane. The picture is built in the frame buffer one bit at a time. If a particular pixel is to be addressed, the corresponding bit in the frame buffer is changed from 0 to 1.

A common implementation of frame buffer is a random access semiconductor memory. Frame buffers can also be implemented using shift registers. Each shift register contributes one pixel in a horizontal scan line. Figure. 6.14 shows a shift register implementation of frame buffer. Shift register frame buffer memory has the disadvantage of low levels of interactivity. In workstations, it is efficient to add a graphics processor with a separate frame buffer memory. The system performance can be improved by this architecture as it meets the update requirements of the frame buffer.

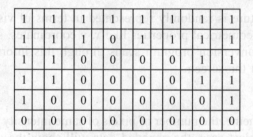

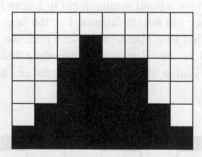

Fig. 6.14 Shift register implementation of frame buffer.

6.6 RENDERING

Rendering is a general term describing the overall process of representation of a 3D object to a shaded 2D projection on the screen of the computer. This involves a number of processes.

- Generation of a data structure for polygon models that will contain the information required for the shading process
- Applying transformations
- Scan converting polygons
- Hidden surface removal
- Shading individual pixels

6.6.1 Rasterizing Polygons

Filling a polygon means finding pixel coordinates of interior points and assigning to these a value calculated using one of the incremental shading schemes. This requires the construction of an edge list for each polygon. There are two approaches to solid area scan conversion of polygons: scan conversion and seed filling.

Scan converting polygons

The scan conversion of a polygon involves finding all pixels that lie inside the polygon boundaries and generating a display by appropriately setting the intensities of these pixels. From a description of the outline of a polygon it is possible to decide whether a given point lies inside or outside the polygon by counting intersections of the boundary with an imaginary line extending from the point to some other point far outside the polygon. If an odd number of intersections is encountered, the point lies inside the boundary otherwise it lies outside.

A scan line intersects a polygon at one or more points. The intersection divides the scan line into a number of regions. Referring to Fig. 6.15, $P_1 - P_2 - P_3 - P_4$ is a polygon. It has a vertex list of P_1, P_2, P_3, P_4 and an edge list P_1P_2, P_2P_3, P_3P_4 and P_4P_1. The intersections are at pixels 4 and 8 along the scan line n. In determining the intensity of shade or colour, the intersections are considered in pairs. For intervals between pairs of intersections, the colour will be that of the background.

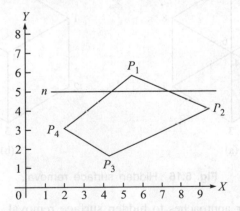

Fig. 6.15 Scan converting polygon.

For the scan line considered from pixel 4 to 8 the colour is that of the polygon and beyond 8, the colour is the background again. Several algorithms are available to create ordered edge lists.

These take into consideration the intersection of the scan line with the vertices of the polygon also.

Seed fill algorithm

The scan algorithm fills the polygon in the scan line order. The seed fill algorithm is designed on the assumption that at least one pixel interior to a polygon is known. Interior or boundary defined regions can be classified into 4 connected or 8 connected. Algorithms can be designed to fill both the regions. An 8 connected algorithm apart from left, right, up and down direction gives four more additional diagonal directions. More efficient algorithm uses a combination of scan line and seed fill approaches.

6.6.2 Hidden Surface Elimination

One of the most difficult problems in computer graphics is the removal of hidden parts from images of solid objects. In real life, the opaque material of these objects obstructs the light rays from the hidden parts and prevents us from seeing them. In the computer generation of an image no such automatic elimination takes place when objects are projected onto the screen coordinate system. Instead all parts of every object, including many parts that should be invisible are displayed. In order to remove these parts to create a more realistic image, we must apply a hidden line or hidden surface algorithm to the set of objects.

In Fig. 6.16(a), an opaque cube is shown in wire frame representation. Edges 15, 48, 37, 14, 43, 12, 23, 58 and 87 are visible whereas edges 56, 57 and 26 are not visible. Correspondingly surfaces 1265, 2673 and 5678 are not visible. Since the object is opaque, the actual representation of the cube must be as shown in Fig. 6.16(b).

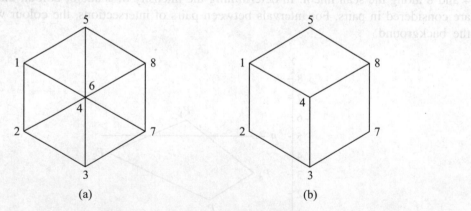

Fig. 6.16 Hidden surface removal.

There are two popular approaches to hidden surface removal. These are scan line based systems and z-buffer based systems. Other important approaches are area subdivision and depth list schemes.

6.7 Z-BUFFER ALGORITHM

This scheme was developed by Catmull using Sutherland's classification scheme. In this technique, pixels interior to a polygon are shaded and their depth is evaluated by interpolation from z-values of the polygon vertices after the viewing transformation has been applied. For every pixel (x,y) values are the pixel coordinates and z value is the viewing space depth. For each interior polygon point a search is carried out to determine the minimum z value. This search is conveniently implemented using a z-buffer that holds for a current point (x,y) the smallest z value so far encountered. The z-buffer algorithm has the advantage of simplicity. It handles scenes of any complexity. There is also no computation required for depth sort. The storage space required, however, is large. This could be reduced by preprocessing, so that polygons nearest the viewpoint are processed first.

6.7.1 Scan Line z-buffer Algorithm

This is a special case of z-buffer algorithm. In this algorithm for each scan line the frame buffer is initialized to the background and z buffer to the minimum z. The intersection of the scan line with the two-dimensional projection of each polygon is found. The depth of each pixel on the scan line between each pair of the intersection is determined. If the pixel depth is greater than that in the z-buffer then this line segment is currently visible.

A spanning scan line algorithm, instead of solving the hidden surface removal on a pixel by pixel basis using incremental z calculation uses spans along the scan line over which there is no depth conflict. Consider the three-dimensional screen space shown in Fig. 6.17. A scan line algorithm moves a scan line plane down the y axis. This plane parallel to the xoz plane, intersects

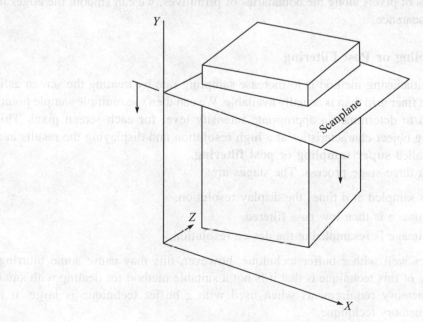

Fig. 6.17 Spanning scan line algorithm.

the objects in the scene and reduces the hidden surface problem to a 2D space. The line segments obtained through the intersection are then analyzed to detect hidden surfaces. This is done by considering the spans which form part of a line segment that is contained between edge intersections of all active polygons.

6.8   ANTIALIASING

Displayed primitives generated by the computer have a jagged or stair step appearance because the sampling process digitizes coordinate points on an object to discrete integer pixel positions. This distortion of information due to low frequency sampling (under sampling) is called **aliasing**. We can improve the appearance of displayed raster lines by applying antialiasing methods that compensate for the undersampling process. One method is to increase the sample rate by increasing the resolution of the raster. Another method is to calculate the raster at higher resolution and display it at lower resolution by averaging the pixel and attribute at the lower resolution. But even at the highest resolution possible with current technology, the jaggies will be apparent to some extent. There is a limit to how big we can make the frame buffer and still maintain the refresh rate at 30 to 60 frames per second. And to represent objects accurately with continuous parameters, we need arbitrarily small sampling intervals. Therefore, unless hardware technology is developed to handle arbitrarily large frame buffers increased screen resolution is not a complete solution to the aliasing problem.

With raster systems that are capable of displaying more than two intensity levels (Colour or gray scale), we can apply antialiasing methods to modify pixel intensities. By appropriately varying the intensities of pixels along the boundaries of primitives, we can smooth the edges to lessen the jagged appearance.

6.8.1   Super Sampling or Post Filtering

A straight forward antialiasing method is to increase sampling rate by treating the screen as if it were covered with a finer grid than is actually available. We can then use multiple sample points across this finer grid to determine an appropriate intensity level for each screen pixel. This technique of sampling object characteristics at a high resolution and displaying the results at a lower resolution is called **super sampling** or **post filtering**.

This method is a three-stage process. The stages are:

(a)   The image is sampled at n times the display resolution.

(b)   The sample image is then low pass filtered.

(c)   The filtered image is resampled at the device resolution.

This method goes well with z buffer technique, however, this may cause some blurring. Another disadvantage of this technique is that it is not a suitable method for dealing with small objects. Since the memory requirements when used with z buffer technique is large, it is essentially a virtual memory technique.

6.8.2 Area Sampling or Prefiltering

An alternative to super sampling is to determine pixel intensity by calculating the areas of overlap of each pixel with the objects to be displayed. Antialiasing by computing overlap areas is referred to as area sampling or prefiltering. Pixel overlap areas are obtained by determining where object boundaries intersect individual pixel boundaries. Since this method involves considerable amount of computation, several modifications to this algorithm have been developed.

6.8.3 Pixel Phasing

Raster object can also be antialiased by shifting the display location of pixel areas. This technique, called **pixel phasing**, is applied by "micro positioning" the electron beam in relation to object geometry.

On raster systems that can address subpixel positions with in the screen grid, pixel phasing can be used to antialias objects stair steps along a line path or object boundary are smoothed out by moving (micro positioning) the electron beam to more nearly approximate positions specified by the object geometry. Systems incorporating this technique are designed so that individual pixel positions can be shifted by a fraction of a pixel diameter. The electron beam is typically shifted by ¼, ½ or ¾ of a pixel diameter to plot points closer to the true path of a line or object edge. Some systems also allow the size of individual pixels to be adjusted as an additional means for distributing intensities.

6.8.4 Stochastic Sampling

This method is a two-stage process:

(a) Sample the image using a sampling grid where the position of each sampling point has been subjected to random perturbation

(b) Use these sample values with a reconstruction filter to determine the pixel intensities to which the unperturbed sample positions correspond.

The problem with this method is that it is only easily incorporated where the image synthesis uses an in coherent sampling method. Since the method splits the objects into micro polygons, it is suitable to objects consisting of parametric bicubic patches.

6.9 REFLECTION

When we view an opaque non-luminous object, we see reflected light from the surfaces of the object. The total reflected light is the sum of the contributions from light sources and other reflecting surfaces in the scene (Fig. 6.18). Thus, a surface that is not directly exposed to a light source may still be visible if nearby objects are illuminated. Sometimes, light sources are referred to as light-emitting sources and reflecting surfaces such as the walls of a room are termed as **light-reflecting sources**. We will use the term light source to mean an object that is emitting radiant energy, such as a light bulb or the sun.

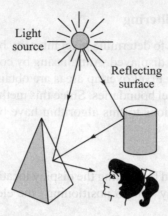

Light
source

Reflecting
surface

Fig. 6.18 Light viewed from an opaque non-luminous surface is in general a combination of reflected light from a light source and reflections of light from other surfaces.

A luminous object in general can be both a light source and a light reflector. For example, a plastic globe with a light bulb inside both emits and reflects light from the surface of the globe. Emitted light from the globe may then illuminate other objects in the vicinity.

The simplest model for a light emitter is a point source. Rays from the source then follow radially diverging paths from the source position as shown in Fig. 6.19. This light source model is a reasonable approximation for sources whose dimensions are small as compared to the size of objects in the scene. Sources such as the sun, that are sufficiently far from the scene can be accurately modelled as point sources. A nearby source such as the long fluorescent light in Fig. 6.20 is more accurately modelled as a distributed light source. In this case, the illumination effects cannot be approximated realistically with a point source, because the area of the source is not small as compared to the surfaces in the scene. An accurate model for the distributed source is one that considers the accumulated illumination effects of the points over the surface of the source.

Fig. 6.19 Diverging ray paths from a point light source.

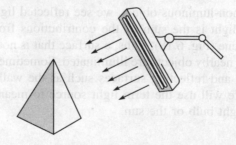

Fig. 6.20 An object illuminated with a distributed light source.

When light is incident on an opaque surface, part of it is reflected and part is absorbed. The amount of incident light reflected by a surface depends on the type of material. Shiny materials reflect more of the incident light and dull surfaces absorb more of the incident light. Similarly, for an illuminated transparent surface, some of the incident light will be reflected and some will be transmitted through the material.

Surfaces that are rough or grainy tend to scatter the reflected light in all directions. This scattered light is called **diffuse reflection**. A very rough matte surface produces primarily diffuse reflections so that the surface appears equally bright from all viewing directions.

Figure 6.21 illustrates diffuse light scattering from a surface. What we call the colour of an object is the colour of the diffuse reflection of the incident light. A blue object illuminated by a white light source, for example, reflects the blue component of the white light and totally absorbs all other components. If the blue object is viewed under a red light, it appears black since all of the incident light is absorbed.

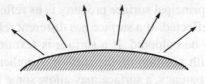

Fig. 6.21 Diffuse reflections from a surface.

In addition to diffuse reflection, light sources create high lights or bright spots called **specular reflection**. This high lighting effect is more pronounced on shiny surfaces than on dull surfaces. An illustration of specular reflection is shown in Fig. 6.22.

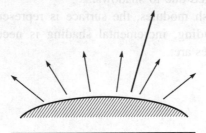

Fig. 6.22 Specular reflection superimposed on diffuse reflection vectors.

A technique to model reflection from an object based on specular reflection has been proposed by Phong. This model assumes that

(a) Light sources are point sources.
(b) All geometry except the surface normal is ignored.
(c) Diffuse and specular components are modelled as local components.
(d) The model to simulate the specular term is empirical.
(e) The colour of specular reflection is that of the light source.
(f) The ambient lighting is constant.

6.10   SHADING

The realism of a raster scan image of a three-dimensional scene depends on the successful simulation of shading effects. Once visible surfaces have been identified by a hidden surface algorithm, a shading model is used to compute the intensities and colours to display for the surface.

The shading model does not precisely simulate the behaviour of light and surfaces in the real world but only approximates actual conditions. In this respect, the shading model is similar to the geometric model. The design of the model is a compromise between precision and computing cost. Trade-offs in a shading model are especially difficult because the properties of the human visual system influence the perception of realism. We must avoid approximations in the model that lead to confusing perceptions by the viewer.

The shading model has two main ingredients: properties of the surface and properties of the illumination falling on it. The principal surface property is its reflectance, which determines how much of the incident light is reflected. If a surface has different reflectances for light of different wavelengths, it will appear to be coloured. If a surface is textured or has a pattern painted on it, the reflectance will vary with position on the surface. Another surface property that plays a role in shaded pictures is transparency, a surface may allow some light to be transmitted through it from behind.

An object illumination is as important as its surface properties in computing its intensity. The scene may have some illumination that is uniform from all directions called **diffuse illumination**. In addition, there may be point sources of light in the scene. They differ from diffuse lighting in that specular reflections or highlights appear on the surfaces. Finally the illumination of an object may be partially blocked due to shadows.

In simple polygonal mesh modules, the surface is represented by constant shading. To introduce more realistic shading, incremental shading is necessary. Two commonly used incremental shading techniques are:

- Gouraud shading
- Phong shading

6.10.1   Gouraud Shading

This intensity-interpolation scheme developed by Gouraud and generally referred to as Gouraud shading, renders a polygon surface by linearly interpolating intensity values across the surface. Intensity values for each polygon are matched with the values of adjacent polygons along the common edges, thus eliminating the intensity discontinuities that can occur in flat shading.

Each polygon surface is rendered with Gouraud shading by performing the following calculations:

- Determine the average unit normal vector at each polygon vertex.
- Apply an illumination model to each vertex to calculate the vertex intensity.
- Linearly interpolate the vertex intensities over the surface of the polygon.

At each polygon vertex, we obtain a normal vector by averaging the surface normals of all polygons sharing that vertex as shown in Fig. 6.23. Thus, for any vertex position V, we obtain the unit vertex normal with the calculation:

$$N_V = \frac{\sum\limits_{k=1}^{n} N_k}{\left| \sum\limits_{k=1}^{n} N_k \right|}$$

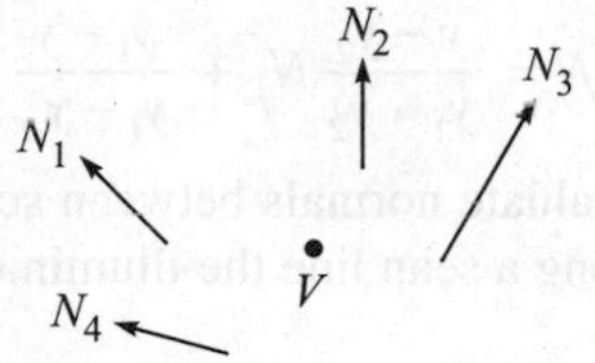

Fig. 6.23 The normal vector at vertex V is calculated as the average of the surface normals for each polygon sharing that vertex.

Once we have the vertex normals, we can determine the intensity at the vertices from a lighting model.

Gouraud shading removes the intensity discontinuities associated with the constant shading model, but it has some other deficiencies. High lights on the surface are sometimes displayed with anomalous shapes and the linear intensity interpolation can cause bright or dark intensity streaks called **mach bands**, to appear on the surface. These effects can be reduced by dividing the surface into a greater number of polygon faces or by using other methods such as Phong shading that require more calculations.

6.10.2   Phong Shading

A more accurate method for rendering a polygon surface is to interpolate normal vectors and then apply the illumination model to each surface point. This method, developed by Phong Bui, is called **Phong shading** or **normal vector interpolation shading**. It displays more realistic highlights on a surface and greatly reduces the Mach-band effect.

A polygon surface is rendered using Phong shading by carrying out the following steps:

- Determine the average unit normal vector at each polygon vertex.
- Linearly interpolate the vertex normals over the surface of the polygon.
- Apply an illumination model along each scan line to calculate projected pixel intensities for the surface points.

Interpolation of surface normals along a polygon edge between two vertices is shown in Fig. 6.24. The normal vector n for the scan line intersection point along the edge between vertices 1 and 2 can be obtained by vertically interpolating between edge end point normals.

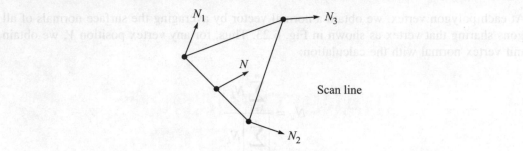

Fig. 6.24 Interpolation of surface normals along a polygon edge.

$$N = \frac{y - y_2}{y_1 - y_2} N_1 + \frac{y_1 - y}{y_1 - y_z} N_2$$

Incremental methods are used to evaluate normals between scan lines and along each individual scan line. At each pixel position along a scan line the illumination model is applied to determine the surface intensity at that point.

REVIEW QUESTIONS

6.1 What are the characteristics of computer graphics?

6.2 Explain the basic principle of the DDA algorithm.

6.3 Explain the basic principle of Bresenham's line algorithm.

6.4 Explain the basic principle of Bresenham's circle algorithm.

6.5 Briefly describe the different techniques used for scan conversion in computer graphics.

6.6 What is rendering? What are the different stages of rendering an image?

6.7 Explain the concept of antialiasing. Discuss the techniques used for antialiasing.

6.8 Describe an algorithm for the removal of hidden lines.

6.9 Discuss hidden surface removal algorithm.

6.10 Discuss the reflection models used in computer graphics.

6.11 Compare the techniques of Phong shading and Gauraud shading.

6.12 Write shortnotes on frame buffer.

6.13 Write z-buffer algorithm. List the advantages and disadvantages.

6.14 Explain the scan line algorithm for polygon filling.

6.15 Mention the applications of computer graphics.

TRANSFORMATION SYSTEMS

INTRODUCTION

With the procedure for displaying output primitives and their attributes, we can create a variety of pictures and graphs. In many applications, there is also a need for altering or manipulating displays design applications and facility layouts are created by arranging the orientations and sizes of the component parts of the scene. Changes in orientation, size and shape are accomplished with geometric transformations that alter the coordinate descriptions of objects. The basic geometric transformations are translation, rotation and scaling other transformations that are often applied to objects include reflection and shear.

7.1 TRANSFORMATION PRINCIPLES

A graphics system should allow the programmer to define pictures that include a variety of transformations. For example, he should be able to magnify a picture so that detail appears more clearly, or reduce it so that more of the picture is visible. He should also be able to apply transformations to symbols. It is also useful to be able to change the scale of a symbol and to rotate it through some angle. Two aspects of the formulation of transformations should be emphasized:

(a) A transformation is a single mathematical entity and as such can be denoted by a single name or symbol.

(b) Two transformations can be combined or concatenated to yield a single transformation with the same effect as the sequential application of the original two. Thus, transformation A might be a translation and transformation B a scaling. The concatenation property allows us to determine a transformation $C = AB$ whose effect is to translate and then scale.

7.2 TWO-DIMENSIONAL GEOMETRIC TRANSFORMATIONS

In computer graphics, drawings are created by a series of primitives which are represented by the co-ordinates of their end points. Certain changes in these drawings can be made by

141

performing some mathematical operations on these coordinates. The basic transformations are scaling, translation and rotation.

7.2.1 Scaling

A scaling transformation alters the size of an object. This operation can be carried out for polygons by multiplying the coordinate values (x, y) of each vertex by scaling factors S_x and S_y to produce the transformed coordinates (x', y')

$$x' = xS_x \quad \text{and} \quad y' = yS_y \tag{7.1}$$

Scaling factor S_x scales objects in the x direction while S_y scales in the y direction. Equation (7.1) can also be written in the matrix form:

$$\begin{bmatrix} x' \\ y' \end{bmatrix} = \begin{bmatrix} S_x & 0 \\ 0 & S_y \end{bmatrix} = \begin{bmatrix} x \\ y \end{bmatrix} \tag{7.2}$$

Any positive numeric values can be assigned to the scaling factors S_x and S_y. Values less than 1 reduce the size of objects and values greater than 1 produce an enlargement. Specifying the values of 1 for both S_x and S_y leaves the size of objects unchanged. When S_x and S_y are assigned the same value, a uniform scaling is produced that maintains relative object proportions. Unequal values for S_x and S_y result in a differential scaling that is often used in design applications, where pictures are constructed from a few basic shapes that can be adjusted by scaling and positioning transformations (see Fig. 7.1).

(a)   (b)

Fig. 7.1 Turning a square (a) into a rectangle (b) with scaling factors $S_x = 2$ and $S_y = 1$.

Objects transformed with Eq. (7.2) are both scaled and repositioned scaling factors with values less than 1 move values greater than 1 move coordinate positions further from the origin.

7.2.2 Translation

A translation is applied to an object by repositioning it along a straight line path from one coordinate location to another. We translate a two-dimensional point by adding translation distances t_x and t_y to the original coordinate position (x, y) to move the point to a new position (x', y') (Fig. 7.2).

$$x' = x + t_x \quad \text{and} \quad y' = y + t_y \tag{7.3}$$

The translation distance pair (t_x, t_y) is called a **translation vector** or **shift vector**.

We can express Eq. (7.3) as a single matrix equation by using column vectors to represent coordinate positions and the translation vector

$$P = \begin{bmatrix} x_1 \\ x_2 \end{bmatrix} \quad P' = \begin{bmatrix} x'_1 \\ x'_2 \end{bmatrix} \quad T = \begin{bmatrix} t_x \\ t_y \end{bmatrix} \tag{7.4}$$

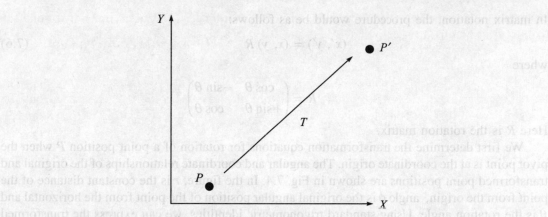

Fig. 7.2 Translating a point P to position P' with translation vector T.

This allows us to write the two-dimensional translation equations in the matrix form:

$$P' = P + T \tag{7.5}$$

Translation is a rigid-body transformation that moves objects without deformation, that is, every point on the object is translated by the same amount. A straight line segment is translated by applying Eq. (7.5) to each of the line end points and redrawing the line between the new end point positions.

7.2.3 Rotation

A two-dimensional rotation is applied to an object by repositioning it along a circular path in the xy plane. To generate a rotation, we specify a rotation angle θ and the position (x_r, y_r) of the rotation point (or pivot point) about which the object is to be rotated (see Fig. 7.3).

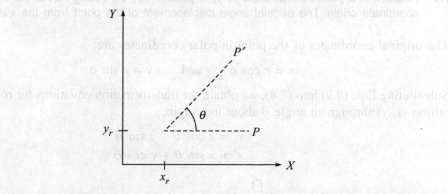

Fig. 7.3 Rotation of an object through angle θ about the pivot point (x_r, y_r).

Positive values for the rotation angle define counterclockwise rotations about the pivot point as in Fig. 7.3 and negative values rotate objects in the clockwise direction. This transformation can also be described as a rotation about a rotation axis that is perpendicular to the xy plane and passes through the pivot point.

In matrix notation, the procedure would be as follows:

$$(x', y') = (x, y)\, R \qquad (7.6)$$

where

$$R = \begin{pmatrix} \cos\theta & -\sin\theta \\ +\sin\theta & \cos\theta \end{pmatrix}$$

Here R is the rotation matrix.

We first determine the transformation equations for rotation of a point position P when the pivot point is at the coordinate origin. The angular and coordinate relationships of the original and transformed point positions are shown in Fig. 7.4. In the figure, r is the constant distance of the point from the origin, angle ϕ is the original angular position of the point from the horizontal and θ is the rotation angle. Using standard trigonometric identities, we can express the transformed coordinates in terms of angle θ and ϕ as

$$x' = r \cos(\phi + \theta) = r \cos\phi \cos\theta - r \sin\phi \sin\theta \qquad (7.7)$$
$$y' = r \sin(\phi + \theta) = r \cos\phi \sin\theta + r \sin\phi \cos\theta \qquad (7.8)$$

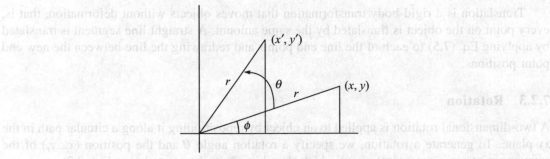

Fig. 7.4 Rotation of a point from position (x, y) to position (x', y') through an angle θ relative to the coordinate origin. The original angle displacement of the point from the x-axis is ϕ.

The original coordinates of the point in polar coordinates are:

$$x = r \cos\phi \quad \text{and} \quad y = r \sin\phi \qquad (7.9)$$

Substituting Eqs. (7.9) into (7.8), we obtain the transformation equations for rotating a point at position (x, y) through an angle θ about the origin.

$$x' = x \cos\theta - y \sin\theta \qquad (7.10)$$
$$y' = x \sin\theta + y \cos\theta$$

7.3 THREE-DIMENSIONAL GEOMETRIC TRANSFORMATIONS

Methods for geometric transformations in three dimensions are extended from two-dimensional methods. By including considerations for the z coordinate, we now translate an object by specifying a three-dimensional translation vector, which determines how much the object is to be moved in each of the three coordinate directions. Similarly, we scale an object with three coordinate scaling factors. The extension for three-dimensional rotation is less straightforward. In three-dimensional space, we can now select any spatial orientation for the rotation axis. Most graphics packages handle three-dimensional rotations as a composite of three rotations, one for each of the three cartesian axes. Alternatively, a user can easily set up a general rotation matrix, given the orientation of the axis and the required rotation angle.

7.3.1   Scaling

The matrix expression for the scaling transformation of a position $P = (x, y, z)$ relative to the coordinate origin can be written as:

$$\begin{pmatrix} x' \\ y' \\ z' \\ 1 \end{pmatrix} = \begin{pmatrix} S_x & 0 & 0 & 0 \\ 0 & S_y & 0 & 0 \\ 0 & 0 & S_z & 0 \\ 0 & 0 & 0 & 1 \end{pmatrix} \begin{pmatrix} x \\ y \\ z \\ 1 \end{pmatrix} \tag{7.11}$$

where scaling parameters S_x, S_y and S_z are assigned any positive values. Explicit expressions for the coordinate transformations for scaling relative to the origin are:

$$x' = x \cdot s_x, \ y' = y \cdot s_y, \ z' = zs_z \tag{7.12}$$

Scaling an object with Eq. (7.11) changes the size of the object and repositions the object relative to the coordinate origin. Also, if the transformation parameters are not all equal relative dimensions in the object are changed. We preserve the original shape of an object with a uniform scaling ($S_x = S_y = S_z$). For equal values of S_x, S_y and S_z, the scaling is linear, otherwise distortions will be introduced into the system.

7.3.2   Translation

In a three-dimensional homogeneous coordinate representation, a point is translated from position $p(x, y, z)$ to $p'(x', y', z')$ with the matrix operation.

$$\begin{pmatrix} x' \\ y' \\ z' \\ 1 \end{pmatrix} = \begin{pmatrix} 1 & 0 & 0 & t_x \\ 0 & 1 & 0 & t_y \\ 0 & 0 & 1 & t_z \\ 0 & 0 & 0 & 1 \end{pmatrix} \begin{pmatrix} x \\ y \\ z \\ 1 \end{pmatrix} \tag{7.13}$$

Parameters t_x, t_y and t_z specifying translation distances for the coordinate directions x, y, and z are assigned any real values. The matrix representation in Eq. (7.13) is equivalent to the three equations:

$$x' = x + t_x, \ y' = y + t_y, \ z' = z + t_z \tag{7.14}$$

An object is translated in three dimensions by transforming each of the defining points of the object.

7.3.3 Rotation

To generate a rotation transformation for an object, we must designate an axis of rotation (about which the object is to be rotated) and the amount of angular rotation. Unlike two-dimensional applications, where all transformations are carried out in the xy plane, a three-dimensional rotation can be specified around any line in space. The easiest rotation axes to handle are those that are parallel to the coordinate axes. Also, we can use combinations of coordinate axis rotations to specify any general rotation.

By convention, positive rotation angles produce counter clockwise rotations about a coordinate axis, if we are looking along the positive half of the axis, if we are looking along the positive half of the axis toward the coordinate origin.

The two-dimensional z-axis rotation equations are easily extended to three dimensions:

$$\left.\begin{array}{l} x' = x\cos\theta - y\sin\theta \\ y' = x\sin\theta + y\cos\theta \\ z' = z \end{array}\right\} \tag{7.15}$$

Parameter θ specifies the rotation angle. In homogeneous coordinate form, the three-dimensional z-axis rotation equations are expressed as:

$$\begin{pmatrix} x' \\ y' \\ z' \\ 1 \end{pmatrix} = \begin{pmatrix} \cos\theta & -\sin\theta & 0 & 0 \\ \sin\theta & \cos\theta & 0 & 0 \\ 0 & 0 & 1 & 0 \\ 0 & 0 & 0 & 1 \end{pmatrix} \begin{pmatrix} x \\ y \\ z \\ 1 \end{pmatrix} \tag{7.16}$$

which we can write more compactly as:

$$P' = R_z(\theta)\, P \tag{7.17}$$

Transformation equations for rotations about the other two coordinate axes can be obtained with a cyclic permutation of the coordinate parameters x, y and z in Eqs. (7.15). That is we use the replacements.

$$x \rightarrow y \rightarrow z \rightarrow x \tag{7.18}$$

Substituting permutations given by Eq. (7.18) in Eqs. (7.15), we get the equations for an x-axis rotation.

$$\left.\begin{array}{l} y' = y\cos\theta - z\sin\theta \\ z' = y\sin\theta + z\cos\theta \\ x' = x \end{array}\right\} \tag{7.19}$$

which can be written in the homogeneous coordinate form:

$$\begin{pmatrix} x' \\ y' \\ z' \\ 1 \end{pmatrix} = \begin{pmatrix} 1 & 0 & 0 & 0 \\ 0 & \cos\theta & -\sin\theta & 0 \\ 0 & \sin\theta & \cos\theta & 0 \\ 0 & 0 & 0 & 1 \end{pmatrix} \begin{pmatrix} x \\ y \\ z \\ 1 \end{pmatrix} \tag{7.20}$$

which we can write more compactly as:

$$P' = R_x (\theta) P \tag{7.21}$$

Cyclically permuting coordinates in Eqs. (7.19) give us the transformation equations for a y-axis rotation.

$$\left. \begin{array}{l} z' = z \cos \theta - x \sin \theta \\ x' = z \sin \theta + x \cos \theta \\ y' = y \end{array} \right\} \tag{7.22}$$

The matrix representation for y-axis rotation is:

$$\begin{pmatrix} x' \\ y' \\ z' \\ 1 \end{pmatrix} = \begin{pmatrix} \cos \theta & 0 & \sin \theta & 0 \\ 0 & 1 & 0 & 0 \\ -\sin \theta & 0 & \cos \theta & 0 \\ 0 & 0 & 0 & 1 \end{pmatrix} \begin{pmatrix} x \\ y \\ z \\ 1 \end{pmatrix} \tag{7.23}$$

which we can write more compactly as $P' = R_y(\theta) P$

7.4 LINEAR TRANSFORMATIONS

Three-dimensional transformations are similar to those in two dimensions. We will concentrate on the following:

Scaling
Translation
Rotation (about one or more axes)

For three-dimensional data $[x, y, z]$ the transformation can be represented by

$$X' = XT$$

where $X = [x, y, z, 1]$ the original coordinates
   $X' = [x', y', z', 1]$ the transformed coordinates

and T can be represented by a single or concatenated 4×4 matrix.

$$T = \begin{pmatrix} r_{11} & r_{12} & r_{13} & 0 \\ r_{21} & r_{22} & r_{23} & 0 \\ r_{31} & r_{32} & r_{33} & 0 \\ t_1 & t_2 & t_3 & 0 \end{pmatrix}$$

where r_{ij} arc the terms of the rotation matrices and t_j the translation offset.

(a)   Rotation about z-axis (see Fig. 7.5)

$$R_z = \begin{pmatrix} \cos \theta_z & -\sin \theta_z & 0 & 0 \\ \sin \theta_z & \cos \theta_z & 0 & 0 \\ 0 & 0 & 1 & 0 \\ 0 & 0 & 0 & 1 \end{pmatrix} \tag{7.24}$$

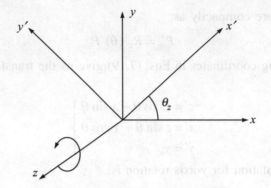

Fig. 7.5 Rotations about z-axis.

(b) Rotation about y-axis (see Fig. 7.6)

$$R_y = \begin{pmatrix} \cos\theta_y & 0 & \sin\theta_y & 0 \\ 0 & 1 & 0 & 0 \\ -\sin\theta_y & 0 & \cos\theta_y & 0 \\ 0 & 0 & 0 & 1 \end{pmatrix} \qquad (7.25)$$

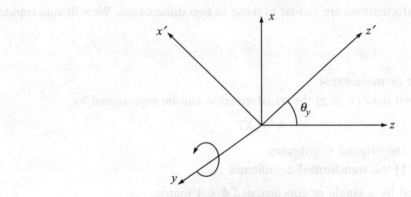

Fig. 7.6 Rotation about y-axis.

(c) Rotation about x-axis (see Fig. 7.7)

$$R_x = \begin{pmatrix} 1 & 0 & 0 & 0 \\ 0 & \cos\theta_x & -\sin\theta_x & 0 \\ 0 & \sin\theta_x & \cos\theta_x & 0 \\ 0 & 0 & 0 & 1 \end{pmatrix} \qquad (7.26)$$

R_x, R_y, R_z can be concatenated to give a general rotation matrix R. However, the order in which the individual matrices are combined will affect the results thereby achieved. The scaling transformation is:

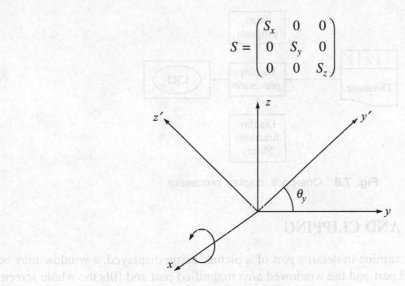

$$S = \begin{pmatrix} S_x & 0 & 0 \\ 0 & S_y & 0 \\ 0 & 0 & S_z \end{pmatrix}$$

Fig. 7.7 Rotation about *x*-axis.

For equal values of S_i, the scaling is linear, otherwise distortions will be introduced into the system.

These transformations allow data to be repositioned anywhere and in any orientation in space. The translation, rotation and scaling effects can all be combined into a single transformation matrix by concatenating the individual matrices. However, the same effects can also be achieved by varying the viewing parameters used to define the object-observer geometry in a perspective transformation.

7.5 DISPLAY

CAD involves displaying a large number of pictures and much time can be consumed in converting structured data into display signals. The display file, may be regarded as a table of instructions to be executed by the display processor. The display file, with hardware consisting of a digitiser and storage tube may be the same as the workspace for a two-dimensional graphics system. Since there is no need for a conventional display memory in the case of a three-dimensional system, intermediate files may be used to advantage.

The display processor consists of a program and a number of subroutines built into a library which are loaded to execute the instructions contained in the display file. This can be depicted in the form of flowchart in Fig. 7.8. In the simplest data structure the instructions are stored in $[I, x, y]$ records with I representing the command code with respect to the display procedure and x, y, the data. Compiling of the graphics language is simple, for instructions are usually generated in a sequential manner. Efficient algorithms for converting display instructions to signals driving the CRT are particularly important for fast display of data.

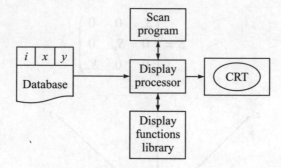

Fig. 7.8 One-pass display processor.

7.6 WINDOWING AND CLIPPING

When it is necessary to examine in detail a part of a picture being displayed, a window may be placed around the desired part and the windowed area magnified part and fills the whole screen. This involves scaling the data which lies within the window so that the window fills the entire screen. Data that lies outside the window must be eliminated so that the data required for display is processed. This process is known as **clipping**. Some hardware devices have the feature of automatic scissoring in which the window and the display vectors may be larger than the display raster.

It is necessary to define two viewing areas: a viewport and a window. It is usual to make the viewport equal in size to the screen to take advantage of the maximum screen area.

The limits of the window are determined by the coordinates on the bottom left-hand corner taken as [0,0] and the dimensions of the required frame. The window is set up by digitising the [0,0] coordinates and moving the cursor on the digitiser to the right until the window encloses the desired area. Once the window is defined the data outside the window is clipped before scaling to the screen coordinates which considerably reduces the amount of data before display signals are generated.

7.7 DISPLAY FILES FOR THREE-DIMENSIONAL DATA

The introduction of a third dimension results in an increase in the number of processes involved in converting data from a basic structure to drive a display terminal. Two major stages are involved: compilation and processing (see Fig. 7.9).

In the compilation stage the data file is first scanned by a viewing algorithm, an output routine which generates function calls to the appropriate subroutines (scaling, perspective, windowing, clipping). Then in response to these function calls the display compiler interprets the stored data and creates a display file. The cumulative effect of these two processes is to generate a second data structure from the first.

In the second stage, the display file, which contains simple commands for the display is scanned by the display processor, which then generates the drive signals for the CRT. The instructions are simple *x-y* drive signals, but the processor may incorporate some hardware character or symbol generation facilities.

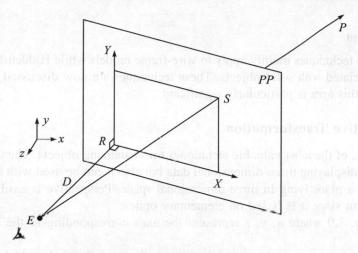

Fig. 7.9 Projections on plane *PP*.

These two stages are very similar, both involving a scanner and an interpreter and can be combined into one, so that a set of data serves both purposes. With a storage tube a conventional memory for storing the display file is not essential, since the image is only displayed once. Because of the lack of selective erasure facilities the whole screen must be redisplayed to incorporate any changes. However, the user of a separate display file does have the advantage of enabling which reduces the amount of compilation involved each time the picture is selectively modified.

It is usual to work with two files so that the data and display files are stored, for example, in FILE 1 and FILE 2 respectively. FILE 2 does not fall within the exact definition of a display file because it contains more than simple instructions for the display processor and it also contains data that lies off the screen area. The main purpose of this file is to permit faster display at different window sizes without the need to repeat viewing transformations. It also permits interactive editing of the database by a parallel search technique.

7.8 VISUALIZATION OF THREE-DIMENSIONAL DATA

The presentation of three-dimensional graphic data on a plane is important since many people have great difficulty in understanding engineering drawings. When an object is rotated, for example, if the new view is not clearly presented then the viewer could be confused. Such confusion might arise from a 'wire cage' drawing where it might be difficult to distinguish between lines representing the front and the rear.

In order to help the CAD system user to interpret static two-dimensional projections of three-dimensional objects a range of visual aids may be adopted. There are several ways in which the visualization of objects can be improved:

- Perspective transformation
- Brightness modulation
- Hidden-line removal

- Shading
- Movement

The first two techniques mainly apply to wire-frame models while Hidden-line removal and shading are associated with solid objects. These techniques are now discussed since the use of the computer in this area is particularly significant.

7.8.1 Perspective Transformation

Perspective is one of the most valuable techniques for visualizing objects. The transformation is often used when displaying three-dimensional data but also it can be used with two-dimensional data to represent a plane lying in three-dimensional space. Perspective is easily represented in mathematical form since it is based on elementary optics.

Consider Fig. 7.9 where x, y, z represent the axes corresponding to the spatial cartesian coordinates.

The system of reference axes used here is in the conventional right-handed reference directions. If the xy plane is made to coincide with the screen, the positive z direction is out of the screen towards the observer.

For the purpose of the projection, a plane PP is defined with coordinates (X, Y). The projection maps the point $\overline{P}$ to a point $\overline{S}$ on the projection plane PP. The projection plane in this case corresponds to the viewing screen, with X to the right and Y upwards.

The observer's eye is situated at the point $\overline{E}[E_x, E_y, E_z]$ and the line of sight makes angles α, β and γ with the spatial x-, y- and z-axes respectively. The angles α, β and γ can be specified by their direction cosines or by a point defined as the centre of attention, from which the direction cosines can be calculated. Normally the centre of attention is made to coincide with the origin or the volumetric centre of the object being observed.

Consider the plane of projection PP at a distance D from E, normal to the line of sight. For an observer with the eye position at $\overline{E}[E_x, E_y, E_z]$ direction of line of sight at angles $[\alpha, \beta, \gamma]$ to the major axes, the point $\overline{P}[P_x, P_y, P_z]$ in space is mapped onto the projection plane PP, by the following relationships:

$$X = \frac{(S_x - R_x)\cos\gamma - (S_x - R_z)\cos\alpha}{\sin\beta}$$

$$Y = \frac{(S_y - R_y)}{\sin\beta}$$

where $[X, Y]$ are the coordinates of the projected position of the point $\overline{P}$. $\overline{S}[S_x, S_y, S_z]$ is the point of intersection of the line of sight on the projection plane and $R[R_x, R_y, R_z]$ the foot of the normal from the observer eye to the plane.

Isometric or orthogonal projections are often preferred to perspective drawings in engineering because measurements can be related more easily to the drawing. However, from aesthetics the perspective view looks right and, since it is easy to produce from a CAD system, its use may well increase.

7.8.2 Brightness Modulation

With this technique parts of the picture near to the observer are bright while those far away are dim.

When this is required on a view an extra routine in the program is entered just before the vector generation and this selects the required brightness levels for the display file as it is being constructed. When a picture has been constructed, the maximum and minimum z coordinates are noted. The z range is then divided into n regions where n are the visible brightness levels available in the display system. The picture is then displayed with the appropriate brightness level corresponding to its z region.

This technique is easy to implement and is very effective when displayed on the screen. It is difficult to obtain a hard copy version unless an electrostatic plotter is used.

7.8.3 Hidden-line Removal

If a complex three-dimensional drawing is fully displayed then the large number of lines usually render the picture impossible to perceive. The main problem lies in that the lines which are normally hidden by the object, are all displayed and this can lead to confusion. The hidden-lines can be removed by the computer but large amounts of computing time are usually required. Computation approximately increases as the square of the number of edges. Therefore, for moderately complex situations computation can become prohibitive on a small computer.

It is not easy to establish reliable algorithms to identify the lines to be removed. In general, the geometric calculations are straightforward if objects are convex polyhedra. But if the three-dimensional bodies are not rectilinear, the problem can be very difficult. A number of successful algorithms have been devised to perform hidden-line removal.

7.8.4 Shading

Shading techniques have been developed extensively in work related to hidden-line removal, particularly at the University of Utah and at the CAD centre at Cambridge. The technique is based on the recognition of distance and shape as a function of illumination.

The technique is similar to finite elements. The surface of a solid is divided into patches and in regions of large curvature, the patches are decreased in size. Each patch or element is then tested for visibility and the degree of shading required. It should be remembered that hidden-line removal is a prerequisite for any shading algorithm.

The amount of shading required is determined by calculating the angle between the normal to the plane of the element and the vector direction of the propagation of the light. The normal vector can be calculated from the cross-product of two vectors on the plane or from the equation of the plane. The angle of incidence is given by the dot-product of the normal and the line of incidence. Brightness and visibility increase as the angle of incidence increases from 0 to 90°. For the visibility test only the sign of the dot-product is required to determine whether the plane is fairing the light. This is a preliminary test to eliminate all planes facing in the wrong direction. Gouraud used a method where the intensity at the point, where the elements meet, is calculated. The intensity is then interpolated to provide smooth shading of the surface. For illumination purposes a point or parallel beam source of light may be used. The surface can also be given a reflective index to make it shiny or dull or even transparent. The output is obtained on devices that can scan at different intensity levels, either by drawing a series of parallel lines or by overwriting. Results obtained by Parker, Gouraud and Newell are impressive but they do rely on powerful, special-purpose hardware. The work at Imperial College by Yi on shading was based on the use of a Calcamp microfilm plotter.

7.8.5 Movement

Movement leads to improved recognition of displayed objects. As an object is rotated or translated, ambiguities that arise due to the superposition of points are eliminated and the geometrical properties of the object are revealed by the interaction of the points defining the object.

The storage tube is not really suitable for displaying movement since the whole picture must be erased before a new one is displayed. This can take time so that real time movement is not possible. However, it is possible to photograph a series of pictures from the screen of a storage tube so that they can be replayed to give real-time movement. This is particularly useful for a simulation exercise since it is an inexpensive method for obtaining a simple animated line test.

The use of movement in CAD will become increasingly important as new cheap refreshed displays are developed. The use of moving blueprints for the simulation of engineering dynamics is an exciting prospect for engineers. Movement, coupled with colour, will be an extremely powerful tool in CAD.

In summarizing the discussion on displaying three-dimensional objects, the main drawback of most of these techniques is the large amount of computing power required together with the requirement for specialized hardware in some cases. Many of these techniques place considerable strains on minicomputer-based systems and skilled programming is required in order to obtain a viable system. Probably the most suitable technique for the minicomputer and storage tube combination is that of perspective.

The advent of microprocessors and new types of refreshed displays is bringing near the time when three-dimensional graphics in colour with real-time movement will be a practical reality in low cost CAD systems.

REVIEW QUESTIONS

7.1 Explain the various graphic transformations required for manipulating the geometric information.

7.2 What is the need for concatenation of transformations?

7.3 Write an algorithm for scaling transformation.

7.4 Discuss the importance of 2D and 3D transformations in any CAD system.

7.5 What do you understand by homogeneous coordinates? Develop the homogeneous transformation.

7.6 Give a brief account on 3D rotations. Show the sequence of transformations for rotating an object about an axis that is parallel to the x-axis.

7.7 Prove that any two successive 3D rotations about a given rotation axis is commutative.

7.8   Explain the method to derive the transformation matrix for rotating an object about any arbitrary axis arbitrary that does not coincide with the coordinate axis x, y and z.

7.9   Explain shear transformations.

7.10 Write the transformation matrix to rotate a point (x, y, z) about z-axis through an angle θ in the clockwise direction.

7.11 Derive the transformation that rotates an object point θ degrees about the x-axis. Write the matrix representation for this rotation.

7.12 What is meant by composite transformations? How does it achieved?

7.13 Derive the composite transformation matrix which magnifies an object by S units in x and y directions about its centre $C(h,k)$.

7.14 Explain about the following 3D transformations:

(a) Scaling   (b) Translation
(c) Rotation about x-axis

7.15 Prove that the 2D rotation and scaling is commute if $S_x = S_y$.

7.16 Derive the relationship for geometric rotation in yz plane.

7.17 Explain the concept of obtaining a rotation about an arbitrary point in xy plane.

PRACTICE PROBLEMS

7.1 Show that a 2D reflection through the x-axis, followed by a 2D reflection through the line $y = -x$, is equivalent to a pure rotation about the origin.

7.2 For the position vectors P_1 [1 1], P_2 [3 1], P_3 [4 2], P_4 [2 3] that define a 2D polygon, develop a signal transformation matrix that
Reflects about the line $x = 0$
Translates by -1 in both the x and y directions
Rotates about the origin by $180°$
Using this transformation, determine the transformed position vectors. Plot both the original and the transformed polygon on the same graph.

7.3 Show that the 2×2 matrix

$$[T] = \begin{bmatrix} \dfrac{1 - t^2}{1 + t^2} & \dfrac{2t}{1 + t^2} \\ \dfrac{-2t^2}{1 + t^2} & \dfrac{1 - t^2}{1 + t^2} \end{bmatrix}$$

represents a pure rotation.

7.4 A unit square is transformed by a 2×2 transformation matrix. The resulting position vectors are:

$$\begin{bmatrix} 0 & 0 \\ 2 & 3 \\ 8 & 4 \\ 6 & 1 \end{bmatrix}$$

What was the transformation matrix?

7.5 Show that for $x = t^2$, $y = t$, the transformation

$$[x \quad y \quad 1] \begin{bmatrix} 0 & -2 & 2 \\ -2 & 2 & -2 \\ 1 & 0 & 1 \end{bmatrix}$$

yields points that lie on a unit circle.

7.6 A line is defined by its end points (0,0) and (2,3) in a two-dimensional graphics system. Express the line in matrix notation and perform the following transformations on this line:

    (a) Scale the line by a factor of 2.0.

    (b) Scale the original line by a factor of 3.0 in the x direction and 2.0 in the y direction.

    (c) Translate the original line by 2.0 units in the x direction and 2.0 units in the y direction.

    (d) Rotate the original line by $45°$ about the origin.

7.7 A triangle is defined in a two-dimensional ICG system by its vertices (0,2) (0,3) and (1,2). Perform the following transformations on this triangle:

    (a) Translate the triangle in space by 2 units in the x direction and 5 units in the y direction.

    (b) Scale the original triangle by a factor of 1.5.

    (c) Scale the original triangle by 1.5 in the x direction and 3.0 in the y direction.

    (d) Rotate the original triangle by $45°$ about the origin.

7.8 A line is defined in two-dimensional space by its end points (1,2) and (6,4). Express this in matrix notation and perform the following transformations in succession on this line:

    (a) Rotate the line by 900 about the origin.

    (b) Scale the line by a factor of ½.

7.9 A cube is defined in three-dimensional space with edges which are one unit in length. The corners of the cube are located at (0,0,0), (0,0,1), (0,1,0) (0,1,1), (1,0,0), (1,0,1), (1,1,0), (1,1,1). Determine the locations of the corners if the cube is first translated by 2.0 units in the x direction and then scaled by a factor of 3.0.

7.10 A line in two-dimensional space has end points defined by (1,1) and (1,3). It is desired to move this line by a series of transformations so that its end points will be at (0,1) and (0,5).

    (a) Describe the sequence of transformations required to accomplish the movement of the line as specified.

    For each transformation.

Chapter

WIRE FRAME MODELLING

8

INTRODUCTION

One of the principal requirements of a CAD/CAM system is the graphical representation of information. Graphical information is always convenient from the standpoint of understanding. This applies to the design stage as well as the manufacturing stage. Early CAD/CAM systems focused on improving the productivity of draftsmen. More recently, they have focused on modelling engineering objects. As a result, geometric models that once were more than adequate for drafting purposes are not acceptable for engineering applications. A basic requirement, therefore, is that a geometric model should be an unambiguous representation of its corresponding object. The model should be unique and complete to all engineering functions from documentation to engineering analysis to manufacturing.

8.1 IMPORTANCE OF GEOMETRIC MODELLING

Geometric modelling to CAD/CAM is as important as governing equilibrium equations to classical engineering fields as mechanics and fluidics. From an engineer's point of view modelling of objects is by itself unimportant. Rather, it is a means to enable useful engineering analysis and judgement. As a matter of fact, the amount of time and effort a designer spends in creating a geometric model cannot be justified unless the resulting database is utilized by the applications module.

8.2 REQUIREMENTS OF GEOMETRIC MODELLING

What does geometric modelling mean? What is it expected to provide? The concept of a product originates in the designer's mind. If it is to be translated into reality, he needs to present it in a relevant form for the manufacturing engineer to understand and carryout the necessary operations on it for its production. Traditionally product drawings were made together with the prototypes for passing the information across. However, in a computerized environment, the information a designer generates can form the basic unit to be accessed by a number of other elements of a CAM system. Hence it is important that the geometric model generated should be as clear and comprehensive as possible so that the other modules of the modelling and manufacturing system are able to use this information in the most optimal way. The functions that are expected of geometric modelling are: design analysis, drafting, manufacturing, production engineering, inspection and quality control.

8.3 TYPES OF MODELS

In general, there are three types of models in common use to represent a physical object in CAD/CAM systems as represented in Fig. 8.1. They are:

- Wire frame model
- Surface model
- Solid model

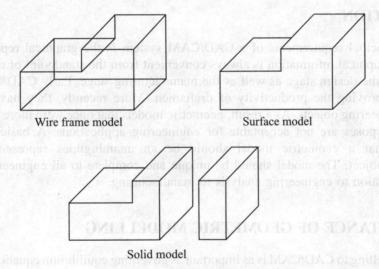

Wire frame model   Surface model

Solid model

Fig. 8.1 Types of geometric models.

8.3.1 Wire Frame Modelling

Wire frame models use networks of interconnect lines (or wires) to represent the edges of the physical objects being modelled. They are also called **edge-vertex** or **stick-figure models**. They are the simplest method of modelling and are most commonly used to define computer models of parts especially in computer assisted draughting systems. The reasons are that wire frame models are simple and easy to create, and that they require relatively little computer time and memory.

Although wire frame models provide accurate information about the location of surface discontinuities on the part, they do not give a complete description of the part very often. The model will subsequently be used as a basis for automatically generating cutter paths to drive an NC machine tool to manufacture the part. As wire frame models contain little, if any, information about the surfaces of the part, they do not distinguish the inside from the outside of part surfaces. They are inadequate for this purpose.

An example of the procedure for creating a wire frame model of a rectangular block is depicted in Fig.8.2. The procedures for creating models may vary from one CAD/CAM system to another depending on its capabilities and the individual technique of the user. In general, most CAD/CAM systems use a split screen approach, as shown in Fig. 8.2, in which multiple views, usually three orthogonal and one isometric view are displayed and manipulated simultaneously. In Fig.8.2(a) four points are indicated on the top view to represent the vertices of one of the faces of the rectangular block to be modelled. The points are then joined up as straight lines to represent the edges of the top face of the rectangular block as shown in Fig. 8.2(b). Then in Fig. 8.2(c), the image is projected into the other three views. Finally, the face is projected into the third dimension to give the model a thickness producing a rectangular block as a result in Fig. 8.2(d).

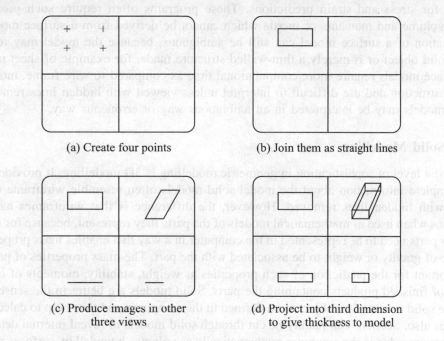

(a) Create four points   (b) Join them as straight lines

(c) Produce images in other   (d) Project into third dimension
three views   to give thickness to model

Fig. 8.2 A typical procedure for creating a wire frame model.

8.3.2 Surface Modelling

A higher level of sophistication in geometric modelling is the surface model which can overcome many of the ambiguities of wire frame models. Surface models are created by connecting various surface elements to represent part geometry. A surface model can be built by defining the surfaces on the wire frame model. They hold the description of an object in terms of points, edges and faces between the edges. Surface modelling systems are able to calculate surface intersections and surface areas and some systems are capable of producing shaded images and removing hidden lines automatically.

Surface models may be constructed using a large variety of surface features often provided by many CAD/CAM systems. The plane is the most basic feature to represent a surface element.

More complex shapes can be defined by tabulated cylinders, ruled surfaces of revolution, sweep surfaces and fillet surfaces.

The most general and complex surface representation are sculptured surfaces which are also known as **curve-mesh surfaces**, **B-surfaces** and **cubic-patch surfaces**. A sculptured surface may be considered as a surface produced from combining two families of curves that intersect one another in a criss-cross manner, creating a network of interconnected patches.

Surface modelling can be used to optimize component weight for both geometry and stress. A model composed of surface elements can easily be made to appear as if it is solid through the technique of hidden line removal, although surface models represent only an envelope of part geometry in computer memory. As surface models do not actually represent the solid nature of parts because they contain no information describing what lies with in the part interior, they cannot be used as a base for engineering analysis programs such as finite element and modal analysis for stress and strain predictions. These programs often require such properties as weight, volume and moments of inertia which cannot be derived from a surface model alone. Interpretation of a surface model can still be ambiguous, because the model may represent a totally solid object or is merely a thin-walled structure made, for example of sheet metal.

Surface models require more computational time as compared to wire frame, more skill in their construction and use difficult to interpret unless viewed with hidden lines removed. The surface models may be interpreted in an ambiguous way or erroneous way.

8.3.3 Solid Modelling

The highest level of sophistication in geometric modelling is 3D modelling. It provides the user with complete information about the model solid models often resemble wireframe or surface models with hidden lines removed. However, the difference is that wireframes have severe limitations when used as mathematical models of the parts they represent, because for modelling purposes parts need to be represented in the computer in a way that enables mass properties such as centre of gravity or weight to be associated with the part. The mass properties of part models are important for the prediction of such properties as weight, stability, moments of inertia and volume of finished products containing the parts. Solid models are better in the sense that they allow the solid nature of an object to be defined in the computer, and thus help to calculate mass properties also. Sections can readily be cut through solid models to reveal internal details. Solid models are recorded in the computer mathematically as volumes bounded by surfaces rather than as stick figure structures. As a result, it is possible to calculate mass properties of the parts which are often required for engineering analysis such as finite element methods and kinematic studies for interference checking.

To give an example such as a cylinder, a wireframe cylinder is defined in a computer as two circles connected by two line segments, whereas the solid model of a cylinder is represented as a 3D object that contains a volume. If it is to determine the volume of the wireframe cylinder, the formula for cylinder volume can be used. However, this formula is only valid for cylinders. For other types of volumes, different formula would have to be programmed into the computer. Obviously, it would be difficult to calculate volumes in this manner for complex shapes because each volume would have to be determined using a different formula. Another drawback is that the computer programmer would have to know the shape of the part in advance before the appropriate formula could be used to program the computer. On the other hand, to calculate the volume of a solid model, it is much easier because the computer can employ a general numerical integration that can be applied to solids of any shape. For example, the volume of a solid may be determined by dividing one face of the solid into a rectangular grid, then tracing the rectangular shapes back through the solid until it reaches the back edges of the model. The volume of the part is then obtained by adding the volumes of all the parallepipeds. This technique is commonly used in solid modelling, and is often referred to as the approximating sum integration method.

Despite the benefits of using solid models, they have their disadvantages. For one thing they use much more memory to represent a model than either wire frame or surface models, for another they require extensive processing for their manipulations owing to the more complicated data structure and associated mathematics.

8.4   WIRE FRAME ENTITIES

Wire frame models consist of only points and lines which are the basic entities of these models. Basic knowledge of how to represent these entities in a computer database is provided in section 8.8, which may help CAD users increase their productivity. It is not uncommon to ask how to draw an ellipse, how to create a circle that is tangent to a line or what a beizer curve is and how to draw it. Furthermore, this will serve as the basis for further discussion. Normally wire frame entities are divided into two categories: analytic and synthetic entities.

8.5   REPRESENTATION OF CURVES

Let us first examine why there is a requirement for alternative geometric representations to those of classical geometry. Most of us will be familiar with the following expressions:

$$y = mx + c \tag{8.1}$$

$$ax + by + c = 0 \tag{8.2}$$

and
$$ax^2 + by^2 + 2kxy + 2fx + 2gy + d = 0 \tag{8.3}$$

which are the explicit equation of a straight line and the implicit equations of a straight line and a conic section curve, respectively.

Why are these not adequate for CAD?

There is an evident problem with the explicit expression for a straight line in that the slope *m* of the line is infinite for a line parallel to the *y*-axis. Near vertical lines will have very large slopes and these may be difficult to define using a computer because very large real numbers may lead to numerical problems. The implicit forms deal with curves of any slope, but have limitations (in common with the explicit forms):

- They represent unbounded geometry. Equation (8.2) defines an infinite line, and Eq. (8.3) defines a complete conic section curve in CAD. However, geometric representation would normally be of a line between two points and often a part of an ellipse or an arc of a circle.

- Curves are often multivalued. For example, for a given value of *x* in Eq. (8.3), there are two values of *y*. Ideally a unique point on the curve would be defined by a single value of a variable defining the curve.

- It is often necessary in CAD to evaluate an orderly sequence of points on a geometric entity. Implicit equations do not offer a natural procedure for evaluating points on a curve (consider the conic section curve again incrementing x by equal intervals gives a very uneven sequence of points on the curve. Some values of x may not yield points on the curve).
- The equation for the curve will depend upon the coordinate system used.

A curve may be described or represented by a set of equations. These equations can be categorized into two forms:

Generic form

The generic form in which any generic point (x, y, z) satisfies a relationship in implicit form in x, y and z, i.e., $f(x, y, z) = 0$. A single constraint generally describes a surface while two constraints considered together can be thought of as a curve which is the intersection of two surfaces. This may be expressed in an explicit form in the following manner:

$$x = g_1(y, z)$$
$$y = g_2(x, z)$$
$$z = g_3(x, y)$$

Parametric form

A parametric curve is one whose defining equations are in terms of a simple, common independent variable known as **parametric variable**. In the parametric form, the representation is done by a set of functions. A curve may be represented by

$$x = x(u)$$
$$y = y(u)$$
$$z = z(u)$$

where x, y, z are coordinates of the points on the curve which are functions of some parameter u and the parametric variable is constrained in the interval $u \, \varepsilon(0,1)$. For example, a point (x, y) is located at an angle 0 from $+x$ axis on a circle with centre at $(0,0)$ and radius $= 1$ can be described in parametric form as:

$$x = \cos \theta \quad \text{and} \quad y = \sin \theta$$

where θ is the parameter.

Parametric design is very popular in CAD for a variety of reasons listed as follows:

- Possibility of separation of variables
- Each variable is treated alike
- More degrees of freedom/control
- Parametric equations can be transformed directly
- Infinite slopes can be handled without computational breakdown
- Easy to express as vectors
- Amenable to plotting and digitizing
- Inherently bounded

8.6 DESIGN OF CURVED SHAPES

Design of curved shapes should satisfy the following requirements:

(a) It should be possible to represent the shape mathematically.

(b) The modelling should involve minimum computation.

(c) It should be possible to generate a CNC program to machine the surfaces (2, 3, 4 and 5 axis machining) or to prepare a mould or die to make the part (as in plastic moulding or casting or automobile panel pressing).

A component can be designed using the curves and shapes which can be mathematically described, e.g. arc, circle, conics, ellipsoid, hyperbolic, paraboloid, sphere, cone, cylinder, linear, conical and circular swept surfaces. However, very often the designer starts with specifying a few points which roughly describe the shape.

Two approaches are available to designers to model curves and surfaces in such cases: interpolation and approximation. The interpolation essentially tries to pass a curve on a surface called **interpolant** through all these points.

Approximation tries to fit a smoother curve on surface which may be close to these points but may not actually pass through each of them. Figure 8.3 illustrates the difference between interpolation and approximation.

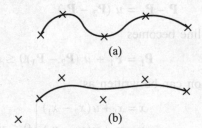

(a)

(b)

Fig. 8.3 (a) Interpolation and (b) approximation.

One of the popular methods of interpolation is to use the lagrange polynomial, which is the unique polynomial of degree N passing through $N + 1$ points. However, Lagrange polynomial is unsuitable in modelling of curves because of:

(a) Large number of computations involved

(b) Tendency for the curve to oscillate between data points when the data points are large

Another approach is to look for a polynomial of fewer degrees than N passing through these $N + 1$ data points. This is done by combining polynomials of lesser degree passing through several consecutive data points and smoothly piercing several such curve segments together by blending. In general, the modelling of curves is done to satisfy the following:

- Axis independence
- Global and local control
- Smoothness of curves
- Versatility
- Continuity between adjoining segments of curve

8.7 ANALYTIC CURVES

Analytic entities include points, straight lines, arcs, circles, ellipses, parabolas and hyperbolas. The properties of these entities and the techniques for manipulating them are well studied and usually taught in high school mathematics courses. The methods for defining these entities in a computer are easy to understand. The most common methods used by CAD/CAM systems to create wire frame entities are shown in parametric representation of analytic curves and methods of defining synthetic curves. Different CAD/CAM systems provide different methods but they are mostly extensions or subsets of the above.

8.8 PARAMETRIC REPRESENTATION OF ANALYTIC CURVES

8.8.1 Line

A line connects two points P_1 and P_2 (see Fig. 8.4). Define a parameter u such that it has values 0 and 1 at P_1 and P_2 respectively. Utilizing the ΔOPP_1, the following equation is obtained as:

$$P = P_1 + (P - P_1)$$

However, the vector $(P - P_1)$ is proportional to the vector $(P_2 - P_1)$ such that

$$P - P_1 = u (P_2 - P_1)$$

Thus, the equation of the line becomes

$$P_1 = P_1 + u (P_2 - P_1)0 \leq u \leq 1$$

In scalar form, this equation can be written as:

$$\left. \begin{array}{l} x = x_1 + u(x_2 - x_1) \\ y = y_1 + u(y_2 - y_1) \\ z = z_1 + u(z_2 - z_1) \end{array} \right\} 0 \leq u \leq 1$$

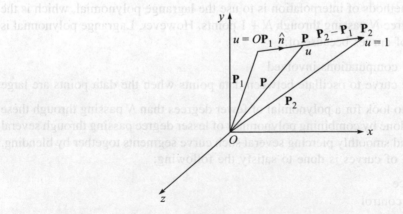

Fig. 8.4 Representation of line.

Any other point on the line or its extension has a certain value u which is proportional to the point location. The tangent vector of the line is given by

$$\mathbf{P}' = \mathbf{P}_2 - \mathbf{P}_1$$

or in a scalar form,

$$x' = x_2 - x_1$$
$$y' = y_2 - y_1$$
$$z' = z_2 - z_1$$

The independence of the tangent vector from u reflects the constant slope of the straight line. For a 2D line, the known infinite (vertical line) and zero (horizontal line) slope conditions can be generated. The unit vector $\hat{n}$ is the direction of the line and is given by

$$\hat{n} = \frac{\mathbf{P}_2 - \mathbf{P}_1}{L}$$

where L is the length of the line

$$L = |\mathbf{P}_2 - \mathbf{P}_1| = \sqrt{(x_2 - x_1)^2 + (y_2 + y_1)^2 + (z_2 + z_1)^2}$$

Regardless of the user input to create a line, a line database stores its two end points and additional information such as its font, width, colour and layer.

8.8.2 Circles

Circles and circular arcs are the most common entities used in the wire frame modelling. Circles and circular arcs together with straight lines are sufficient to construct a large percentage of existing mechanical parts and components in practice.

Besides other information, a circle database stores its centre and radius as its essential geometric data. If the plane of the circle cannot be defined from the user input data while specifying a centre and a radius, it is assumed by the software to be the xy plane of the current WCS at the time of construction. Regardless of the user input information to create a circle, such information is always converted into a radius and centre by the software.

The basic parametric equation of a circle is written as:

$$\left. \begin{array}{l} x = x_c + R \cos u \\ y = y_c + R \sin u \\ z = z_c \end{array} \right\} \quad 0 \le u \le 2\pi$$

assuming the plane of circle as xy plane.

In this equation, the parameter u is the angle measured from the x-axis to any point P on the circle (see Figure 8.5). This parameter is used by commercial software to locate points at certain angles on the circle for construction purposes.

For display purposes, points are generated on the circle circumference by incrementing u from 0 to 360. These points are, in turn, connected with line segments to display the circle. However, this is an inefficient way due to computing the trigonometric functions in the equation for each point. A less computational method is to assume that there is an increment Δu between

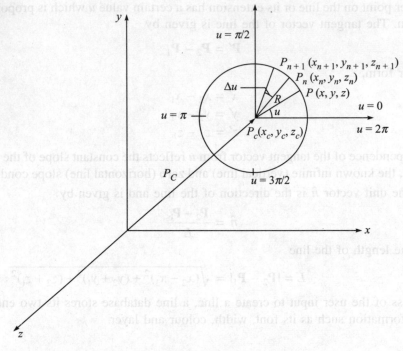

Fig. 8.5 Representation of circle.

two consecutive points $P(x_n, y_n, z_n)$ and $P(x_{n+1}, y_{n+1}, z_{n+1})$ on the circle circumference. The following recursive relationship can be written as:

$$x_n = x_c + R \cos u$$
$$y_n = y_c + R \sin u$$
$$x_{n+1} = x_c + R \cos (u + \Delta u)$$
$$y_{n+1} = y_c + R \sin (u + \Delta u)$$
$$z_{n+1} = z_n$$

Expanding x_{n+1} and y_{n+1} equations give

$$x_{n+1} = x_c + (x_n - x_c) \cos \Delta u - (y_n - y_c) \sin \Delta u$$
$$y_{n+1} = y_c + (y_n - y_c) \cos \Delta u + (x_n - x_c) \sin \Delta u$$
$$z_{n+1} = z_n$$

Thus, the circle can start from an arbitrary point and successive points with equal spacing can be calculated recursively. The values of $\cos \Delta u$ and $\sin \Delta u$ have to be calculated only once, which eliminates computation of trigonometric functions for each point. This algorithm is useful for hardware implementation to speed up the circle generation and display.

A circular arc can be written as:

$$\left. \begin{array}{l} x = x_c + R \cos u \\ y = y_c + R \sin u \\ z = z_c \end{array} \right\} u_s \leq u \leq u_e$$

where u_s and u_e are the starting and ending angles of the arc respectively. An arc database includes its centre and radius, as a circle, as well as its starting and ending angles. The arc is always connected by its beginning and ending points in a counter-clockwise direction.

8.8.3 ELLIPSE

Ellipse is a curve generated by a point moving in space such that at any position the sum of its distances from two fixed points (foci) is constant and equal to the major diameter. Each focus is located on the major axes of the ellipse at a distance from its centre equal to $\sqrt{A^2 - B^2}$, where A and B are the major and minor axes radii. Circular holes and forms become ellipses when they are viewed obliquely relative to their planes.

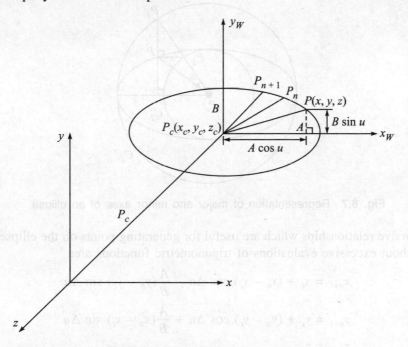

Fig. 8.6 Representation of ellipse.

The development of the parametric equation and other related characteristics of ellipses, elliptic arcs and fillets are similar to those of circles, circular arcs and fillets. Four conditions are required to define the geometric shape of an ellipse. The default plane of an ellipse is the xy plane of the current WCS (world coordinate system) at the time of construction if the user input is not enough to define the ellipse plane, as in the case of inputting the centre, half of the length of the major axes, the half of the length of the minor axes. The database of an ellipse usually stores user input as a centrepoint, half the length of the major axes, half the length of the minor axes and other information (orientation, starting and ending angles, font, layer, name, colour, etc.).

The parametric equation of an ellipse is written as:

$$x = x_c + A \cos u$$
$$y = y_c + B \sin u \left.\right\} 0 \le u \le 2\pi$$
$$z = z_c$$

The parameter u is the angle. To find point P on the ellipse that corresponds to an angle u, the two concentric circles C_1 and C_2 are constructed with centres at P_c and radii of A and B respectively (refer Fig. 8.7). A radial line is constructed at an angle u to intersect both circles at points P_1 and P_2 respectively. If a line parallel to the minor axes is drawn from P_1 and a line parallel to the major axes is drawn from P_2, the intersection of these two lines defines the point P.

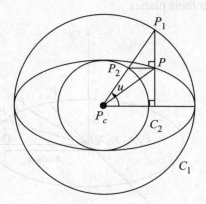

Fig. 8.7 Representation of major and minor axes of an ellipse.

The recursive relationships which are useful for generating points on the ellipse for display purposes without excessive evaluations of trigonometric functions are:

$$x_{n+1} = x_c + (x_n - x_c) \cos \Delta u - \frac{A}{B}(y_n - y_c) \sin \Delta u$$

$$y_{n+1} = y_c + (y_n - y_c) \cos \Delta u + \frac{A}{B}(x_n - x_c) \sin \Delta u$$

$$z_{n+1} = z_c$$

If the ellipse major axes is inclined with an angle α relative to the x-axis, ellipse equation becomes

$$x = x_c + A \cos u \cos \alpha - B \sin u \sin \alpha$$
$$y = y_c + A \cos u \sin \alpha + B \sin u \cos \alpha \left.\right\} 0 \le u \le 2\pi$$
$$z = z_c$$

and

$$x_{n+1} = x_c + A \cos (u_n + \Delta u) \cos \alpha - B \sin (u_n + \Delta u) \sin \alpha$$
$$y_{n+1} = y_c + A \cos (u_n + \Delta u) \cos \alpha + B \sin (u_n + \Delta u) \sin \alpha$$
$$z_{n+1} = z_n$$

where $u_n = (n-1)u$. The first point corresponds to $n = 0$ which lies at the end of the major axes. There is computational saving.

8.8.4 Parabola

The parabola shown in Fig. 8.8 is defined mathematically as a curve generated by a point which moves such that its distance from a fixed point (the focus P_F) is always equal to its distance to a fixed line (the directrix).

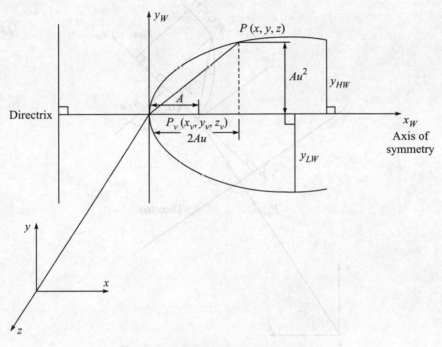

Fig. 8.8 Representation of parabola.

The vertex P_v is the intersection point of the parabola with its axis of symmetry. It is located midway between the directrix and focus. The focus lies on the axis of symmetry. Useful applications of the parabolic curve in engineering design include its use in parabolic sound and light reflectors, radar antennas and in bridge arches.

Three conditions are required to define a parabolic curve or a parabolic arc. The default plane of a parabola is the xy plane of the current WCS at the time of construction. The database of a parabola usually stores the coordinates of its vertex, distances y_{HW} and y_{LW}, that define its end points, the distance A between the focus and the vertex (the focal distances), and the orientation angle α. The two-end points determine the amount of the parabola to be displayed.

Assuming the local coordinate system of the parabola, its parametric equation can be written as:

$$\left.\begin{array}{l} x = x_v + Au^2 \\ y = y_v + 2Au \\ z = z_v \end{array}\right\} \ 0 \le u \le \infty$$

If the range of the y coordinate is limited to y_{HW} and y_{LW} for +ve and –ve values respectively, the corresponding values of u become

$$u_H = \frac{y_{HW}}{2A}$$

$$u_L = \frac{y_{LW}}{2A}$$

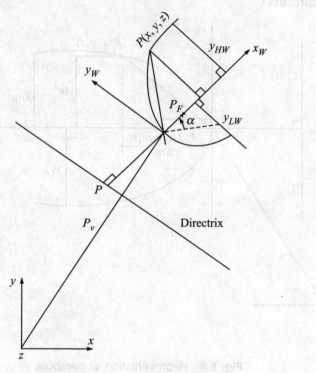

Fig. 8.9 Representation of parabola inclined at an angle α.

The recursive relationships to generate points on the parabola are obtained by substituting $u_n + \Delta u$ for points $n + 1$. This gives

$$x_{n+1} = x_n + (y_n - y_v)\Delta u - A\,(\Delta u)^2$$
$$y_{n+1} = y_n + 2A\Delta u$$
$$z_{n+1} = z_n$$

If the parabolic axis of symmetry is inclined with an angle α, its equation becomes

$$x = x_v + Au^2 \cos\alpha - 2Au \sin\alpha$$
$$y = y_v + Au^2 \sin\alpha + 2Au \cos\alpha$$
$$z = z_v$$

and the recursive relationships reduce to

$$x_{n+1} = x_n \cos \alpha + (1 - \cos \alpha)x_v + (\Delta u \cos \alpha - \sin \alpha)(y_n - y_v) + A\Delta u (\Delta u \cos \alpha - 2 \sin \alpha)$$

$$y_{n+1} = (\cos \alpha + \Delta u \sin \alpha)y_n + (1 - \cos \alpha - \Delta u \sin \alpha)y_v + (x_n - x_v) \sin \alpha + A\Delta u (\Delta u \sin \alpha + 2 \cos \alpha)$$

$$z_{n+1} = z_n$$

If the $x_W y_W$ and the xy planes coincide, the transformation matrix becomes

$$[T] = \begin{bmatrix} \cos \alpha & \sin \alpha & 0 & x_v \\ -\sin \alpha & \cos \alpha & 0 & y_v \\ 0 & 0 & 1 & z_v \\ 0 & 0 & 0 & 1 \end{bmatrix}$$

If the two planes are different then the unit vectors $\hat{n}_{xW}$, $\hat{n}_{yW}$ and $\hat{n}_{zW}$ that define the orientation of the parabolic local coordinate system relative to MCS must be calculated.

8.8.5 Hyperbola

A hyperbola is described mathematically as a curve generated by a point moving such that at any position the difference of its distances from the fixed points (foci) F and F is a constant and equal to the transverse axis of the hyperbola.

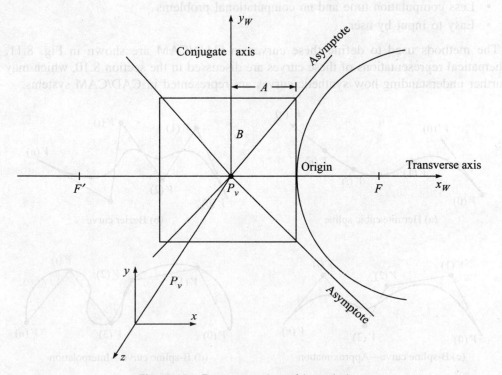

Fig. 8.10 Representation of hyperbola.

The parametric equation of the hyperbola is given by

$$x = x_v + A \cosh u$$
$$y = y_v + B \sinh u$$
$$z = z_v$$

This equation is based on the non-parametric implicit equation of the hyperbola which can be written as:

$$\frac{(x - x_v)^2}{A^2} - \frac{(y - y_v)^2}{B^2} = 1 \quad (\text{using } \cosh^2 u - \sinh^2 u = 1)$$

8.9 SYNTHETIC CURVES

In addition to simple analytic entities that we discussed in the previous section 8.8, some more general methods for representing curves are needed to meet geometric design requirements of mechanical parts and various engineering applications. Much research has been done during the past couple of decades on developing better methods for representing general curves and shapes and easier ways to manipulate them by computer. A good representation of engineering objects have the following properties:

- Easy to control the continuity of the curves to be designed
- Requires less storage to represent a curve
- Less computation time and no computational problems
- Easy to input by user

The methods used to define these curves in CAD/CAM are shown in Fig. 8.11. The mathematical representations of these curves are discussed in the section 8.10, which may help in further understanding how synthetic curves are represented in CAD/CAM systems.

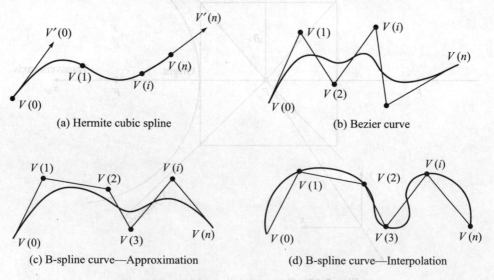

(a) Hermite cubic spline (b) Bezier curve

(c) B-spline curve—Approximation (d) B-spline curve—Interpolation

Fig. 8.11 Methods of defining synthetic curves.

8.10 PARAMETRIC REPRESENTATION OF SYNTHETIC CURVES

Analytic curves as described in the section 8.8, are usually well defined and well studied. However, in the geometric design of mechanical parts, some other curves are required. They do not belong to any analytic curves we presented in the section 8.8. One familiar example is the car body. The curves of the car body not only have to envelop inside the mechanical parts but also help reduce air resistance. It is unusual for a well-designed car body to consist only of analytic curves or surfaces such as circles or cylinders. Other examples include the fuselage, wings and propeller blades of airplanes, whose shape is based on aerodynamic and fluid flow simulations.

How are those curves or surfaces created in CAD/CAM systems? What are their mathematical representations? To answer these questions, we provide a detailed discussion of various types of synthetic curves in the sections 8.10.1 to 8.10.3.

First we introduce the concept of continuity. Intuitively, continuity means the smoothness of the connection of two curves or surfaces at the connection points or edges. Normally three types of continuity C^0, C^1 and C^2 are defined to characterize the smoothness of connection of two curves.

C^0 continuity implies simply connecting two curves in which the gradients of two curves at the point of joining as well as their curvatures may be different (Fig. 8.12a).

In C^1 continuity, the gradients at the point of joining must be same (Fig. 8.12b).

However, C^2 implies curvature continuity that is not only the gradient but also the centre of curvature is the same, as shown in Fig. 8.12c. Curves that are constructed by many curve segments are called **synthetic curves**. Different types of curve segments can be used to construct synthetic curves with certain continuity requirements.

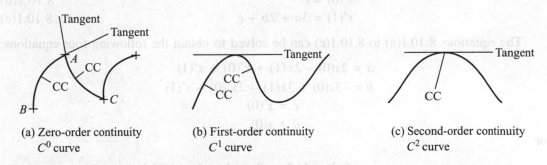

| (a) Zero-order continuity | (b) First-order continuity | (c) Second-order continuity |
| C^0 curve | C^1 curve | C^2 curve |

Fig. 8.12 Types of continuities.

Synthetic curves usually have some good properties for example it is easy to control the continuity of the curve to be designed and computation requirements are low. The types of synthetic curves normally provided by major CAD/CAM systems include

- Hermite cubic spline
- Bezier curves
- B-spline curves
- Rotational β-splines
- Non-uniform rational B-splines.

8.10.1 Cubic Spline Curves

Splines are functions that are used for fitting a curve through a number of data points. A cubic spline has the property that the three coordinate functions $x(u)$, $y(u)$ and $z(u)$ are each cubic polynomials in the variable u.

$$x(u) = au^2 + bu2 + cu + d$$
$$y(u) = eu^3 + fu^2 + gu + h$$
$$z(u) = ju^3 + ku^2 + lu + m$$

Parametric curves are generally considered to be defined over an infinite range of parameter values of u. It is always necessary to limit these to a finite interval when modelling shapes on a computer. The domain u is therefore taken as $0 < u \leq 1$ in the discussion that follows:

A spline passes through two end points and satisfies differentiality condition at these end points. These four conditions require a polynomial degree of at least 3. The derivatives of cubic splines for x, y and z are similar. The difference will be only in the end point coordinates and derivatives at the end points. Therefore only the derivatives of $x(u)$ are considered here.

Let $x(0)$, $x(1)$, $x'(0)$ and $x'(1)$ are the end points and the derivatives at the end points respectively.

Since

$$x(u) = au^3 + bu^2 + cu + d \qquad \text{8.10.1(a)}$$

we get

$$x(0) = d \qquad \text{8.10.1(b)}$$
$$x(1) = a + b + c + d \qquad \text{8.10.1(c)}$$
$$x'(0) = c \qquad \text{8.10.1(d)}$$
$$x'(1) = 3a + 2b + c \qquad \text{8.10.1(e)}$$

The equations 8.10.1(a) to 8.10.1(e) can be solved to obtain the following four equations:

$$a = 2x(0) - 2x(1) + x'(0) + x'(1)$$
$$b = -3x(0) + 3x(1) - 2x'(0) - x'(1)$$
$$c = x'(0)$$
$$d = x(0)$$

or

$$\begin{pmatrix} a \\ b \\ c \\ d \end{pmatrix} = \begin{pmatrix} 2 & -2 & 1 & 1 \\ -3 & 3 & -2 & -1 \\ 0 & 0 & 1 & 0 \\ 1 & 0 & 0 & 0 \end{pmatrix} \begin{pmatrix} x(0) \\ x(1) \\ x'(0) \\ x'(1) \end{pmatrix}$$

The coefficients of cubic spline polynomials $y(u)$ and $z(u)$ are similarly obtained by replacing x data by y and z data.

In a vectorial form, the parametric cubic curve can be expressed as:

$$P(u) = au^3 + bu^2 + cu + d$$

i.e

$$p(u) = au^3 + bu^2 + cu + d$$

where $u\varepsilon\ (0,1)$

or $$p(u) = [u^3\ u^2\ u\ 1][a\ b\ c\ d]^T$$

Let $$u = [u^3\ u^2\ u\ 1] \quad \text{and} \quad A = [a\ b\ c\ d]$$

or $$P = UA$$

Curves are defined by interpolating points and tangents at these points. The parametric cubic curve can be designed using four vectors $\mathbf{p(0)}$, $\mathbf{p(1)}$, $\mathbf{p'(0)}$ and $\mathbf{p'(1)}$. Each one of these vectors has 3 components

$$P(0) = d$$
$$P(1) = a + b + c + d$$
$$P'(0) = c$$
$$P'(1) = 3a + 2b + c$$

$$P = (2u^3 - 3u^2 + 1)\ p(0) + (-2u^3 + 3u^2)\ P(1) + (u^3 - 2u^2 + u)\ p'(0) + (u^3 - u^2)\ P'(1)$$

Let $$F_1 = 2u^3 - 3u^2 + 1$$
$$F_2 = -2u^3 + 3u^2$$
$$F_3 = u^3 - 2u^2 + u$$
$$F_4 = u^3 - u^2$$

Then $$P = F_1\ P + F_2\ P + F_3\ P' + F_4\ P'$$

Let $$F = [F_1\ F_2\ F_3\ F_4] \quad \text{and} \quad B = [P(0)\ P(1)\ P'(0)\ P'(1)]$$

Then $$[P] = [F]\ [B]$$

F_1, F_2, F_3, F_4 are called **blending functions** or **Hermite polynomial basis functions**. B is the matrix of geometric coefficients. F is the blending function matrix.

The blending function matrix F can be written as the product of two matrices.

$$F = UM$$

where $$M = \begin{pmatrix} 2 & -2 & 1 & 1 \\ -3 & 3 & -2 & -1 \\ 0 & 0 & 1 & 0 \\ 1 & 0 & 0 & 0 \end{pmatrix}$$

M is called **universal transformation matrix**.

Therefore, $$P = UMB$$
or $$P = UA$$
Hence $$A = MB$$
and conversely $$B = M^{-1} A$$

where $$M^{-1} = \begin{pmatrix} 0 & 0 & 0 & 1 \\ 1 & 1 & 1 & 1 \\ 0 & 0 & 1 & 0 \\ 3 & 2 & 1 & 0 \end{pmatrix}$$

For a given set of end points $p(0)$ and $p'(1)$ and slopes $p'(0)$ and $p'(1)$ several sets of curves can be generated by varying the magnitude of the tangent vectors (unit tangent vector $t = \dfrac{pu}{(pu)}$).

Figure 8.13 shows a cubic spline connecting four points. There are several other formulations of cubic splines. Ferguson used cubic polynomials for curve and surface design in aircraft. However, these methods generally suffer from the fact that they do not allow the user to change the smoothness of the curve.

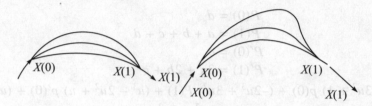

Fig. 8.13 Cubic spline.

8.10.2 Bezier Curves

Bezier curves developed by P. Bezier at Renault Automobile Company and P.De Casteljau at Citreon, France is used for curve and surface design of automobile panels. A Bezier curve is blended at the joints and is completely defined by four consecutive points (x_0, y_0), (x_1, y_1), (x_2, y_2), (x_3, y_3). The curve passes only through the first and fourth points. The two intermediate points are used to define the slope of the curve at the end points. If x_0, x_1, x_2 and x_3 are the x-coordinates of the control points, it is assumed that

(a) The curve passes through the end points x_0 and x_3.
(b) The slope at the points are:

$$x'(0) = 3\,(x_1 - x_0)$$
$$x'(1) = 3\,(x_3 - x_2)$$

Using these properties the coefficients of Bezier curve for x term can be expressed as:

$$\begin{pmatrix} a \\ b \\ c \\ d \end{pmatrix} = \begin{pmatrix} -1 & 3 & -3 & 1 \\ 3 & -6 & 3 & 0 \\ -3 & 3 & 0 & 0 \\ 1 & 0 & 0 & 0 \end{pmatrix} \begin{pmatrix} x_0 \\ x_1 \\ x_2 \\ x_3 \end{pmatrix}$$

The cubic polynomial $x(t)$ in the Bezier form can be expressed as:

$$x(t) = \begin{pmatrix} u^3 & u^2 & u & 1 \end{pmatrix} \begin{pmatrix} -1 & 3 & -3 & 1 \\ 3 & 0 & 3 & 0 \\ -3 & 3 & 0 & 0 \\ 1 & 0 & 0 & 0 \end{pmatrix} \begin{pmatrix} x_0 \\ x_1 \\ x_2 \\ x_3 \end{pmatrix}$$

The advantages of Bezier curve over cubic spline is that the direction of the curve at the joints can be defined and changed simply by specifying the position of the second and third data points. Changing a control point not only affects the shape of the curve near the control point but has an influence throughout the curve. This lack of local control is a major weakness of Bezier curve. Figure 8.14 shows Bezier cubic segments for two sets of values of x.

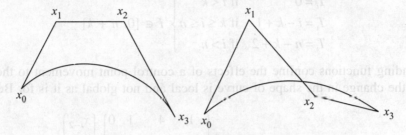

Fig. 8.14 Bezier cubic segments for two sets of values of x.

The Bezier cubic segment discussed earlier is a special case of the more general Bezier polynomial curve segment.

$$P(u) = \sum_{i=0}^{3} P_i B_{i \cdot 3} (u)$$

With blending functions $\qquad B_{i,n}(u) = C(n, i)\, u^i\, (1-u)^{n-i}$

where n is the binomial coefficient.

and $\qquad\qquad\qquad\qquad C(n,\ i) = \dfrac{n!}{i'(n-1)!}$

and P are the control points and there are $(n + 1)$ of them defining the vertices of the characteristic polygon.

8.10.3 B-spline Curves

Neither the Bezier nor the cubic spline curve formulations allow local modification of curves, and Bezier polynomials are in addition somewhat constrained in the number of points that they may approximate without the degree of the curve becoming inconveniently high. Both of these limitations are overcome by a generalization of the Bezier approach known as the **B-spline method**, which again uses blending functions to combine the influence of a series of control or track points in an approximate curve.

This form of cubic segments uses a third set of basis functions different from the types discussed earlier. A cubic B-spline curve is a special case of spline curve. The equation for this curve can be written as:

$$P(u) = \sum_{i=0}^{n} P_i N_{i \cdot k} (u) \qquad u \in [0,(n-k+2)]$$

This generates what is known as **piecewise polynomial**. The $N(u)$ blending functions are defined recursively as:

$$N(u) = 1 \qquad \text{if } t_i = u \le t_{i+1} = 0$$

Otherwise
$$N_{i,k}(u) = \frac{(u - t_i) N_{i,k-1}(u)}{t_{i+k-1} - t_i} + \frac{(t_{i+k} - u) N_{i+1,k-1}(u)}{t_{i+k} - t_{i+1}}$$

The resulting parametric polynomials are of degree $(k-1)$. The ts are called **knot values**. For an open curve, they are chosen as follows:

$$\left. \begin{array}{ll} T_i = 0 & \text{if } i < k \\ T_i = i - k + 1 & \text{if } k \le i \le n \\ T_i = n - k + 2 & \text{if } i > n \end{array} \right\} I \in [0, n + k]$$

The blending functions confine the effects of a control point movement to the immediate local. Thus, the change in the shape of curve is local and not global as it is for Bezier curves.

$$x(u) = \begin{pmatrix} 1 & u & u^2 & u^3 \end{pmatrix} \begin{Bmatrix} 1 & 4 & 1 & 0 \\ -3 & 0 & 3 & 0 \\ 3 & -6 & 3 & 0 \\ -1 & 3 & -3 & 1 \end{Bmatrix} \begin{Bmatrix} x_{i-2} \\ x_{i-1} \\ x_i \\ x_{i+1} \end{Bmatrix}$$

When the control points are distinct, this curve is continuous in slope and curvature between successive segments but it does not pass through any of the intermediate control points. The cubic B-spline has the advantage that the control points may be moved without affecting slope and curvature continuity and only four spans of the overall curve will be affected by the change. Moreover, by allowing two control points to coincide, it is possible to create a curvature discontinuity. A slope discontinuity similarly can be introduced by choosing three successive control points to be coincident.

It is possible to represent complex curve shapes by considering composite curves constructed from individual segments, in the case of cubic spline Bezier and B-spline techniques.

8.11   LIMITATIONS OF WIRE FRAME MODELLING

Although wire frame modelling is simple and straightforward in concept, it has a number of limitations:

(a)   From the point of view of engineering applications, it is not possible to calculate volume and mass properties of a design. Other applications such as NC path generation, cross sectioning and interference detection, also encounter problems when wire frame modelling is used because the wire frame model database contains only low-level information such as points and lines, the wire frame methods are limited in scope when high level information is required by particular applications such as process planning.

(b)   In the wire frame representation the virtual edges (profile or silhouette) are not usually provided. For example, a cylinder is represented by three edges, that is, two circles and one straight line. But the straight line is not enough to represent the profile or silhouette of a cylinder.

(c) The creation of wire frame models usually involves more user effort to input necessary information than that of solid models, especially for large and complex parts. For example, consider the creation of a 3D model of a simple cube.

In the wire frame model, we need to draw 12 lines, whereas with solid modelling methods a cube may be created by providing the positions of three corner points. Furthermore the latter model provides more information such as the inside or outside of the part.

8.12   CURVE FITTING TECHNIQUES

8.12.1   Displaying

Displaying curves provide the designer with a means of visualizing geometric models within the units of the wire frame modelling technique. Various colours can be assigned to various curves for identification and other purposes. To display a curve many closely spaced points on it are generated for various permissible values of the parameter u. The curve parametric equation is used to generate the coordinates of these points. The display processing unit receives these coordinates and changes them to 2D coordinates relative to the device coordinate system of the display terminal. The resulting points are displayed and are connected by short-line segments. Line-generation hardware exists for creating and displaying these line segments.

To display the straight line, the two end points are fed to the line generation hardware to generate intermediate points and connect them by short-line segments. Equations of curves, such as circle and conic sections, that involve trigonometric functions are usually rewritten to minimize computation time to generate points on these curves for display purposes.

8.12.2   Blending

Given two curve segments $P_1(u_1)$ $0 < u_1 < a$, $P_2(u_2)$, $0 < u_2 < b$, the condition for the two segments to be continuous at the joint is that the upper limits u, and u_9 are taken to be a and b and not 1. Three classes of continuity can be considered at the joint.

The first is C^0 continuity, i.e., the ending point of the first curve and the starting point of the second curve are the same. This gives

$$P_1(a) = P_2(0)$$

If the two segments are to be C continuous as well, they must have slope continuity

$$\left.\begin{array}{l} P_1'(a) = \alpha_1 \mathbf{T} \\ P_2'(0) = \alpha_2 \mathbf{T} \end{array}\right\} \tag{8.1}$$

where α_1 and α_2 are constants and $\mathbf{T}$ is the common unit tangent vector at the joint. For slope continuity at the joint, the last segment of the control polygon of the former and the first segment of the control polygon of the latter must be collinear.

The third useful class of continuity is of the curvature (C^2 continuity) which has to be continuous at the joint in addition to position and slope. To achieve curvature continuity is less straightforward and requires the binomial vector to a curve at a point. On a curve segment, $\mathbf{T}$ is the tangent unit vector, $\mathbf{N}$ is the normal unit vector, O is the centre of curvature and ρ is the radius of curvature at point P.

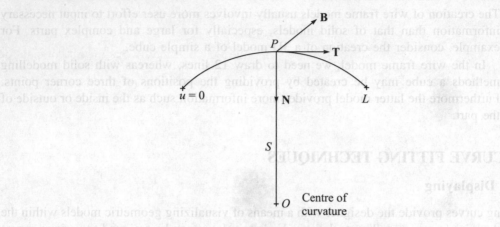

Fig. 8.15 Binomial vector to a curve.

The curvature at P is defined as $1/\rho$. The binomial vector $\mathbf{B}$ is defined as:

$$\mathbf{B} = \mathbf{T} \times \mathbf{N}$$

The curvature is related to the curve derivatives through the vector $\mathbf{B}$ by the equation:

$$\frac{1}{\rho}\mathbf{B} = \frac{\mathbf{P}' \times \mathbf{P}''}{|\mathbf{P}'|^3}$$

where $\mathbf{P}''$ is the second derivative w.r.t. to the parameter u. The condition for curvature continuity at the joint is:

$$\frac{\mathbf{P}_1'(a) \times \mathbf{P}_1''(a)}{|\mathbf{P}_1'(a)|^3} = \frac{\mathbf{P}_2'(0) \times \mathbf{P}_2''(0)}{|\mathbf{P}_2'(0)|^3} \tag{8.2}$$

Substituting Eq. (8.1) in Eq. (8.2), we get

$$\mathbf{T} \times \mathbf{P}_1''(a) = \left(\frac{\alpha_1}{\alpha_2}\right)^2 \mathbf{T} \times \mathbf{P}_2''(0)$$

This equation can be satisfied if

$$\mathbf{P}_2''(0) = \left(\frac{\alpha_2}{\alpha_1}\right)^2 \mathbf{P}_1''(a)$$

In general, if

$$\mathbf{P}_2''(0) = \left(\frac{\alpha_2}{\alpha_1}\right)^2 \mathbf{P}_1''(0) + \gamma \mathbf{P}_1''(a)$$

where γ is an arbitrary scalar which is chosen as zero for practical purposes.

8.12.3   Segmentation

Segmentation or curve splitting is defined as replacing one existing curve by one or more curve segments of the same curve type such that the shape of the composite curve is identical to that of the original curve. Segmentation is a very useful feature for CAD/CAM system. It is implemented as a 'divide entity' command. Model clean-up for drafting and documentation purposes is an example where an entity (curve) might be divided into two at the line of sight of another (normally the two entities do not intersect in space). One of the resulting segments is then removed and blanked out.

Mathematically, curve segmentation is a reparametrization or parameter transformation of a curve. Splitting lines, circles and cones is a simple problem. To split a line connecting two points P_0 and P_1 at a point P_2, the requirement is to define two new lines connecting the point pairs (P_0, P_2) and (P_2, P_1). For circles and cones, the angle corresponding to the splitting point together with the starting and ending angles of the original curve defines the proper range of the parameter u for the resulting two segments.

Polynomial curves such as cubic splines, Bezier curves and B-spline curves require a different parameter transformation. If the degree of the polynomial defining a curve is to be unchanged, which is the case of segmentation, the transformation must be linear. Let us assume that a polynomial curve is defined over the range $u = u_0$, u_m as shown in Fig. 8.16. To split the curve at a point defined by $u = u_1$ means that the first and the second segments are to be defined over the range $u = u_0$, u, and $u = u$, u_m respectively. A new parameter v is introduced for each segment such that its range is $v = 0$, 1. The parameter transformation takes the form:

$$u = u_0 + (u, - u_0)\ v \text{ for the first segment}$$
$$u = u_1 + (u_m - u_1)\ v \text{ for the second segment}$$

$v = 0$, 1 corresponds to the proper u values as required. To facilitate substitution in the curve polynomial equation where u is raised to different powers, the following equation is used:

$$u_n = \sum_{\gamma=0}^{n} \binom{n}{\gamma} u_0^r (\Delta u_0 v)^{n-\gamma}$$

where $\Delta u_0 = u_1 - u_0$. A similar equation can be written for the second segment and so on.

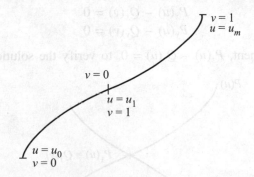

Fig. 8.16 Reparametrization of a segmental curve.

8.12.4 Trimming

Trimming curves or entities is a very useful function provided by all CAD/CAM systems. Trimming can truncate or extend a curve. Trimming is mathematically identical to segmentation but the only difference between the two is that the result of trimming a curve is only one segment of the curve bounded by the trimming boundaries. All the mathematical treatment of segmentation applies here. The trimming function requires evaluating the equation of the desired segment of the curves, deleting the original curve, and then storing and displaying the desired segment. Figure 8.17 shows the reparametrization of a trimmed curve.

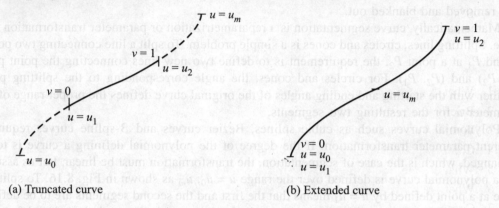

(a) Truncated curve   (b) Extended curve

Fig. 8.17   Reparametrization of a trimmed curve.

8.12.5   Intersection

The intersection point P_1 of two parametric curves $P(u)$ and $Q(v)$ in 3D space shown in Fig. 8.18 requires the solution of the following equation in the parameters u and v:

$$P(u) - Q(v) = 0$$

This equation represents three scalar equations that take the polynomial non-linear form generally with two parameters unknown. One way to find u and v is to solve the x and y components of the equation simultaneously, i.e.,

$$P_x(u) - Q_x(v) = 0$$
$$P_y(u) - Q_y(v) = 0$$

and then use the z-component, $P_z(u) - Q_z(u) = 0$, to verify the solution.

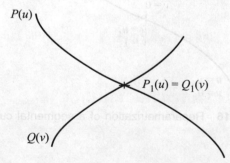

Fig. 8.18   Intersection of two parametric curves in space.

The roots of the equation can be found by numerical analysis method such as the Newton–Raphson method.

8.12.6   **Transformation**

- Translate
- Rotate
- Mirror
- Scale

REVIEW QUESTIONS

8.1   What does geometric modelling mean?

8.2   Briefly explain what wire frame model is and give its advantages and disadvantages.

8.3   Write on the importance of studying geometric modelling in CAD.

8.4   Compare 2D and 3D wire frame models.

8.5   What are the limitations in wire frame modelling? Explain with a suitable example.

8.6   What are the various types of geometric models available in CAD/CAM systems?

8.7   Comment on the characteristics of Bezier and B-spline curves.

8.8   What is non-parametric representation of curves and what are their limitations?

8.9   Define parametric representation. Discuss how the limitations of the non-parametric representation are overcome here.

8.10   With suitable figures explain the difficulty in displaying holes and curved ends in wire frame modelling.

8.11   What are entities? Explain the methods of defining lines and circles in wire frame modelling?

8.12   Explain how the curves are represented in (a) generic form; (b) parametric form.

8.13   Distinguish between interpolation and approximation approaches used in design of curves.

8.14   Explain the basic curve fitting techniques.

8.15   With examples, explain the terms, analytic and synthetic curves.

8.16   With a simple example, explain the need for hidden line removal in wire frame modelling.

8.17   Explain the significance of curves in wire frame modelling.

8.18   Explain the different types of synthetic curves.

8.19   Explain how a Bezier curve is superior to a cubic spline curve from the designer point of view.

PRACTICE PROBLEMS

8.1   Write down the implicit form of a line passing through (50, 50) as its mid point.

8.2   Find the intersection points of a line passing through (30, 30) and (100, 80) and a circle at (100, 50) and radius 75 mm.

8.3   Find the equation for a line passing through (80, 60) and (30, 30). Find the equation of a line that is perpendicular to the above line and passing through a point (60, 30).

8.4 Find the equation for a line passing through (80, 60) and (30, 30). Find the equation of a line that is parallel to the above line and passing through a point (40, 50).

8.5 Represent a circle with centre (0, 0) and a radius of 50 mm through the implicit form as well as parametric form.

8.6 Write down the implicit and parametric form of circle with a centre at (50, 50) and radius 100 mm.

8.7 Find the radius and centre of a circle that is passing through three points (30, 30), (60, 30) and (50, 40).

8.8 Find the intersection points of a line whose equation is given by $Y = 5X + 30$, and a circle with centre (0, 0) and radius 65 mm.

8.9 Two lines are passing through (20, 30), (120, 80) and (80, 20), (40, 90). Find the equation of a circle that is tangent to the above lines and having a radius of 40 mm.

8.10 Find the equation of a line that is tangent to a circle whose equation is $x^2 + y^2 = 25$ and passing through the point (8, 1).

8.11 Find the equation of a Bezier curve which is defined by the four control points as (80, 30, 0), (100, 100, 0) and (250, 30, 0).

8.12 A cubic Bezier curve is defined by the control points as (30, 30), (50, 80), (100, 100) and (150, 30). Find the equation of a curve and its mid point.

8.13 Calculate the points on cubic spline at $t = 0.25, 0.5, 0.75$ defined by the following points.

$$V_0(0) = [1, 2] \qquad V_0'(0) = [2, 4]$$
$$V_1(1) = [8, 10] \qquad V_1' = [4, 0]$$

8.14 Compare the middle point of cubic spline having same initial and final points but different tangents.

	10 curve	2 curve
Initial point $A = [2, 4]$	Slope at $A = 60°$	Slope at $A = 30°$
Final point $B = [8, 12]$	Slope at $B = 30°$	Slope at $B = 60°$

8.15 Fit a Bezier curve with the following control points:

$$P_0(1, 2) \ P_2(2, 2) \ P_3(3, 1) \text{ and } P_4(4, 0)$$

8.16 Fit a Bezier curve and find out its middle point.

$$P_0(1, 2) \ P_2(3, 5) \ P_3(4, 2) \text{ and } P_4 = P_0$$

8.17 Construct a Bezier curve and obtain points at $t = 0.25, 0.5$ and 0.75.

$$P_0(1, 3) \ P_2(3, 5) \ P_3(4, 2) \text{ and } P_4(5, 6)$$

8.18 Calculate the coefficients for the functions $x = x(u)$, $y = y(u)$ for a Hermite interpolation parametric cubic curve through the points:

$$P_1 = (1, 2) \qquad P_2 = (5, 6)$$
$$P_1' = (1, 1) \qquad P_2' = (1, 0)$$

Sketch the curve and the blending functions for the defining points and vectors of a parametric cubic curve.

8.19 A cubic Bezier curve is defined by the points (1, 1), (2, 3), (4, 4) and (6, 1). Calculate the coordinates of the parametric mid-point of this curve, and verify that its gradient (dy/dx) is 1/7 at this point. Use this information to sketch the curve.

8.20 Sketch the second-derivative continuous cubic spline curve that fits points $p_0 = (50, 50)$, $p_1 = (100, 50)$, $p_2 = (100, 100)$, $p_3 = (150, 100)$, $p_4 = (150, 50)$, $p_5 = (200, 50)$, $p_6 = (200, 100)$, and $p_7 = (250, 100)$. What information is required in addition to the point values to enable a second derivative cubic curve to be defined, and why is this additional information required? Suggest a parameter value for each control point in the curve.

8.21 Using the blending curves, sketch the B-spline curve that interpolates the point set $p_0 = (1, 1)$, $p_1 = (4, 1)$, $p_2 = (5, 4)$, $p_3 = (7, 2)$, $p_4 = (7, 2)$, $p_5 = (7, 2)$, $p_6 = (9, 4)$, and $p_7 = (12, 4)$. What is the range of the parameter u over which p_4 influences the shape of the curve? Estimate the coordinates of the curve at $u = 2.5$. What would be the shape of the curve if a B-spline of order 2 were fitted to the same point set?

8.22 Why is the rational form of polynomial functions used in geometric modelling? Complete the parametric mid-point of a rational Bezier curve through points $p_0 = (0, 0)$, $p_1 = (50, 50)$, and $p_2 = (10, 0)$, if the weights for points p_0 and p_2 is 1, and for p_1 is 2.

8.23 A lame's oval is described by

$$\frac{x^n}{c_1} + \frac{y^n}{c_2} = 1$$

Determine the equation of the oval through the three points P_1 [0 6], P_2 [5 5], P_3 [6 0]. Determine the intersection of this oval with the circle of radius 3 centered at $x = 6$, $y = 0$.

8.24 Calculate an equal number of points on the perimeter of the quarter circle in the first quadrant using a non-parametric explicit $y = +\sqrt{1 - x^2}$ $0 \le x \le 1$, a trigonometric parametric $P(\theta) = [x \ y] = [\cos \theta \ \sin \theta]$ $0 \le \theta \le 2\pi$ and the alternate parametric

$$P(t) = \left[\frac{1-t^2}{1+t^2} \ \frac{2t}{1+t^2}\right] \quad \theta \le t \le 1 \text{ analysis. Plot and compare the results.}$$

8.25 What type of conic section is represented by each of the following equations:

$$x^2 + y^2 + 4x - 2y = 10$$
$$x^2 + y^2 + xy - x + y = 3$$
$$x^2 - 4xy + 4y^2 + 3x = 6$$
$$2x^2 - y^2 + 4xy + 2x - 3y = 6$$
$$x^2 - 4xy + 3y^2 + 5y = 7$$
$$x^2 - y^2 = 1$$

8.26 Show that the following quadratic equation is a parabola and draw the segment between $x = 1$ and $x = 3$ for y_{min}:

$$16x^2 - 24xy + 9y^2 - 60x - 80y + 20 = 0$$

8.27 The first quadrant of a unit circle is parametrically represented by

$$x = \frac{(1-\sqrt{2})t^2 + (\sqrt{2}-2)t + 1}{(2-\sqrt{2})t^2 + (\sqrt{2}-2)t + 1}$$

$$y = \frac{(1-\sqrt{2})t^2 + \sqrt{2}\,t}{(2-\sqrt{2})t^2 + (\sqrt{2}-2)t + 1}$$

Determine the value of y for $x = 0.5$.

[**Hint:** Use an interactive technique to first t for $x = 0.5$].

SURFACE MODELLING

INTRODUCTION

In wire frame modelling, we take advantage of the simplicity of certain surfaces. For example, a plane is represented by its boundaries. We say nothing about the middle of the plane, which is fine because we know the middle of a plane is still a plane. However this assumption is not applicable to general cases. It is common knowledge that the shapes of cars, air planes, ships and so forth do not simply consist of standard geometry such as planes and cylinders. Therefore, it is not easy to represent them by wire frame models. To overcome the problems inherent in wire frame modelling, a different geometric modelling scheme called **surface modelling** is available to engineers.

Surface modelling is a widely used modelling technique in which objects are defined by their bounding faces. Surface modelling systems contain definition of surfaces, edges and vertices. Surface modelling goes one step further than wire frame modelling. For example, it contains not only the information of a wire frame model but also other information such as the connection of two surfaces. A surface model of an object can be used to determine the cutter path, whereas a wire frame usually cannot. In such surface modelling systems, a user may input the vertices and edges of a work piece in a manner that outlines or bounds one face at a time. Surface modelling systems offer better graphic interaction, although the models are more difficult to create than wire frame models. Some time an intermediate model must be created.

Although a surface modelling scheme is better than a wire frame modelling scheme, it still has some drawbacks. It does not provide information about the topology of the entities, such as the concept of the component's inside and outside. As a result with surface models, a user may still not be able to distinguish the interior and exterior of an object on the monitor. As a matter of fact, a surface modelling system may not guarantee that the user has designed a realizable object that is the collection of surfaces may not define a physical port.

9.1 SURFACE MODELLING

A higher level of sophistication in geometric modelling is the surface model which can overcome many of the ambiguities of wire frame models. Surface models are created by connecting various surface elements to represent part geometry. A surface model can be built by defining the surfaces on the wire frame model. They hold the description of an object in terms of points, edges and faces between the edges. Surface modelling systems are able to calculate surface intersections and surface areas and some systems are capable of producing shaded images and removing hidden lines automatically.

Surface models may be constructed using a large variety of surface features often provided by many CAD/CAM systems. The plane is the most basic feature to represent a surface element. More complex shapes can be defined by tabulated cylinders, ruled surfaces of revolution, sweep surfaces and fillet surfaces.

The most general and complex surface representation are sculptured surfaces which are also known as **curve-mesh surfaces**, **B-surfaces** and **cubic-patch surfaces**. A sculptured surface may be considered as a surface produced from combining two families of curves that intersect one another in a criss-cross manner, creating a network of interconnected patches.

Surface modelling can be used to optimize component weight for both geometry and stress. A model composed of surface elements can easily be made to appear as if it is solid through the technique of hidden line removal, although surface models represent only an envelope of part geometry in computer memory. As surface models do not actually represent the solid nature of parts because they contain no information describing what lies within the part interior, they cannot be used as a base for engineering analysis programs such as finite element and modal analysis for stress and strain predictions. These programs often require such properties as weight, volume and moments of inertia which cannot be derived from a surface model alone. Interpretation of a surface model can still be ambiguous, because the model may represent a totally solid object or is merely a thin-walled structure made, for example, of sheet metal.

Surface models require more computational time as compared to wire frame, more skill in their construction and use difficult to interpret unless viewed with hidden lines removed. The surface models may be interpreted in an ambiguous way or erroneous way.

9.2 SURFACE ENTITIES

A brief discussion of various surface entities that are needed to construct a surface model is given hereinafter.

9.2.1 Plane Surface

A plane surface is the simplest surface that is defined by three non-coincident points or its variation. It is the most basic surface in engineering design. Two plane surfaces are shown in Fig. 9.1.

9.2.2 Ruled (Lofted) Surface

A ruled (lofted) surface can be defined as a linear interpolation between two different generating or edge

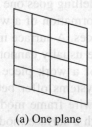

(a) One plane

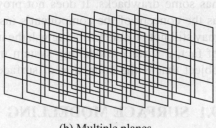

(b) Multiple planes

Fig. 9.1 Plane surface.

curves. Informally speaking, the straight lines connecting the two rails (general curves) form the surface. This is a special case of using interpolation between two general curves. Figure 9.2 shows two space curves "connected" by straight lines that form a surface.

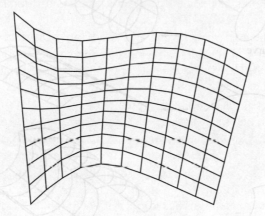

Fig. 9.2 Ruled surface.

9.2.3 Surface of Revolution

This surface is generated by rotating a planar curve in space about an axis at a certain angle. This surface is particularly useful when modelling turned parts or parts which possess axial symmetry. Commercial CAD systems will often provide the facility to revolve multiple curves in a profile with a single command. In Fig. 9.3, a straight line revolves about an axis and forms a revolution surface.

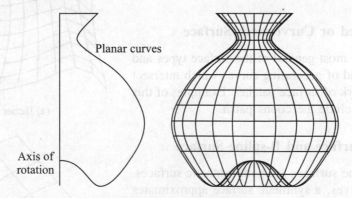

Planar curves

Axis of rotation

Fig. 9.3 Surface of revolution.

9.2.4 Tabulated Cylinder

This surface is generated by sweeping a planar curve in space in a certain direction at a certain distance. In Fig. 9.4, a straight line sweeps along a path that is a circle and forms a surface. The straight line is called a **generatrix** and the circle a **directrix**.

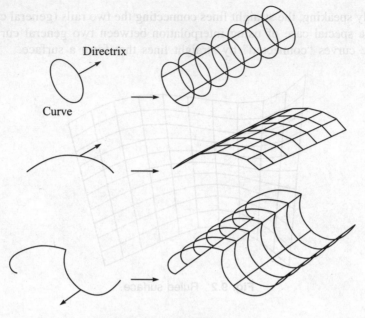

Fig. 9.4 Tabulated cylinder.

9.2.5 Swept Surface

This surface in a sense is an extension of the surface of revolution where the defining curve is swept along an arbitrary spine curve instead of a circular arc.

9.2.6 Sculptured or Curve-mesh Surface

This is among the most general of the surface types and is defined by a grid of generating curves which intersect to form a patchwork of surface patches. Examples of this type of surface include the coons patch.

9.2.7 Bezier Surface and B-spline Surface

Bezier and B-spline surfaces are both synthetic surfaces. Like synthetic curves, a synthetic surface approximates given input data (an array of given points in 3D space). Bezier and B-spline surfaces are general surfaces that permit twists and kinks. The difference between them, also similar to the case of curves, is that the local control is possible for the B-spline surface but not for the Bezier surface. Figure 9.5 shows a B-spline surface.

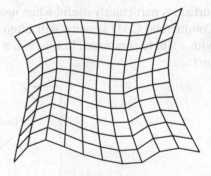

(a) Bezier surface

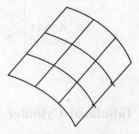

(b) B-spline surface

Fig. 9.5 Bezier surface and B-spline surface.

9.2.8 Coons Patch

The above surfaces are used with either open boundaries or given data points. The coons patch is used to create surface using curves that form closed boundaries as shown in Fig. 9.6.

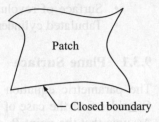

Fig. 9.6 Coons patch.

9.2.9 Fillet Surface

This is a B-spline surface that blends two surfaces together as shown in Fig. 9.7. The two original surfaces may or may not be trimmed.

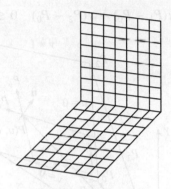

Fig. 9.7 Fillet surface.

9.2.10 Offset Surface

Existing surfaces can be offset to create new ones identical in shape but may have different dimensions. It is a useful surface to use to speed up the surface construction. For example, to create hollow cylinder, the outer or inner cylinder can be created using a cylinder command and other one can be created by an offset command as shown in Fig. 9.8.

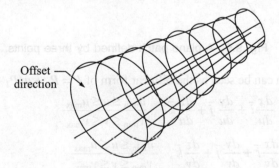

Fig. 9.8 Offset surface.

9.3 PARAMETRIC REPRESENTATION OF ANALYTIC SURFACES

- Plane surface
- Ruled surface

- Surface of revolution
- Tabulated cylinder

9.3.1 Plane Surface

The parametric equation of a plane can take different forms depending on the given data. Consider first the case of a plane defined by three points P_0, P_1, and P_2 as shown in Fig. 9.9. Assume that the point P_0 defines $u = 0$ and $v = 0$ and the vectors $(P_1 - P_0)$ and $(P_2 - P_0)$ define the u and v directions respectively. Assume also that domains for u and v are [0,1]. The position of any point P on the plane can be now written as:

$$P(u, v) = P_0 + u(P_1 - P_0) + v(P_2 - P_0) \quad 0 \leq u \leq 1, 0, v \leq 1 \quad (9.1)$$

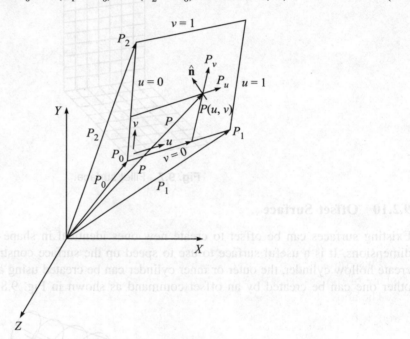

Fig. 9.9 A plane patch defined by three points.

The above equation can be seen as the bilinear form of $P = P_1 + u(P_1 - P_2)\, 0 \leq u \leq 1$. Utilizing

Eqs. $p_u(u, v) = \dfrac{dp}{du} = \dfrac{dx}{du}\bar{i} + \dfrac{dy}{du}\bar{j} + \dfrac{dz}{du}\bar{k}$, $\quad \begin{matrix} u_{min} \leq u \leq u_{max} \\ v_{min} \leq v \leq v_{max} \end{matrix}$

and $p_v(u, v) = \dfrac{dp}{dv} = \dfrac{dx}{dv}\bar{i} + \dfrac{dy}{dv}\bar{j} + \dfrac{dz}{dv}\bar{k}$, $\quad \begin{matrix} u_{min} \leq u \leq u_{max} \\ v_{min} \leq v \leq v_{max} \end{matrix}$

$$P_u(u, v) = P_1 - P_0, \quad P_v(u, v) = P_2 - P_0, \quad 0 \leq u \leq 1, \quad 0 \leq v \leq 1 \quad (9.2)$$

and the surface normal is:

$$\hat{n}(u, v) = \frac{(P_1 - P_0) \times (P_2 - P_0)}{\left|(P_1 - P_0) \times (P_2 - P_0)\right|}, \; 0 \leq u \leq 1, 0 \leq v \leq 1 \quad (9.3)$$

which is constant for any point on the plane. As for the curvature of the plane, it is equal to zero because all the second fundamental coefficients of the plane are zero.

Another case of constructing a plane surface is when the surface passes through a point P_0 and contains two directions defined by the unit vectors **r** and **s** as shown in Fig. 9.10. Similar to the above case, the plane equation can be written as

$$P(u, v) = P_0 + uL_u\hat{\mathbf{r}} + vL_u\hat{\mathbf{s}}, \quad 0 \leq u \leq 1, 0 \leq v \leq 1 \tag{9.4}$$

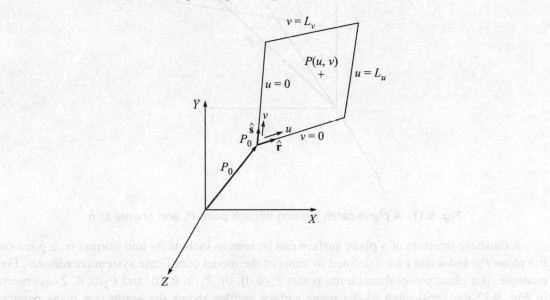

Fig. 9.10 A plane patch defined by a point and two directions.

This equation is also considered as the bilinear form of $P = P_1 + L\hat{\mathbf{n}}$. The equation assumes a plane of dimensions L_u and L_v that may be set to unity.

The above two cases can be combined to provide the equation of a plane surface that passes through two points P_0 and P_1 and is parallel to the unit vector $\hat{\mathbf{r}}$. In this case, we can write

$$P(u, v) = P_0 + u(P_1 - P_0) + vL_v\hat{\mathbf{r}}, \quad 0 \leq u \leq 1, 0 \leq v \leq 1 \tag{9.5}$$

The last case to be considered is for a plane that passes though a point P_0 and is perpendicular to a given direction $\hat{\mathbf{n}}$. Figure 9.11 shows this case. The vector $\hat{\mathbf{n}}$ is normal to any vector in the plane. Thus,

$$(P - P_0)\cdot\hat{\mathbf{n}} = 0 \tag{9.6}$$

which is non-parametric equation of the plane surface. A parametric equation can be developed by using Eq. (9.6) to generate two points on the surface, which can be used with P_0, in Eq. (9.1). Planes that are perpendicular to a current working coordinate system are special cases of Eq. (9.6). For example, in the case of plane perpendicular to the x axis, $\overline{n}$ is (1,0,0) and the plane equation is $x = x_0$.

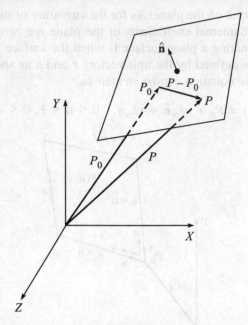

Fig. 9.11   A plane patch passing through point P_0 and normal to $\hat{\mathbf{n}}$.

A database structure of a plane surface can be seen to include its unit normal $\hat{\mathbf{n}}$, a point on the plane P_0, and u and v axes defined in terms of the model coordinate system coordinates. For example, if a plane passes through the points $P_0(0, 0, 0)$, $P_1(2, 0, 0)$, and $P_2(0, 0, 2)$ as shown in Fig. 9.12 the verification of the plane surface entitles shows the entity is a plane passing through $P_0(0, 0, 0)$ and has a unit normal of $(0, -1, 0)$. In addition, the u and v axes are defined by the coordinates $(1, 0, 0)$ and $(0, 0, 1)$ respectively.

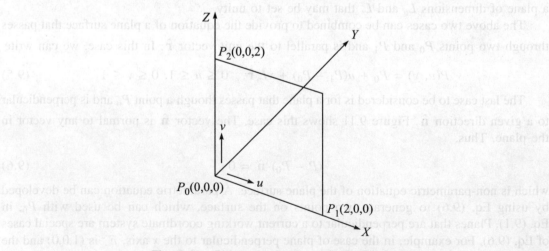

Fig. 9.12   Plane passing through P_0, P_1, P_2.

9.3.2 Ruled Surface

A ruled surface is generated by joining corresponding on two space curves $G(u)$ and $Q(u)$ by straight lines (also called rulings or generators), as shown in Fig. 9.13. The main characteristic of a ruled surface is that there is that at least one straight line passing through the point $P(u, v)$ and lying entirely in the surface. In addition, every developable surface is a ruled surface. Cones and cylinders are examples of ruled surfaces.

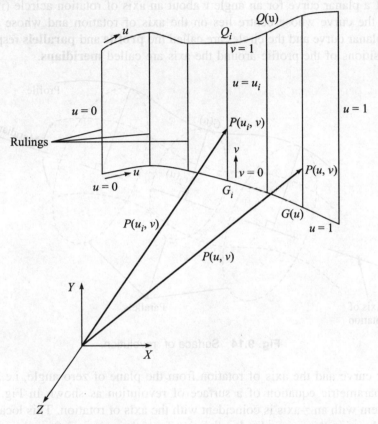

Fig. 9.13 Paramedic representation of a ruled surface.

To develop the parametric equation of a ruled surface, consider the ruling $u = u_i$ joining points G_i and Q_i on two space curves $G(u)$ and $Q(u)$ respectively. Using $P = P_1 + u(P_1 - P_2)$ the equation of the ruling becomes

$$P(u_i, v) = G_i + v(Q_i - G_i) \qquad (9.7)$$

where v is the parameter along the ruling. Generalizing Eq. (9.8) for any ruling, the parametric equation of a ruled surface defined by two space curves is:

$$P(u, v) = G(u) + v[Q(u) - G(u)] = (1 - v)\, G(u) + vQ(u),\ 0 \le u \le 1,\ 0 \le v \le 1 \qquad (9.8)$$

Holding the u value constant in the above equation produces the rulings given by Eq. (9.8) in the v direction of the surface, while holding the v value constant yields curves in the u

direction, which are a linear blend of the two space curves. Infact, of $G(u)$ and $Q(u)$ are $P(u, 0)$ and $P(u, l)$ respectively. Therefore, the closer the value of v to zero, the greater the influence of $G(u)$ and the less the influence of $Q(u)$ on the v = constant curve. Similarly, the influence of $Q(u)$ on the ruled surface geometry increases when the v value approaches unity.

9.3.3 Surface of Revolution

The rotation of a planar curve for an angle v about an axis of rotation acircle (if $v = 360°$) for each point on the curve whose centre lies on the axis of rotation and whose radius $r(u)$ is variable. The planar curve and the circles are called the **profile** and **parallels** respectively while the various positions of the profile around the axis are called **meridians**.

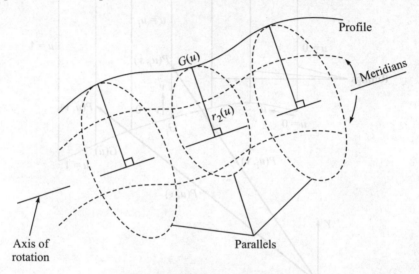

Fig. 9.14 Surface of revolution.

The planar curve and the axis of rotation from the plane of zero angle, i.e., $v = 0$ is used to derive the parametric equation of a surface of revolution as shown in Fig. 9.14. A local coordinate system with an z-axis is coincident with the axis of rotation. This local system L can be created by choosing the perpendicular direction from the point $u = 0$ on the profile as the X_L axis and the intersection point between X_L and Z_L (as the origin of the local system). The Y_L axis is automatically determined by the right-hand rule. Consider a point $G(u) = P(u, 0)$ on the profile that rotates an angle v about Z_L when the profile rotates the same angle. The parametric equation of the surface of rotation can be written as:

$$P(u, v) = r_z(u) \cos v \; n_1 + r_z(u) \sin v \; n_2 + Z_L(u) \; n_3, \; 0 < 1, \; 0 < v < 2\pi \qquad (9.9)$$

Choose $Z_L(u) = u$ for each point on the profile, equation gives the local coordinates (X_L, Y_L, Z_L) of a point $P(u, v)$ as $[r_z(u) \cos v, r_z(u) \sin v, u]$. The local coordinates are transformed to MCS-Model Coordinate System coordinates before displaying the surface, where the rotation matrix is formed from $\hat{n}_1$, $\hat{n}_2$ and $\hat{n}_3$ and the position of the origin of the local system is given by P_L, see Fig. 9.15.

The database of a surface of revolution must include its profile, axis of rotation and the angle of rotation as starting and ending angles. Whenever the user requests the display of the surface with a mesh size $m \times n$, the u range is divided equally into $(m - 1)$ divisions and m values of u are obtained. Similarly, the v range is divided equally into n values and is used to generate points on the surface.

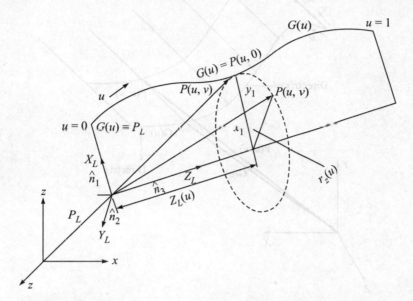

Fig. 9.15 Parametric representation of a surface of revolution.

9.3.4 Tabulated Cylinder

A tabulated cylinder has been defined as a surface that results from translating a space planar curve along a given direction. It can also be defined as a surface that is generated by moving a straight line (called generatix) along a given planar curve (called directrix). The straight line always stays parallel to a fixed given vector that defines the v direction of the cylinder as shown in Fig. 9.16. The planar curve $G(u)$ can be any wire frame entities. The position vector of any point $P(u, v)$ on the surface can be written as

$$P(u, v) = G(u) + vn\hat{\mathbf{n}}_v \quad 0 \le u \le u_{max}, 0 \le v \le v_{max} \tag{9.10}$$

From a user point of view, $G(u)$ is the desired curve the user digitizes to form the cylinder, v is the cylinder length, and $\hat{\mathbf{n}}_v$ is the cylinder axis. The representation of $G(u)$ is already available in the database at the time of creating it. The cylinder length v is input in the form of lower and higher bounds where the difference between them gives the length. A zero value of the lower bound indicates the plane of the directrix. The user inputs the cylinder axis as two points that are used to determine $\hat{\mathbf{n}}_v$, which is the unit vector along the axis.

As seen from Eq. (9.10), the database of a tabulated cylinder includes its directrix, the unit vector $\hat{\mathbf{n}}_v$, and the lower and upper of the cylinder. The display of a tabulated cylinder with a mesh $m \times n$ follows the same approach as discussed with surfaces of revolution.

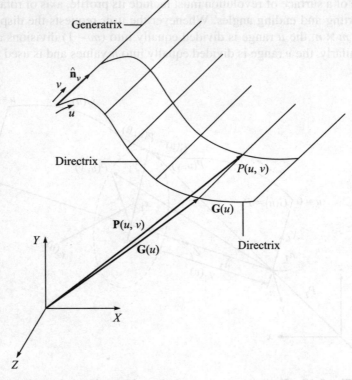

Fig. 9.16 Paramedic representation of a tabulated cylinder.

9.4 SURFACE REPRESENTATIONS

Similar with the discussion of curves, we start our discussion of surfaces with their representation equation. Recalling that a general curve can be defined by either an implicit or an explicit equation. A general surface can also be defined by implicit and explicit equations as follows:

9.4.1 Implicit Equation

$$F(x, y, z) = 0 \qquad (9.11)$$

Its geometric meaning is that the locus of the points that satisfy the constraint equation defines the surface.

In parametric form, a surface may be represented as:

$$x = x(u, v)$$
$$y = y(u, v)$$
$$z = z(u, v)$$

where x, y and z are suitable functions of two parameters u and v. For example, the parametric representation of the surface of a sphere whose centre is at the origin of coordinates and of radius R is:

$$X = x(0, \phi) = R \sin \phi \cos \theta$$

$$Y = y\,(0,\,\phi) = R\,\sin\,\phi\,\sin\,\theta$$
$$Z = z\,(0,\,\phi) = R\,\cos\,\phi$$

9.4.2 Explicit Equation

$$V = (x,\,y,\,z)^T = [x,\,y,\,F(x,\,y)]^T \qquad (9.12)$$

where V is the position vector of a variable point on the surface. In this equation, we directly define the variable point coordinates x, y, z. The z coordinates of the position vector of the variable points are defined by x, y through a suitable function $[f(x,\,y)]$ as shown in Fig. 9.17.

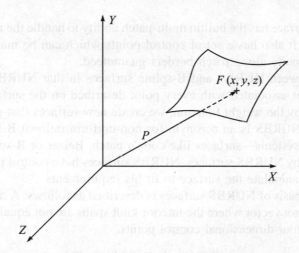

Fig. 9.17   Explicit equation in surface representation.

Comparing Eqs. (9.11) and (9.12) with those for the curves we can see the difference between a space curve and a surface mathematically. The points on a space curve have one degree of freedom, the points on a surface have two.

The general form of the function $f(x,\,y)$ for a surface to pass through all the given data points is a polynomial with respect to x and y. Usually, an arbitrary surface is defined by an x-y grid of size $(p + 1)\,(q + 1)$ points.

9.5   PARAMETRIC EQUATION OF A SURFACE

There are no extra parameters in Eqs. (9.11) and (9.12), and the surface representation equations of this form are called **non-parametric equations**. Equations that utilize parameters are called **parametric equations**. Because the points on the surface have two degrees of freedom, it is natural to write the parametric representation of a surface as:

$$V(s,\,t) = (x,\,y,\,z)^T = [x\,(s,\,t),\,y\,(s,\,t),\,z\,(s,\,t)]^T$$
$$S_{\min} \le S \le S_{\max},\, t_{\min} \le t \le t_{\text{tmax}}$$

where x, y and z are functions of the two parameters S and t.

9.6 SURFACE REPRESENTATION SCHEME

Two important surface representation schemes exist that extend the control of shape beyond movement of control vertices. These are NURBS (Non-Uniform Rational B-splines) and B-splines. In the case of NURBS, a local vertex is extended to a four-dimensional coordinate, the extra parameter being a weight that allows a subtle form of control which is different in effect to moving a control vertex. In the simplest form of B-spline control, two global parameters (bias and tension) are introduced which effect the whole curve.

9.6.1 NURBS

In general a B-spline surface has the builtin multi-patch ability to handle the interpolation through a mesh of data points. It also has a set of control points which can be manipulated to edit the surface, with the continuity along patch borders guaranteed.

The difference between NURBS and B-spline surfaces in that NURBS are rational. This means there is a weight associated with every point described on the surface. By dividing the surface point function by the weight function, we create new surfaces that are not possible with non-rational surfaces. NURBS is an acronym for non-uniform rational B-spline. Other widely used surfaces in CAD systems—surfaces like coon's patch, Beizer or B-spline surfaces can all be exactly represented by NURBS surfaces. NURBS surfaces have control points that a designer may use to pull and manipulate the surface to fit his requirements.

The mathematical basis of NURBS surfaces is described as follows: A non-uniform B-spline curve is defined on a knot vector where the interior knot spans are not equal. A rational B-spline is defined by a set of four-dimensional control points.

$$P = (w_i \, x_i, \, w_i \, y_i, \, w_i \, z_i, \, w_i)$$

The perspective map of such a curve in 3D space is called a **rational B-spline curve**.

$$P(u) = H \left[\sum_{i=0}^{n} P_i^w N_{i,k}(u) \right]$$

$$= \frac{\displaystyle\sum_{i=0}^{n} P_i W_i N_{i,k}(u)}{\displaystyle\sum_{i=0}^{n} W_i N_i k(u)}$$

$$= \sum_{i=0}^{n} P_i R_i k(u)$$

where

$$R_i k(u) = \frac{N_i k(u) \, W_i}{\displaystyle\sum_{i=0}^{n} N_j k(u) \, W_j}$$

Rational B-splines have the same analytical and geometric properties as non-rational B-splines. The W_i associated with each control point is called a **weight,** and can be viewed as an extra shape parameter. It affects the curve only locally and can be interpreted geometrically as a coupline factor. The curve is pulled towards a control point if W_i increases.

9.6.2 B-splines

B-splines are obtained from B-splines by introducing two new degrees of freedom: bias and tension, which can be applied uniformly or non-uniformly.

A singular disadvantage of B-spline is that it does not pass through any control point.

9.7 TECHNIQUES FOR SURFACE MODELLING

The general form of the parametric representation of a surface was observed to be:

$$x = x(u, v) \qquad y = y(u, v) \qquad z = z(u, v) \tag{9.13}$$

and some of the more important polynomial representations of curves were described. The most widely used techniques for the modelling of free-form surfaces are not surprisingly, extensions into the second parametric dimension of the polynomial curve techniques, and the resultant surface types share many of the characteristics of the curve forms.

9.7.1 The Surface Patch

Just as the curve segment is the fundamental building block for curve entities, so the patch is the fundamental building block for surfaces. Also just as the parametric variable u varies monotonically along the segment so the two variables u and v vary across the patch—the patch may be termed as **biparametric**. The parametric variables often lie in the range 0 to 1, although other parametric intervals may be used if appropriate. Fixing the value of one of the parametric variables results in a curve on the patch in terms of the other variable (known as an **isoparametric curve**). The result of doing this for a variety of values of u and v is an intersecting mesh of curves on the patch. Figure 9.18, for example, shows a surface with u and v in the range 0 to 1 with curves at intervals of u and v of 0.1.

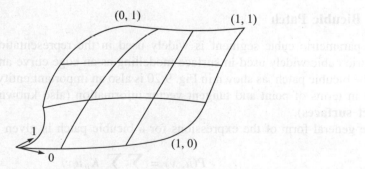

Fig. 9.18 Surface patch.

9.7.2 The Coons Patch

The more general surface forms—the sculptured surfaces—often involve interpolation across an intersecting mesh of curves that in effect comprise a rectangular grid of patches, each bounded by four boundary curves. A variety of techniques have been developed for interpolating between such boundary curves of which perhaps the simplest is the linearly blended coons patch. Figure 9.19 shows surface patches defined using this formulation. Linear blending has limitations, and higher order blending functions such as cubics are used for formulations that allow tangency continuity between adjacent patches. The term *coons patch* is used generically to include other patches which are blendings of arbitrary boundaries.

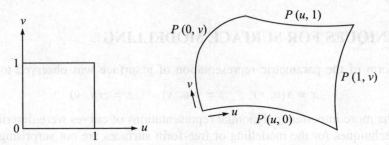

Fig. 9.19 Coons patch.

The linearly blended coons patch

Let us first consider the case of a bilinearly blended coons patch which interpolates to the four boundary curves of arbitrary parametric form $C_i(u)$ and $D_j(v)$ (see Fig. 9.19) with the linear blending functions.

$$F(t) = 1 - t \qquad g(t) = t \qquad (9.14)$$

Simply applying the blending functions to the curves would give incorrect results, especially at the corners of the patch and, therefore, functions of the corner points P_{ij} must also be introduced into the equation:

$$P(u,v) = C_0(u)\, f(v) + C_1(u)\, g(v) + D_0(v)\, f(u) + D_1(v)\, g(u) - P_{00}\, f(u)\, f(v)$$
$$- P_{01} F(u)\, g(v) - P_{10}\, g(u)\, f(v) - P_{11}\, g(u)\, g(v) \qquad (9.15)$$

9.7.3 Bicubic Patch

As the parametric cubic segment is widely used in the representation of curves, so is the parametric cubic widely used in surface modelling as an edge curve and the equivalent surface form 'the bicubic patch' as shown in Fig. 9.20 is also an important entity for surface descriptions defined in terms of point and tangent vector information (also known as **tensor** or **cartesian product surfaces**).

The general form of the expressions for a bicubic patch is given by

$$P(u,\ v) = \sum_{i=0}^{3} \sum_{j=0}^{3} K_{ij} u_i v_j \qquad (9.16)$$

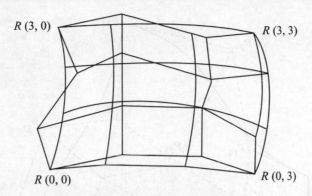

Fig. 9.20 Bicubic patch.

Equation (9.16) is a vector equation with 16 unknown parameters K_{ij}. These might be found by Lagrange interpolation through points, in which case 16 points, for example, in a 4×4 grid would be required. Hermite interpolation through points and tangent vectors may also be used, as for curves but in this case simply using the four corner points and the tangent vectors at the corners (two at each point) does not provide sufficient constraints. The remaining four boundary conditions are normally supplied by the cross-derivative vectors $\partial^2 \rho / \partial u \partial v$ at each corner. These partial derivatives are commonly known as the **twist vectors** because they are said to be a measure of 'twist' in the surface. However, although mathematically convenient, they are not a quantity for which the designer will generally have much feel, and therefore implementations of bicubic patches in the Hermite basis generally have to interpolate the twist vectors in some fashion or to assume that they are zero and both of these routes may lead to unsatisfactory results.

9.7.4 Bezier Surfaces

In the same way that the Bezier curve uses the more tractable control polygon in place of control points and tangent vectors so too does the Bezier surface formulation use a characteristic polygon (also called a **characteristic mesh**) in place of points and tangent and twist vectors. Points on the Bezier surface are given by simple extension of the expression for a curve:

$$P(u, v) = \sum_{i=0}^{m} \sum_{j=0}^{n} B_{i,m}(u) \; B_{j,n}(v) \; P_{ij} u, \; v \in [0, 1] \qquad (9.17)$$

where P_{ij} are the vertices of the characteristic polygon and $B_{i,m}$ and $B_{j,n}$ are the blending functions as defined for curves.

Examples of Bezier surfaces and their respective control polygons are shown in Fig. 9.21. These show that the surfaces share a number of characteristics with Bezier curves, in particular, that the surfaces pass through the corner points of the characteristic polygon only and have edge curves that are tangential to the edges of the characteristic polygon at the corner points. Furthermore the surfaces are variation diminishing and have a convex hull property.

The Bezier surface is superior to a bicubic surface in that it does not require tangent or twist vectors to define the surface flower, its main limitation is the lock of local control changing one

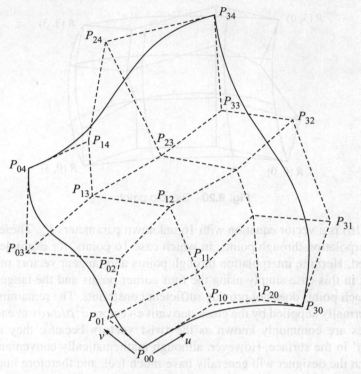

Fig. 9.21 Bezier surface.

or more control point affects the shape of the whole surface. Therefore, the user cannot selectively change the shape of part of the surface.

9.7.5 B-spline Surfaces

The limitations of the Bezier surface are largely overcome by the B-spline surface formulation. This formulation shares certain characteristics with the Bezier scheme and also with the B-spline curve scheme. The surface again approximates a characteristic polygon (as shown in Fig. 9.21) and (generally) passes through the corner points of the polygon. Where its edges are tangential to the edges of the polygon (The word generally is used here because B-splines may also be defined using periodic blending functions that are all the same shape. Figure 9.22 shows the effect of using a periodic blending function in combination with a closed control polygon, in which control points on two edges of a mesh are connected). Just as the control point of a curve only influences the shape of the curve over a limited range of the parametric variable u so a control point of the surface influences the surface only over a limited (rectangular) portion of the parametric space of variables u and v. The extent of the influence of a control point and hence the degree of approximation of the control polygon may be varied by varying the order of the B-spline blending curves employed.

The expression for the B-spline surfaces is again a straightforward extension of the curve:

$$P(u, v) = \sum_{i=0}^{m} \sum_{j=0}^{n} N_{i,k}(u) N_{j,l}(v) P_{ij} \tag{9.18}$$

where P_{ij} are the vertices of the defining polygon and N_i, K and $N_{j,l}$ are blending functions of the same form as those for B-spline curves.

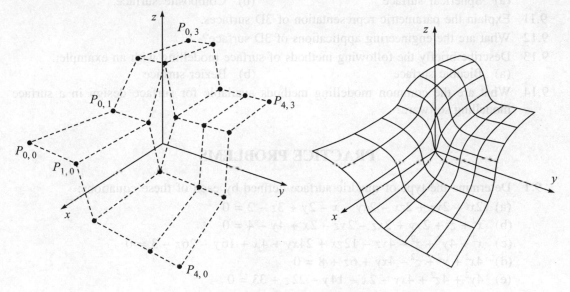

Fig. 9.22 B-spline surface.

9.8 LIMITATIONS

(a) It does not give information regarding inner details of a component.
(b) It does not calculate mass properties like MI, CG and volume etc.

REVIEW QUESTIONS

9.1   Discuss the importance of surface modelling in computer-aided graphics and design.
9.2   Give a classification of the different surfaces that can be used in geometric modelling applications.
9.3   What are the various surface entities that are needed to construct a surface model?
9.4   Explain the coons and Bezier surfaces. What are the differences and applications for which these are used?
9.5   Describe the parametric equation of a composite surface.
9.6   Discuss the modelling guidelines to be followed by the user while constructing a surface model on a CAD/CAM system.
9.7   Briefly discuss about the composite surface and Bezier surface.
9.8   Derive a parametric equation for a cubic Bezier and cubic B-spline surface whose boundary curves and cross boundary derivatives are given.
9.9   What do you mean by blending function? Explain reparametrisation of a surface.

9.10 Write the mathematical representation, application and limitations of the following surfaces:
    (a) Spherical surface   (b) Composite surface.

9.11 Explain the parametric representation of 3D surfaces.

9.12 What are the engineering applications of 3D surface?

9.13 Describe briefly the following methods of surface modelling with an example:.
    (a) Bicubic surface   (b) Bezier surface

9.14 What are the common modelling methods available for surface design in a surface modelling software?

PRACTICE PROBLEMS

9.1 Determine the type of quadric surface defined by each of these equations:
    (a) $2x^2 - 2yz + 2zx - 2xy - x - 2y + 3z - 2 = 0$
    (b) $x^2 + z^2 + 2xy + 2xz - 2yz - 2x + 4y - 4 = 0$
    (c) $x^2 + 4y^2 + z^2 - 4yz - 12zx + 24xy + 4x + 16y - 26z - 3 = 0$
    (d) $4x^2 + y^2 + z^2 - 4xy + 6z + 8 = 0$
    (e) $4y^2 + 4z^2 + 4xy - 2x - 14y - 22z + 33 = 0$

9.2 Using a parametric representation, generate a hyperbolic parabolic.

9.3 Draw the individual component mapped surfaces, i.e., Q (u, v, x), Q (u, v, y), Q (u, v, z), (see Fig. 6.24), and the complete surface Q (x, y, z), for a point, a line and a plane.

9.4 For the ruled surface with edge curves defined by

$$z = 0 \quad y = 0.25 \cos (2\pi x) \quad 0 \le x \le 1$$
$$z = 1 \quad y = 0.25 \sin (2\pi x) \quad 0 \le x \le 1$$

Determine the surface point at Q (0.5, 0.5).

9.5 Construct the bilinear surface corresponding to the four points P (0, 0) = [0.25   0], $P(1, 0)$ = [0.75   0], $P(0,1)$ = [0.75   0.9], $P(0, 0)$ = [0.28   0.8].

9.6 Construct the bilinear surfaces that correspond to the four points defined by the opposite ends of the diameters of two unit circles lying in parallel planes 1 unit apart as they rotate in opposite directions through 180° at 15° intervals.

9.7 Show by example that a planar Coons bicubic surface results when the position, tangent and twist vectors all lie in the same plane.

9.8 Find the matrix $[P]$ that yields a cylindrical Coons bicubic surface patch.

9.9 Find the matrix $[P]$ that yields a general ruled Coons bicubic surface patch.

9.10 Determine the surface point at $u = w = 0.5$ for the Coons bicubic surface with the following position. $P(0, 0)$ = [−100   0   100], $P(0, 1)$ = [100   −100   100], $P(1, 1)$ = [−100   0   −100], $P(1, 0)$ = [−100   −100   −100]; u-tangent $P^{1,0}(0, 0)$ = [10   10   0], $P^{1,0}(0, 1)$ = [−1   −1   0], $P^{1,0}(1, 1)$ = [−1   1   0], $P^{1,0}(1,0)$ = [1   −1   0]; w-tangent $P^{0,1}(0, 0)$ = [0   −10   −10], $P^{0,1}(0, 1)$ = [0   1   −1], $P^{0,1}(1, 1)$ = [0   1   1], $P^{0,1}(1, 0)$ = [0   1   1]; and twist vectors given by $P^{1,1}(0, 0)$ = [0   0   0], $P^{1,1}(0, 1)$ = [0.1   0.1   0.1],

$P^{1,1}(1, 1) = [0 \;\; 0 \;\; 0]$, $P^{1,1}(1, 0) = [-0.1 \;\; -0.1 \;\; -0.1]$. Systematically vary the tangent and twist vectors, and compare the results.

9.11 Extend the analysis of parabolically blended curves to parabolically blended surfaces (end conditions for periodic B-spline curves). What is the continuity of individual patches? What degree of continuity can be maintained across patch boundaries? What is the extend of the influence of moving a single point on a surface built up of multiple patches?

[**Hint:** Use parabolically blended curves for four boundary curves and a Coons linear surface style formulation with blending functions given by $F_1(t) = 1 - 3t^2 + 2t^3$, $F_2(t) = 3t^2 - 2t^3$. Note that the resulting surface is controlled by the four corner points and by two additional pseudo vertices at each corner that control the tangent vectors at the ends of the boundary curves.

9.12 Show for a 4×4 bicubic Bezier surface that for $u - w = 0$,

$$Q_{uw} (0, 0) = 9[(B_1, 1 - B_0, 1) - (B_1, 0 - B_0, 0)]$$

Compare the result to the appropriate term in Eq. (6-60).

9.13 Show that neither a hyperbolic parabolic nor a hyperbola of one sheet is a developable surface.

9.14 Compute the fourth order B-spline surface for the 4×5 defining polygon net given by

$B_{1,1}$ [0 0 100]	$B_{2,1}$ [25 0 150]	$B_{3,1}$ [50 0 100]
	$B_{4,1}$ [75 0 50]	$B_{5,1}$ [100 0 100]
$B_{1,2}$ [0 33 150]	$B_{2,2}$ [25 33 200]	$B_{3,2}$ [50 33 100]
	$B_{4,2}$ [75 33 50]	$B_{5,2}$ [100 33 50]
$B_{1,3}$ [0 66 50]	$B_{2,3}$ [25 66 25]	$B_{3,3}$ [50 66 100]
	$B_{4,3}$ [75 66 150]	$B_{5,3}$ [100 66 150]
$B_{1,4}$ [0 100 100]	$B_{2,4}$ [25 100 50]	$B_{3,4}$ [50 100 100]
	$B_{4,4}$ [75 100 150]	$B_{5,4}$ [100 100 100]

Also

(a) Determine by hand calculation the point in the centre of an 11×15 parametric surface net.

(b) Write a program to compute the two diagonal parametric lines $u = 2w$ and $u = 2(1 - w)$.

(c) List the points from (0, 0) to (2, 1) for $u = 2w$ and from (0, 1) to (2, 0) for $u = 2(1 - w)$.

(d) List 11 values for each diagonal.

(e) Compute and display the 11×15 parametric surfaces. List 15 u and 11 w values.

Use an appropriate viewing transformation.

Suggestion: Output the results to a file in either standard graphics format (.sgf) or standard polygon format (.spf) and use the three-dimensional manipulator developed in a three-dimensional manipulator program that accepts input data in the form of either an .sgf or .spf file to display the surface.

SOLID MODELLING

10

INTRODUCTION

The wire frame and surface modelling approaches have limited engineering applications. Solid modelling finds wide applications that cut across functional boundaries, such as the use of solid models with finite element analysis and fluid flow analysis in the conceptual design of products. Furthermore solid models can be used to evaluate the size, shape and weight of products early during the conceptual design phase.

In order to develop a product right from the conceptual stage designers should have at their disposal, efficient means of representing three-dimensional objects. Solid modelling is an efficient tool for the real 3D representation of an object. This method enables the designer to visualize an object in terms of its basic geometry primitives joined together by Boolen operations.

A solid model is an unambiguous and informationally complete mathematical representation of the shape of a physical object.

In a solid modelling system, objects are defined directly by primitive shapes called **building blocks**, instead of the surfaces, lines and points used in wire frame and surface modelling. This means that an independent surface, line or point does not have any meaning in solid modelling.

Solid modelling is a natural extension from the use of essentially 'one-dimensional' entities (curves) or 'two-dimensional' entities (surfaces) to the modelling of shapes using three-dimensional solids.

10.1 SOLID MODELLING

The highest level of sophistication in geometric modelling is 3D modelling. It provides the user with complete information about the model. Solid models often resemble wire frame or surface models with hidden lines removed. However, the difference is that wire frames have severe limitations when used as mathematical models of the parts they represent, because for modelling purposes parts need to be represented in the computer in a way that enables mass properties such as centre of gravity or weight to be associated with the part. The mass properties of part models are important for the prediction of such properties as weight, stability, moments of inertia, volume, etc., of finished products containing the parts. Solid models are better in the sense that they allow the solid nature of an object to be defined in the computer and thus help to calculate mass properties. Also, sections can readily be cut through solid models to reveal internal details. Solid models are recorded in the computer mathematically as volumes bounded by surfaces rather than as stick figure structures. As a result, it is possible to calculate mass properties of the parts which are often required for engineering analysis such as finite element methods and kinematic studies for interference checking.

To give an example such as a cylinder, a wire frame cylinder is defined in a computer as two circles connected by two line segments, whereas the solid model of a cylinder is represented as 3D object that contains a volume. If it is to determine the volume of the wire frame cylinder, the formula for cylinder volume can be used. However, this formula is only valid for cylinders. For other types of volumes, different formula would have to be programmed into the computer. Obviously it would be difficult to calculate volumes in this manner for complex shapes because each volume would have to be determined using a different formula. Another drawback is that the computer programmer would have to know the shape of the part in advance before the appropriate formula could be used to program the computer. On the other hand, to calculate the volume of a solid model, it is much easier because the computer can employ a general numerical integration that can be applied to solids of any shape. For example, the volume of a solid may be determined by dividing one face of the solid into a rectangular grid, then tracing the rectangular shapes back through the solid until it reaches the back edges of the model. The volume of the part is then obtained by adding the volumes of all the parallelopipeds. This technique is commonly used in solid modelling, and is often referred to as the *approximating sum integration method*.

Despite the benefits of using solid models, they have their disadvantages. For onething, they use much more memory to represent a model than either wire frame or surface models, for another they require extensive processing for their manipulations owing to the more complicated data structure and associated mathematics.

10.2 SOLID ENTITIES

Solid modelling packages have a CSG—compatible user input, and therefore, provide users with a certain set of building blocks, often called **primitives**. Primitives are simple basic shapes and are considered the solid modelling entities which can be combined by a mathematical set of boolean operations to create the solid. The user usually positions primitives as required before applying boolean operations to construct the final solid.

The four most commonly used primitives are: block, cylinder, cone and sphere. These are based on the four natural quadrics, planes, cylinders, cones and spheres. Planar surfaces result from rolling, chamfering and milling; cylindrical surfaces from turning or filleting; spherical surfaces from cutting with a ball end cutting tool; conical surfaces from turning as well as from dull tips and counter sinks. Natural quadrics are distinguished by the fact that they are combinations of linear motion and rotation. Other surfaces, except the torus, require atleast dual axis control.

A primitive requires a set of location data, a set of a specific system syntax, orientation data to define it completely. Location data entails a primitive local coordinate system and an input point defining its origin. Geometrical data differs from one primitive to another and are user-input. Orientation data is typically used to orient primitives properly relative to the MCS (Model

Coordinate System) or WCS (World Coordinate System) of the solid model under construction. Primitives are usually translated and/as rotated to position and orient them properly before applying Boolean operations.

Block

This is a box whose geometrical data is its width, height and depth. Its local coordinates system is $X_L \, Y_L \, Z_L$ as shown in Fig. 10.1. Point P defines the origin of the $X_L \, Y_L \, Z_L$ system. The signs of W, H and D determine the position of the block relative to its coordinate system.

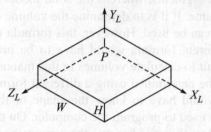

Fig. 10.1 Block.

Cylinder

This primitive is a right circular cylinder whose geometry is defined by its radius (or diameter) R and length H as shown Fig. 10.2. The length H is taken along the direction of the Z_L axis. H can be positive or negative.

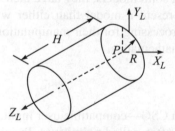

Fig. 10.2 Cylinder

Cone

This is a right circular cone or a frustrum of a right circular cone as shown in Fig. 10.3 whose base radius R, top radius (for truncated cone) and height H are user defined.

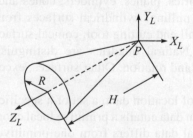

Fig. 10.3 Cone

Sphere

This is defined by its radius or diameter as shown in Fig. 10.4 and is centered about the origin of its local coordinate system.

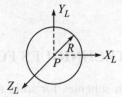

Fig. 10.4 Sphere

Wedge

This is a right-angled wedge as shown in Fig. 10.5 whose height H, width W and base depth D form its geometric data.

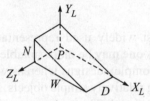

Fig. 10.5 Wedge

Torus

This primitive is generated by the revolution of a circle about an axis lying in its plane. The torus geometry as shown in Fig. 10.6 can be defined by the radius (or diameter) of its body R_1 and the radius (or diameter) of the centre line of the torus body R_2, or the geometry can be defined by the inner radius (or diameter) R_I and outer radius (or diameter) R_0.

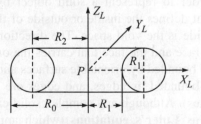

Fig. 10.6 Torus

For all the above primitives, there are default values for the data defining their geometries. Most package use default values of 1. Some packages assume that the origin P of the local coordinate system is coincident with that of the MCS or WCS and require the user to translate the primitive to the desired location, thus eliminating the input of point P by the user.

Two or more primitives can be combined to form the desired solid. To ensure the validity of the resulting solid, the allowed combinational relationships between primitives are achieved via

Boolean (or set) operations. The available Boolean operators are union (U or +), intersection (n or I) and difference (−). The union operator is used to combine or add together two objects or primitives. Intersecting two primitives gives a shape equal to their common volume. The difference operator is used to subtract one object from the other and results in a shape equal to the difference in their volumes.

10.3 SOLID MODELLING TECHNIQUES FOR VOLUME MODELLING

There are a number of representation schemes for solid modelling, which includes

- Boundary representation (B-rep)
- Constructive solid geometry (CSG)
- Sweep representation
- Primitive instancing
- Cell decomposition
- Analytical solid modelling

BREP and CSG are the most widely used representation schemes for solid modelling. However, for different applications one may be more suitable than the other. For example, BREP is more suitable for representing complex designs whereas with CSG, models are easy to create but are usually used in representing relatively simple objects. In some modellers, a hybrid scheme employing both BREP and CSG is used. These representation schemes are discussed in more detail in the following sections.

10.4 BOUNDARY REPRESENTATION

Boundary representation is essentially a scheme that, as the name implies, describes the geometry of an object in terms of its boundaries, namely vertices, edges and surfaces An example of BREP is shown in Fig. 10.7. In order to represent a solid object by its surfaces, we require the orientation of each surface that defines the inside or outside of the object. Usually the inside is the material part and the outside is the void space. The direction of the face normal is usually used as the orientation of the face and the face that carries the orientation information is called **orientable surface**. A solid is bounded by orientable surfaces and we can define a solid by a set of faces. A face is generally bounded by edges, and edges are bounded by vertices (with the exception of spheres and circles). Although any complex solid can be represented by faces, the system of equations known as **Euler's equations** (which appear as syntax of CAD/CAM systems) is used to ensure the validity of BREP models, that is to ensure that a real object is formed or bounded. For example, three planes will not form a solid object. Similarly, a face is bounded by edges and a face will not be bounded by two straight lines.

Theoretically, BREP's can be used to describe any object. However, some BREPs are restricted to planar and polygonal boundaries. It is worth mentioning that curved surfaces are difficult to represent and, in general, are approximated by polygons.

10.4.1 Basic Entities for BREP

To create a model of a part using the boundary representation scheme, as shown in Fig. 10.7 designers try to define the part faces explicitly. In turn, the faces are defined by edges and the edges are defined by vertices. Face, edge and vertex are usually regarded as the basic elements of BREP.

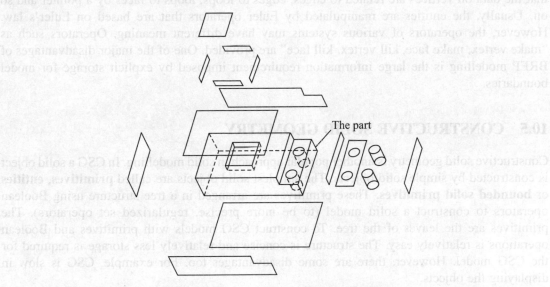

The part

Fig. 10.7 Boundary representation.

10.4.2 Validation of BREP Model Using Euler's Law

Euler's law states that a polyhedron is topologically valid if the following equation is satisfied:

$$F - E + V = 2 \qquad (10.1)$$

which means that to construct a valid polyhedron, the number of faces (F), edges (E) and vertices (V) must satisfy this equation.

For example, the simplest polyhedron, a tetrahedron consists of 4 faces, 6 edges and 4 vertices. Similarly, a cube has 6 faces, 12 edges and 8 vertices.

The generalized version of Euler's law is

$$F - E + V - L = 2 (B - G) \qquad (10.2)$$

where F, E, V, L, B and G are the numbers of faces, edges, vertices, faces inner loops, bodies and general (such as torus, through-hole) respectively. This law is used for the purpose of validation of polyhedral objects with passage way sand holes. It may not ensure the validity of such solids with passage ways in all cases. For example, for objects with curved surfaces such as cylinders, spheres and cones, it is not so easy to apply Euler's law. In such cases, we have to define the structure of the curved objects.

From this discussion, we know that the validity of a polyhedral object is ensured via Euler's law. In order to create a valid BREP model, we have to follow the construction syntax of the software which is mainly based on Eq. (10.2).

Both topological and geometric information is necessary to define a solid. While we provide the connectivity of entities, the related geometric information is also attached, such as the equations of curves and surfaces and the coordinates of points. The information is stored in an ordered structure, the object–body–genus–face–loop–edge–vertex structure.

This means that when we look closely at the database of a model of an object, we will find that the data on vertices are related to edges, edges to loops, loops to faces by a pointer and so on. Usually, the entities are manipulated by Euler operators that are based on Euler's law. However, the operators of various systems may have different meaning. Operators such as "make vertex, make face, kill vertex, kill face" are provided. One of the major disadvantages of BREP modelling is the large information requirement imposed by explicit storage for model boundaries.

10.5 CONSTRUCTIVE SOLID GEOMETRY

Constructive solid geometry is another popular approach in solid modelling. In CSG a solid object is constructed by simple solid objects. The simplest solid objects are called **primitives**, **entities** or **bounded solid primitives**. These primitives are arranged in a tree structure using Boolean operators to construct a solid model (to be more precise, regularized set operators). The primitives are the leaves of the tree. To construct CSG models with primitives and Boolean operations is relatively easy. The structure is concise and relatively less storage is required for the CSG model. However, there are some disadvantages too. For example, CSG is slow in displaying the objects.

It is usually converted internally into a BREP to display the model or generate wire frame drawing. This is why many systems provide both BREP and CSG. CSG provides easier input and BREP provides faster display and line drawing.

10.5.1 Solid Entities (Primitives)

We construct a wire frame model by inputting data consisting of points and lines. A solid model of an object is constructed by inputting data consisting of simple basic shapes that are considered as solid modelling entities, such as block, cylinder, cone and sphere, as shown in Fig. 10.8. These entities are not simply objects represented by a few lines as in wire frame modelling. For example, a block entity is not a simple object consisting of 12 lines and 8 vertices, it is an object that occupies space. The consequence is that when two blocks intersect, a more complex combined object is created due to intersection of lines, whereas the wire frame model creates only 24 lines and a 16 vertex set, which represents nothing. Various entities are provided to users by various CAD/CAM systems. The entities provided also depend on the application orientation of the system. Some CAD/CAM systems give users some degree of freedom to define entities that are not predefined in the system.

Generally, the four entities most commonly available to user of most systems are block, cylinder, cone and sphere. They can be used to construct complex objects or user-defined entities. These entities usually represent certain machining operations. For example, a block of material may be removed by milling, which may result in a step, a pocket or a slot. A cylinder of material may be removed by drilling, which may be used to construct a hole in an object.

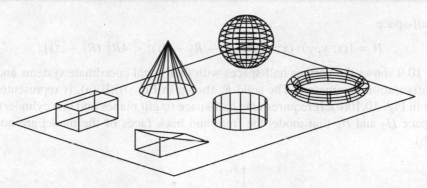

Fig. 10.8 Solid modelling entities.

10.5.2 Half-spaces

In fact, the primitives themselves may be viewed as combinations of even simpler entities termed as **unbounded half-spaces**, which provide a basis for mathematical understanding. In this section, we present this concept and how half-spaces are used in CSG.

In general, a half-space is considered the basic element of primitives. By applying set operations to combine half-spaces, various primitives may be constructed. Mathematically, a half-space is defined as a regular point set in E^3 space as follows:

$$H = \xi v: f(v) < 0, v \ \xi \ E^3 \qquad (10.3)$$

where v is a point set in E^3 (3D space) and $f(v) = 0$ defines the surface equation of the half-space boundaries. The point set v that satisfies $f(v) < 0$ is considered a solid part, whereas the point set v that validates the $f(v) > 0$ equation is considered empty.

Although $f(v) < 0$ may represent any surface boundaries, the most commonly used half-spaces are planar, cylindrical, spherical, conical and toroidal. Using these half-spaces, we may form many primitives by set operations. For example, a block can be formed by six planar half-spaces using AND operators.

The most widely used half-spaces are:

Planar half-space

$$H = \{(x, y, z): z < 0\}$$

Cylindrical half-space

$$H = \{(x, y, z): x^2 + y^2 < R^2\}$$

Spherical half-space

$$H = \{(x, y, z): x^2 + y^2 < R^2\}$$

Conical half-space

$$H = \left\{(x, y, z): x^2 + y^2 \left[\tan\left(\frac{x}{2}\right)z\right]^2\right\}$$

Toroidal half-space

$$H = \{(x,\ y,\ z)\colon (x^2 + y^2 + z^2 - R_2^2 - R_1^2)^2 < 4R_2^2\ (R_1^2 - z^2)\}$$

Figure 10.9 shows the various half-spaces with their local coordinate systems and the limits on that configuration parameters. The solid R, shown in Fig. 10.10(a), is represented by using half-spaces in Fig. 10.10(b). It requires nine half-space (eight planes and one cylinder) H_1 to H_{10}. The half-space H_7 and H_8 that model the front and back faces of the model are not shown in Fig.10.10(b).

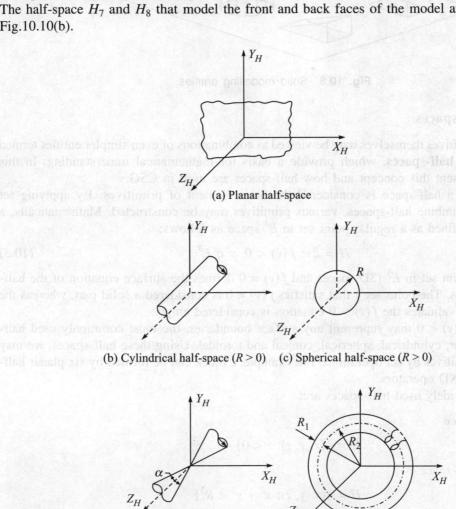

(a) Planar half-space

(b) Cylindrical half-space ($R > 0$)   (c) Spherical half-space ($R > 0$)

(d) Two-sheet conical half-space
(0 < α < π)

(e) Torodial half-space
($R_1 > 0, R_2 > 0, R_2 > R_1$)

Fig.10.9 Unbounded half-spaces.

Half-spaces are lower level primitives and any complex object can be modelled as half-spaces combined by the set of operations. One form of CSG can be based on unbounded half-spaces. The main advantage of half-spaces is its conciseness in representing objects compared to other

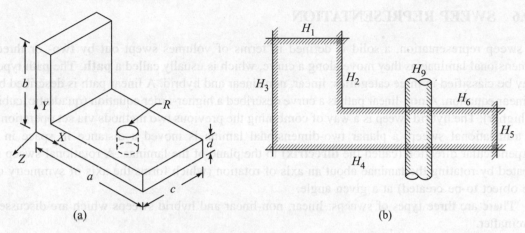

Fig. 10.10 Solid, half-space representation.

schemes such as CSG. However, this representation can lead to unbounded solid models resulting in missing faces and abnormal shaded images. Modelling by half-spaces is also cumbersome for designers. It is useful only for research purposes.

The basic element of entities is the half-space. However, the block, cylinder, cone and sphere are usually considered the basic entities of CSG without further concern about half-spaces. This is analogous to another situation. When we program using high-level languages, we are not concerned that the high-level languages are supported by some low-level languages.

10.5.3 Regularized Set Operations and Their Geometric Meaning

Boolean set operations are utilized in CSG to combine various primitives. We know that the set operations include union, intersection and difference. The geometric meaning of these operators is shown in Fig. 10.11. A binary tree is formed to represent the complete object data structure.

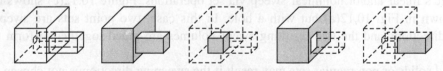

Fig. 10.11 The geometric meaning of set operations.

First, let us understand that there are problems with set operations when they are used in geometric modelling. It is normal to define a point by the intersection of two straight lines. This can be represented using set operations as point $A = $ line l_1 n line l_2. Extending this concept to solid modelling and using two solid cubes, we have the following situation with set operation intersection, we may generate a null set, a point, a line, a surface or a cube. However, points, lines and surfaces are not objects or parts in the real world.

Furthermore, their presence leads to problems in solid modelling. Because we need set operations a modification of the set operations is necessary. The modified set operation is called **regularized set operation**. The set operations used to construct a solid are not unique. For example, you can use other sequences of set operations to construct a solid.

10.6 SWEEP REPRESENTATION

In sweep representation, a solid is defined in terms of volumes swept out by two- or three-dimensional laminae as they move along a curve, which is usually called a **path**. The path types may be classified in three categories: linear, non-linear and hybrid. A linear path is described by a linear equation. A non-linear path is a curve described a higher-order equation (quadratic, cubic or higher). The hybrid sweep is a way of combining the previous two methods via set operations. In translational sweep, a planar two-dimensional lamina is moved a distance in space in a perpendicular direction (called the **directrix**) to the plane of the laminae. A rotational sweep is created by rotating the laminae about an axis of rotation (which forms the axis of symmetry of the object to be created) at a given angle.

There are three types of sweeps: linear, non-linear and hybrid sweeps which are discussed hereinafter.

Linear sweep

In linear sweep, the path is a linear or circular vector described by a linear, most often parametric. Linear sweep can be divided into translational and rotational sweep. In translational sweep, a planar two-dimensional point set described by its boundary (or contour) can be moved at a given distance in space in a perpendicular direction (called the **directrix**) to the plane of the set [see Fig. 10.12(a)]. For rotational sweep, the planar two-dimensional point set is rotated about the axis of rotation by a given angle [see Fig. 10.12(a)].

Non-linear sweep

In non-linear sweep, the path is a curve described by a higher-order equation (quadratic, cubic or higher) as shown in Fig.10.12(b).

Hybrid sweep

In combine's linear and/or nonlinear sweep via set operations. Figure 10.12(c) shows the same object shown in Fig. 10.12(a) but with a hole. In this case, two point sets are swept in two different directions and the two resulting swept volumes are glued together to form the final object.

Invalid solids or non-regular sets may result if the sweeping direction is not chosen properly as shown in Fig. 10.12(d).

Sweeping operations are useful in engineering applications that involve swept volumes in space. Two widely known applications are simulations of material removal due to machining operations and interference detection of moving objects in space.

Sweep representation is useful once it develops. Its modelling domain can be extended beyond 2 ½D objects if non-linear sweep is available. Non-linear sweep may be useful in creating non-rigid objects and studying their deformation as they travel in space. Complex mechanical parts such as screws, springs and other components that require helical and special loci can be represented by sweeping.

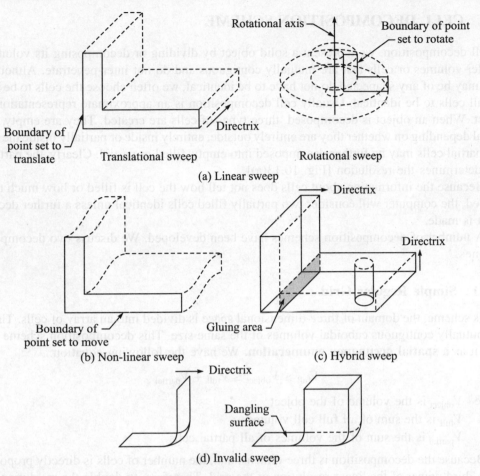

Fig. 10.12 Types of sweep.

10.7 PRIMITIVE INSTANCING METHOD

The pure primitive instancing technique is based on considering an object that has the same topology as a potential primitive (also called **generic primitive**) but different geometry. By predefining the topology of the primitives, the user is required to provide only the geometry of an object and the family to which the object belongs. For example, a bolt can be defined by bolt (number of sides, length, pitch, diameter) the other dimensions are generated automatically and the topology is predefined. This scheme is basically the philosophy of group technology. It is easier to use and create unambiguous and unique solids.

The main disadvantage is clearly the limited domain of objects that can be handled. The method is restricted to the primitives predefined in the system and they are quite limited. Another problem is lack of generality to develop any algorithm to infer the properties of the represented solid.

10.8 CELL DECOMPOSITION SCHEME

In cell decomposition, we represent a solid object by dividing or decomposing its volume into smaller volumes or cells that are mutually contiguous and do not inter-penetrate. Although the cells may be of any shape and do not have to be identical, we often choose the cells to be cuboid and all cells to be identical. Usually cell decomposition is an approximate representation of an object. When an object is decomposed, three types of cells are created. They are empty, full or partial depending on whether they are entirely outside, entirely inside or partially inside the object. The partial cells may be further decomposed into empty, filled or partial. Clearly, the partial cell size determines the resolution [Fig. 10.13(a)].

Because the information about cells does not tell how the cell is filled or how much the cell is filled, the computer will consider two partially filled cells identical unless a further decomposition is made.

A number of decomposition schemes have been developed. We discuss two decomposition schemes.

10.8.1 Simple Regular Grid

In this scheme, the domain of three-dimensional space is divided into an array of cells. The cells are mutually contiguous cuboidal volumes of the same size. This decomposition scheme is also known as a **spatial occupancy enumeration.** We have the following equation:

$$V_{full} \leq V_{object} \leq V_{full} + V_{partial}$$

where V_{object} is the volume of the object

V_{full} is the sum of all full cell volumes

$V_{partial}$ is the sum of the volumes of all partial cells

Because the decomposition is three-dimensional, the number of cells is directly proportional to the third power of the linear resolution of the grid. Therefore, to double the resolution of the grid, each cell is divided into eight smaller identical cells, leading to an eight-fold increase in the number of cells and hence in the amount of storage space required, not to mention the computational time. For simplicity, we show the decomposition scheme in two dimensions in Fig. 10.13(a), where the total number of cells is proportional to the second power of the linear resolution of the grid.

10.8.2 Octree Adaptive Grid

In this scheme, instead of slicing the three-dimensional space into an array of equal sized and regularly spaced cells, a hierarchical subdivision scheme is used. At the initial level, a simple regular grid is formed and at each subsequent level further subdivision may be performed on the partial cells. This gives rise to smaller and smaller cell sizes, leading to higher and higher spatial resolution. The 3D space decomposition results in eight subdivisions, hence the name *Octree*. In this scheme, the number of cells at the boundary of the object increases rapidly, so the amount of computer storage is proportional to the surface area of the object. This is, however, more economical than the simple grid decomposition scheme in which the computer storage is proportional to the volume of the object. Figure 10.13(b) shows a quadtree, which is the 2D version of an Octree. It is evident from Figs. 10.13(a) and (b) that the quadtree requires a fewer cells than the simple regular grid decomposition scheme for the same resolution.

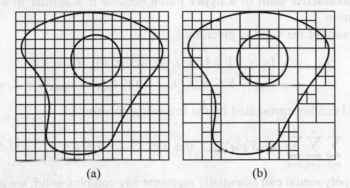

(a)   (b)

Fig. 10.13 Simple regular grid decomposition and quadtree adaptive grid decomposition.

10.9 ANALYTICAL SOLID MODELLING

In analytical solid modelling (ASM), as shown in Fig. 10.14 the tensor product method that is used to represent surfaces is extended to three-dimensional parametric space with parameters, say *s*, *t* and *u*. This is similar to representing a curve by one-dimensional parametric space with one parameter (say *t*) and a surface by two-dimensional parametric space with two parameters

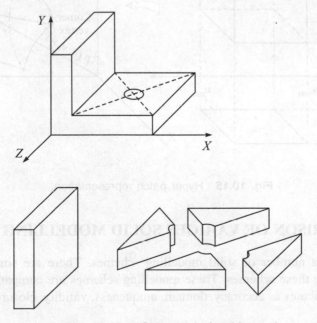

Fig. 10.14 ASM model of solids.

(say s and t). The techniques for creating spline or patch curve segments or surface patches are valid in ASM. As in surface representation a general solid described by x, y and z in cartesian space is mapped into three-dimensional parametric space via the tensor product formulation. This solid is called a **parametric solid** or a **hyper patch** because it is similar to a surface patch in surface representation.

The variable point of the solid is given by

$$V_{(s,\ t,\ u)} = [x,\ y,\ z] = [x(s,\ t,\ u),\ y(s,\ t,\ u),\ z(s,\ t,\ u)]$$
$$S_{min} \le S \le S_{max},\ t_{min} \le t \le t_{max},\ \le u_{min} \le u \le u_{max} \qquad (10.4)$$

A general solid can be represented by the following polynomial:

$$V_{(s,\ t,\ u)} = \sum_{i=1}^{4} \sum_{j=1}^{4} \sum_{k=1}^{4} C_{ijk} S^i t^j u^k, \qquad 0 \le s \le 1, \qquad 0 \le t \le 1, \qquad 0 \le u \le 1 \quad (10.5)$$

Although this polynomial can essentially represent any complex solid, we prefer to combine simple hyper patches and manipulate them to create complex objects ASM originated from finite element analysis applications. So it is natural that this method of modelling is appealing in design and analysis applications.

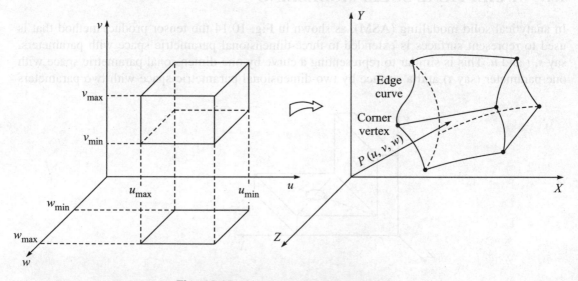

Fig. 10.15 Hyper patch representation.

10.10 COMPARISON OF VARIOUS SOLID MODELLING SCHEMES

We have presented a number of solid modelling schemes. There are similarities as well as dissimilarities among these schemes. These modelling schemes are compared in Table 10.1 on the basis of such attributes as accuracy, domain, uniqueness, validity, closure, and compactness.

Table 10.1 Comparison of Various Solid Modelling Schemes

	BREP	CSG	Sweep	Primitives instancing	Cell decomposition	ASM
Accuracy	Curved surface support is needed to reach high accuracy	Non-polyhedral primitives are needed to reach high accuracy	High accuracy	Similar to sweep	Usually used in low accuracy applications	Good in accuracy
Domain	Theoretically it can support any object	Similar to BREP	Limited to certain domain	Similar to sweep	Almost any object	With computational effort it is fairly good
Uniqueness	No	No	Need careful definition	Similar to sweep	Simple grid and Octree generally generate unique object	No
Validity	Hard to check	Easy to check	Easy to check	Easy to check	Easy to check	Easy to check
Closure	Fairly good	Fairly good	Hard to achieve	Hard to achieve	Fairly good	Fairly good
Compactness and efficiency	Good at internal process	Good for user interface	Good for interface	Good for interface	Good for interface	Good at internal use

REVIEW QUESTIONS

10.1 What are the common modelling methods available for solid design in a solid modelling software?

10.2 Why the sweep representations are useful in creating solid models of 2D objects? Explain.

10.3 Discuss with the help of neat sketches, the most commonly used solid entities.

10.4 Describe the scheme of boundary representation to create solid models of physical objects.

10.5 What is solid modelling? Compare Bezier representation with CSG representation scheme.

10.6 What are the desirable properties of any solid modelling scheme?

10.7 Describe how a solid model is generated through cell composition and spatial occupancy enumeration.

10.8 How do you define a solid model? Explain various solid modelling schemes with their applications and limitations.

10.9 Differentiate between solid modelling and surface modelling methods and representation schemes.

10.10 Why the sweep representation is useful in creating solid models of 2½D objects? Explain.

10.11 Distinguish between C.S.G. and B-rep models.

10.12 Explain the concept of the three basic Boolean operations used in solid modelling. Give neat sketches showing the effect of these operators on any two basic primitives.

PRACTICE PROBLEMS

10.1 A valid solid is defined as a point set that has an interior and a boundary as given by the equation $S = iS \cup bS$.

A valid boundary must be in contact with the interior. Sketch a few 2D and 3D solids and identify iS and bS for each one. Is iS always joint for any S? Can bS be disjoint? What is your conclusion?

10.2 Three point sets in E^2 define three valid polygonal solids S_1, S_2 and S_3. The three solids are bounded by three boundary sets bS_1, bS_2, and bS_3 given by their corner points as: $bS_1 = (2, 2), (5, 2), (5, 5), (2, 5)$, $bS_2 = (3, 3), (7, 3), (7, 6), (3, 6)$, and $bS_3 = (4, 1),$ (6, 1), (6, 4), (4, 4). Find $S_1 \cup S_2 \cup S_3$, $S_1 \cup S_2 \cap S_3$, and $S_1 - S_2 - S_3$.

10.3 Using the set membership classification, classify the line L with respect to the solid shown in the following figure, if the solid is given as a B-rep and a CSG.

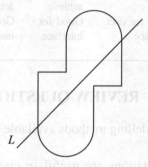

10.4 A solid S is defined as the union of two intersecting spheres, that is, $S = SP_1 \cup SP_2$. Sketch the solid S. What is $SP_1 \cap SP_2$ if the spheres have the same radius, and if they have different radii? Verify the Euler's equation for the solid S.

10.5 Solids in E^2 (that is, 2D) and their solid modellers are valuable in understanding many of the concepts needed to handle solids in E^2 (that is, 3D solids). A 2D solid is a point set of ordered pairs (x, y). A 2D solid modeler based on half-spaces is to be developed utilizing linear and disk half-spaces. Find the equation of the two half-spaces. Develop a parametric equation for their intersection.

10.6 We wish to develop bounded primitives for a 2D solid modeler based on the CSG scheme. A plate (rectangular plate and trip late) and disk primitives are to be developed. Find the mathematical definitions of these primitives. Develop intersection equations (refer to Problem 10.5).

10.7 How can you use a cylinder primitive to generate a sphere?

10.8 How can you generate a torus using other natural quadrics?

CAD

Design Aspects of Industrial Components

Chapter

FINITE ELEMENT MODELLING AND ANALYSIS

11

INTRODUCTION

The availability of sophisticated computers over the last three decades has enabled engineers to take up challenging tasks and solve intractable problems of earlier years. Now-a-days rapid decrease in hardware cost has enabled every engineering firm to use a desktop computer or a microprocessor. Moreover they are ideal for engineering design because they easily provide an immediate access and do not have the system jargon associated with large computer system. It is to be expected that the software is to be sold or leased and the hardware is to be supplied with software.

Industrial applications of FEA were pioneered in the 1950s and 1960s by the aeronautical industry to aid creation of designs with high strength to weight ratio. FEA is now extensively used in automobile industry for optimizing the design and to improve fuel efficiency. While the predominant application of FEA has been in stress analysis the technique is increasingly used to solve design problems associated with fluid flow, heat transfer, natural frequencies, electronics, radiation analysis, acoustics and magnetics.

Finite element structural analysis is a method of predicting the behaviour of a real structure under specified load and displacement conditions. The finite element modelling is a generalization of the displacement or matrix method of structural analysis to two-dimensional and three-dimensional problems. The basic concept of FEM is that the structure to be analyzed is considered to be an assemblage of discrete pieces called **elements** that are connected together at a finite number of points or nodes. The finite element is a geometrically simplified representation of a small part of the physical structure.

11.1 GENERAL PROCEDURE FOR FINITE ELEMENT ANALYSIS

The solution of continuum problem by the finite element analysis usually follows an orderly step-by-step process. The following steps show in general how the finite element analysis works:

1. *Discretize the given continuum.* The essence of the finite element analysis is to divide a continuum, that is, problem domain into quasi-disjoint non-overlapping elements. This

can be achieved by replacing the continuum by a set of key points, called **nodes,** which when connected properly produce the elements. The collection of nodes and elements forms the finite element mesh. A variety of element shapes and types are available. The analyst can mix element types to solve one problem. The number of nodes and elements that can be used in a problem lies in the hands of the designer's judgement. As a general rule, the larger, the number of nodes and elements, the more accurate the finite element solution, but also the more expensive the solution is more memory space is needed to store the finite element model and more computer time is needed to obtain the solution. Figure 11.1(a) shows an example of discretizing a cantilever beam made of steel and supporting a concentrated load at its free end. Figure 11.1(b) shows node generation and numbering. Figure 11.1(c) shows two types (four-node and six-node) of a quadrilateral (element shape) element.

2. *Select the solution approximation.* The variation of the unknown (called **field variable**) in the problem is approximated within each element by a polynomial. The field variable may be a scalar (e.g., temperature) or a vector (e.g., horizontal and vertical displacements).

   Polynomials are usually used to approximate the solution over an element domain because they are easy to integrate and differentiate. The degree of the polynomial depends on the number of nodes per element, the number of unknowns (components of field variable) at each node and certain continuity requirements along element boundaries.

3. *Develop element matrices and equations.* The finite element formulation involves transformation of the governing equilibrium equations from the continuum domain to the element domain. Once the nodes and material properties of a given element are defined its corresponding matrices (stiffness matrix, mass matrix, etc.) and equations can be derived.

4. *Assemble the element equations.* The individual element matrices are added together by summing the equilibrium equations of the elements to obtain the global matrices and system of algebraic equations. The structure is analyzed for a given set of boundary conditions.

5. *Solve for the unknowns at the nodes.* The global system of algebric equations is solved via Gauss elimination methods to provide the values of the field variables at the nodes of the finite element mesh. Values of field variables and their derivatives at the nodes form the complete finite element solution of the original continuum problem before discretization.

6. *Interpret the results.* The final step is to analyze the solution and results obtained from the previous step to make design decisions. The correct interpretation of these results requires a sound background in both engineering and FEA. The process is continued until an optimum design is evolved.

In a typical finite element problem, the solution is usually derived in four separate phases in the following order:

1. *Formulation phase*: The physical region of interest is divided into elements and element types and appropriate interpolation functions are selected.

2. *Evaluation phase*: The contribution of each element to the problem formulation is computed, and this generally involves evaluating stiffness matrices and generalized force vectors.

3. *Assembly phase*: The contributions from the individual elements are assembled to produce a large system of equations for solution.

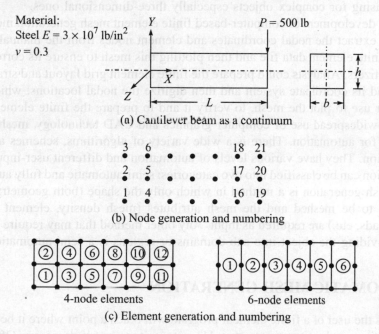

(a) Cantilever beam as a continuum

(b) Node generation and numbering

(c) Element generation and numbering

Fig. 11.1 Continuum discretization for FEA.

4. *Solution phase:* These equations are solved to find the primary unknowns of the problem. These unknowns are generally values of some physical quantity (such as displacement) at the nodes of the elements.

11.2 MESH GENERATION TECHNIQUES

In many cases, the user of a finite element program will reach a point where it becomes desirable to use an existing mesh generating system or to write to one if necessary. Often it is difficult to write a program to generate a mesh for complex structures and the user has to spend many hours drawing meshes in a drawing office and measuring the coordinates of each nodal point. So, in some problems where the number of nodal points is very large, the user frequently has to spend many hours mostly involving data checking and rechecking by tedious graphical methods.

Mesh generation forms the backbone of the FEA. Mesh generation refers to the generation of nodal coordinates and elements. It also includes the automatic numbering of nodes and elements based on a minimal amount of user-supplied data. Automatic mesh generation reduces errors and saves a great deal of user time, therefore, reducing the FEA cost.

Before the existence of preprocessors, finite element meshes were generated manually. In manual mesh generation, the analyst discretizes the simplified geometry of the object to be studied, that is, the geometric model of the object into nodes and elements. Nodes are defined by specifying their coordinates while element connectivity (connecting nodes) defines the elements. Manual meshing is inefficient, error-prone, and meshing data can grow rapidly and become confusing for complex objects especially three-dimensional ones.

The early development of computer-based finite element mesh generation methods began by attempting to extract the nodal coordinates and element nodes from the manually prepared (by the analyst) finite element data file and then plotting this mesh to ensure its correctness with the advent of digitizers. Analysts could prepare the finite element grid layout and strategy, define the grid origin and its coordinate system and then digitize the nodal locations, which are stored in a file for later use to plot the mesh, to verify it and to prepare the finite element data file.

With the widespread use of computer graphics and CAD technology, mesh generation has been a target for automation. There is a wide variety of algorithms, schemes and methods for mesh generation. They have various levels of automation and different user-input requirements. Mesh generation can be classified into two categories: semi-automatic and fully automatic. A fully automatic mesh generation is a method in which only the shape (both geometry and topology) of the object to be meshed and the mesh attributes (mesh density, element type, boundary conditions, loads, etc.) are required as input. Any other method that may require additional input such as subdividing the object into sub-domains or regions is a semi-automatic one.

11.3 AUTOMATIC MESH GENERATION

In many cases the user of a finite element program will reach a point where it becomes desirable to use an existing mesh generating system [1] or to write to one if necessary. Often it is difficult to write a program to generate a mesh for complex structures and the user has to spend many hours drawing meshes in a drawing office and measuring the coordinates of each nodal point. So, in some problems where the number of nodal points is very large, the user frequently has to spend many hours mostly involving data checking and rechecking by tedious graphical methods.

Using a CAD system in the data generation process has considerable advantages over most other methods because the user is able to see the element connections and position of each element directly on a display, as element generation is in progress. Also, if the system is interactive, the user is able to change a mesh instantaneously to arrive at the best mesh arrangement to suit a particular problem by adding or deleting elements. The CAD system can also be used after the finite element analysis to present data graphically so that the results can quickly be assessed.

If a CAD system is to be used in the generation and presentation of data in finite element analysis, it should ideally be capable of providing the following facilities:

- Generation of mesh for two- or three-dimensional structures
- Ability to represent curved edges and surfaces
- Ability to control element density and generation of non-uniform meshes
- Facility for concentrating and grading the mesh over any region. Speedy node and element numbering system which will lead to computational efficiency.
- Facility to display idealization

- Ability to rotate the idealization from any desired angle
- Option to display any portion of the model
- On-line interactive modification of data for alternative idealizations
- Level of automatic mesh generation with manual override
- Preparation of input data for analysis programs
- User-orientated and easy to use with minimum of input
- Economical with respect to both computer time and manual effort

11.4 MESH REQUIREMENTS

Before we describe the various existing mesh generation methods, it is important to list the requirements that make a mesh valid, that is, produces the correct FEA results and economical. Some of the requirements are (some are necessary while others are described):

Nodal locations

Nodes must lie inside or on the boundaries of the geometric model to be meshed. Nodes that are very close to the boundaries must be pulled to lie on them to accurately mesh the model. Some generation methods offset (shrink) the model boundary by a small amount and generate the nodes based on the offset boundary and then pull the boundary nodes to the original boundary of the model.

Element type and shape

It is desirable if various elements (large element library) can be generated to provide users with the required flexibility to meet the compatibility and completeness requirements.

Mesh gradation

This usually refers to mesh grading and density control. Most often objects on which FEA is performed may have holes or sharp corners. It is usually required that mesh density (number of nodes and elements) is increased around these regions to capture the rapid change (e.g., stress variation around holes and sharp corners) of the field variable.

Some generation methods allow users to specify various mesh densities for various regions. Mesh gradation is usually encountered in transition regions. A transition region is one that connects two neighbouring regions with either different types of elements or the same type but with different numbers of nodes. Transition elements are usually employed in the transition region to connect the meshes in the other two regions.

Mesh conversion

It may be desirable to convert a mesh of a given type of element to another mesh of a different element type. In two-dimensional meshes, for example, it is always possible to convert a triangular element into three quadrilateral elements (a tetrahedron can be subdivided into four hexahedral) or combine two triangular elements to produce a quadrilateral element. A quadrilateral element mesh may be converted into a triangular element mesh by splitting each quadrilateral into two triangles. Mesh conversion must be done with care as poorly formed elements (especially in three dimensions) may result.

Element aspect ratio

For geometric invariance, it is important to keep the aspect ratio of any element close to 1, that is, all sides of an element are equal in length.

Mesh geometry and topology

As the object to be meshed has geometry and topology, so does its mesh. Mesh geometry refers to the coordinates of nodal points and the connectivity information of elements. Mesh topology refers to the mesh orientation relative to object topology. Object topology always determines the mesh topology as shown in Fig. 11.2.

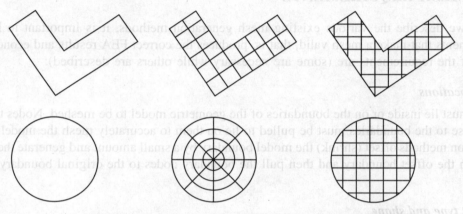

Fig. 11.2 Geometry and topology of finite element meshes.

Compatibility with representation schemes

A mesh generation method is inherently related to the type of geometric model to be meshed whether it is a wire frame surface or solid model. For example, generation methods based on wire frames or surfaces can never be fully automatic due to their lack of topological information.

Cost effectiveness

The time it takes to generate the mesh and the time it takes to perform the FEA are crucial. To reduce both it is important that the mesh generation method optimizes the mesh and minimizes the number of nodes and elements that comprise the mesh and yet meets the conversion requirements.

11.5 THREE-DIMENSIONAL SHAPE DESCRIPTION AND MESH GENERATION

Three-dimensional problems were described by utilizing a 2½D technique which is based on describing a three-dimensional mesh by a mesh digitized in two dimensions and then giving the mesh a uniform depth with the increasing power of computers. The finite element technique is becoming increasingly used, not only for stress analysis but also in areas such as heat transfer and reactor physics. This increase in utilization of the finite element method has increased the requirement for true three-dimensional shape definition and mesh generation.

The graphical definition in three dimensions has been thoroughly described in Chapter 5. As we know, there are a number of ways of entering three-dimensional data into a computer. The first method is based on digitizing two or more orthogonal views which are subsequently assembled within the computer to give the required three-dimensional shape. The second method consists of digitizing contours in two dimensions with the third dimension normal to the surface, being added numerically. A surface patch technique is then used to interpolate the contour data in order to produce data of sufficient quality for the graphical representation. Finally, a third method may be used which is based on photogrammetry where two perspective drawings or photographs are digitized and the data converted into a three-dimensional format.

Of these three methods of graphical data description, the first method was found to be more suited to complicated shapes. The method was varied slightly in that a complex shape is always divided into a number of basic building blocks which are assembled at a later stage. The use of the building block technique may seem a tedious method but it is the most efficient one for computers for small core sizes. Each building block can be described in core and the mesh generated in the building block without large numbers of disc transfers. The data generated can be stored in a series of files which can easily be edited and finally assembled into a single file.

So far only two groups of input have been discussed: first the numerical input of the number of the input points to be digitized and secondly the actual points digitized on the table. Once the number of 3D points digitized on the table is equal to that specified by the user, the program jumps out of the digitizing input loop and requests the user to certify the acceptability of these stored points. If any of the points are not acceptable to the user, for instance, if the program requested point number three in the xy plane and the user digitized point number five in the particular input shape, the program jumps into an editor phase guiding the user to correct the point which has been digitized in error. At this stage all the input points can be changed and if the initial number of input points specified is less than that required, more points could be added. The program can only get out of the editor mode when the user requests for a null point to be edited. Once the user is satisfied with the input points, the program requests for the macroblock number, to call the right overlay into core for the building block required by the user.

Before loading the appropriate service overlay for the required building block into core, 3D mesh stacks its third entry point on the overlay stack, so that control of the system is returned to it after the mesh is developed by the service routine. All the routines used for the generation of meshes within the separate building blocks are written in such a way that, if the number of input points specified at first entry to 3D mesh is not compatible with the routine, control is automatically handed over to 3D mesh at the second entry point where the user can correctly input the required macro block. The user can exit from 3D mesh by supplying a macro block number of six—a dummy macroblock number used for exit purposes. When a mesh generating routine is successfully executed and control handed over to 3D mesh, the element connection generating routine is loaded only if a coordinate is stored on the system disc. After the element connections are generated and a file is created, or if there is no coordinate file, 3D mesh will ask the user for more macro blocks or an exit from the program. If the addition of more building blocks to the work space is requested, 3D mesh places its first entry point on the overlay stack and calls the overlay responsible for displaying the three orthogonal views. After displaying the three orthogonal views, 3D mesh is loaded into core and the first request from the program 'NUMBER OF INPUT POINTS' appears on the VDU.

The use of the overlay 3D mesh as a coordinating program for the whole 3D mesh generation cannot be overemphasized. It is responsible for loading and coordinating four separate mesh generation overlays, and element connection data generating overlay and last but most important itself.

11.6 NATURAL COORDINATES

11.6.1 One-dimensional Elements

Linear model: (Fig. 11.3)

$$\begin{Bmatrix} 1 \\ x \end{Bmatrix} = \begin{Bmatrix} 1 & 1 \\ x_1 & x_2 \end{Bmatrix} \begin{Bmatrix} L_1 \\ L_2 \end{Bmatrix}$$

or

$$\begin{Bmatrix} L_1 \\ L_2 \end{Bmatrix} = \frac{1}{(x_2 - x_1)} \begin{bmatrix} x_2 & -1 \\ -x_1 & 1 \end{bmatrix} \begin{Bmatrix} 1 \\ x \end{Bmatrix} = \frac{1}{1} \begin{bmatrix} x_2 & -1 \\ -x_1 & 1 \end{bmatrix} \begin{Bmatrix} 1 \\ x \end{Bmatrix}$$

$$\varphi(x) = [N]\vec{\phi}^e = [N_1 \ N_2] \, \varphi^{(e)}$$

$$N_i = a_1^{(i)} L_1 + a_2^{(i)} L_2, \ i = 1, 2$$

where

$$N_1 = \begin{cases} 1 \text{ at node 1} & (L_1 = 1, L_2 = 0) \\ 0 \text{ at node 2} & (L_1 = 0, L_2 = 1) \end{cases}$$

$$N_2 = \begin{cases} 0 \text{ at node 1} & (L_1 = 1, L_2 = 0) \\ 1 \text{ at node 2} & (L_1 = 0, L_2 = 1) \end{cases}$$

$$\therefore \quad N_1 = L_1, \ N_2 = L_2$$

$$\vec{\phi}^e = \begin{Bmatrix} Q_1 \\ Q_2 \end{Bmatrix}^{(e)} = \begin{Bmatrix} \phi(x = x_1) \\ \phi(x = x_1) \end{Bmatrix}^{(e)} = \begin{Bmatrix} \phi(\text{at } L_1 = 1, L_2 = 0) \\ \phi(\text{at } L_1 = 0, L_2 = 1) \end{Bmatrix}^{(e)}$$

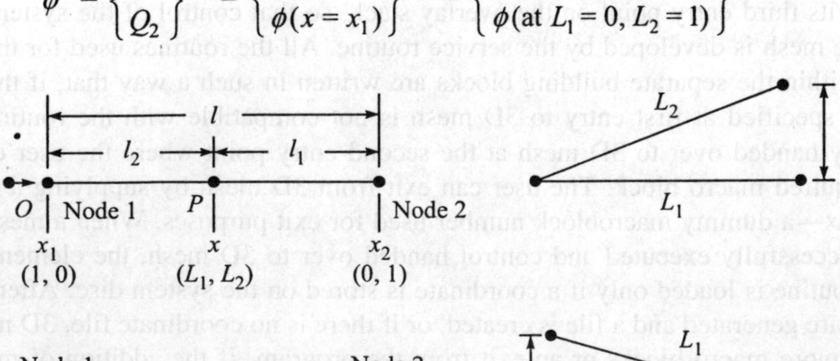

Fig. 11.3 One-dimensional element in natural coordinates.

Quadratic model: (Fig. 11.4)

$$\phi(x) = [N_1 \quad N_2 \quad N_3]\overrightarrow{\varphi^{(e)}}$$

where
$$N_1 = L_1 - 2L_1 L_2$$
$$N_2 = 4L_1 L_2$$
$$N_3 = L_2(2L_2 - 1)$$

$$\phi(e) = \begin{Bmatrix} \phi_1 \\ \phi_2 \\ \phi_3 \end{Bmatrix}^{(e)} = \begin{Bmatrix} \phi(\text{at } L_1 = 1, L_2 = 0) \\ \phi\left(\text{at } L_1 = \frac{1}{2}, L_2 = \frac{1}{2}\right) \\ \phi(\text{at } L_1 = 0, L_2 = 1) \end{Bmatrix}^{(e)}$$

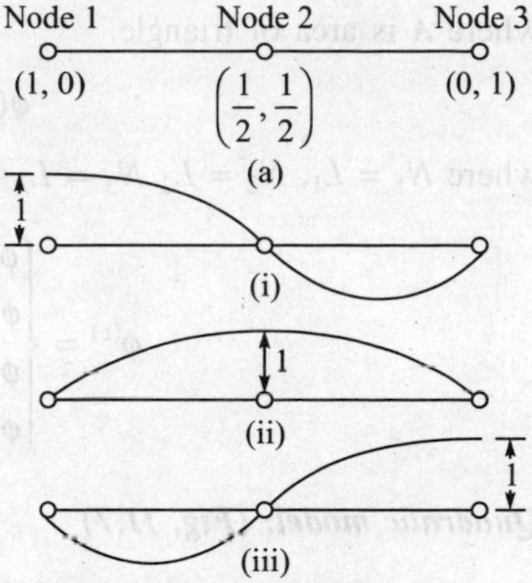

Cubic model: (Fig. 11.5)

$$\phi(x) = [N_1 \quad N_2 \quad N_3 \quad N_4]\phi^{(e)}$$

where
$$N_1 = L_1\left(1 - 9L_1\frac{L_2}{2}\right)$$

$$N_2 = -\frac{9}{2}(L_1 L_2)(1 - 3L_1)$$

$$N_3 = 9(L_1 L_2)\left(1 - 3\frac{L_1}{2}\right)$$

$$N_2 = -\frac{9}{2}(L_1 L_2)(1 - 4)$$

$$\phi(e) = \begin{Bmatrix} \phi_1 \\ \phi_2 \\ \phi_3 \\ \phi_4 \end{Bmatrix} = \begin{Bmatrix} \phi(\text{at } L_1 = 1, L_2 = 0) \\ \phi\left(\text{at } L_1 = \frac{2}{3}, L_2 = \frac{1}{3}\right) \\ \phi\left(\text{at } L_1 = \frac{1}{3}, L_2 = \frac{2}{3}\right) \\ \phi(\text{at } L_1 = 0, L_2 = 1) \end{Bmatrix}^{(e)}$$

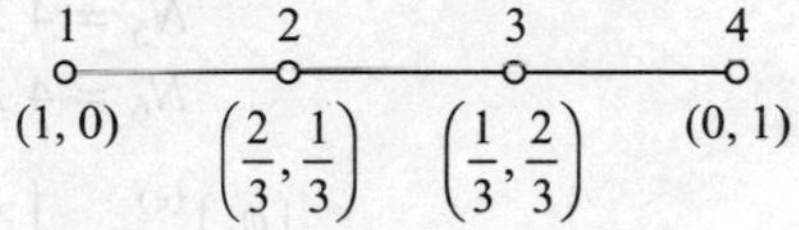

Node 1 (1, 0) Node 2 $\left(\frac{1}{2}, \frac{1}{2}\right)$ Node 3 (0, 1)

(a)

(i)

(ii)

(iii)

Fig. 11.4 Quadratic model.

1 (1, 0) 2 $\left(\frac{2}{3}, \frac{1}{3}\right)$ 3 $\left(\frac{1}{3}, \frac{2}{3}\right)$ 4 (0, 1)

Fig. 11.5 Cubic model.

11.6.2 Two-dimensional Triangular Elements

Linear model: (Fig. 11.6)

$$\begin{Bmatrix} 1 \\ x \\ y \end{Bmatrix} = \begin{bmatrix} 1 & 1 & 1 \\ x_1 & x_2 & x_3 \\ y_1 & y_2 & y_3 \end{bmatrix} \begin{Bmatrix} L_1 \\ L_2 \\ L_3 \end{Bmatrix}$$

or
$$\begin{Bmatrix} L_1 \\ L_2 \\ L_3 \end{Bmatrix} = \frac{1}{2A} \begin{bmatrix} x_2 y_3 - x_3 y_2 & y_2 - y_3 & x_3 - x_2 \\ x_3 y_1 - x_1 y_3 & y_3 - y_1 & x_1 - x_3 \\ x_1 y_2 - x_2 y_1 & y_1 - y_2 & x_2 - x_1 \end{bmatrix} \begin{Bmatrix} l \\ x \\ y \end{Bmatrix}$$

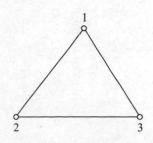

Fig. 11.6 Linear model.

where A is area of triangle.

$$\phi(x, y) = [N_1 \quad N_2 \quad N_3]\phi^{(e)}$$

where $N_1 = L_1$, $N_2 = L_2$, $N_3 = L_3$

$$\phi^{(e)} = \begin{Bmatrix} \phi_1 \\ \phi_2 \\ \phi_3 \\ \phi_4 \end{Bmatrix}^{(e)} = \begin{Bmatrix} \phi(\text{at } L_1 = 1, \ L_2 = L_3 = 0) \\ \phi(\text{at } L_2 = 1, \ L_1 = L_3 = 0) \\ \phi(\text{at } L_3 = 1, \ L_1 = L_2 = 0) \end{Bmatrix}^{(e)}$$

Quadratic model: (Fig. 11.7)

$$\phi(x, y) = [N_1 \quad N_2 \quad \cdots \quad N_6]\phi^{(e)}$$

where $N_i = L_i(2(L_i - 1)$, $i = 1, 2, 3$

$$N_4 = 4\,L_1 L_2$$
$$N_5 = 4\,L_2 L_3$$
$$N_6 = 4\,L_1 L_3$$

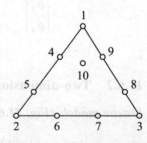

Fig. 11.7 Quadratic model.

$$\phi^{(e)} = \begin{Bmatrix} \phi_1 \\ \phi_2 \\ | \\ | \\ | \\ \phi_6 \end{Bmatrix}^{(e)} = \begin{Bmatrix} \phi(\text{at } L_1 = 1, \ L_2 = L_3 = 0) \\ \phi(\text{at } L_2 = 1, \ L_1 = L_3 = 0) \\ \cdots\cdots\cdots\cdots\cdots\cdots\cdots \\ \cdots\cdots\cdots\cdots\cdots\cdots\cdots \\ \cdots\cdots\cdots\cdots\cdots\cdots\cdots \\ \phi\left(\text{at } L_1 = L_3 = \dfrac{1}{2}, \ L_2 = 0\right) \end{Bmatrix}^{(e)}$$

Cubic model: (Fig. 11.8)

$$\phi(x, y) = [N_1 \quad N_2 \quad \cdots \quad N_{10}]\phi^{(e)}$$

where $N_i = \frac{1}{2} L_i(3L_i - 1)(3L_i - 2)$, $i = 1, 2, 3$

$$N_4 = \left(\frac{9}{2}\right)(L_1 L_2)(3L_1 - 1)$$

$$N_5 = \left(\frac{9}{2}\right)(L_1 L_2)(3L_2 - 1)$$

$$N_6 = \left(\frac{9}{2}\right)(L_2 L_3)(3L_2 - 1)$$

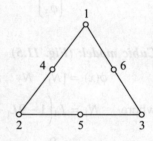

Fig. 11.8 Cubic model.

$$N_7 = \left(\frac{9}{2}\right)(L_2 L_3)(3L_3 - 1)$$

$$N_8 = \left(\frac{9}{2}\right)(L_1 L_3)(3L_3 - 1)$$

$$N_9 = \left(\frac{9}{2}\right)(L_1 L_3)(3L_1 - 1)$$

$$N_{10} = 27L_1L_2L_3$$

$$\phi^{(e)} = \begin{Bmatrix} \phi_1 \\ \phi_2 \\ | \\ | \\ | \\ \phi_{10} \end{Bmatrix} = \begin{Bmatrix} \phi(\text{at } L_1 = 1, \ L_2 = L_3 = 0) \\ \phi(\text{at } L_2 = 1, \ L_1 = L_3 = 0) \\ \dots\dots\dots\dots\dots\dots\dots \\ \dots\dots\dots\dots\dots\dots\dots \\ \dots\dots\dots\dots\dots\dots\dots \\ \phi_6\left(\text{at } L_1 = L_2 = L_3 = \dfrac{1}{3}\right) \end{Bmatrix}^{(e)}$$

11.6.3 Three-dimensional Tetrahedron Elements

Linear model: (Fig. 11.9)

$$\begin{Bmatrix} 1 \\ x \\ y \\ z \end{Bmatrix} = \begin{bmatrix} 1 & 1 & 1 & 1 \\ x_1 & x_2 & x_3 & x_4 \\ y_1 & y_2 & y_3 & y_4 \\ z_1 & z_2 & z_3 & z_4 \end{bmatrix} \begin{bmatrix} L_1 \\ L_2 \\ L_3 \\ L_4 \end{bmatrix}$$

$$\begin{bmatrix} L_1 \\ L_2 \\ L_3 \\ L_4 \end{bmatrix} = \frac{1}{6V} \begin{bmatrix} a_1 & b_1 & c_1 & d_1 \\ a_2 & b_2 & c_2 & d_2 \\ a_3 & b_3 & c_3 & d_3 \\ a_4 & b_4 & c_4 & d_4 \end{bmatrix} \begin{Bmatrix} 1 \\ x \\ y \\ z \end{Bmatrix}$$

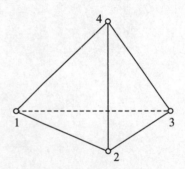

Fig. 11.9 Linear model.

where V is the volume of tetrahedron.

$$\phi(x, y, z) = [N_1, N_2, N_3, N_4]\phi^{(e)}$$

where $N_i = L_i$, $i = 1$ to 4

$$\phi(e) = \begin{Bmatrix} \phi_1 \\ \phi_2 \\ \phi_3 \\ \phi_4 \end{Bmatrix} = \begin{Bmatrix} \phi(\text{at } L_1 = 1, L_2 = 0) \\ \phi\left(\text{at } L_1 = \dfrac{2}{3}, L_2 = \dfrac{1}{3}\right) \\ \phi\left(\text{at } L_1 = \dfrac{1}{3}, L_2 = \dfrac{2}{3}\right) \\ \phi(\text{at } L_1 = 0, L_2 = 1) \end{Bmatrix}^{(e)}$$

Quadratic model: (Fig. 11.10)

$$\phi(x, y, z) = [N_1, N_2, \dots N_{10}]\phi^{(e)}$$

where $N_i = L_i(2L_i - 1)$, $i = 1$ to 4

$$N_5 = 4L_1L_2$$
$$N_6 = 4L_2L_3$$
$$N_7 = 4L_1L_3$$
$$N_8 = 4L_1L_4$$

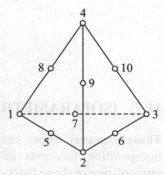

Fig. 11.10 Quadratic model.

$$N_9 = 4L_2L_4$$
$$N_{10} = 4L_3L_4$$

$$\phi^{(e)} = \begin{Bmatrix} \phi_1 \\ \phi_2 \\ | \\ | \\ | \\ \phi_{10} \end{Bmatrix}^{(e)} = \begin{Bmatrix} \phi(\text{at } L_1 = 1, L_2 = L_3 = L_4 = 0) \\ \phi(\text{at } L_2 = 1, L_1 = L_3 = L_4 = 0) \\ \cdots\cdots\cdots\cdots\cdots\cdots \\ \cdots\cdots\cdots\cdots\cdots\cdots \\ \cdots\cdots\cdots\cdots\cdots\cdots \\ \phi\left(\text{at } L_3 = L_4 = \dfrac{1}{2}, L_1 = L_2 = 0\right) \end{Bmatrix}$$

Cubic model: (Fig. 11.11)

$$\phi(x, y, z) = [N_1, N_2, \ldots N_{20}]\phi^{(e)}$$

where $N_i = \frac{1}{2} L_i (3L_i - 1)(3L_i - 2)$, $i = 1$ to 4

$$N_5 = \left(\frac{9}{2}\right)(L_1L_2)(3L_1 - 1)$$

$$N_6 = \left(\frac{9}{2}\right)(L_1L_2)(3L_2 - 1)$$

$$N_7 = \left(\frac{9}{2}\right)(L_2L_3)(3L_2 - 1)$$

$$N_8 = \left(\frac{9}{2}\right)(L_2L_3)(3L_3 - 1)$$

$$N_{17} = 27 L_1L_2L_4$$

$$N_{18} = 27 L_2L_3L_4$$

$$N_{19} = 27 L_1L_2L_3$$

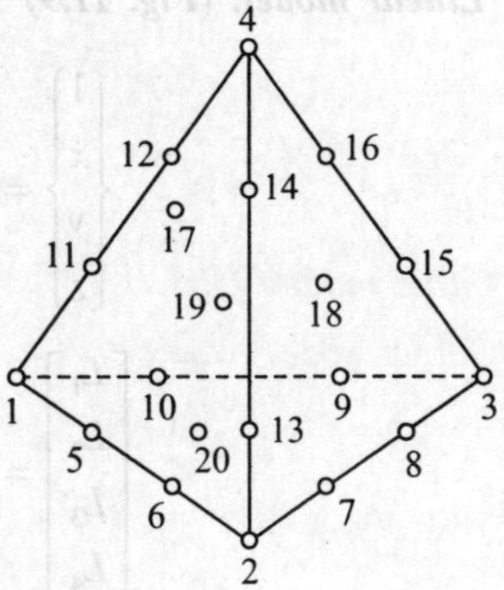

Fig. 11.11 Cubic model.

$$\phi^{(e)} = \begin{Bmatrix} \phi_1 \\ \phi_2 \\ | \\ | \\ | \\ \phi_{20} \end{Bmatrix}^{(e)} = \begin{Bmatrix} \phi(\text{at } L_1 = 1, L_2 = L_3 = L_4 = 0) \\ \phi(\text{at } L_2 = 1, L_1 = L_3 = L_4 = 0) \\ \cdots\cdots\cdots\cdots\cdots\cdots \\ \cdots\cdots\cdots\cdots\cdots\cdots \\ \cdots\cdots\cdots\cdots\cdots\cdots \\ \phi\left(\text{at } L_1 = L_2 = L_3 = L_4 = \dfrac{1}{3}\right) \end{Bmatrix}^{(e)}$$

11.7 ISOPARAMETRIC ELEMENTS

Those elements whose shape (or geometry) and field variables are described by the same interpolation functions of the same order are known as **isoparametric elements**. An interpolation or shape function is a function which has unit value at one nodal point and zero value at all other nodal points.

11.7.1 Rectangular Elements

Linear model: (Fig. 11.12)

$$N_i = \frac{1}{4}(1 + \xi_0)(1 + \eta_0),\ i = 1 \text{ to } 4$$

where $\qquad \xi_0 = \xi\,\xi_i \quad$ and $\quad \eta_0 = \eta\eta_i$

Quadratic model: (Fig. 11.13)

Corner nodes:

$$N_i = \frac{1}{4}(1 + \xi_0)(1 + \eta_0)(\xi_0 + \eta_0 - 1)$$

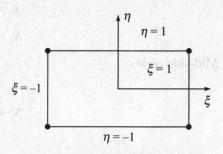

Fig. 11.12 Linear model.

Mid-side nodes:

$$\left[\begin{array}{l} \xi_i = 0,\ N_i = \frac{1}{2}(1 - \xi^2)(1 + \eta_0) \\[2mm] \eta_i = 0,\ N_i = \frac{1}{2}(1 + \xi_0)(1 - \eta^2) \end{array}\right]$$

Cubic model: (Fig. 11.14)

Corner nodes:

$$N_i = \frac{1}{32}(1 + \xi_0)(1 + \eta_0)[-10 + 9(\xi^2 + \eta^2)]$$

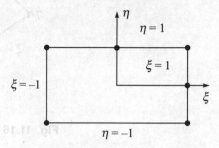

Fig. 11.13 Quadratic model.

Mid-side nodes:

$$\xi_i = \pm 1 \quad \text{and} \quad \eta_i = \pm\frac{1}{3}$$

$$N_i = \frac{9}{32}(1 + \xi_0)(1 + \eta^2)(1 + 9\eta_0)$$

11.7.2 Quadrilateral Elements

Linear model: (Fig. 11.15)

$$N_i = \frac{1}{4}(1 + \varepsilon\varepsilon_i)(1 + \eta\eta_i)$$

$$x = \sum_{i-1}^{n} N_i(\varepsilon, \eta)x$$

$$y = \sum_{i=1}^{n} N_i(\varepsilon, \eta)y$$

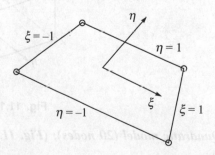

Fig. 11.14 Cubic rectangular model.

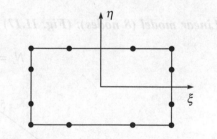

Fig. 11.15 Linear quadrilateral model.

Quadratic model: (Fig. 11.16)

Corner nodes:

$$N_i = \frac{1}{4}(1 + \varepsilon\varepsilon_i)(1 + \eta\eta_i) - \frac{1}{4}(1 + \varepsilon^2)(1 + \eta\eta_i) - \frac{1}{4}(1 + \varepsilon\varepsilon_i)(1 + \eta^2)$$

Mid-side nodes:

$$\varepsilon_i = 0, \quad \eta_i = \pm 1$$

$$N_i = \frac{1}{2}(1 + \varepsilon^2)(1 + \eta\eta_i)$$

Mid-side nodes:

$$\varepsilon_i = \pm 1, \quad \eta_i = 0$$

$$N_i = \frac{1}{2}(1 + \varepsilon\varepsilon_i)(1 + \eta^2)$$

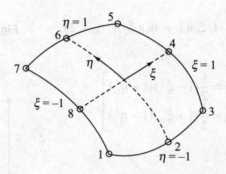

Fig. 11.16 Quadratic quadrilateral model.

11.7.3 Hexahedron Elements

Linear model (8 nodes): (Fig. 11.17)

$$N_i = \frac{1}{8}(\varepsilon\varepsilon_i)(1 + \eta\eta_i)(1 + \xi\xi_i)$$

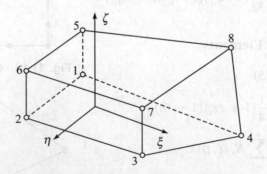

Fig. 11.17 Eight-node hexahedron.

Quadratic model (20 nodes): (Fig. 11.18)

Corner nodes:

$$N_i = \frac{1}{8}(1 + \varepsilon\varepsilon_i)(1 + \eta\eta_i)(1 + \xi\xi_i)(\varepsilon\varepsilon_i + \eta\eta_i + \xi\xi_i - 2)$$

Mid-side nodes:

$$\varepsilon_i = 0, \ \eta_i = \pm1, \ \xi_i = \pm1$$
$$N_i = \frac{1}{4}(1 + \varepsilon^2)(1 + \eta\eta_i)(1 + \xi\xi_i)$$

Mid-side nodes:

$$\varepsilon_i = \pm1, \ \eta_i = \pm1, \ \xi_i = 0$$
$$N_i = \frac{1}{4}(1 + \varepsilon\varepsilon_i)(1 + \eta\eta_i)(1 + \xi^2)$$

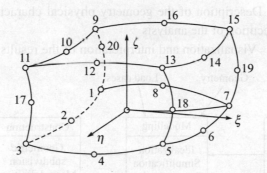

Fig. 11.18   20-node quadratic model.

11.8   CAD APPLICATION TO FEM

The idealization of the structure by suitable elements of regular shape manually is a time consuming process and subject to considerable errors. Automating this process can greatly enhance the use of the technique. Using a CAD system in the data generation process has considerable advantages as the user can see the element connections and position of each element directly on the display as element generation is in progress. The user is also able to change a mesh to suit a particular problem. After the analysis the data can be presented graphically to assess the results.

A CAD system for automatic mesh generation and presentation of data in FEM should provide the following facilities:

1. Generation of mesh for 2D or 3D structures
2. Ability to represent curved edges and surfaces
3. Ability to control element density and generation of non-uniform meshes
4. Facility for concentrating and grading the mesh over any region
5. Speedy node and element numbering system which will lead to computation efficiency
6. Facility to display idealization
7. Ability to rotate the idealization from any desired angle
8. Option to display any portion of the model
9. Facility for modification of data for alternative idealization

10. Preparation of input data for analysis programs
11. Easy to use with minimum of input
12. Economically feasible
13. User-orientated and easy to use with minimum of input

11.9 FINITE ELEMENT MODELLING

Following are the essentially three stages involved in applying the FEM to engineering problems:

1. *Pre-processing* Description of the geometry physical characteristics and the mesh
2. *Processing* Execution of the analysis
3. *Post-processing* Visualization and interpretation of the results

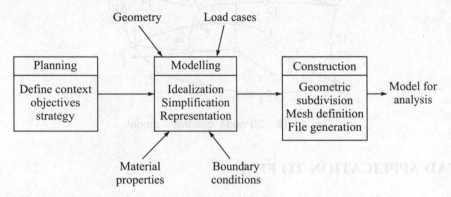

Fig. 11.19 Stages in finite element pre-processing.

11.9.1 Pre-processing

The pre-processing module must accomplish the following three functions: (a) Description of the geometry of the object, (b) mesh generation, (c) definition of the regions and the boundaries. The stages in finite element pre-processing are shown in Fig. (11.19). A commonly applied technique for the sub-division of shapes (called **mesh generation**) is to divide the overall shape of the part to be meshed into a number of large fairly regular regions, which are then divided by software into individual nodes and elements. The nodes are defined by their coordinates while the elements are characterized by their type and a list of their nodes. The meshing software normally has facilities for filling various sorts of shape—triangles, trapezia, blocks, wedges and so on, with a variety of element types. The finite element model also contributes to data preparation, checking the correctness of large data files by plotting the mesh, by element shrinking and finding such things as missing elements and inspecting the boundary conditions.

11.9.2 Processing

The execution of the analysis can be carried out to find the solution of system of equations. For non-linear problems, any one of the following methods may be used to find their solution:

    (a) Newton–Raphson method

    (b) Continuation methods

    (c) Minimization methods

    (d) Perturbation methods

Two large classes of methods are used to solve sets of equations: point or block methods of relaxation or global matrix methods.

11.9.3 Post-processing

The post-processor performs two tasks:

    (a) Extraction of significant information

    (b) Synthetic presentation of the numerical data via graphics facilities

This makes the data more understandable and easier to interpret.

11.9.4 Mesh Generation

Mesh generation refers to the generation of nodal coordinates and elements. It also includes the automatic numbering of nodes and elements based on a minimal amount of user supported data. Automatic mesh generation reduces errors and saves a great deal of user time, thereby reducing the FEA cost.

    With the widespread use of computer graphics and CAD technology, mesh generation has been a target for automation. There is a wide variety of algorithms, schemes and methods for mesh generation. Broadly these methods may be classified into semi-automatic and fully automatic. A fully automatic mesh generation is a method in which only the shape (both geometry and topology) or the object to be meshed and the mesh attributes (mesh density, element type, boundary conditions, loads, etc.) are required as input.

11.10 GENERAL STRUCTURE OF A FINITE ELEMENT ANALYSIS PROCEDURE

The analysis of a structure during its design process is accomplished by the solution of the partial differential equations which describe the given model. This involves the following three steps (Fig. 11.20):

    (a) The description of the geometry the physical characteristics and the mesh (pre-processing)

    (b) The application of the finite element analysis (solution)

    (c) The visualization and interpretation of the results of the solution (post-processing)

    These three steps are quite distinct and correspond to creating, on the programming level, the three distinct modules:

    (i) The module to enter the data

    (ii) The module to perform the analysis

    (iii) The module to interpret and display the results

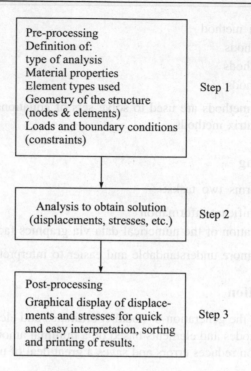

Fig. 11.20 Steps in finite element analysis.

11.10.1 Pre-processing Module

The pre-processing module is used for entering all the information necessary to define the problem. This data relates to the discretization of the structure and the representation of its physical behaviour.

The pre-processing module must accomplish the following three functions:

- Description of the geometry of the object in terms of the chosen element types
- Mesh generation
- Definition of loading and boundary condition

The mesh generation consists of forming a collection of nodes and elements which form an acceptable discretization of the structure to be analyzed. Such a discretization must conform to the boundaries of the component and the interfaces between different regions. The shape of the elements must not be too irregular (elongated) and should, as much as possible, resemble the standard elements (triangles or tetrahedra, squares or cubes, etc.).

The nodes are defined by their coordinates while the elements are characterized by their type and a list of their nodes. Certain formulations involve boundary integrals. In this case the designer must define not only interior finite elements (volume elements in three dimensions, surface elements in two) but also boundary finite elements (surface elements in three dimensions, line elements in two) on the corresponding boundaries.

It is also necessary to describe the physical characteristics of the elements, such as description of material properties (e.g., thermal conductivity, modulus of elasticity, density,

Poisson's ratio); description of heat sources, description of boundary conditions, and description of initial conditions (for time dependent problems).

Generally, this information is entered region by region of the component. The description of the geometry is sometimes implicitly linked to the meshing, however the trend at present is to separate the two. The description of the geometry is done first and then the mesh is generated. The most extreme case of this separation is the use of two separate specialized programs: a solid modeller for the geometrical input and a mesh generator for the discretization.

More than one type of element can be used in a model, including two-dimensional elements where all forces and displacements act in a plane; axisymmetric elements for solids of revolution that are also two-dimensional but have node displacements in radial and axial directions; three-dimensional solid elements where forces and displacements act in all three directions or when a structure has a complex geometry that does not allow two-dimensional analysis. Finally, specialty elements such as plate and beam types are used where sections of a structure behave according to conventional plate and beam theory.

Elements with nodes at vertices generally model only a linear variation of displacement with constant strain. Elements are also available with nodes at edges. These elements can model high-order variations in displacements and strain within an element. Many elements with nodes along their sides are *Isoparametric* where additional nodes allow element sides to form curved boundaries. These elements also increase accuracy. Isoparametric models can cut model construction times but require more processing time per element.

11.10.2 Pre-processing from CAD Model

In computer aided engineering the product development starts with a CAD model. This model is then subjected to several types of analysis (FEA, mechanisms analysis, etc.) to verify the model and explore modifications to design. It requires the transfer of CAD database to the finite analysis software. This can be carried out broadly in two ways:

(a) *Using a preprocessor package*

Commercial packages are available which will carry out pre-processing of CAD model, so that time consuming tasks of node generation and element generation can be avoided. For example, if the CAD model is made by a PRO/Engineer, CADDS 5 or I-DEAS software package, the FEA model can be created by a finite element modelling interface software. The automatic mesh generation facility (volume meshing or area meshing as the case may be) takes out the tedium of meshing. The model is more accurate as the redefinition of the geometry is avoided.

There are several ways in which finite element models are created by this technique. Different mesh types are used for creating the models. These are:

(i) Shell meshing. Shell meshing is performed using triangular mesh elements. This is designed to mesh surfaces. Surfaces are designated for shell meshing in three ways:

(a) *Pair meshing* In this case, part surfaces are paired so that material is sandwiched between selected surfaces. These surfaces are then compressed together to a single surface using thickness as a property. The shell meshing is done in the compressed plane.

  (b) *Boundary meshing*   The surface of a part is shell meshed using this option.

  (c) *Quilt meshing*   Two-dimensional features are shell meshed using this option.

(ii) Solid meshing. Solid meshing of interiors of solid is done with tetrahedral elements (Fig. 11.21)

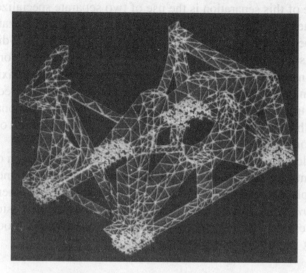

Fig. 11.21   Tetrahedral element meshing.

(iii) Mixed meshing. Shell and tetrahedral meshing technique can be used advantageously for many parts. If the model contains thin and thick parts, thin parts can be shell meshed and thick parts can be meshed with tetrahedral elements. The corresponding tetrahedral mesh elements are forced to coincide to provide analytical continuity.

11.10.3   Procedure for Creation of Finite Element Model from Solid Model

The following procedure is recommended to create a finite element model from a solid model:

  (a) Simplify the model by removing features which are unnecessary to the finite element analysis. Features like rounds, chamfers and small holes can be suppressed. Dimensions can be modified to align edges and surfaces are very nearly aligned. Another way to simplify the model is to use symmetry of the part to be modelled. Only a portion of a symmetrical object needs modelling.

  (b) Add loads and boundary conditions to surfaces, edges and datum points on the model.

  (c) If the model is to be shell meshed, define shell model. As mentioned earlier a shell model represents solid features with a network of surfaces formed by pairing and compressing part surfaces together. Pairing of surfaces can be done automatically or manually.

  (d) Assign material properties to shell pairs and solid elements.

  (e) Use mesh control to determine fineness of the mesh.

  (f) Mesh the model.

(g) Goto Step 1 if needed.

(h) Output to a specially formatted file for the finite element analysis.

11.10.4 Use of Neutral Files

A neutral file like IGES file can be used to transfer the geometry to FEA. Here the CAD model is converted to an IGES file which is then input to the FEA software using a utility software. The geometry can then be used for meshing and additional information necessary can be input to complete the model.

11.10.5 Analysis

The analysis part computes the unknowns in the finite element problem, i.e., it solves the linear or non-linear system of equations based on the variational or the projective formulation.

Input to analysis module is the finite element model, the physical characteristics and the boundary conditions (pre-processor file). Its output is the value of the unknown quantity at each of the nodes of the grid.

Two large classes of methods are used to solve these sets of equations: point or block methods of relaxation or global matrix methods. The latter, more popular today, requires several steps:

(a) Creation of sub-matrices and sub-vectors corresponding to each individual finite element

(b) Assembly of these elementary matrices and vectors to build the system matrix and right-hand vector

(c) Solution of the linear system of equations

The solution of linear algebraic systems can be done in several ways: by direct methods (Gauss, Choleski), semi-direct methods or block iterative methods (Gauss–Seidel).

When the system of equations is non-linear, these operations are repeated in an iterative scheme (Gauss–Seidel, Newton–Kantorovich, Newton–Raphson). When the problem is time dependent, these steps must be repeated for each time step (implicit or explicit finite difference methods, Crank–Nicholson, Predictor–Corrector, Guyan reduction techniques, and Householder method).

11.10.6 Post-processor

The pre-processor describes the problem which is then solved by the computational modules. However, the results of this solution are not always directly useful for the following reasons:

- The state variables, computed at the finite element nodes, describe the state of the system in mathematical form. However, sometimes the physical meaning is not always clear (e.g., vector potential for an electromagnetic problem).
- The large amount of data coming from the solution phase (several thousands of nodal values) are often too much to be understood without further processing.

The post-processor then performs two tasks:

(i) *Extraction of significant information.* The information may be related to local quantities (magnetic flux density, stress distribution) or global quantities (heat flux, electromagnetic forces, etc).

(ii) *Synthetic presentation of the numerical data via graphics facilities.* This makes the data more understandable and easier to interpret (stress plots, displacement plots, isothermal plots, temperature vs time curves, magnetic field along a line, etc.).

REVIEW QUESTIONS

11.1 Explain what you understand by the finite element method.

11.2 Write a short note on automatic mesh generation with an illustrative example.

11.3 Briefly explain the steps to be followed in manually carrying out the finite element solution to a physical problem.

11.4 Give details of various types of element shapes usually employed for modelling components.

11.5 What do you understand by the finite element model? Give an example of modelling a component.

11.6 What are the steps to be carried out for solving a physical problem with the help of a FEM software?

11.7 Explain the function served by a pre-processor in FEM.

11.8 Explain the functions served by a post-processor in FEM.

11.9 What different methods are available for processing while using a CAD system in conjunction with a FEM software?

11.10 How CAD systems can be applied to FEM? Explain briefly.

11.11 What considerations are taken into account while discretizing the domain for FEM? Explain briefly.

11.12 What are shape functions? How would you determine shape functions for one-dimensional cubic model?

11.13 What are isoparametric elements? Determine the shape functions for a quadrilateral isoparametric element.

11.14 Explain the procedure of formulating plane stress problems by the finite element method.

11.15 Why interpolating functions in the form of polynomials are used for the field variation? List the considerations to be taken while choosing the order of the polynomial.

11.16 Discuss the direct methods of formulating characteristic matrices and vectors for the FEM.

11.17 How would you determine the characteristic matrix corresponding to the global coordinates from the local coordinates?

PRACTICE PROBLEMS

11.1 A symmetrical indeterminate truss subjected to a pair of parting forces is shown in Fig. 11.22. All bars have the same axial rigidity EI = 2 × 10E7. There is no connection between the two diagonal bars. Find the displacement at all joints and axial force for each bar.

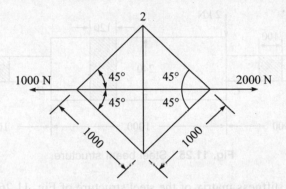

Fig. 11.22 Indeterminate truss.

11.2 A three-bar steel truss is shown in Fig. 11.23. All bars have the same area of cross-section of 100 mm. If the temperature of bar 1–2 is increased above a certain uniform temperature of the other elements truss, find the displacement of joint and the axial force in the three bars.

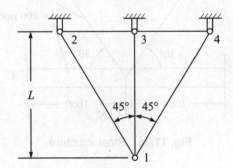

Fig. 11.23 Three-bar steel truss.

11.3 Derive the stiffness matrix for a beam element with constant cross-section.

11.4 Using the beam elements to model at a steel beam structure as shown in Fig. 11.24. Find the deflected shape and the reaction forces.

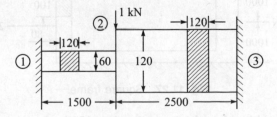

Fig. 11.24 Steel beam structure.

11.5 Using three elements to model the steel beam structure as shown in Fig. 11.25. Find the deflected shape and the reaction forces at the ends.

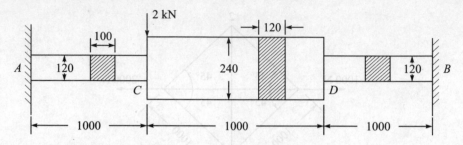

Fig. 11.25 Steel beam structure.

11.6 Formulate the stiffness matrix of the steel structure of Fig. 11.26. A beam connected to a truss is shown.

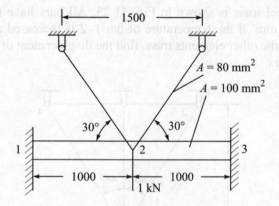

Fig. 11.26 Steel structure.

11.7 A square CI frame subjected to a pair of forces is shown in Fig. 11.27. Find the deflected shape.

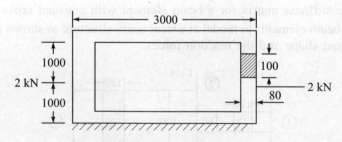

Fig. 11.27 Square frame.

11.8 A portal frame with side columns inclined at 45° is shown in Fig. 11.28. The cross-section of the members is a square with 100 mm side. A horizontal load 2 kN is applied at ②. Find the deflected shape.

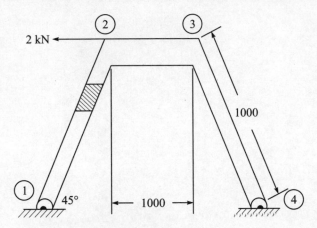

Fig. 11.28 A portal frame.

11.9 Find centre of deflection of the steel ring shown in Fig. 11.29.

11.10 Obtain stiffness in buckling for a plane frame element oriented arbitrarily in a two-dimensional plane.

11.11 Obtain stiffness equation for the 8-DOF rectangular element in the state of plane stress (Fig. 11.30).

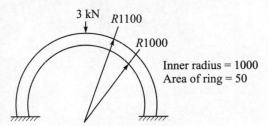

Inner radius = 1000
Area of ring = 50

Fig. 11.29 Partial ring.

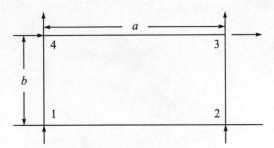

Fig. 11.30 Rectangular element.

11.12 Derive the shape function from Fig. 11.31.

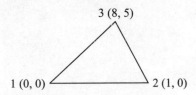

Fig. 11.31 Triangular elements.

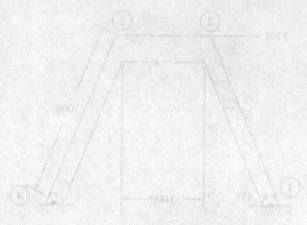

CAM

Numerical Control
Production Systems

NUMERICAL CONTROL IN PRODUCTION SYSTEMS

INTRODUCTION

The introduction of numerical control (NC) in 1952 opened a new era in automation. NC is based on digital computer principles, which was a new technology at that time. The new era of automation, which started with the introduction of NC machine tools, was undoubtedly stimulated by the digital computer. Digital technology and computers enabled the design of more flexible automation systems, namely systems which can be adapted by programming to produce a new product in a short time. The advent of computers has introduced the concept of automation in industries. The concept of automation was introduced to reduce the role of the operation in conventional machines and cope with changes in product design.

The main components of automation are manufacturing/processing machines, computer, sensors, drives and controller.

The concept of numerical control evolved during the Second World War as a consequence of the need for advanced manufacturing techniques to machine complex aircraft parts. Numerical control is essentially an application of the digital technology to control a machine.

12.1 WHAT IS NUMERICAL CONTROL?

Numerical control (NC) is the technique of giving instructions to a machine in the form of a code which consists of numbers, letters of the alphabet, punctuation marks and certain other symbols. Controlling a machine tool by means of a prepared program is known as **numerical control**.

NC equipment has been defined by the Electronic Industries Association (EIA) as:

"A system in which actions are controlled by the direct insertion of numerical data at some point. The system must automatically interpret at least some portion of this data".

Instructions are supplied to the machine as blocks of information. A block of information is a group of commands sufficient to enable the machine to carry out one individual machining operation. Each block is given a sequence number for identification.

A set of instructions forms an NC program. When the instructions are organized in a logical manner they direct the machine tool to carry out a specific task. It is thus termed as **part program**.

In a typical NC system, the numerical data which is required for producing a part is maintained on a punched tape and is called the **part program**. The part program is arranged in the form of blocks of information, where each block contains the numerical data required to produce one segment of the work piece. The punched tape is moved forward by one block each time the cutting of a segment is completed.

Preparing the part program for a NC machine tool requires a part programmer. The part programmer must possess knowledge and experience of tools, cutting fluids, machinability data and fixture design techniques.

Part programmers must be familiar with the function of NC machine tools and machining processes and have to decide on the optimal sequence of operations. Part programs are written manually or by using a computer-aided language, such as automated program tool (APT).

12.2 THE NC PROCEDURE

To utilize numerical control in manufacturing, the following steps must be accomplished:

(a) *Process planning:* The engineering drawing of the workpart must be interpreted in terms of the manufacturing process to be used. This step is referred to as the process planning and it is concerned with the preparation of a route sheet. The route sheet is a listing of the sequence of operations which must be performed on the workpart. It is called a **route sheet** because it also lists the machines through which the part must be routed in order to accomplish the sequence of operations. We assume that some of the operations will be performed on one or more NC machines.

(b) *Part programming:* A part programmer plans the process for the portions of the job to be accomplished by NC. Part programmers are knowledgeable about the machining process and they have been trained to program for numerical control. They are responsible for planning the sequence of machining steps to be performed by NC and to document these in a special format. There are two ways to program for NC:

Manual part programming
Computer-assisted part programming

In manual part programming, the machining instructions are prepared on a form called a **part program manuscript.** The manuscript is a listing of the relative cutter/workpiece position which must be followed to machine the part. In computer-assisted part programming, much of the tedious computational work required in manual part programming is transferred to the computer. This is especially appropriate for complex work piece geometries and jobs with many machining steps. Use of the computer in these situations results in significant savings in part programming time.

(c) *Tape preparation:* A punched tape is prepared from the part programmer's NC process plan. In manual part programming, the punched tape is prepared directly from the part program manuscript on a typewriter like device equipped with tape punching capability. In computer-assisted part programming, the computer interprets the list of part programming instructions, performs the necessary calculations to convert this into a detailed set of machine tool motion commands, and then controls a tape punch device to prepare the tape for the specific NC machine.

(d) *Tape verification:* After the punched tape has been prepared, a method is usually provided for checking the accuracy of the tape. Sometimes the tape is checked by running it through a computer program which plots the various tool movements (or table movements) on paper. In this way, major errors in the tape can be discovered . The "acid test" of the tape involves trying it out on the machine tool to make the part. A form or plastic material is sometimes used for this tryout. Programming errors are not uncommon, and it may require about three attempts before the tape is correct and ready to use.

(e) *Production:* The final step in the NC producer is to use the NC tape in production. This involves ordering the raw workparts, specifying and preparing the tooling and any special fixturing that may be required, and setting up the NC machine tool for job. The machine tool operator's function during production is to load the raw workpart in the machine and establish the starting position of the cutting tool relative to the work piece. The NC system then takes over and machines the part according to the instructions on tape. When the part is completed, the operator removes it from the machine and loads the next part.

12.3   NC COORDINATE SYSTEMS

In order for the part programmer to plan the sequence of positions and movements of the cutting tool relative to the work piece, it is necessary to establish a standard axis system by which the relative positions can be specified. Using an NC drill press as an example, the drill spindle is in a fixed vertical position, and the table is moved and controlled relative to the spindle. However, to make things easier the programmer adopt the view point that the work piece is stationary while the drill bit is moved relative to it. Accordingly, the coordinate system of axes is established with respect to the machine table.

Two axes, x and y, are defined in the plane of the table, as shown in Fig. 12.1. The z-axis is perpendicular to this plane and movement in z direction is controlled by the vertical motion of

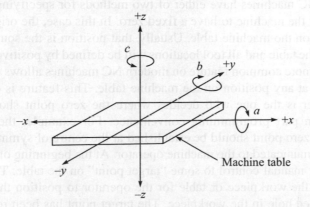

Fig. 12.1   NC machine tool axis system per milling and drilling operations.

the spindle. The positive and negative directions of motion of tool relative to table along these axes are as shown in Fig. 12.2. NC drill presses are classified as either two-axis or three-axis machines, depending on whether they have the capability to control the z-axis.

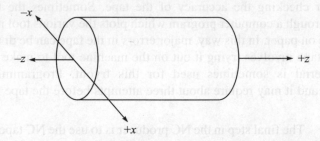

Fig. 12.2 NC machine tool axis system for turning operation.

A numerical control milling machine and similar machine tools (boring mill, for example) use an axis system similar to that of the drill press. However, in addition to the three linear axes, these machines may possess the capacity to control one or more rotational axes. Three rotational axes are defined in NC: the a, b and c axes. These axes specify angles about the x, y and z axes, respectively. To distinguish positive from negative angular motions, the right-hand rule can be used. Using the right hand with the thumb pointing in the positive linear axis direction (x, y or z), the fingers of the hand or curled to point in the positive rotational direction.

For turning operations, two axes are normally required to command the movement of the tool relative to the rotating work piece. The z axis is the axis of rotation of the workpart, and x axis defines the radial location of the cutting tool. This arrangement is illustrated in Fig. 12.2.

The purpose of the coordinate system is to provide a means of locating the tool in relation to the workpiece. Depending on NC machine, the part programmer may have different options available for specifying this location.

12.3.1 Fixed Zero and Floating Zero

The programmer must determine the position of the tool relative to the origin (zero point) of the coordinate system. NC machines have either of two methods for specifying the zero point. The first possibility is for the machine to have a fixed zero. In this case, the origin is always located at the same position on the machine table. Usually, that position is the southwest corner (lower left-hand corner) of the table and all tool locations will be defined by positive x and y coordinates.

The second and more common feature on modern NC machines allows the machine operator to set the zero point at any position on the machine table. This feature is called **floating zero**. The part programmer is the one who decides where the zero point should be located. The decision is based on part programming convenience. For example, the work part may be symmetrical and the zero point should be established at the centre of symmetry. The location of the zero point is communicated to the machine operator. At the beginning of the job, the operator moves the tool under manual control to some "target point" on the table. The target point is the convenient place on the workpiece or table for the operator to position the tool. For example, it might be a predrilled hole in the workpiece. The target point has been referenced to the zero point by the part programmer. In fact, the programmer may have selected the target point as the zero point for tool positioning. When the tool has been positioned at the target point, the machine operator presses a "zero" button on the machine tool console, which tells the machine where the origin is located for subsequent tool movements.

12.3.2 Absolute Positioning and Incremental Positioning

Another option available to the part programmer is to use either an absolute system of tool positioning or an incremental system. Absolute positioning means that the tool locations are always defined in relation to the zero point. If a hole is to be drilled at a spot that is 8 in. above the x axis and 6 in. to the right of the y axis, the coordinate location of the hole would be specified as $x = +6.000$ and $y = +8.000$. By contrast, the incremental positioning means that the next tool location must be defined with reference to the previous tool location. If in our drilling example, suppose that the previous hole had been drilled at an absolute position of $x = +4.000$ and $y = +5.000$. Accordingly, the incremental position instructions would be specified as $x = 2.000$ and $y = +3.000$ in order to move the drill to the desired spot. Figure 12.3 illustrates the difference between absolute and incremental positioning.

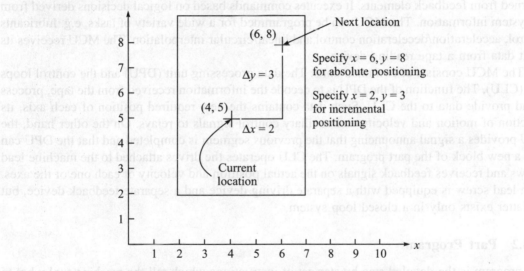

Fig. 12.3 Absolute versus incremental positioning.

12.4 ELEMENTS OF NC SYSTEM

A typical NC system consists of part program, machine control unit and machine tool which are depicted in Fig. 12.4.

12.4.1 Machine Control Unit

The machine control unit (MCU) is the brain of an NC machine. The information contained in the part program is read by the MCU which in turn generates signals to appropriately control the tool movement. The functions that it performs vary with the complexity of the machine tool and

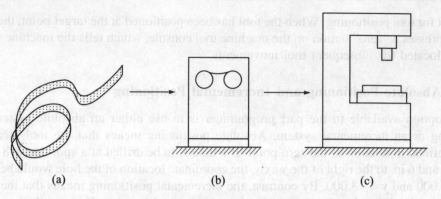

(a)   (b)   (c)

Fig. 12.4 Three basic components of a numerical (NC) System.

the tasks that it is required to execute. The MCU converts the coded information in the part program into voltage or current pulses of varying frequency or magnitude, which are fed to the drive motors or relays to control the operation of the machine. The MCU decodes input data returned from feedback elements. It executes commands based on logical decisions derived from all system information. The MCU can be programmed for a wide variety of tasks, e.g. lubricants control, acceleration/deceleration control and linear/circular interpolation. The MCU receives its input data from a tape reading device.

The MCU consists of two main units: The data processing unit (DPU) and the control loops unit (CLU). The function of the DPU is to decode the information received from the tape, process it and provide data to the CLU. Such data contains the new required position of each axis, its direction of motion and velocity and auxiliary control signals to relays. On the other hand, the CLU provides a signal announcing that the previous segment is completed and that the DPU can read a new block of the part program. The CLU operates the drives attached to the machine lead screws and receives feedback signals on the actual position and velocity of each one of the axes. Each lead screw is equipped with a separate driving device and a separate feedback device, but the latter exists only in a closed loop system.

12.4.2 Part Program

Part program is the detailed step by step set of instructions which tell the machine tool what to do. It is coded in numerical or symbolic form on some type of input medium that can be interpreted by the machine control unit. Input medium used are punched tape, punched cards, magnetic tape and 35 mm motion picture film. The input to NC system is entered to the machine control unit either manually or by means of direct link with computer. The part program is prepared by part programmer.

12.4.3 Machine Tool

The machine tool consists of worktable and spindle as well as motors and controls necessary to drive them. It also contains cutting tools, work fixtures and other auxiliary equipment needed in machining operation. Machine tool takes the instructions from machine control unit and functions accordingly, and sends feedback signals so that the instructions are properly executed by the machine.

12.5 CLASSIFICATION OF NC SYSTEMS

The classification of NC machine tool systems can be done in four ways:

- According to the type of machine: Point-to-point, straight-cut and continuous path
- According to the programming method: Absolute and incremental
- According to the type of control system: Open-loop and closed-loop

12.5.1 Point-to-Point

Point-to-point machines move only in straight lines. They are limited to drilling, reaming, boring, etc. and straight milling cuts parallel to a machine axis. When making an axis move, all affected drive motors run at the same speed. When one axis motor has moved the instructed amount, it stops while the other motor continues until its axis has reached its programmed location. The point-to-point positioning NC system is illustrated in Fig. 12.5.

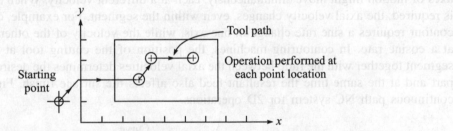

Fig. 12.5 Point-to-point (positioning) NC system.

The simplest example of a point-to-point (PTP) NC machine tool is a drilling machine. In a drilling machine, the work piece is moved along the axes of motion until the centre of the hole to be drilled is exactly beneath the drill. Then the drill is automatically moved towards the work piece, the hole is drilled and the drill moves out in a rapid traverse feed. The work piece moves to a new point and the above sequence of actions are repeated.

12.5.2 Straight-cut NC

Straight-cut control systems are capable of moving the cutting tool parallel to one of the major axes at a controlled rate suitable for machining. It is, therefore, appropriate for performing milling operation to fabricate work pieces of rectangular configurations. With this type of NC system it is not possible to combine movements in more than a single axis direction. Therefore, angular cuts on the work piece would not be possible. An example of straight-cut operation is shown in Fig. 12.6. An NC machine capable of straight-cut movements is also capable of PTP movements.

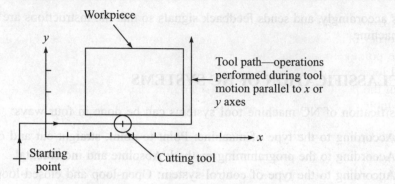

Fig. 12.6 Straight-cut system.

12.5.3 Continuous Path

A continuous path machine has the ability to move its drive motors at varying rates of speed while positioning the machine which facilitates cutting of arc segments and angles. The most common type of continuous path operations are milling and lathe operations. In continuous path machine, the tool is cutting while the axes of motion are moving, as for example, in a milling machine. All axes of motion might move simultaneously, each at a different velocity. When a non-linear path is required, the axial velocity changes, even within the segment. For example, cutting a circular contour requires a sine rate change in one axis, while the velocity of the other axis is changed at a cosine rate. In contouring machines, the position of the cutting tool at the end of each segment together with the ratio between the axial velocities determines the desired contour of the part and at the same time the resultant feed also affects the surface finish. Figure 12.7 shows continuous path NC system for 2D operations.

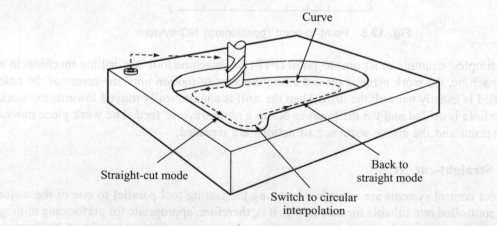

Fig. 12.7 Contouring (continuous path) NC system for two-dimensional operations.

12.5.4 Absolute Programming

Absolute positioning is another type of programming system. In this system, the tool locations are always defined in relation to point zero. The position commands are given as absolute distances from the reference point. The reference point can be defined outside the work piece or at a corner of the work piece. The reference point or point zero could be fixed or floating. When the point zero is fixed, the origin is always located at the same position on the machine table. All locations must be defined by positive x and y coordinates relative to that fixed origin.

When the point zero is floating, the operation can set the point zero at any position on the machine table. This point zero is decided based on part programming convenience.

Advantages of absolute programming

- In cases of interruptions that force the operator to stop the machine, the cutting tool automatically returns to previous position. Since it always moves to the absolute coordinate called for and the machining proceeds from the same block where it was interrupted.
- Possibility of easily changing the dimensional data in the part program whenever required.
- When describing contours and positions, it is always preferable to employ absolute dimensioning, because the first incorrect dimensioning of an individual point has no effect on the remaining dimensions and the absolute system is easier to check for errors.

12.5.5 Incremental Programming

Incremental positioning is a programming system used to define the position of the tool in NC machines. In an incremental system, the next tool location must be defined with reference to the previous tool location. The dimensional data applied to the system will be a distance increment measured from the preceding point at which the axis of motion was present.

Advantages of incremental programming

- If manual programming is used with incremental systems the inspection of the part program, before punching the tape is easy. Since the end point, when machining a part is identical to the starting point, the sum of the position, commands (for each axis separately) must be zero. A non-zero sum indicates that an error exists.
- The performance of the incremental system can be checked by a closed-loop tape. The last position command on the tape the table to return to the initial position.
- Mirror-image programming is facilitated with the incremental systems.
- Incremental dimension programming is advantageous for certain individual partial contours in a work piece are repeated several times, and the associated program sections can be employed several times without a coordinate shift.

12.5.6 Open-loop and Closed-loop System

In NC system, every control system may be designed as an open or a closed-loop control. The term *open-loop control* means that there is no feedback in the total system and the action of the controller has no information about the effect of the total system and the command signals that it produces. The controller produces commands for actions of the motions of the NC machine tool.

The open-loop NC systems are of digital type and use stepping motors for driving the slides. A stepping motor is a device whose output shaft rotates through by a fixed angle in response to an input pulse. The stepping motors are the simplest way for converting electrical pulses into proportional movement. Each pulse drives the stepping motor by a fraction of one revolution called the **step angle**. Since there is no feedback from the slide position, the system accuracy is solely dependent on the ability of the motor and accuracy of the mechanical parts.

The closed-loop control measures the actual position and velocity of the axis and compares them with the help of a comparator. The comparator is a device that compares the output signal with the signal received from the feedback device. The difference between the actual and the desired values is the error. The control system is designed in such a way as to eliminate or reduce to a minimum, the error, namely the system is of a negative feedback type.

In NC system both the input to the control loop and the feedback signals may be a sequence of pulses. Each pulse representing one BLU, i.e., 0.01 mm. The digital comparator correlates the two sequences and gives, by means of a digital-to-analog converter (DAC), a signal representing the position error of the system and the output of DC drives the DC motor. A closed loop system uses position sensors attached to the machine table to measure its position relative to the input value for the axis.

12.6 ADVANTAGES OF NC SYSTEMS

The primary motivation for the development of NC systems for machine tools was the demand for high accuracy in manufacturing of complicated parts, combined with the desire to shorten production time. The NC is superior to conventional manufacturing in a number of ways. These are as follows:

- Inspection time is reduced, since all the parts in a batch would be identical provided proper care is taken about tool compensation in part program and operation. NC systems with the use of inspection probes in the case of some advanced CNC controllers, the measurement function also becomes part of the program for improving the required tolerances.
- The problem of operator fatigue does not come into picture in the manufacturing of a part. The NC machine tool can be utilized continuously and moreover in NC, the accuracy is repeatable.
- The need for certain types of form tools is completely eliminated in NC machines. This is because the profile to be generated can be programmed, even if it involves in three dimensions.
- In NC system, tools can be utilized at optimum feeds and speeds which can be programmed easily.
- The need for a highly skilled and experienced operator is avoided because of automation and the absence of interrelated human factors.
- In NC system, it is possible to reduce the non-productive time by adopting the following methods:
  o By reducing the number of set-ups
  o By reducing set-up time

- o By reducing work piece-handling time
- o By reducing tool-changing time
- In NC system, the operator has free time since his involvement in part manufacture is reduced to a minimum. This time may be used for looking after other machining operations.
- The need for expensive jigs and fixtures is reduced or eliminated since the part programs take care of the geometry generated.
- Any modification in part design can be very easily translated into manufacture by the simple changes in part programs.
- Machining times and costs can be predicted to a greater accuracy, since all the elements involved in manufacturing would have to be analyzed properly before a part program is prepared.

12.7 DISADVANTAGES OF NC SYSTEMS

- There is a very high initial investment. The cost of NC machine is nearly 5 to 10 times more as compared to a conventional machine. The cost of tooling is also high.
- A properly trained part programmer is needed to write instructions in desired languages for the machines on the shop floor. They have to be acquainted with the manufacturing process.
- A highly skilled machine operator is needed in view of the complex and sophisticated technology involved. Similarly, machine operators have to be trained to adopt for the new technology.
- A special maintenance team is required and needed to be trained in all the subsystems present such as mechanical, hydraulic, pneumatic and electronics. Though the latest machines are equipped with latest diagnostic facilities, maintenance is more complicated.
- There is a relatively higher running cost involved due to the automatic operation of NC machines.

REVIEW QUESTIONS

12.1 Briefly explain the meaning of numerical control.

12.2 When was numerical control introduced?

12.3 What are the various elements of a NC system?

12.4 Why were the earlier NC machines referred to as "tape-controlled" machines?

12.5 Explain the function of NC machines with the help of a neat sketch.

12.6 Explain the function of DPU.

12.7 Explain the function of CLU.

12.8 Discuss NC positioning systems.

12.9 Enumerate the advantages of NC systems.

12.10   List out the disadvantages of NC systems.

12.11   Derive the performance parameters of a NC positioning system.

12.12   What are the advantages of absolute programming?

12.13   What are the advantages of incremental programming?

12.14   What are the various types of control systems designed in NC system?

12.15   Explain the designation of axes in a NC system.

12.16   What is the difference between absolute and incremental programming?

12.17   Explain the numerical control procedure.

12.18   List out the input media which have been used in NC. What are the material used for punched tapes?

12.19   Discuss the types of tape-coding systems used in the NC industry. Compare the similarities and differences between the two systems.

12.20   Explain the various types of formats used on NC tapes.

12.21   What are the factors that contributed to the development of numerical control?

12.22   Briefly explain the functions that are expected to be served by numerical control in machine tools.

12.23   Explain the principle of numerical control.

12.24   What are the applications where numerical control is most suitable?

Chapter

COMPUTER CONTROL OF NC MACHINES

Chapter

13

INTRODUCTION

Competition between manufacturing firms is increasingly dictated by quality, cost, and variety and servicing. Each one of these attributes of a successful product can only be produced by achieving the highest possible efficiency in manufacturing. Computer numerical control has revolutionized the manufacturing automation during the last 40 years. The movement of the tool and the operation of the various components of a machine tool through a program bring in the kind of flexibility to manufacturing which is not possible by means of the hard automation. Use of the digital computer has also permitted substantial improvements to be made in the controls for the numerical control. Computer numerical control involves the replacement of the conventional hard-wired numerical controller unit by a small computer (minicomputer or microcomputer). The small computer is used to perform some or all of the basic numerical control functions by programs stored in its read/write memory. One of the distinguishing features of computer numerical control is that one computer is used to control one machine tool.

A computer numerical control system, therefore, replaces some or all of the hardware functions previously performed by the machine control unit with a dedicated computer i.e a computer assigned to control a single numerical machine. With computer numerical control, the program is entered once and then stored in the computer memory. Thus, the tape reader is used only for the original loading of the part program and data. Computer numerical control offers additional flexibility and computational capability over numerical control.

13.1 WHAT IS COMPUTER NUMERICAL CONTROL?

Computer numerical control (CNC) retains the fundamental concepts of NC but utilizes a dedicated stored-program computer within the machine control unit. CNC is largely the result of technological progress in microelectronics rather than any radical departure in the concept of NC. The first step in the process of implementing automation in any industry is to manufacture parts or components through automation using machines and machine tools, with little or no human intervention. Machines or machine tools of CNC can be operated automatically using computers.

CNC attempts to accomplish as many of the MCU functions as possible within the computers software which is programmed into the computerized control unit. CNC control units, like the computers on which they are based, operate according to a stored program held in computer memory. This means that part programs are now able to become totally resident within the memory of the control unit, prior to their execution.

Therefore, CNC is self-contained NC system single machine tools including a dedicated mini-computer or all of the basic NC functions. Figure 13.1 depicts general configuration of CNC system.

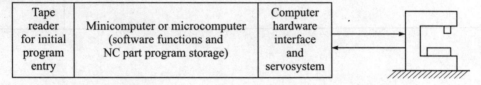

Tape reader for initial program entry	Minicomputer or microcomputer (software functions and NC part program storage)	Computer hardware interface and servosystem	

Fig. 13.1 General configuration of computer numerical control (CNC) system.

Components of CNC system

CNC machine has the following components:

- Computer
- Controller unit
- Machine tool

13.2 FUNCTIONS OF CNC

The principal functions of CNC are:

- Machine tool control
- In-process compensation
- Diagnostics

13.2.1 Machine Tool Control

This involves conversion of the part program instructions into machine tool motions through the computer interface and servosystem. The capability to conveniently incorporate a variety of control features into the software controller unit is the main advantage of CNC. Some of the control functions, such as circular interpolation can be accomplished more efficiently with hardwired circuits than with the computer. Therefore, there are two alternative controller designs in CNC:

- Hybrid CNC
- Straight CNC

In hybrid CNC, the controller consists of the software computer plus hardwired logic circuits. The hardwired components perform the functions such as feed rate generation and circular interpolation. The computer performs the remaining control functions. Use of these hardwired circuits saves the computer from performing the calculation chores. Hence a less expensive component is required in the hybrid CNC controller. Figure 13.2 illustrates hybrid CNC.

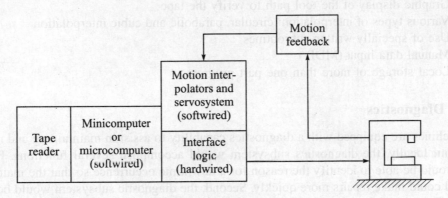

Fig. 13.2 Hybrid CNC.

The straight CNC system uses a computer to perform all the NC functions. The only hardwired elements are those required to interface the computer with the machine tool and the operators console. Interpolation, tool position feedback and all other functions are performed by computer software. The computer required in straight CNC system must be powerful. The advantage of straight CNC configuration is its flexibility. It is possible to make changes in the interpolation programs, whereas the logic contained in the hardwired circuits of hybrid CNC cannot be altered. Figure 13.3 illustrates straight CNC.

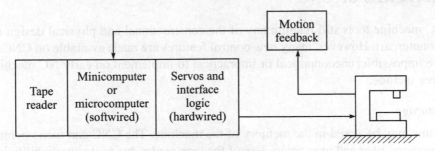

Fig.13.3 Straight CNC.

13.2.2 In-process Compensation

This involves the dynamic correction of the machine tool motions for changes or errors which occur during process. Some of the CNC in-process compensation are:

- Adjustments for errors sensed by in-process inspection probes and gauges
- Recompilation of axis positions when an inspection probe is used to locate a datum reference on a work part
- Offset adjustments for tool radius and length
- Adaptive control adjustments to speed and/or feed

- Computation of predicted tool life and selection of alternative tooling when indicated
- Improved programming and operating features: (software)
- Editing of part programs at the machine permits correction and optimization of program
- Graphic display of the tool path to verify the tape
- Various types of interpolation: circular, parabolic and cubic interpolation
- Use of specially written subroutines
- Manual data input (MDI)
- Local storage of more than one part program

13.2.3   Diagnostics

CNC machines are equipped with a diagnostics capability to assist in maintaining and repairing the system. Ideally, the diagnostics subsystem would accomplish several functions. First, the system would be able to identify the reason for a downtime occurrence so that the maintenance personnel could make repairs more quickly. Second, the diagnostic subsystem would be alert to signs that indicate imminent failure of a certain component. Hence maintenance personnel could replace the faulty component during a scheduled downtime, thus avoiding an unplanned interruption of production.

A third possible function is certain amount of redundancy of components which are considered unreliable. When one of these components fails, the diagnostics subsystem would automatically disconnect the faulty component and activate the redundant component. Repairs could thus be accomplished without any breaks in normal operations.

13.3   FEATURES OF CNC

Many CNC machine tools still retain many of the constructional and physical design aspects of their NC counterpart. However, many new control features are made available on CNC machines which were impossible, uneconomical or impractical to implement on early NC machines. Such new features include:

Program storage

Part programs may be stored in the memory of the machine. The CNC can then operate directly from this memory, over and over again. Use of the tape reader due to its unreliability is virtually eliminated. For long production runs, the part program may be retained in memory, even when the power is removed, by the use of battery back-up facilities that keep only the memory supplied with power. Often, more than one program may be resident in the control unit memory at one time with the ability to switch between them.

Online part programming and editing

In NC, online programming is more or less impractical as it is highly time consuming. In CNC, the program can be easily entered in word address format or machine languages from the alphanumeric keyboard of the operator panel of the controller very often. The imperfections and flows in a new part program becomes visible when it is implemented on the shop floor. Editing can be carried out on the part program held in memory. Thus, errors, updates and improvements can be attended to at the machine. Such editing can be stored in computer memory and override the tape information as read in. A new and corrected tape may then be punched directly from the CNC control unit. For that read the program can be output on any desired media provided the relevant output device has been interfaced in the CNC system. Alternatively punched tape or floppy or cassette can be used offline to put the program on other media with the help of relevant I/O devices.

Enhanced cutter compensation

When a part program is written, it is normally done with a particular type and size of cutter in mind. The positioning of the cutter relative to the work piece will need to take account of the dimensions of the cutter. It may be the case that when the part program comes to be run on the machine, the particular cutter specified is not available. CNC control units allow compensations and offsets to be made for the differences in dimensions between the actual cutter and the specified cutter. Thus the part program is now independent of the cutter specified when writing the program. Because of availability of semiconductor memory several tool compensations can be easily entered.

Flexibility to modify control functions

In CNC systems, re-programming of the dedicated control programs can modify the control function to meet the desired requirements. This gives rise to great flexibility of adapting to several new options without adding to the controller's components.

Programming features

Some standard programming features include:

- Absolute/Incremental programming
- Decimal point programming
- Diameter/Radius programming
- Linear, circular and helical interpolation

Easier maintenance of CNC via diagnostic software

Most modern CNC machines are well equipped with comprehensive diagnostic software checking of its electronic operation. A unique advantage of CNC systems is that a diagnostic routine can be executed on its computer for effective maintenance inbuilt to check the operating conditions. These routines continue to check input/output and then the machine operation ceases.

Tool path generation

Programs have been developed which can help online checking of the part program without running the machine. The program interprets the part program in question and displays the path which would be taken by the tool by simulation.

Program proving facilities

Many modern control systems contain software that will process the resident part program information and indicate the component shape that will produce before machining takes place. This is often displayed graphically on a visual display unit (VDU) on the operating console.

Advanced hardware architecture

Custom built very large scale integrated circuits (VLSI) are used with several functions in a single chip. This way the benefits of reduced component count and improved reliability are realized.

Background programming

It is a common facility available in CNC systems. The user can run edit and simulate another program while a program is being executed on the machine tool.

- Multiple part program storage
- Sequence number search
- Feed/min or Feed/rev programming
- Dwell programming
- Position preset

Complex interpolation

In addition to the linear and circular interpolations which were possible in NC system also, parabolic, helical and cubic interpolation control programs are also available in CNC systems. In fact any type of interpolation program to meet the user's requirements can be developed and inbuilt into the CNC controller.

Custom built or user built macros/subroutines manufacturers offer several types of macros built around the customer's requirements. Repetitive operation can be written as subroutines and main program can call these subroutines. This, of course, requires the training in writing the macros which like subroutine writing would involve the use of variables, intervariable calculations trigonometry, branching and main part program connection techniques.

Special canned cycles

This is a powerful control feature intended to shorten the length of the part program and consequently the length of tape. Canned cycles are single block multipass fixed cycles. For example, canned cycles for rough turning, circular pocket and rectangular pocket milling, and drilling, boring and tapping in machining centres.

Software modularity

Each function is written as a discrete module (structured programming). Additional software features can be added at any time to upgrade the control without affecting the software.

Programmable logic controller features

Programmable logic controller is a soft wire oriented machine interface between the CNC system and machine tool functions such as miscellaneous functions (coolant on/off, pallet operation, spindle speed functions and tool functions). Signals from CNC to PLC are on/off type. The programmable logic controllers available today are microprocessor-based units which are integrated into the CNC system.

13.4 ADVANTAGES OF CNC

The development of CNC system has progressed as a result of the rapidly improving capabilities and coupled with developments in MCU technology, semiconductor technology and integrated circuits technology. The trend away from conventional NC to the computer control system means a change from purely hardware based NC to a software based system. This brings the user a number of advantages which are listed as follows:

- An increase in flexibility
- A reduction in hardware circuits and simplification of the remaining hardware as well as the availability of automatic diagnostic programs brings a subsequent need for fewer maintenance personnel.
- A reduction in inaccuracies in manufacturing due to the reduced use of tape reader.
- An improvement in the possibilities for correcting errors in part programs, the editing feature, the possibility of using the computer's peripheral equipment for debugging the edited part program.
- Extremely good accuracy and repeatability of the components produced enables a greater uniformity of production.
- CNC machines offer complete control of all axes, under optimum cutting conditions.
- Extremely short set-up times are possible since standard tooling is all that is required. The need for jigs and fixtures is almost eliminated.
- The quality of the finished job is no longer under the control of the operator but under the control of a computer-run part program. This ultimately translates into lower costs per part and much-reduced lead times.
- Part programming is often carried out by specialist part programmers, away from the machine. The facility to prepare new jobs away from the machine means that the machine tool reduces nonproduction time.
- Extended tool life since optimum cutting conditions are realized.
- The number of components produced is controlled by the machine and not by the operator, therefore, delivery is accurately determined.
- Accuracy is controlled by the machine, therefore, better interchangeability (for spares), easier fitting and assembly and reduced inspection are achieved.

13.5 INDUSTRIAL APPLICATIONS OF CNC

The most exciting developments in CNC applications are taking place in large-scale manufacturing. In many industries, computer integration of the entire manufacturing process is taking place.

With the latest generation of CNC controls, manufacturers are making available the option to include network adapter cards in the controls. The network card allows the CNC control to plug into a company's local area network (LAN). The control can send and receive information over the company's LAN. Software available for these controls allows plant manager or manufacturing supervisor to view the production status of work running on the CNC equipment from his or her desktop PC. Continuous monitoring of the production process can be accomplished allowing any necessary adjustments to the process to be accomplished before a problem develops to the point of producing bad parts.

13.6 DNC CONCEPTS

Direct numerical control can be defined as a manufacturing system in which a number of machines are controlled by a computer through direct connection and in real time.

The tape reader is omitted in DNC, thus relieving the system of its least reliable component. Instead of using the tape reader, the part program is transmitted to the machine tool directly from the computer memory. In principle, one large computer can be used to control more than 100 separate machines. The DNC computer is designed to provide instructions to each machine tool on demand. When the machine needs control commands, they are communicated to it immediately. DNC also involves data collection and processing from the machine tools back to the computer.

The computer is connected between the tape reader and the NC machine thereby passing the tape reader. This system was therefore called as **behind the tape reader system (BTR)**. With the development of CNC, DNC concept was extended to CNC machines also mainly for part program management. The DNC computer could serve as a number of CNC machines in shop floor.

The DNC computer stores all the part programs and transfers the part programs to the CNC machines in response to the requests of the operators. DNC systems are generally designed for 4, 8 or 16 CNC machines. However with the wide spread acceptance of the local area network concept, the possibility of connecting more CNC machines in a DNC network has become a reality. The concept of internet, intranet and extranet has further enlarged the scope of distributed numerical control.

13.7 OBJECTIVES OF DNC

DNC is the first step towards the factory future. DNC serves many purposes and is now considered as an essential for the efficient management of CNC machine tools in the shop floor. The main objectives of implementing DNC are:

- To provide for uploading, i.e., entry of part programs from CNC systems to the control computer.
- To transfer or download directly the part programs from central computer to individual machine control units of machines.
- Easy editing of the existing programs.
- Eliminating the use of punched tape in the old generation of NC machines.
- To provide software for processing of the shop floor data for management purposes.
- Eliminating the need for manually punching the program at the keyboard.
- To establish two-way communication among the various subsystems. The essential communication links in direct numerical control are between the following components of DNC system:

  Central computer and machine tools

  Central computer and NC part programmer terminal

  Central computer and bulk memory

- To post process part programs which are written in higher languages like APT.
- To provide for integration of the central computer with other system like CAD/CAM system shop floor control system, corporate data processing computer (shown in Fig. 13.5), remote maintenance diagnostics system and other computer-automated systems in the plant.
- Tool length offsets from tool pre-setters can be transferred directly to machine tool controls. It is also possible to connect coordinate measuring machines to DNC networks.

13.8 COMPONENTS OF DNC SYSTEM

DNC consists of four basic components. Figure 13.4 illustrates the configuration of the basic DNC system.

- Central computer
- Bulk memory which stores the NC part programs
- Telecommunication lines
- Machine tools

The computer calls the part program instructions from bulk storage and sends them to the individual machines as the need arises. It also receives data back from the machine. This two-way information flow occurs in real time, which means that each machine's request for instructions must be satisfied almost instantaneously. Similarly, the computer must always be ready to receive information from the machines and to respond accordingly. The remarkable feature of the DNC system is that the computer separates machine tools, all in real time.

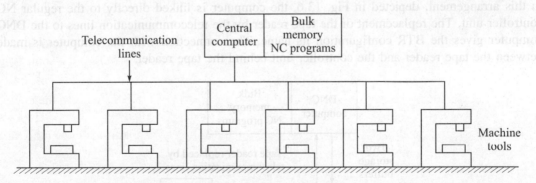

Fig.13.4 General configuration of direct numerical control (DNC) system.

Depending on the number of machines and the computational requirements that are imposed on the computer it is sometimes necessary to make use of the satellite computers as shown in Fig. 13.5. These satellites are minicomputers and they serve to take some of the burden off the central computer. Each satellite controls several machines. Groups of part program instruction are received from the central computer and stored in buffers. They are then dispensed to the individual machine as required. Feedback data from the machines are also stored in the satellite's buffer before being collected at the central computer.

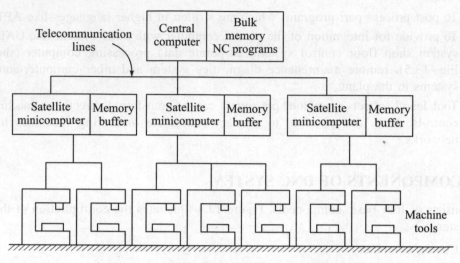

Fig.13.5 DNC with satellite minicomputers.

13.9 TYPES OF DNC

There are two alternative system configuration by which the communication link is established between the computer and the machine tool. One is behind the tape reader system and the other configuration make use of a specialized machine control unit.

13.9.1 Behind the Tape Reader System (BTR)

In this arrangement, depicted in Fig. 13.6, the computer is linked directly to the regular NC controller unit. The replacement of the tape reader by the telecommunication lines to the DNC computer gives the BTR configuration its name. The connection with the computer is made between the tape reader and the controller unit behind the tape reader.

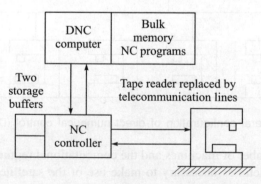

Fig. 13.6 DNC with behind the tape reader (BTR) configuration.

Except for the source of the command instructions, the operation of the system is similar to conventional NC. The controller unit uses two temporary storage buffers to receive blocks of instructions from the DNC computer and convert them into machine action. While one buffer is receiving a block of data, the other is providing control instructions to the machine tool.

13.9.2 Special Machine Control Unit

The other strategy in DNC is to eliminate the regular NC controller altogether and replace with a special MCU. The configuration is illustrated in Fig. 13.7. This special MCU is a device that is specifically designed to facilitate communication between the machine tool and the computer. This communication link is important in circular interpolation of the cutter path. The special MCU configuration achieves a superior balance between accruing of the interpolation and the fast metal removal rates.

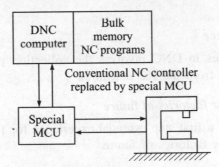

Fig.13.7 DNC with special machine control unit (MCU).

The special MCU is softwired while the conventional NC controller is hardwired. The advantage of softwired MCU is its flexibility. Its control functions can be altered with relative ease to make improvements. It is much more difficult to make changes in the regular NC controller because rewiring is required.

13.10 ADVANTAGES OF DNC

Elimination of punched tapes and tape readers

DNC eliminates the least reliable element in the conventional NC system. This makes part program input convenient and error-free preparation. Storage and handling of paper tape become very difficult.

Greater computational capability and flexibility

The large DNC computer provides the opportunity to perform the computational and data processing functions. Because these functions are implemented with software rather than with hardwired devices, there exists the flexibility to alter and improve the method by which these functions are carried out.

Convenient storage of NC part programs

The huge memory of DNC computer makes it possible to store very long part programs thus making free the memory of controller. Further if same programs are required to be run on different identical machines then these can be recalled or used from the DNC computer.

Easier and effective part programming

Programming software at the DNC computer makes part programming and program providing easier and effective. Further programming of highly complex jobs which need to be in high level language at APT is facilitated because of post-processing facilities which can be made available at the DNC computer.

Programs stored as CLFILE

Storage of part programs in DNC is generally in the form of cutter path data rather than post-processed programs for specific machine tools. Storing of the programs in this more general format affords the flexibility in production scheduling to process a job on any of several different machine tools.

Reporting of shop performance

One of the important features in DNC involves the collection processing and reporting of production performance data from the NC machines.

Establishes the framework for factories of future

The DNC concept represents a first step essential experience for future expansions leading to fully automated production or factories of future.

13.11 ADAPTIVE CONTROL

Adaptive control refers to the in-process sensing of tool wear, component size and component quality in order to provide automatic compensation of machining process parameters to maintain consistent accuracy and quality of output. In-process sensing means that sensing is carried out while actual machining is taking place.

The most important component in a CNC machine tool is the tool tip. If the tool tip breaks down then the whole system is rendered ineffective and the cost of production increases. The reasons for increase in production costs are:

- Tool wear can result in oversize components being produced, necessitating rework operations.
- Tool wear can reduce optimum cutting conditions resulting in components of inferior quality.
- Tool wear can cause premature tool breakage.
- Tool breakage can scrap expensive components, damage the machine tool and cause serious injuries to the operator.
- Tool breakage can disrupt machining and production schedules.
- Adaptive control implies two things: Firstly that the tool and component characteristics of size and quality can be sensed while machining is taking place and, secondly, that there are process parameters which can be automatically adjusted in a responsive manner.

In conventional machining situations, the operator provides the adaptive control.

The experienced machinist is able to assess the quality of the cutting tool by sight, sound, touch and smell, and can then affect control by adjusting the speed, feed, depth of cut or by replacing the tool. In automatic adaptive control, speeds, feeds and depth of cut are still the process variables which are controlled. Monitoring of tool and component condition, however, has to be done by some form of sensing.

13.12 TOOL WEAR SENSING

Tool wear can be predicted using well-tried tool life calculations. A tool may thus be replaced after x-minutes cutting or when y-components have been machined. This method is unsuitable in the modern manufacturing environment for the following reasons:

- Certain assumptions have to be made regarding the formulae used.
- Ideal cutting conditions are assumed in the calculations.
- New tool materials are constantly being introduced.
- Difficult-to-machine work piece materials are often not accommodated.
- The method is based on "trial and error" formulae.
- A more reliable method is required.

The majority of adaptive control systems sense the onset of tool wear. This may be done in a variety of ways. In all cases an effective system must

- Properly reflect the degree of wear at the tool-tip.
- Have quick response.
- Be sensitive to sudden changes in cutting conditions.
- Be robust and adaptable for use on the machine tool.
- Be reliable and require little maintenance.
- Be available at a realistic cost.

Sensing methods may be contact or non-contact. They may also be either direct, measuring dimensional, volumetric or other changes in the tool itself, or indirect via other conditions that can be more conveniently measured. Three classes of adaptive control are:

13.12.1 Feedback from the Tool

Electrical sensors monitor the change in electrical resistance across the tool-work piece junction when the tool is in contact with the work piece during cutting. As the tool wears, its contact area with the work piece increases and the electrical resistance across the junction decreases.

Pneumatic sensors monitor air back-pressure formed between a nozzle set in the clearance face of the cutting tool and the work piece surface. As the tool wears, the gap between the clearance face of the tool and the work piece surface decreases resulting in an increase in back-pressure.

Optical sensor illuminates the wear zone of the tool tip when it is not in contact with the work piece. The light beam is reflected to impinge on a photo transistor. The output signal from the phototransistor can be correlated to tool wear. This system can only be used when the cutting tool is not continuously in contact with the work piece, for example in milling, when the cutting edge can be exposed to a light beam during that part of its rotation when it is not in contact with the work piece.

13.12.2 Feedback from the Work Piece

Contacting or non-contacting transducers measure and compare actual work piece dimensions with commanded work piece dimensions. Tool wear is directly related to dimensional changes in the component. Inspection (touch trigger) probes held in the tool spindle are a common way of implementing this approach.

Fibre optic sensors probe the same path as the cutting tool and measure the reflectivity of the newly machined surface. Reflectivity of the surface varies with the roughness of the surface texture. The deterioration in the quality of the surface of the work piece is taken as an indication of tool wear.

Electronic feeler displacement transducers measure the distance between the work piece and the tool post. The output signal of the transducer varies as the displacement varies and this is proportional to tool wear.

13.12.3 Feedback from the Cutting Process

Thermocouple transducers measure the cutting temperature at the tool/work piece junction. Cutting temperature will increase as tool wear increases. Infrared sensors can also be focused onto the cutting zone to measure the cutting temperature.

Piezoelectric transducers located near the cutting tool edge respond to changes in vibration. An increase in vibration indicates the probability of the onset of tool wear.

Power and torque transducers (watt meters and accelerometers, respectively) mounted with in the machine spindle detect increase in power and torque respectively. As the tool wears, rubbing takes place and the power and torque values increase.

These values need to be preset prior to machining provide reference levels. This will probably be accomplished by cutting a one-off with new, sharp tools and recording torque and power levels. The programmer then has to decide what increase in level constitutes a worn tool. Tool wear sensing is important since tool wear is probably the biggest cause of tool breakage.

The control system software will need to be programmed to take appropriate action conditional on the output from the adaptive control system. This may be to cease operation and provide an alarm system requesting attention from an operator or automatically initiate a tool change for a sister tool.

Broken tool detection

A complementary facility to an adaptive control system is that of broken tool detection. It is normally provided as an emergency check rather than a routine since it takes valuable time to implement. It may be accomplished in a number of ways.

One method is to provide a light spring-loaded switch mounted to the side of the machine. At various times during the machining cycle, the tool is traversed to negotiate this switch. If no signal is received by the control system, it is assumed that the tool is either missing or broken.

A second method involves passing the tool through a caliper to break either a light beam or an air jet. With intelligent application this method can also be used to automatically set the tool length offset of a replacement tool.

13.13 ADAPTIVE CONTROL MACHINING

For a machining operation the term *adaptive control* denotes a control system that measures certain output process variables and uses these to control speed and or feed. Some of the process variables that have been used in adaptive control machining system include spindle deflection or force torque, cutting temperature, vibration amplitude and horsepower.

Nearly all the metal cutting variables that can be measured have been tried in experimental adaptive control systems. The motivation for developing an adaptive machining system lies in trying to operate the process more efficiently. The typical measures of performance in machining have been metal remove rate and cost per volume of metal removed. Adaptive control implies that the CNC system is responsive to adapt itself to operate at those machining parameters which result into higher productivity. A part programmer specifies cutting speeds, feed rates, depth of cut, etc. on the basis of knowledge and experience to achieve desired levels of manufacturing objectives, say, accuracy, surface finish and productivity. Figure 13.8 shows a schematic diagram illustrating the operation of the AC system during the machining process.

The programmed parameters are chosen offline and the machine operator by his online intervention further attempts to achieve the desired objectives. Let us think of a control system where in real time the operating parameters, viz. cutting speeds, feeds, depth of cut go on automatically and constantly adjusting themselves so as to

- use the available spindle power to maximum, or
- limit the deflection of the cutter, or
- limit the cutting tool temperature, or
- limit the vibration amplitude of the cutter.

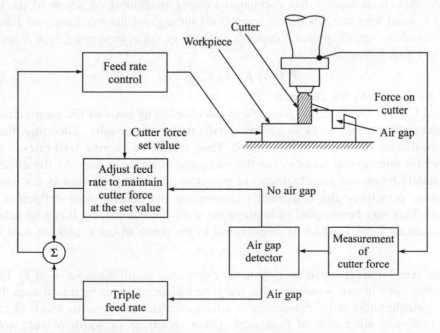

Fig. 13.8 Configuration of typical adaptive control machining system that uses cutter forces as the measured process variable.

The control system with constraints as defined earlier are named as adaptive control with constraints or ACC. The other type of adaptive control known as **adaptive control optimization** (ACO) determines operating parameters to optimize some production performance criteria or index in the light of constraints. Usually this criterion is the ratio of metal removal (MRR) to tool wear rate (TWR).

$$\text{Performance index, PI} = \frac{\text{MRR}}{\text{TWR}}$$

13.13.1   ACC Type of Adaptive Control

The ACC type of adaptive control requires constant measurement of relevant manufacturing constraints like cutting forces, torques, spindle motor power/current/tool wear/tool deflection, cutting temperature, vibrations, etc. Accordingly sensors/transducers/instruments like dynamometers, strain gauges, accelerometers and contact type probes are employed. The measured value at no time exceeds the permitted one. The AC software is essentially a machinability program which is written as a macro and can be integrated with the main CNC system programs. The program calculates those values of machining parameters which would give the desired values (extreme values objective), say, maximum metal removal rate. The programmed values of machining parameters like that of the feed rate are continuously replaced by the new calculated values. The real time effect of operating these values is checked by measuring the constraints values. The value of operating parameter is accordingly adjusted so as to remain within the imposed constraints.

For example, consider milling on a CNC system, the programmed value of feed rate is usually based on the maximum depth it has encountered during machining of whole of the job. This results into a small feed rate f which is maintained throughout the machining of job. It is so because the available spindle power p depends on depth of cut a, feed rate f, rpm N and milling tool diameters d *as* follows:

$$P = K_1 \ K_2 \ K_3 \ K_4 \ af^k ND$$

where K_1, K_2, K_3 and K_4 are constants.

In the ACC type, the operating feed rate is calculated. The basis of the programmed depth of cut continuously changes so as to use the spindle motor power fully. Therefore, the spindle power is a constraint which cannot be violated. Thus, it leads to varying feed rates in real time without operator intervention so as to use the maximum of motor power. As the effect of new rate is constantly monitored measurements of power available at the motor is not exceeded.

On similar principles, the constraint of maximum permissible tool deflection may be implemented. This may be intended to improve the work piece accuracy. It can be achieved by limiting the cutting force F which is proportional to the depth of cut a and the feed rate f as

$$F = K_1 A f^K$$

The feed rate is thus changed so as to never exceed the permissible value of F. The actual value of cutting force in turn is monitored in real time by force measuring transducers. It is clear that AC is suitable under those conditions in which enough variations in width of cut exists permitting profitable adaptation of feed rates. Other variations in width of cut, work piece material, hardness, work piece stiffness, tool wear rate, etc. also present suitable situation where AC may be applied.

13.13.2 ACO Type of Adaptive Control

This type of control measures manufacturing process output variables such as cutting torque, tool temperature, tool wear and machine vibrations. These sensor measurements along with programmed feed rate, spindle speed and constraints are fed into CNC system and processed by the ACO software. The processing computes real time performance criteria or index (PI) on the basis of MRR and TWR. This index is compared with the desired value by another program in the AC software which recalculates the cutting process variables like feed rate and speed rate so as to keep performance index at highest level without violating any of the constraints. The process variables are thus continuously changed so as to obtain optimum value of the performance index. Figure 13.9 shows the adaptive control system for a CNC machine tool.

These systems may be advantageously used to:

- Increase productivity
- Maintain high levels of work piece precision
- Raise the level of automation by reducing operator intervention
- Increase cutter life by reducing tool loads
- Increase work piece accuracy and CNC machine life by maintaining low levels of vibration

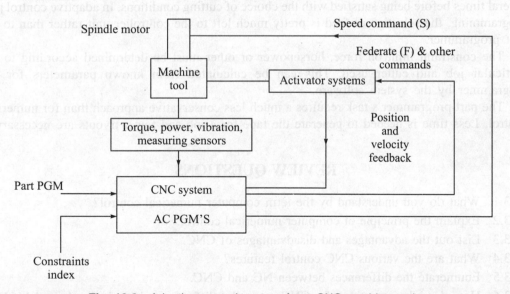

Fig. 13.9 Adaptive control system for a CNC machine tool.

13.14   BENEFITS OF ADAPTIVE CONTROL MACHINING

Increased production rates

Productivity improvement was the motivating force behind the development of adaptive control machining. Online adjustments allow for variations in work geometry material, and tool wear provide the machine tool with the capability to achieve the highest metal removal rates that are consistent with existing conditions. This capability translates into more parts per hour.

Increased tool life

In addition to higher production rates, adaptive control will generally provide a more efficient and uniform use of the cutter throughout its tool life. Because adjustments are made in the feed rate to prevent severe loading of the tool, fewer cutters will be broken.

Greater part protection

Instead of setting the cutter force constraint limit on the basis of maximum allowable cutter and spindle deflection, the force limit can be established on the basis of work size tolerance. In this way, the part is protected against an out-of-tolerance condition and possible damage.

Less operator intervention

The advent of adaptive control machining has transferred the control of the process even further out of the hands of the operator and into the hands of management via the part programmer.

Easier part programming

With ordinary numerical control, the programmer must plan the speed and feed for the worst conditions that the cutter will encounter. The programmer may have to try out the program several times before being satisfied with the choice of cutting conditions. In adaptive control part programming, the selection of feed is pretty much left to the controller unit rather than to the part programmer.

The constraint limit on force, horsepower or other must be determined according to the particular job and cutter used. This can be calculated from known parameters for the programmer by the system software.

The part programmer's task requires a much less conservative approach than for numerical control. Less time is needed to generate the tape for a job and fewer tryouts are necessary.

REVIEW QUESTIONS

13.1 What do you understand by the term computer numerical control?

13.2 Explain the principle of computer numerical control.

13.3 List out the advantages and disadvantages of CNC.

13.4 What are the various CNC control features?

13.5 Enumerate the differences between NC and CNC.

13.6 How does the operation of a CNC controller differ from that of a hardwired MCU?

13.7 What is the need of DNC? Explain?

13.8 List out the advantages of DNC.

13.9 Discuss the elements of DNC system.

13.10 Explain the functions of DNC system.

13.11 Enumerate the differences between CNC and DNC.

13.12 Explain the working of BTR type of interface in DNC system.

13.13 What do you understand by the term adaptive control machining?

13.14 Draw the block diagram of adaptive control with constraints and briefly discuss.

13.15 Explain the concept of adaptive control systems, its types and associated features.

13.16 List out the advantages of adaptive control systems.

13.17 Explain neatly with a sketch adaptive control with optimization.

13.18 Under what circumstances, the adaptive control machining system is used? Discuss briefly.

13.19 Distinguish between adaptive control with optimization and adaptive control with constraints.

13.20 Discuss the applications of adaptive control system.

13.21 What are the process variables in the adaptive control system?

13.22 What is mirror imaging? Why is it used?

NC PART PROGRAMMING AND COMPUTER AIDED PART PROGRAMMING

Chapter

14

INTRODUCTION

NC part programming is concerned with the planning and documentation of the sequence of processing steps to be performed on a numerical control machine. It is usually accomplished by a part programmer. The planning portion of part programming requires a knowledge of machining as well as geometry and trigonometry. The sequence of processing steps in NC involves a series of movements of the processing head w.r.t the machine table and workpart. Part programming can be carried out either manually or with the aid of computers. Part programs for simple components can be easily done manually. However, if the component has complex features which require too many repetitive and/or tedious calculations for preparing its program for cutter path description then it is recommended that the computer aided programming be resorted to.

The necessary data for producing a part program may be classified as follows:

- Information taken directly from a drawing: dimensions such as length, width, height, radius, etc.; segment shape; linear, circular or parabolic; diameter of holes to be drilled. The tool path is calculated based on this information.

- Machining parameters, which depend on surface quality, required tolerances and type of work piece and cutting tool, feeds, cutting speeds and auxiliary functions such as turning on and off the coolant.

- Data determined by the part programmer such as the cutting direction and changing of tools. Part programmers establish the optimal sequence of operations which required to produce the part program. Therefore, they must be familiar with manufacturing processes and have detailed knowledge of the characteristics of the particular NC system.

- Information depending on the particular NC system such as acceleration and deceleration intervals or programming a two-spindle machine. When using computer languages for programming, those functions are performed by a post-processor program.

Programming for NC/CNC machines can be done in three ways:

1. Manual part programming
2. Computer assisted programming
3. Generation of program using CAD/CAM packages

14.1 MANUAL PART PROGRAMMING

One should know about the machine tools, cutting tools and fixtures to be used in the manufacture of the components to write a manual part program. Shop experience is essential for a good programmer as only careful process planning can lead to satisfactory programs. A programmer is supposed to study the drawing of the components in detail and prepare a process plan showing the list of operations to be carried out in sequence, the machine and the tools to be used and the speeds and feeds to be maintained during various operations. The reference axes should be chosen so that the coordinates for various features can be determined. (Fig. 14.1)

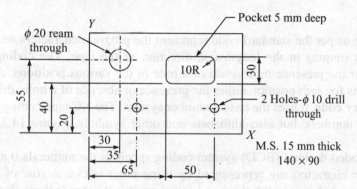

Fig. 14.1 A typical component.

These details enable the programmer to know the operation motions which he is to write in code form, i.e. the program. This would then be punched on a paper tape to and the tape be fed into the tape reader of the machine controller. The program can also be prepared on microprocessor based terminals or personal computers and then be stored or transmitted directly to machine tools. The programs can also be entered directly on the machine control panel. The programming section should maintain an updated list of tools available and information regarding machining parameters. Figure 14.2 shows the block diagram of manual part programming.

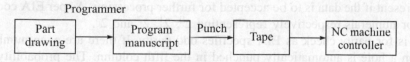

Fig. 14.2 Manual part programming.

Manual part programming jobs can be divided into two categories: point-to-point jobs and contouring jobs. Except for complex work parts with many holes to be drilled, manual programming is ideally suited for point-to-point applications. Except for the simplest milling and turning jobs, manual programming is time-consuming for applications requiring continuous path control of the tool.

14.1.1 The Punched Tape in NC

The part program is converted into a sequence of machine tool operations by means of the input medium, which contains the program and the controller unit is used to interpret the input medium. The punched tape of 1 inch wide is the standardized input medium as per EIA standard. The punched tape uses coded symbols which represent the part program. The information blocks of the program are converted into a punched tape using flexowriter in which the hole patterns corresponds to the various characters in the program. The code according to which these holes get punched has been standardized by EIA, ISO and ASCII.

14.1.2 Tape Coding and Format

Coding

The punched tape as per the standard code represent the program characters on its eight tracks/columns of holes running in the lengthwise direction of the tape. The coding of the tape is provided by either the presence or absence of a hole in the various positions. As there are two possible conditions for each position either the presence or absence of a hole, this coding system is called the **binary code**. It uses the base 2 number system. The NC tape coding system is useful to code not only numbers, but also alphabets and other symbols. Figure 14.3 shows the tape format.

The binary coded decimal (BCD) system coding specifies the numerals 0 and 9 on the first four tracks. Other characters are represented by using other tracks. A row of feeding/driving/timing holes is present between the third and fourth tracks. It extends through the length of the tape, and is used to engage with the teeth of a sprocket which transports the tape. These feed holes have no relationship with the holes which have been punched according to the code.

The punched tape code holes diameter is different from the feed hole or sprocket holes diameter. The thickness of the tape is about 0.004 + 0.003 inches. Although flexowriter, the tape punching device, is least reliable, there is always a possibility that some holes will be erroneously deleted or added. To help detect this error, the codes prescribe either odd number of holes or even number of holes for each character. Thus, EIA code specifies odd number of holes while the ISO code and ASCII code prescribe even number of holes. A circuit known as **parity check** can be incorporated in the MCU (machine control unit) to ensure that the desired number of holes are always present if the data is to be accepted for further processing. As per EIA coding, Tracks 1 to 4 are for numerals respectively representing 2^0, 2^1, 2^2 and 2^3.

Track 5 is for parity check as EIA specifies odd parity. If there are even number of holes in a row then a hole is automatically punched in the fifth column. The probability of error is minimized by parity check. Coordinates are five digit numbers with decimal placed before two least significant digits, e.g. 125.87. Alphabets are coded by dividing them into three groups of 9, 9 and 8 letters.

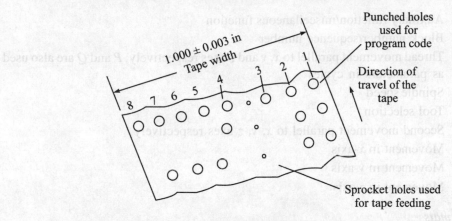

Fig. 14.3 Tape format.

The Group I has holes punched in 6th and 7th tracks.
The Group II has holes punched in 6th track.
The Group III has holes punched in the 7th track.

A number punched along with 6th or 7th or both tracks indicates the alphabets placed in their group.

Group/Number	1	2	3	4	5	6	7	8	9
I	A	B	C	D	E	F	G	H	I
II	J	K	L	M	N	O	P	Q	R
II	S	T	U	V	W	X	Y	Z	

A hole is punched in the 8th track to indicate end of block (EOB).

Format. Word address format is used in part programming. In this format, a letter precedes each word and is used to identify the word type and to address the data to a particular location in the controller unit. These words are essentially alphanumeric, i.e., the first character is an alphabet which is also the address character, followed by the numerical digits pertaining to information for the address character.

Address characters as per DIN 66025:

Character	Meaning
A	Rotation about x-axis
B	Rotation about y-axis
C	Rotation about z-axis
D and E	Rotation about additional axes
F	Feed rate
G	Preparatory function, identifying the action to be executed
I	Interpolation parameter/thread pitch parallel to x-axis
J	Thread pitch parallel to y-axis
K	Thread pitch parallel to z-axis

M	Auxiliary function/miscellaneous function
N	Block number/sequence number
P, Q, R	Thread movement parallel to x, y and z axes respectively; P and Q are also used as parameters in cycles
S	Spindle speed
T	Tool selection
U, V, W	Second movement parallel to x, y, z axes respectively
X	Movement in x-axis
Y	Movement in y-axis
Z	Movement in z-axis

Types of tape formats

The information about an operation which would be necessary to be conveyed to the controller for the machine tool operation would consist of the following:

- Operation number
- Operation code
- Coordinates for position or motion
- Tool information
- Speed and feed for operation
- Miscellaneous information like spindle clockwise/anticlockwise rotation, coolant ON/OFF etc.

In a program, this entire information is called **block**. There are three types of formats for representing the block in punched tape, viz.

- Fixed sequential format
- Word address format
- Tab sequential format

Fixed sequential format. Every instruction contains all the words in the same sequence irrespective of the words being the same as in the previous blocks. Hence, the identifying address letter need not be provided.

For example, if some coordinate values (i.e, x, y and z coordinates) remain constant from one block to next block, these values have to be specified in the next block also. The data must be input in a specified sequence and characters within each word must be of the same length.

N010 G00 X10 Z0 F50 S1000 E0B
N020 G01 X20 Z0 F50 S1000 E0B

Word address format. Each word is preceded and identified by its length address. This format enables instructions which remain unchanged from the preceding block to be omitted from succeeding blocks. This system speeds programming and tape lengths are considerably reduced. This is the format adopted by most CNC machine control units. Detailed format classification is provided by the control system manufacturer.

N010 G00 X10 Z0 TAB F50 S1000 E0B
N020 G01 X20 E0B

Tab sequential format. The words in each instruction/block are always provided in the same sequence but each word is preceded by the TAB character. If instructions remain unchanged in succeeding blocks, the instructions need not be repeated but TAB character must be punched. Here also, the identifying letter address need not be employed.

N010 G00 X10 TAB Z0 TAB F50 TAB S1000 E0B

N020 G01 X20 E0B

Program format

In this format, the standard sequence of words for a two-axis NC system is:

N-word	This is used to identify the block.
G-word	This is used to prepare the controller for instructions.
X-word ⎫	
Y-word ⎭	These give the coordinate position of the tool.
Y-word	
F-word	This specifies the feed rate in machining.
S-word	This specifies the rate at which the spindle rotates.
T-word	This specifies which tool is to be used in the operation.
M-word	This is used to specify certain miscellaneous functions.
EOB	

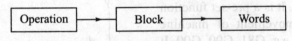

Fig. 14.4 Program format.

An operation is described, in code form, as a block in a program. Figure 14.4 shows the program format. The block consists of the operation number word, data words and the end of block (EOB). The operation number is an 'N' word address and must be programmed at the beginning of every block. As per ISO 2539, the operation number has a minimum of three digits, e.g. N009, N028.

The data word consists of the words for the preparatory functions to be used in the manufacture of the component. A typical ISO format is:

N3G2X +/–43Y + /–43Z+/–43 U..V..W..I..J..K.. F4 S4 T4 M2

where the numerical values indicate the number of digits for the preceding address character, e.g. G, the preparatory function is followed by two-digit information, say, G00 or G81. 43 indicates that four digits before the decimal and three after it are needed to describe any of the coordinates $(x, y, z, \ldots k)$.

As per the standards followed, decimal sign should not be given, its position being defined by format specifications. Since each function is indicated by its address character, the order of writing words in a block is not important except that the letter N should come right in the beginning and the end of block (EOB) should be placed where the information for that block is completed.

14.1.3 Sequence Number

The first word in a block of information is usually the sequence number and it is used to identify each particular block of information. The sequence number is usually a three-digit numerical word (though it can also be four digits) preceded by the letter code Nxxxx is normally used to represent block number. The letter N is generally used to identify all other sequence numbers ranging from N000 to N9999. Sequence numbers are especially valuable when it becomes necessary to edit a tape and make corrections or revisions to a program. They can be used to search the tape or program for a particular sequence number, stop when it is found for editing purpose.

If sequence numbers are assigned with flexibility in mind during programming, they can be very useful if it becomes necessary to add, revise or correct a program. It is not considered good practice to assign numbers in numerical order (for example, N001, N002, N003) because this approach provides no flexibility if revisions or additions must be made to a program. If a correction or revision must be made to the program, all following sequence numbers must be made to the program, all following sequence numbers must be renumbered. The most common method used to assign sequence numbers is in progression of 10 (for example, N010, N020, N030). This allows enough room between each sequence number to insert as many as nine blocks of new information before any sequence number would have to be renumbered.

14.1.4 Preparatory Function

This is denoted by 'G'. It is a pre-set function associated with the movement of machine axes. It has two digits, e.g. G81, G90, G00. It is possible to include more than one 'G' address in one block, provided these functions are not mutually exclusive. For example, G02 and G03 together in one block are not permissible. This enables rapid movement along the axes involved for achieving the position programmed. This stays active till cancelled by any other function of its family, i.e. G01, G02, G03 (modal function) and available as soon as the system is switched on or when a new program starts (turn on mode). Figure 14.5 refers the rapid positioning of tool from position A until it achieves position B. Some of the preparatory functions supplied by the EIA are listed in Table 14.1.

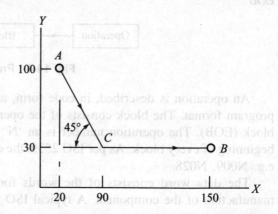

Fig. 14.5 Positioning, preparatory function G00.

14.1.5 Feed Rate Function, F

Once the feed rate is programmed in a block, e.g. F250 then this feed rate of 250 mm per minute remains in force in subsequent blocks too, till it is replaced by another 'F' value. The feed rate units are either mm/min or mm/rev.

TABLE 14.1 Some of the Preparatory Functions

Number	Operation	Definition
G00	Point-to-point	Point-to-point positioning at rapid traverse rate
G01	Linear interpolation	A mode of contouring control which uses the information contained in a block to produce a straight line in which the vectorial velocity is held constant
G02	Arc clockwise circular interpolation (two-dimensional)	An arc generated by the coordinate motion of two axes in which curvature of the path of the tool w.r.t the work piece is clockwise
G03	Arc counterclockwise circular interpolation (two-dimensional)	An arc generated by the coordinate motion of two axes in which curvature of the path of the tool w.r.t the work piece is counterclockwise
G04	Dwell	A time delay of programmed or established duration
G20	Inch data input	Used by British system
G21	Metric data input	Used by Indian system (metric units)
G28	Reference point return	Used for a new operation
G32	Thread cutting	Used for thread cutting (single pass)
G40	Tool nose radius compensation, Cancel	Used to deactivate the controller for tool nose radius compensation
G41	Tool nose radius compensation—left	Cutter on left side of work surface looking from cutter in the direction of relative cutter motion with displacement normal to the cutter path to adjust for the difference between actual and programmed cutter radii
G42	Tool nose radius compensation—right	Cutter on right side of work surface looking from cutter in the direction relative cutter motion with displacement normal to the cutter path to adjust for the difference between actual and programmed cutter radii
G71	Metric programming	Mode for programming in metric units
G73	Pattern repeating	Used for example profiles
G74	Peck drilling in z-axis	Used for only peck drilling operation
G75	Grooving in x-axis	Used for only grooving operation
G76	Thread cutting cycle	Used for multiple thread cutting cycle
G90	Absolute input	A control mode in which the data input is in the form of absolute dimensions
G91	Incremental input	A control mode in which the data input is in the form of incremental dimensions
G92	Absolute preset	
G98	Feed per minute	mm/min
G99	Feed mm/rev	mm/rev

14.1.6 Miscellaneous Function 'M'

Miscellaneous NC functions perform a variety of auxiliary commands, such as stopping the program, starting or stopping the spindle or feed, tool changes, coolant flow, etc. which control the machine tool. They are generally multicharacter ON/OFF codes which select a function controlling the machine tool. Miscellaneous functions are used at the beginning or end of a cycle and are identified by the letter address 'M' followed by a two-digit number.

In most cases, miscellaneous codes such as M00, M01, M02, M06 or M26 are effective only in the specific block in which they are programmed. If they are needed in two successive blocks, they must be programmed in each block. Most other miscellaneous codes do not have to be repeated in succeeding blocks.

Both preparatory and miscellaneous functions are generally classified as either modal or non-modal. When modal miscellaneous functions such as M03 (spindle CW) and preparatory functions such as G81 (canned drill cycle) are programmed, they stay in effect in succeeding blocks until they are replaced by another function code. The modal function or code is changed or cancelled as soon as a new miscellaneous or preparatory function code is programmed.

All non-modal functions such as M00, M01, M02, M06, etc. are valid or operational only in the block programmed. If they are needed in successive blocks, they must be programmed again. Some miscellaneous functions that have been supplied by the EIA are listed in Table 14.2.

TABLE 14.2   Some of the Miscellaneous Functions: FANUC Code

Number	Operation	Definition
M00	Program stop	A miscellaneous function command cancels the spindle and coolant functions and terminate further program execution after completion of other commands in the block
M01	Optional stop	A miscellaneous function command similar to a program stop except that the control ignores the command unless the operator has previously validated the command
M02	End of program	A miscellaneous function indicating completion of work piece stops spindle coolant and feed after completion of all commands in the block
M03	Spindle forward (CW)	Starts spindle rotation to advance a right-handed screw into the work piece
M04	Spindle reverse (CCW)	Starts spindle rotation to retract a right-handed screw from the work piece
M05	Spindle stop	Stops spindle in normal most efficient manner
M06	Tool change	Stops spindle and coolant and retracts tool to full retract position. Should be coded in last block of information in which a given tool is used
M08	Coolant ON	Turns coolant ON
M09	Coolant OFF	Turns coolant OFF
M10	Chuck open	
M11	Chuck close	Can pertain to machine slides work piece, fixtures, spindle, etc.
M13	Spindle forward (CW) and coolant ON	Starts spindle rotation and turns the coolant ON
M14	Spindle reverse (CCW) and coolant ON	Starts spindle rotation and turns the coolant ON
M16	Special tool call	Calls for a special tool which is not used frequently
M19	Spindle orientate	Spindle orientation
M20	Spindle indexing	
M21	Spindle indexing	Spindle indexing
M22	Spindle indexing	
M23	Spindle indexing	

TABLE 14.2 Some of the Miscellaneous Functions: FANUC Code (Contd.)

Number	Operation	Definition
M25	Quill extend	Extends the quill (tailstock)
M26	Quill retract	Retracts the quill (tailstocks)
M29	DNC mode	
M30	Program reset and rewind	Stops the spindle, turns the coolant OFF terminates the CNC program and resets it
M38	Door open	Opens the door
M39	Door close	Closes the door
M40	Parts catcher	Extend
M41	Parts catcher	Retract
M43	Swarf conveyor	Forward
M44	Swarf conveyor	Reverse
M45	Swarf conveyor	Stop
M48	Lock feed	Speed at 100%
M49	Cancel M48	(default)
M52	Threading pull	Out angle(90)
M53	Cancel M52	
M56	Internal chucking	
M57	External chucking	
M68	Turret indexing	Only at home position
M69	Turret indexing	Anywhere
M70	Mirror in x ON	
M80	Mirror in x OFF	
M98	Subprogram call	
M99	Subprogram exit	

14.1.7 Spindle Speed Function 'S'

In some of the NC machines, speeds are set manually and so are not programmed. Some machines have stepped drive which can be programmed as S1, S2, S3,..., and refer to the particular speeds available on the corresponding steps. Most NC machines have step less variable speed drive and these are programmed as the symbol 'S' followed by the speed required, e.g. S4000 means spindle speed is 4000 rpm.

14.1.8 Tool Change Function 'T'

All NC machines are generally equipped with turrets or tool magazines with automatic tool changers (ATC) which enable the positioning of the preset tools in a few seconds. Thus, the ratio of the cutting time to total machine time is increased. For programming this event, the instruction in the program block is given by the word 'T', followed by the tool number, e.g. T07. In the case of a turret, it would index bringing tool 7 into a ready position. If the ATC is present it would withdraw the existing tool in the machine spindle and replace it by tool 7.

14.1.9 Dimensional Words

Dimensional words X, Y and Z are used to specify the movement of the programming axes. Figure 14.6 indicates various dimensional words used in NC.

X, Y and Z words refer to coordinate movement of the machine tool for positioning or machining purposes.

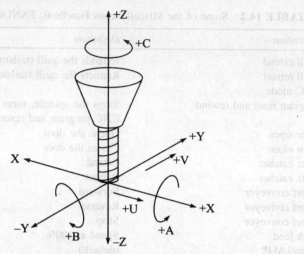

Fig. 14.6 Specification of programming axes.

To avoid confusion and errors when programming, the distance to the right or left of the Y-axis is generally given first. Any X distance to the right of the Y-axis is uniformly referred to as a positive (+) dimension; those to the left are negative (–) dimensions. Any Y distance above the X-axis is referred to as a positive (+) dimension, those below are negative (–) dimensions. The third plane or Z-axis is perpendicular to the plane established by the X and Y axes and on a vertical mill or drill press refers to the movement of the cutting tool.

Towards or into the work piece is a negative (–) motion, while away from the work piece is a positive (+) motion. Table 14.3 describes the various axes and its information.

TABLE 14.3

Address	Information
X,Y,Z	Linear axes
A,B,C	Rotary axes
U,V,W	Axes parallel to x, y, z axes
I,J,K	Axes used as auxiliary of x, y, z axes
R,Q	Axes used as auxiliary of z-axis

Coordinate information must be programmed in the following sequence in order for the MCU to understand the command correctly:

(a) *Axis movement:* X, Y or Z axes must be specified when motion is required on more than one axis, program X first, Y second and Z last.

(b) *Direction movement:* The information must indicate whether the movement is positive (+) or negative (–) from the origin point. The positive movement sign (+) does not have to appear on a program, since it is assumed. Negative dimensions must have the minus (–) sign. For example, the dimension X 5.678 would mean that X = +5.678 mm.

(c) *Dimension movement:* This is normally a seven-digit number with the decimal point fixed in the tape format to allow four places to the right of the decimal. Decimal points are not usually programmed, however, modern controls do allow decimal points to be programmed. For example, the dimension Z-5.678 would mean a 5.678 mm movement below the Z-axis.

(d) *Depth selection (Z-axis)*: Machining operations such as producing holes, milling slots or steps, or cutting into the surface of a work piece generally involve the Z-axis. The Z-axis on any machine tool is usually a line drawn through the centre of the machine spindle. The Z-axis of motion is always parallel to spindle of the machine and perpendicular to the workholding surface on milling, boring, drilling and tapping machines. The spindle is the tool-rotating device on lathes, grinders and other machines. Where the work revolves, the spindle is the work-rotating device. A positive (+) Z movement moves the cutting tool away from the work piece, while a negative (−) Z movement moves the cutting tool into a work piece. The Z motion can be controlled by the operator (manual data input) by preset stops or by programmed NC tape.

The word address letter 'Z' generally consists of a seven-digit number with the decimal point fixed at four places to the left. The Z movement generally involves a rapid move to the programmed gage height or a position above the work piece and then a slower feed rate during the machine operation.

Work plane

The word address letter 'R' refers to either the work surface or the rapid-feed distance (sometimes called the work plane) programmed. The 'R' work surface is set either at the highest surface on the part to be machined or at a specific height or distance from this surface setting is referred to as R000000 or the reference dimension and all programmed depths for cutting tools and surfaces to be machined are taken from the R surface.

14.1.10 Interpolation

The calculation of successive increments in slide position to reach the programmable point is called **interpolation**. Interpolation are of two types: linear interpolation and circular interpolation.

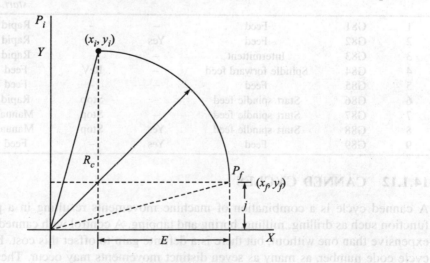

Fig. 14.7 Circular interpolation.

In linear interpolation, the coordinates of the destination are provided in the block along with code G01 and other relevant machining parameters. The data processing unit calculates the slope and intercept for the straight line along which the tool is supposed to move.

Figure 14.7 shows circular interpolation. In circular interpolation, besides the destination coordinates, it is required to provide the coordinates of centre of arc w.r.t starting point of arc. The data processing unit calculates the radius of the circle for which the arc forms the parts that determines the points required for interpolation.

14.1.11 Fixed Cycles

An important feature of the NC system is the facility of canned cycles, which provide the sequence of fixed motions as a built-in aspect in the control.

All manufacturing operations consist of a number of motions, e.g. rapid approach, entering into the material at a preset feed rate and retracting of the cutting tool.

Approach and retraction constitute the non-producing time. In NC, these can be programmed to be carried out at the highest feed rate available along with actual cutting being undertaken at the desired feed rate by an appropriate 'G' function in one block only, there not only improving the productivity but also providing a very simple programming procedure.

The 'R' word is applicable only in the cycle mode. These cycles remain in force till cancelled by the word G80. If the same operation is to be carried out at a number of positions, then in subsequent programming blocks only the coordinates of various positions need to be specified. In the complete cycle, all the positioning moves automatically take place at 'Rapid'. G80 is effective in which block it is commanded. As per ISO recommendations, G81 to G89 are assigned to fixed cycles. Table 14.4 shows the preparatory functions of fixed cycles.

TABLE 14.4 Preparatory Functions (Fixed Cycles)

Fixed cycle			At bottom		Movement	Typical usage
Number	Code	Movement in	Dwell	Spindle	out to feed start	
1	G81	Feed	–	–	Rapid	Drill, spot drill
2	G82	Feed	Yes	–	Rapid	Drill, counterbore
3	G83	Intermittent	–	–	Rapid	Deep hole
4	G84	Spindle forward feed	–	ReV	Feed	Tap
5	G85	Feed	–	–	Feed	Bore
6	G86	Start spindle feed	–	Stop	Rapid	Bore
7	G87	Start spindle feed	–	Stop	Manual	Bore
8	G88	Start spindle feed	Yes	Stop	Manual	Bore
9	G89	Feed	Yes	–	Feed	Bore

14.1.12 CANNED CYCLES

A canned cycle is a combination of machine movements resulting in a particular machining function such as drilling, milling, boring and tapping. A control with canned cycles may be more expensive than one without, but there is a definite gain to offset this cost. By programming one cycle code number, as many as seven distinct movements may occur. These seven movements would normally take at least six blocks of programming on a control without canned cycles. Using canned cycles, it is possible to realize a savings upto one third less data processing time. Tape length can also be reduced by at least one third with canned cycles. Figure 14.8 shows the motion in a canned cycle.

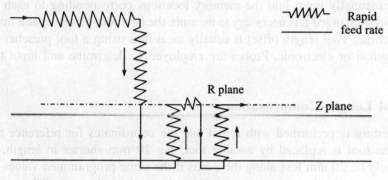

Fig. 14.8 Motion in canned cycle.

Most control manufacturers today have both canned and non-canned cycles as part of their standard control package. Rapid and feed are the two important terms which need to be defined before any further discussion of preparatory functions.

Rapid

Positioning the cutter and work piece into close proximity with one another at a high rate of travel speed, usually 150 to 400 inches per minute (IPM) before the cut is started.

Feed

The programmed or manually established rate of movement of the cutting tool into the work piece for the required machining operation.

G71, G72, G76, etc. are codes for canned cycles.

Figure 14.9 illustrates the G81 drill cycle. The canned cycles need another coordinate to be defined by the R word, which specifies the plane above work surface up to which the tool would position in 'rapid', then proceed at feed rate into the material and then return to the R plane (Fig. 14.8).

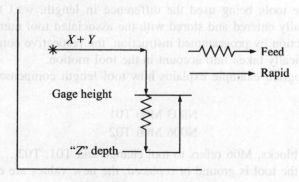

Fig. 14.9 G81—Drill cycle.

14.1.13 Tool Offset

The NC program is written for the movement of the slides whereas it is the path of the tip of the tool which produces the desired surface. Hence it is necessary to alter the coordinate position of the slides to take the length of the tool into account. This is called **tool length offset**.

These are manually input into the memory locations corresponding to each tool position. Whenever a tool is changed it is necessary to measure the new tool offset input it into the memory of the NC machine. Tool length offset is usually measured using a tool presetter which can be mechanical, optical or electronic. Probes are employed to determine and input the tool offset automatically.

14.1.14 Tool Length Compensation

Absolute presetting is performed with a tool and the coordinates for reference are registered. However, if the tool is replaced by another tool, say 20 mm shorter in length, then the tool movement would be 20 mm less along the z-axis if the same programmed values are used. The alternative is to modify the program every time the tool is changed. When a G81 cycle is programmed, the tool will move in rapid in X and/or Y axis, rapid in the Z-axis to gage height feed in the Z-axis to the Z depth and rapid retraction to the gage height (Fig. 14.9). Figure 14.10 shows the tool length compensation.

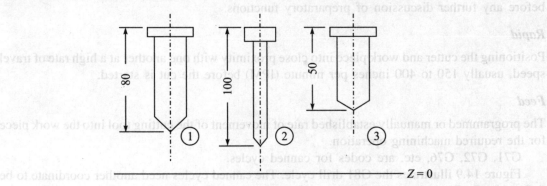

Fig. 14.10 Tool length compensation.

In NC practice, all tools are measured in the assembled state and this information is always kept uptodate. For the tools being used the difference in length, w.r.t the presetting tool is recorded and is manually entered and stored with the associated tool number. Whenever these tools are called into action by programmed instruction, the respective compensation values are activated and automatically taken into account in the tool motion.

The following program example explains how tool length compensation is automatically taken care of:

$$\text{N003 M06 T01}$$
$$\text{N006 M06 T02}$$

In these program blocks, M06 refers to tool change and T01, T02,…, refer to the tools to be loaded. Whenever the tool is ground or replaced, the new values are entered to replace the earlier ones and thus the program remains unchanged. This is an essential facility, without which the multiplicity of the programs for each job/tool combination would be enormous and futile. The values entered compensate for the difference in lengths and, therefore, all tools effectively become independent of tool dimensions, if the dimensions of all the tools are stored. When programming, the tool dimensions are not considered since the compensation values are calculated by the control system itself during manufacturing.

14.1.15 Tool Nose Radius Compensation

In profiling operations, one needs to calculate the tool path for preparing the program. This path refers to the spindle axis which is away from the profile required. Figure 14.11 shows the component and the tool path.

Apart from the problem of calculating, one should examine that whenever the cutter size changes, the program would need editing. However, if a compensation equal to the radius of the cutter is entered and stored in the control system, then the program could be written for the component profile and thus no change in program would be required.

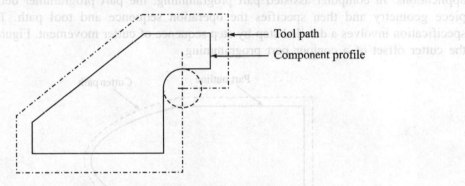

Fig. 14.11 Cutter radius compensation.

The preparatory functions G40, G41 and G42 are used for radius compensation and form one group. These are modal and the one programmed in any block remains active till cancelled by the other.

In some systems, tool nose radius compensation is possible and in these cases, the value of the diameter is entered as the compensation value. The tool radius entry is always positive. If the programmed path determined for a particular size of the cutter, the compensation value would be '+' or '−' depending on whether the cutter used for machining is oversize or undersize.

Preparatory functions G41, G42

N010	G92	X0	Y0 Z50—Absolute preset at a 50 mm above work surface
N020	G90		
N030	M06	T1	
N040	Z-30	M03	S800
N050	G41	G01	X100 Y-60 F100—Tool nose radius compensation ON
N060	X300		
N070	Y-260		
N080	X100		

N090	Y-60		
N100	G40	X0	Y0—Tool nose radius compensation OFF
N110	G00	Z50	T01 M05—Compensation value for tool T01 cancelled
N120	M06	T02	–Different tool is used for internal profile
N130	X200	Y-120	Z200 S600 M03
N140	G01	Z-5	F100
N150	G42	Y-100—Cutter compensation ON	
N160	G02	I 0	J-60
N170	G00	Z50	M05
N180	G40	X0	Y0 T0 M02

14.2 COMPUTER ASSISTED PART PROGRAMMING

Computer assisted part programming is used for complicated point-to-point and contour profile applications. In computer assisted part programming, the part programmer defines the work piece geometry and then specifies the operation sequence and tool path. This tool path specification involves a detailed step-by-step sequence of cutter movement. Figure 14.12 shows the cutter offset of a contour part programming.

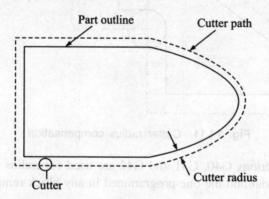

Fig. 14.12   Cutter offset of a contour part programming.

The computer's job in computer assisted part programming consists of the following steps:

- Input translation
- Arithmetic calculations
- Cutter offset computation
- Post-processor

The part programmer enters the program in an NC language. The input translation component converts the coded instructions contained in the program into computer usable form for further processing. The arithmetic calculations unit of the system consists of a comprehensive set of subroutines for solving the mathematics required to generate the part surface. These subroutines are called by the various part programming language statements.

The actual tool path is different from the part outline because the tool path is defined as the path taken by the centre of the cutter. It is at the periphery of the cutter that machining takes place. The purpose of the cutter offset compensations is to offset the tool path from the desired part surface by the radius of the cutter.

The post-processor is a separate computer program that has been written to prepare the punched tape for a specific machine tool. The input to the post processor is the output from the other three components, a series of cutter locations and other instructions. The output of the post-processor is the NC tape written in the correct format for the machine on which it is to be used.

14.3 COMPUTER ASSISTED NC PART PROGRAMMING LANGUAGES

An NC part programming language consists of a software package plus the special rules, conventions and vocabulary words for using that software. Table 14.5 shows a list of part programming languages.

TABLE 14.5 List of Part Programming Languages

Program	Developed by	Features
ART	MIT/IITRI/CAM-I	P, C
ADAPT	IBM	P, C
AUTOSPOT	IBM	P
CINTURN	Cincinnati Milacron	T
COMPACT-II	MDSI	P, C
EXAPT-I	TH-Aachen	P
EXAPT-II	TH-Aachen	T
EXAPT-III	TH-Aachen	C
GENTURN	General Electric	T
MILTURN	Metal Institute, Netherlands	T
NEL 2PL	Ferranti	P
NEL 2C	Ferranti	T
NEL 2CL	Ferranti	C
PROMPT	Weber	C
NEL APT	NEL	C
SPLIT	Sun strand	C
UNIAPT	United computing co	P, C

14.4 APT LANGUAGE

Automated programmed tool (APT) is the most widely used NC part programming language. A great number of computer systems have been developed for NC programming. The APT system is available on many computers and is widely used by many manufacturers of NC systems.

APT language consists of English like statements and it shows how they can be formulated to command the cutting tool through its sequence of machining operations. APT is not only a NC language, it is also the computer program that performs the calculations to generate cutter positions based on APT statements.

APT is a three-dimensional system that can be used to control up to five axes. APT can be used to control a variety of different machining operations. There are over 400 words in the APT vocabulary.

To program in APT, the work part geometry must first be defined. Then the tool is directed to various point locations and along surfaces of the work parts to carry out the machining operations. The viewpoint of the part programmer is that the work piece remains stationary and the tool is instructed to move relative to the part.

There are four types of statements in the APT language:

14.4.1 Geometry Statements

These define the geometric elements that comprise the work part. They are also sometimes called **definition statements**.

When the tool motions are specified, their description is in terms of points and surfaces. Therefore, the points and surfaces must be defined before tool motion commands can be given.

The general form of an APT geometry statement is

$$\text{Symbol} = \text{Geometry type/descriptive data}$$

e.g.
$$\text{P1} = \text{POINT/5.0,4.0,0.0}$$

The statement is made up of three sections: The first is the symbol used to identify the geometric element. A symbol can be any combination of six or fewer alphabetic and numeric characters. At least one of the six must be an alphabetic character.

The second section is an APT vocabulary word that identifies the type of geometry element.

The third section of the geometry statement is the descriptive data that defines the element precisely. These data may include quantitative dimensional and positional data, previously defined geometry elements and other APT words.

To specify a point
$$\text{P2} = \text{POINT/INT OF, L1, L2}$$

To specify a line
$$\text{L3} = \text{LINE/P3, P4}$$
$$\text{L4} = \text{LINE/P5, PARLEL, L3}$$

To specify a plane
$$\text{PL1} = \text{PLANE/P1, P4, P5}$$
$$\text{PL2} = \text{PLANE/}P_2\text{, PARLEL, PL1}$$

To specify a circle
$$\text{C1} = \text{CIRCLE/CENTRE, P1, RADIUS, 5.0}$$

Rules for formulating APT geometry statement

- Coordinate data must be specified in the order x, y, z.
- Any symbol used as descriptive data must have been previously defined.
- A symbol can be used to define only one geometry element.
- Lines defined in APT are of infinite length in both directions. Similarly, planes extend indefinitely and circles defined in APT are complete circles.

14.4.2 Motion Statements

These are used to describe the path taken by the cutting tool.

The general form of a motion statement is: Motion command/descriptive data

e.g.

GOTO/P2

The statement consists of two sections separated by a slash. The first section is the basic motion command, which tells the tool what to do. The second section comprised descriptive data which tells the tool where to go. At the beginning of the motion statements, the tool must be given a starting point. This point is likely to be the target point, the location where the operator has positioned the tool at the start of the job.

FROM/TARG

FROM/–2.0, –2.0,0.0

Point-to-point motions

There are only two basic PTP motion commands: GOTO and GODLTA. The GOTO statement instructs the tool to go to a particular point location specified in the descriptive data.

GOTO/GOTO/P2

GOTO/2.0,7.0,0.0

The GODLTA command is useful in drilling and related operations. The tool can be directed to a particular hole location with the GOTO statement.

Contouring motions

Contouring commands are somewhat more complicated because the tool's position must be continuously controlled throughout the move. Figure 14.13 illustrates the three surfaces in APT contouring motions that guide the cutting tool. To accomplish this control, the tool is directed along three surfaces:

(a) *Drive surface:* This is the surface that guides the side of the cutter.

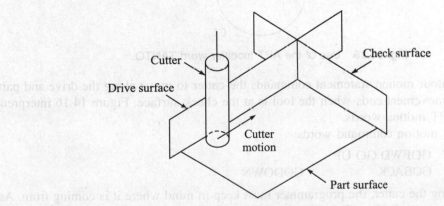

Fig. 14.13 Three surfaces in APT contouring motions that guide the cutting tool.

(b) *Part surface:* This is the surface on which the bottom of the cutter rides. The part surface may or may not be an actual surface of the work part. The part programmer must define this plus the drive surface for the purpose of maintaining continuous path control of the tool.

(c) *Check surface:* This is the surface that stops the movement of the tool in its current direction. It checks the forward movement of the tool.

There are several ways in which the check surface can be used. This is determined by APT modifier words within the descriptive data of the motion statement. The three main modifier words are TO, ON and PAST. Figure 14.14 depicts the use of APT modifier words in motion statements. A fourth modifier word is TANTO. This is used when the drive surface is tangent to a circular check surface. In this case the cutter can be brought to the point of tangency with the circle by use of the TANTO modifier word. Figure 14.15 shows the use of the APT modifier word TANTO.

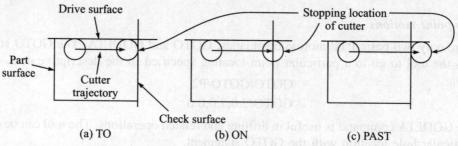

Fig. 14.14 Use of APT modifier words in motion statements.

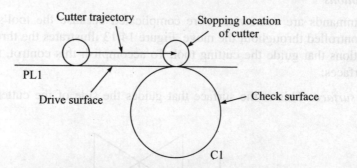

Fig. 14.15 Use of the APT modifier word TANTO.

The APT contour motion statement commands the cutter to move along the drive and part surfaces and the movement ends when the tool is at the check surface. Figure 14.16 interprets the use of the APT motion words.

There are six motion command words:

GOLFT	GOFWD GO UP	
GORGT	GOBACK	GODOWN

In commanding the cutter, the programmer must keep in mind where it is coming from. As the tool reaches the new check surface does the next movement involve a right turn or an upward turn or what? The tool is directed accordingly by one of the six motion words. In the use of these

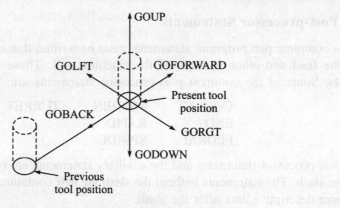

Fig. 14.16 Use of the APT motion words.

words, it is helpful for the programmer to assume the viewpoint that the work piece remains stationary and the tool is instructed to move relative to the piece. To begin the sequence of motion commands, the FROM statement is used in the same manner as for PTP moves. The statement following the FROM statement defines the initial drive surface, part surface and check surface. The sequence is of the following form:

FROM/TARG
GO/TO, PL1,TO,PL2,TO,PL3

Figure 14.17 shows the initialization of APT contouring motion sequence. The symbol TARG represents the target point where the operator has set up the tool. The GO command instructs the tool to move to the intersection of the drive surface (PL1), the part surface (PL2) and the check surface (PL3). The periphery of the cutter is tangent to PL1 and PL3, and the bottom of the cutter is touching PL2. This cutter location is defined by use of the modifier word. To the three surfaces included in the GO statement must be specified in the order: drive surface first, part surface second, and check surface last.

Note that the GO/TO command is different from the GOTO command. GOTO is used only for PTP motions. GO/TO is used to initialize the sequence of contouring motions.

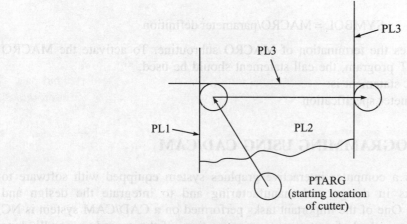

Fig. 14.17 Initialization of APT contouring motion sequence.

14.4.3 Post-processor Statements

To write a complete part program, statements must be written that control the operation of the spindle, the feed and other features of the machine tool. These are called **post-processor statements**. Some of the common post-processor statements are:

COOLNT	MACHIN	TURRET
END	RAPID	
FEDRAT	SPINDL	

The post-processor statements and the auxiliary statements are of two forms: either with or without the slash. The statements without the slash are self contained. The APT words with the slash require descriptive data after the slash.

14.4.4 Auxiliary Statements

The complete APT program must also contain various other statements, called **auxiliary statements**. These are used for cutter size definition, part identification and so on. Some of the common auxiliary statements are:

CLPRNT	INTOL
CUTTER	OUTTOL
FINI	PARTNO

The offset calculation of the tool path from the part outline is based on the CUTTER/ definition. For example, the statement

$$CUTTER/.500$$

would instruct the APT program that the cutter diameter is 0.500 in. Therefore, the tool path must be offset from the part outline by 0.250 in.

The MARCO statement in APT

The MARCO statement in APT is like a subroutine. The MARCO subroutine is defined by a statement of the format.

$$SYMBOL = MACRO/parameter\ definition$$

"TER MAC" signifies the termination of MACRO subroutine. To activate the MACRO subroutine within an APT program, the call statement should be used.

The syntax of CALL statement is:

CALL/symbol, parameter specification

14.5 NC PART PROGRAMMING USING CAD/CAM

A CAD/CAM system is a computer interactive graphics system equipped with software to accomplish certain tasks in design and manufacturing and to integrate the design and manufacturing functions. One of the important tasks performed on a CAD/CAM system is NC part programming. In this method of part programming portions of the procedure usually done by the part programmer are instead done by the computer. Recall that the two main tasks of the part programmer in computer-assisted programming are:

(a) Defining the part geometry
(b) Specifying the tool path

Advanced CAD/CAM systems automate portions of both of these tasks.

Geometry definition using CAD/CAM is to integrate the design engineering and manufacturing engineering functions. Certainly one of the important design functions is to design the individual components of the product. If a CAD/CAM system is used, a computer graphics model of each part is developed by the designer and stored in the CAD/CAM database. That model contains all of the geometric, dimensional and material specifications for the part.

When the same CAD/CAM system or a CAM system that has access to the same CAD database in which the part model resides, is used to perform NC part programming it makes little sense to recreate geometry of the part during the programming procedure. Instead, the programmer has the capability to retrieve the part geometry model from storage and to use that model to construct the appropriate cutter path. The significant advantage of using CAD/CAM in this way is that it eliminates one of the time consuming steps in computer-assisted part programming, geometry definition. After the part geometry has been retrieved, the usual procedure is to label the geometric elements that will be used during part programming. These labels are the variable names (symbols) given to the lines, circles and surfaces that comprise the part. Most systems have the capacity to automatically label the geometry elements of the part and to display the labels on the monitor. The programmer can then refer to those labelled elements during tool path construction.

If the NC programmer does not have access to the database, then the geometry of the part must be defined. This is done by using similar interactive graphics techniques that the product designer would use to design the part. The advantage of using the interactive graphics system over conventional computer-assisted part programming is that the programmer receives immediate visual verification of the definitions being created. This tends to improve the speed and accuracy of the geometry definition process.

14.6 TOOL PATH GENERATION USING CAD/CAM

The second task of the NC programmer in computer-assisted part programming is tool path specification. The first step in specifying the tool path is to select the cutting tool for the operation. Most CAD/CAM systems have tool libraries that can be called by the programmer to identify what tools are available in the tool crib. The programmers must decide which of the available tools is most appropriate for the operation under consideration and specify it for the tool path. This permits the tool diameter and other dimensions to be entered automatically for tool can be specified by the programmer. It then becomes part of the library for future use.

The next step is tool path definition. There are differences in capabilities of the various CAD/CAM systems, which result in different approaches for generating the tool path. The most basic approach involves the use of the interactive graphics system to enter the motion commands one-by-one, similar to computer-assisted part programming. Individual statements in APT or other part programming language are entered and the CAD/CAM system provides an immediate graphic display of the action resulting from the command, thereby validating the statement.

A more-advanced approach for generating tool path commands is to use one of the automatic software modules available on the CAD/CAM system. These modules have been developed to accomplish a number of common machining cycles for milling, drilling and turning. They are subroutines in the NC programming package that can be called and the required parameters given to execute the machining cycle. Figure 14.18 illustrates examples of machining cycles available in automatic programming modules.

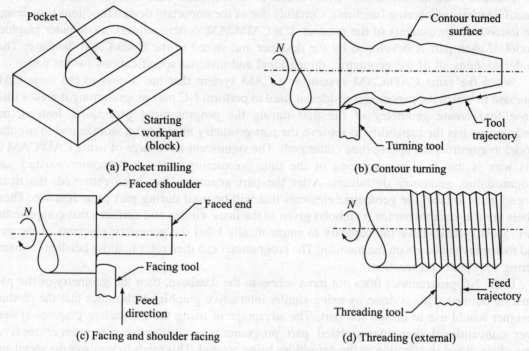

Fig. 14.18 Examples of machining cycles available in automatic programming modules.

When the complete part program has been prepared, the CAD/CAM system can provide an animated simulation of the program for validation purposes.

14.7 COMPUTER-AUTOMATED PART PROGRAMMING

In the CAD/CAM approach to NC part programming, several aspects of the procedure are automated. In the future, it should be possible to automate the complete NC part programming. Given the geometric model of a part that has been defined during product design, the computer-automated system would possess sufficient logic and decision-making capability to accomplish NC part programming for the entire part without human assistance.

This can most readily be done for certain NC processes that involve well-defined, relatively simple part geometries. Examples are point-to-point operations such as NC drilling and electronic component assembly machines. In these processes, the program consists basically of a series of locations in an x-y coordinate system where work is to be performed (e.g. holes are to be drilled or components are to be inserted). These locations are determined by data that are generated during product design. Special algorithms can be developed to process the design data and generate the NC program for the particular system. NC contouring systems will eventually be capable of similar level of automation. Automatic programming of this type is closely related to computer-automated process planning (CAPP).

Some of the popular CAM packages available for generation of part programs are:

Pro Engineer (manufacturing module)

CATIA
UNI GRAPHICS
Master CAM
Surf CAM
God CAM
Edge CAM
Auto CAM
Gibbs CAM

Figure 14.19 shows the procedure involved to generate a machine specific NC program using a CAM.

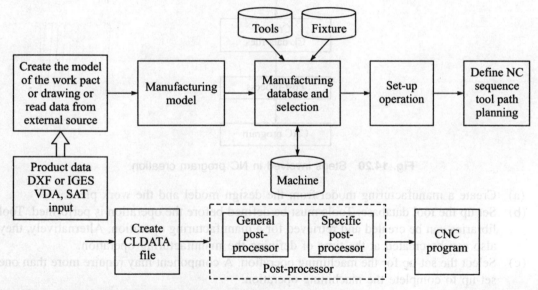

Fig. 14.19 Procedure involved to generate a machine specific NC Program using a CAM.

14.8 TECHNOLOGY OF CAM

The block diagram shown in Fig. 14.20 illustrates the steps involved in creating a NC program using a CAM software package. The starting point of CAM is the CAD file. A common approach is the program creation carried out using solid models or surface models. Data for program creation can also be obtained from SAT (ACIS Solids), IGES, VDA, DXF, CADL, STL and ASCII using suitable translators.

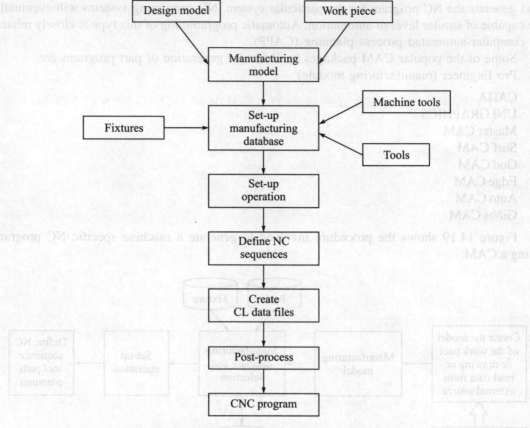

Fig. 14.20 Steps involved in NC program creation.

(a) Create a manufacturing model from the design model and the work piece.

(b) Set up the tool database. Tools must be defined before the operation is performed. Tool libraries can be created and retrieved for a manufacturing operation. Alternatively, they also can be created at the time of defining the manufacturing operation.

(c) Select the set-up for the machining operation. A component may require more than one set-up to complete the machining operation.

(d) Fixtures are necessary at each set-up. Fixtures can be displayed along with the work piece. This is to ensure that the path of the tools does not interfere with the tools. Fixtures consist of the locating and clamping elements to hold the work piece on the machine table during the machining operation. In the case of a turning operation, appropriate chucking devices should be used. Standard fixture elements can be created as a library of parts and assembled before the manufacturing operation is commenced. These are also available as standard libraries.

(e) Create a machinability database. Parameters like spindle speed and feed rate can be selected from the machinability database.

(f) Create the manufacturing operations to generate the CL data. As each manufacturing operation takes place, material is removed from the work piece to simulation is enhanced by the tool path being displayed for each operation.

(g) If needed, the CL data can be modified by modifying the operation parameters or by editing the CL data file.

(h) Create a manufacturing session.

(i) Post-process the CL file to create the NC program.

14.9   HOW COMPUTER ASSISTED PROGRAMMING SYSTEMS WORK?

All programming systems are really two systems working together. The programming system itself is called the **processor**. The APT language is a text-based processing system. Master cam, surface and Auto pro are example of CAD/CAM processing systems. Pro/Engineer is an example of a solid modelling processing system. The purpose of the programming system (programming processor) is to take the programmer's specified information and convert it into a file called a **centre line data file (CL file)**. The CL file contains the cutter centre line locations of the NC program along with commands indicating when to turn the spindle ON or OFF, turn the coolant ON or OFF, turn on cutter diameter compensation and so on.

The second system is called the **post-processor**. The job of the post-processor is to convert the CL files information into the tape commands that the NC control needs. Most NC shops have a variety of NC controls. While only one processor, for example, master cam, may be in use in a given shop, a separate post-processor is required for each machine control.

Regardless of what type of programming system is used (computer aided, CAD/CAM or solid modelling), each system works in the same basic way that is shown in Fig. 14.20. First the NC programmer creates a source file. This file or collection files, contains the part geometry, the tool paths to be developed and any auxiliary statements (turning ON or OFF the spindle, coolant and so on). The NC programming system then processes this information into a CL file, lastly the post-processor converts the CL file into the final tape image file that will be loaded into the NC control.

CNC PROGRAM EXAMPLES

Prepare CNC programs to machine the part given in Figs. 14.21 to 14.25 using lathe.

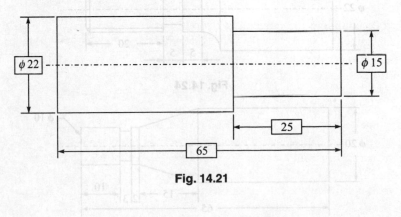

Fig. 14.21

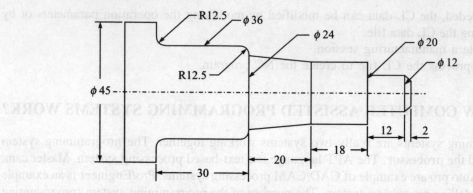

Fig. 14.22

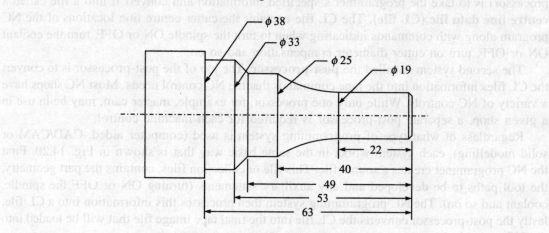

Fig. 14.23

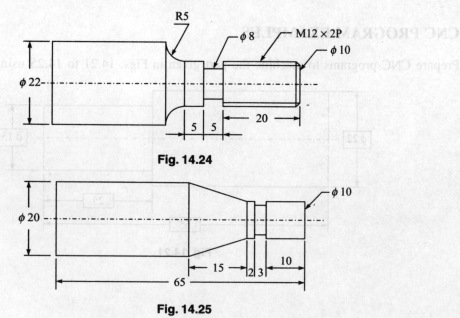

Fig. 14.24

Fig. 14.25

Solution:

(i) The CNC program is as follows:

```
[BILLET X22 Z65
G21 G98
G28 U0 W0
M06 T2
M03 S1200
G00 X22 Z1
G01 X21
G01 Z-25 F45
G00 X22
G00 Z1
G01 X20
G01 Z-25
G00 X22
G00 Z1
G01 X19
G01 Z-25
G00 X22
G00 Z1
G01 X18
G01 Z-25
G00 X22
G00 Z1
G01 X17
G01 Z-25
G00 X22
G00Z1
G01 X16
G00 X22
G00 Z1
G01 X15
G01 Z-25
G00 X22
G00 Z1
G28 U0 W0
M05
M30
```

(ii) The CNC program is as follows: Note that the rod with diameter 45 mm is used for machining.

```
[BILLET X45 Z85
N010 G21 G98
N020 G28 V0 W0
N030 M06 T02
N040 M03 S1200
N0S0 G00 X45 Z1
N060 G71 U1 R1
N070 G71 P80 Q170 U0.1W0.1 F50
N080 G01 X8
N090 G01 Z0
```

```
N100 G01 X12 Z-2
N110 G01 X12 Z-14
N120 G01 X20
N130 G01 Z-32
N140 G01 X24 Z-52
NI50 G03 X30 Z-67 R12.5
N160 G02 X36 Z-82 R12.5
N170 G01 X45
N180 G70 P80 Q170
N190 G28 U0 W0
N200 M05
N210 M30
```

(iii)   The CNC program is as follows:

```
[BILLET X38 280
N010 G21
N020 G28 U0 W0
N030 G50 S2000
N040 M06 T2
N050 G99
N060 G96 S200
N070 G00 X38 Z0
N080 G01 X-2 F.05
N090 G00 X35 Z2
N100 G 71 U1 R1
N110 G71 P120 Q210 00.1 W0.1 E07
N120 G01 X14
N130 G01 Z0
N140 G01 X20 Z-1 F0.05
NI50 G01 Z-22
N160 G01 X19 Z-25
N170 G02 X25 Z-40 R30
N180 G01 Z-49
N190 G01 X33 Z-53
N200 G01 Z-62
N210 G00 G00 X38 Z2
N220 G42
N230 G70 P120 Q210
N240 G40
N2S0 G 28 U0 W0
N260 M05
N270 M30
```

(iv)   The CNC program is as follows: Note that the rod diameter 22 mm is used for machining.

```
[BILLET X22 Z65
N010 G21 G98
N020 G28 U0 W0
N030 M06 T02
N040 M03 S1200
N050 G00 X23 Z2
N060 G 71 U0.5 R1
```

```
N070 G 71 P80 Q120 U0.1 W0.1 F50
N080 G01 X10
N090 G01 Z0
N100 G01 X12 Z-1
N110 G01 Z-30
N120 G02 X22 Z-35 R5
N130 G70 P80 Q120
N140 G28 U0 W0
N1s0 M06 T06
N160 M03 S800
N170 G00 X13 Z-23
N180 G75 X8 Z-25 P500 Q500 F50
N190 G28 U0 W0
N200 M06 T04
N210 M03 S500
N220 G00 X13 Z3
N230 G76 P031560 Q100 R0.15
N240 G76 X9.850 Z-21 Q150 F1.75
N250 G28 U0 W0
N260 M05
N270 M30
```

(v) The CNC program is as follows:

```
[BILLET X20 265
N010 G21 G98 G40
N020 G28 U0 W0
N030 M06 T02
N040 M03 S1500
N050 G00 X21 Z1
N060 G94 X10 Z-1 F50
N070Z-2
N080Z-3
N090Z-4
N100Z-5
N110Z-6
N120Z-7
N130Z-8
N140Z-9
N150Z-10
N160Z-11
N170Z-12
N180Z-13
N190Z-14
N200Z-15
N210 G00 X20 Z-15
N220 G94 X20 Z-15 R-15 F50
N230 X20
N240 X19
N250 X18
N260 X17
N270 X16
```

```
N280 XI5
N290 X14
N300 X13
N310 X12
N320 X11
N330 X10
N340 G28 U0 W0
N350 M06 T06
N360 M03 S800
N370 G00 X10 Z-10
N380 G90 X9.5 2-13 F30
N390 X9
N400 X8.5
N410 X8
N415 G00 X21 21
N420 G00 X21 Z-50
N430 M98 P0204040
N440 G00 X20
N450 G28 U0 W0
N460 M05
N470 M30
[PARTING OFF
[CALLING SUB PROGRAM 4040
04040
G0I U-0.S F30
W-1
U-0.5 F30
W1
M99
```

MILLING EXAMPLES

Programming using Linear and Circular Interpolation

EXAMPLE 14.1 Write a manual part program for contouring operation for the component shown in Fig. 14.26.

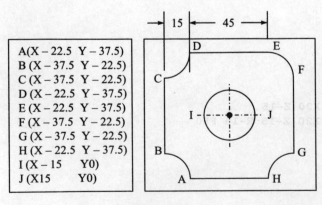

A (X – 22.5 Y – 37.5)
B (X – 37.5 Y – 22.5)
C (X – 37.5 Y – 22.5)
D (X – 22.5 Y – 37.5)
E (X – 22.5 Y – 37.5)
F (X – 37.5 Y – 22.5)
G (X – 37.5 Y – 22.5)
H (X – 22.5 Y – 37.5)
I (X – 15   Y0)
J (X15   Y0)

Fig. 14.26

Solution:

```
O1001
(Fig. 14.26
(G01-LINEAR INTERPOLATION
(G02-CIRCULAR INTERPOLATION CLOCKWISE
(G03-CIRCULAR INTERPOLATION ANTI-CLOCKWISE
```

BILLET X100 Y100 Z10	It defines the billet dimensions
EDGEMOVE X-50 Y-50	This directive sets up the required offset from the program zero position to the middle of the billet.
	This is used for simulation.
TOOLDEF TI D6	Defining tool.
G21 G94	G21 code specifies that program is done in metric units.
	G94 gives the unit of feed in mm/min.
G91 G28 Z0	G28-Go to home position along Z-axis in incremental mode.
G28 X0 Y0	Go to home position along X and Y axes.
M06 T1	Tool change to Tool No.1.
M03 S2000	M03 makes the spindle rotate in clockwise direction.
	S2000 setting the spindle speed at 2000 rpm.
G90 G00 X0 Y-37.5 Z5	G90 absolute mode.
	G00 gives rapid position of the tool to a point X0 Y-37.5 Z5 which is just above the billet. This point is called tool entry point.
G01 Z-1 F50	Giving the depth of cut along Z axis at a feed of 50 mm/min.
G01 X-22.5 Y-37.5	
G03 X-37 Y-22.5 R15	CCW interpolation.
G01 X-37.5	
G01 X-37.5 Y22.5	
G03 X-22.5 Y37.5 R15	
G01 X22.5 Y37.5	
G02 X37.5 Y22.5 R15	CW interpolation.
G01 X37.5 Y-22.5	
G03 X22.5 Y-37.5 R15	
G01 X0 Y-37.5	
G00 Z5	
G00 X-15 Y0	
G01 Z-1 F50	
G02 X15 Y0 R15	
G02 X-15 Y0 R15	
G00 Z5	
G91 G28 Z0	
G28 X0 Y0	Makes the tool to go to home position.
M05	Stops the spindle rotation.
M30	Program stop and rewind.

Contouring with Left Cutter Diameter Compensation

EXAMPLE 14.2 Write a manual part program for contouring operation with left cutter diameter compensation for the component shown in Fig. 14.27.

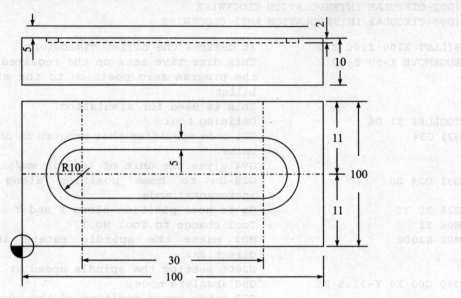

Fig. 14.27

Solution:

```
O1002
(Fig. 14.27
(PROGRAM FOR CUTTER COMPENSATION
(G CODES USED G40, G41
(G40-CUTTER DIAMETER COMPENSATION CANCEL
(G41-CUTTER DIAMETER COMPENSATION LEFT
```

BILLET X100 Y100 Z10	It defines the billet dimensions.
EDGEMOVE X-0 Y-0	This directive sets up the required offset from the program zero position to the lower left hand corner of the billet. This is used for simulation.
TOOLDEF TI D5	Defining tool.
G21 G94	G21 code specifies that program is done in metric units.
	G94 gives the unit of feed in mm/min.
G91 G28 Z0	G28 got to home position along Z-axis in incremental mode.
G28 X0 Y0	Go to home position along X and Y axes.
M06 T1	Tool change to Tool No.1.
M03 S2000	M03 makes the spindle rotate in clockwise direction.
	S2000 setting the spindle speed at 2000 rpm.
G90 G41 G00 X40 Y40 Z5	G90 absolute mode.
	G41 cutter diameter compensation left.
	G00 gives rapid position of the tool to a point

X40 Y40, Z5 which is just above the billet. This point is called tool entry point.

```
G01 Z-1 F50          Giving the depth of cut along Z axis.
G01 X70 Y40
G03 X70 Y60 R10
G01 X-40 Y60
G03 X40 Y40 R10
G00 Z5
G40                  G40 cancelling cutter radius compensation.
G91 G28 Z0           Makes the tool to go to home position.
G28 X0 Y0
M05                  Stops the spindle rotation.
M30                  Program stop and rewind.
```

Contouring with Right Cutter Diameter Compensation

EXAMPLE 14.3 Write a manual part program for contouring operation with right cutter diameter compensation for the component shown in Fig. 14.28.

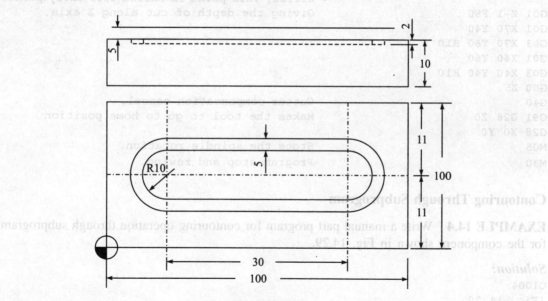

Fig. 14.28

Solution:
```
O1003
(Fig. 14.28
(PROGRAM FOR COMPENSATION
(G CODES USED G40 AND G42
(G40 CUTTER DIAMETER COMPENSATION CANCEL
(G41 CUTTER DIAMETER COMPENSATION RIGHT

BILLET X100 Y100 Z10          It defines the billet dimensions.
EDGEMOVE X0 Y0                This directive sets up the required offset
```

	from the program zero position to the lower left hand corner of the billet. This is used for simulation.
TOOLDEF TI D5	Defining tool.
G21 G94	G21 code specifies that program is done in metric units.
	G94 gives the unit of feed in mm/min.
G91 G28 Z0	G28 go to home position along Z-axis in incremental mode.
G28 X0 Y0	Go to home position along X and Y axes.
M06 T1	Tool change to Tool No.1.
M03 S2000	M03 makes the spindle rotate in clockwise direction
	S2000 setting the spindle speed at 2000 rpm.
G90 G42 G00 X40 Y40 Z5	G90 absolute mode.
	G42 cutter diameter compensation right.
	G00 gives rapid position of the tool to a point X40 Y40, Z5 which is just above the billet. This point is called tool entry point.
G01 Z-1 F50	Giving the depth of cut along Z axis.
G01 X70 Y40	
G03 X70 Y60 R10	
G01 X40 Y60	
G03 X40 Y40 R10	
G00 Z5	
G40	Cutter compensation cancel.
G91 G28 Z0	Makes the tool to go to home position.
G28 X0 Y0	
M05	Stops the spindle rotation.
M30	Program stop and rewind.

Contouring Through Subprogram

EXAMPLE 14.4 Write a manual part program for contouring operation through subprogram for the component shown in Fig. 14.29.

Solution:

```
O1004
(Fig. 14.29
(PROGRAM FOR SUBPROGRAM
(M CODES USED M98, M99
(M98-SUBPROGRAM CALL
(M99-SUBPROGRAM EXIT
(TOTAL DEPTH OF CUT: 5 mm
(DEPTH OF CUT FOR EACH PASS: 0.5 mm
```

BILLET X100 Y100 Z10	It defines the billet dimensions.
EDGEMOVE X0 Y0	This directive sets up the required offset from the program zero position to the lower left hand corner of the billet. This is used for simulation.

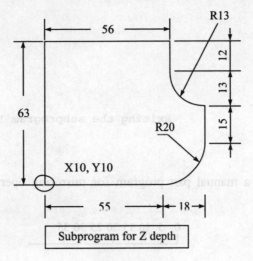

Subprogram for Z depth

Fig. 14.29

```
TOOLDEF TI D5              Defining tool.
G21 G94                   G21 code specifies that program is done in
                          metric units.
                          G94 gives the unit of feed in mm/min.
G91 G28 Z0                G28 got to home position along Z-axis in
                          incremental mode.
G28 X0 Y0                 Go to home position along X and Y axes.
M06 T1                    Tool change to Tool No.1.
M03 S2000                 M03 makes the spindle rotate in clockwise
                          direction.
                          S2000 setting the spindle speed at 2000 rpm
G90 G00 X10 Y10 Z5        G90 absolute mode.
                          G00 gives rapid position of the tool to a
                          point X10 Y10, Z5 which is just above the
                          billet. This point is called tool entry point.
G00 Z0
M98 P0014000              Calling the subprogram '4000' five times.
M98 P0014000
M98 P0014000
M98 P0014000
G00 Z5
G91 G28 Z0                Going to home position.
G28 X0 Y0
M05                       Stops the spindle.
M30                       Program stop and rewind.
04000                     Subprogram for INCREASING DEPTH.
M98 P0014001
G01 X10 Y73 F50
X66
Y61
G03 X83 Y44 R13
```

```
G01 X83 Y29
G02 X65 Y10 R20
G01 X10 Y10
M99
04001
G91 G01 Z-0.5 F50
G90..................
M99                                    Exiting the subprogram '04000'
```

Mirroring

EXAMPLE 14.5 Write a manual part program for mirroring operation for the component shown in Fig. 14.30.

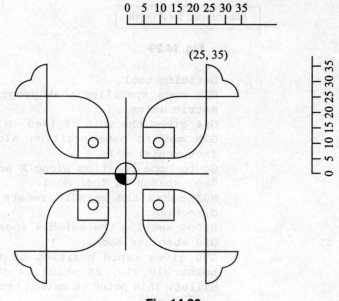

Fig. 14.30

Solution:

```
O1005
(Fig. 14.30
(PROGRAM FOR MIRRORING
(M CODES USED M70, M71, M80,M81
(M70X-AXIS MIRROR ON
(M71Y-AXIS MIRROR ON
(M80X-AXIS MIRROR OFF
(M81X-AXIS MIRROR OFF

BILLET X100 Y100 Z10          It defines the billet dimensions.
EDGEMOVE X-50 Y-50            This directive sets up the required offset
                             from the program zero position to the lower
                             left hand corner of the billet. This is used
                             for simulation.
```

```
TOOLDEF TI D5                  Defining tool.
G21 G94                        G21 code specifies that program is done in
                               metric units.
                               G94 gives the unit of feed in mm/min.
G91 G28 Z0                     G28 got to home position along Z-axis in
                               incremental mode.
G28 X0 Y0                      Go to home position along X and Y axes.
M06 T1                         Tool change to Tool No.1.
M03 S2000                      M03 makes the spindle rotate in clockwise
                               direction.
                               S2000 setting the spindle speed at 2000 rpm.
G90 G00 X0 Y0 Z5               G90 absolute mode.
                               G00 gives rapid position of the tool to a
                               point X0 Y0, Z5 which is just above the
                               billet. This point is called tool entry point.
M98 P0015000                   Calling the subprogram '5000' one time.
M70                            Mirroring about 'X' axis.
M98 P0015000                   Calling the subprogram '5000' one time.
M80                            Cancelling of mirroring about 'X'.
M70                            Mirroring about 'X' axis.
M71                            Mirroring about 'Y' axis.
M98 P0015000                   Calling the subprogram '5000' one time.
M80
M81
M71
M98P00 1500
G00 Z5
G91 G28 Z0                     Going to home position.
G28 X0 Y0
M05                            Stops the spindle.
M30                            Program stop and rewind.
01005
(Fig. 14.30
05000
G00 X5 Y5 Z5                   Rapid traverse to 5,5,5.
G01 Z-1 F50                    Feed rate at 35 mm/min.
X15 Y5
X15 Y15
X5 Y15
X5 Y5
G00 Z5
X10 Y10
G01 Z-1 F50
G00 Z5
X10 Y10
G01 Z-1 F50
G00 Z5
G03 X25 Y15 R15                CCW interpolation.
G01 X25 Y35
G02 X30 Y30 R5
```

```
G02 X35 Y25 R5
G01 X15 Y25
G03 X5 Y15 R15
G00 Z5
G00 X0 Y0
M99                              Exiting the subprogram.
```

Drilling

EXAMPLE 14.6 Write a manual part program for drilling operation for the component shown in Fig. 14.31.

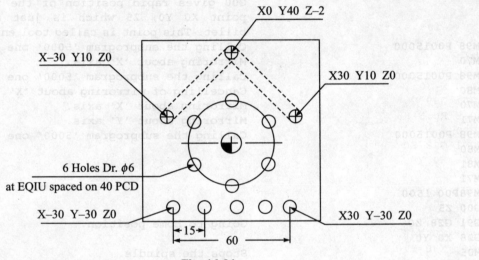

X0 Y40 Z–2

X–30 Y10 Z0

X30 Y10 Z0

6 Holes Dr. ϕ6
at EQIU spaced on 40 PCD

X–30 Y–30 Z0

X30 Y–30 Z0

15

60

Fig. 14.31

Solution:
```
O1006
(Fig. 14.31
(PROGRAM FOR DRILLING
(G CODES USED G73,G83,G98,G99
(G73-HIGH SPEED PECK DRILLING
CYCLE
(G83-PECK DRILLING CICLE
```

```
[BILLET X100 Y100 Z10                It defines the billet dimensions.
[EDGEMOVE X-50 Y-50                  This directive sets up the required
                                     offset from the program zero position to
                                     the middle of the billet. This is used
                                     for simulation.
[TOOLDEF T1 D8 T2 D5                 Defining tool.
G21 G94                              G21 code specifies that program is done
                                     in metric units.
G91 G28 Z0                           G28—Go to home position along Z axis in
                                     incremental mode.
```

G28 X0 Y0	Go to home position along X and Y axes.
M06 T1	Tool change to tool No.1.
M03 S1500	M03 makes the spindle rotate in clockwise direction.
	S2000 setting the spindle speed at 2000 rpm.
G90 G00 X0 Y0 Z5	G90 absolute mode
	G00 gives rapid position of the tool to a point X0 Y0 Z5 which is just above the billet. This point is called tool entry point.
G83 G99 X0 Y20 Z-5 Q.5 R.5 F.50	Peck drilling cycle.
X17.32 Y10	
Y-10	
X0 Y-20	
X-17.32 Y-10	
Y10	
G80	Cancel the canned cycle.
G00 X-30 Y-30	
G91 G99 G73 X10 Y0 Z-5 P100	
Q.5 R0.5 K5 F50	High speed peck drilling cycle.
G80	Cancel the canned cycle
G91 G28 Z0	
G28 X0 Y0	
M06 T2	
M03 S1500	
G90 G00 X-30 Y10 Z0	
G01 Z-1	
G01 X0 Y40	
G01 X30 Y10	
G00 Z5	
G91 G28 Z0	Going to home position.
G28 X0 Y0	
M05	Stop the position
M30	Program stop and rewind

Pocketing

EXAMPLE 14.7 Write a manual part program for pocketing operation for the component shown in Fig. 14.32.

Solution:

O1007
(Fig. 14.32
(PROGRAM FOR POCKETING
(G CODES USEDG170,G171-CIRCULAR
POCKETING
(G172,G173-RECTANGULAR POCKETING

[BILLET X100 Y100 Z10	It defines the billet dimensions.
[EDGEMOVE X-50 Y-50	This directive sets up the required

A(−30, −30)
B(−30, −10)
C(−30, 10)
D(−30, 30)
E(−10, 30)
F(10, 30)
G(30, 30)
H(30, 10)
I(30, −10)
J(30, −30)
K(10, −30)
L(−10, −30)
M(−10, 10)
N(−10, −10)
O(−23, −23)
P(7, 7)

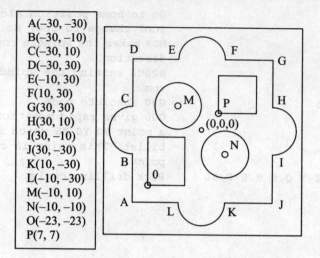

Fig. 14.32 Circular and rectangular pocketing.

offsets from the program zero position to the middle of the billet. This is used for simulation.

```
[TOOLDEF T1 D5
```
Defining tool.

```
G21 G94
```
G21 code specifies that program is done in metric units.

```
G91 G28 Z0
```
G28 HGO to home position along Z-axis in incremental mode.

```
G28 X0 Y0
```
Go to home position along X and Y axes.

```
M06 T1
```
M06—tool change to Tool No.1 (5 mm slot drill)

```
M03 S1500
```
M03 makes the spindle speed at 1500 rpm.

```
G90 G00 X-30 Y-30 Z5
```
G90 absolute mode.

G00 gives rapid position of the tool to a point X−30 Y−30 Z5 which is Just above the billet. This point is called tool entry point.

```
G01 Z-1 F50
G01 X-30 Y-10
G02 X-30 Y10 R10
G01 X-30 Y30
G01 X-10 Y30
G02 X10 Y30 R10
G01 X30 Y30
G01 X30 Y10
G02 X30 Y-10 R10
G01 X30 Y-30
G01 X10 Y-30
G02 X-10 Y-30 R10
G01 X-30 Y-30
```

```
G00 Z5
G00 X-10 Y10
G170 R0P0 Q.5 X-10 Y10 Z-5 I.2 J.2 K9    Circular pocketing (rough operation).
G171 P50 S2000 R35 F45 B2500 J50
G170 R0 P1 Q.5 X10 Y-10 Z-5 I0 J0 K9      Circular    pocketing    (finishing
                                          operation).

G171 P50 S2000 R35 F45 B2500 J50
G172 I15 J15 K0 P0 Q.5 R0 X7 Y7 Z-5       Rectangular    pocketing    (rough
                                          operation).

G173 I.2 K.2 P50 T1 S2000 R35 F45
B2500 J50 Z5
G172 I15 J15 K0 P1 Q.5 R0 X-23            Rectangular pocketing (finishing
Y-23 Z-5                                  operation).

G173 I0 K0 P75 T1 S2000 R40 F60 B2500
J50 Z5
G170 R0 P1 Q.5 X10 Y-10 Z-5 I.1 J.1 K9    Circular    pocketing    (roughing
                                          operation).

G171 P50 S2000 R35 F45 B2500 J50
G172 I15 J15 K0 P0 Q.5 R0 X7 Y7 Z-5       Rectangular   pocketing   (Roughing
                                          operation).

G172 I15 J15 K0 P1 Q.5 R0 X-23 Y-23
Z-5
G173 I0 K0 P75 T1 S2000 R40 F60 B2500
J50 Z5
G00 Z5
G91 G28 X0 Y0 Z0                          Going to home position.
M05                                       Stop the spindle.
M30                                       Program stop and rewind.
```

APT PROGRAMMING EXAMPLES

EXAMPLE 14.1   Write an APT program to cut the profile as shown in Fig.14.33. With the cutter of dia. 0.256 inches and cutting speed of 900 rpm and feed rate of 3 in/min.

Solution:
```
PARTNO/1
MACHIN / MILL01
INTOL / 0.01
OUTOL / 0.01
CUTTER / 0.256
SPINDL / 900
FEDRAT / 3
COOLNT / ON
P0 = POINT / -3.0, -3.0, 0.0
P1 = POINT / 0, 0, 0
P2 = POINT / 6.0, 0, 0
P3 = POINT / 6.0, 1.875, 0
P4 = POINT / 2.0, 3.75, 0.0
```

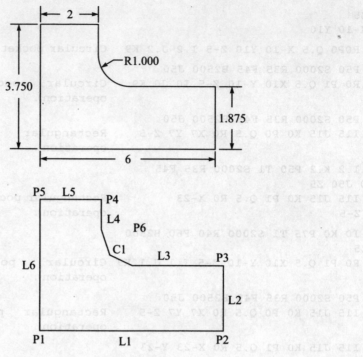

Fig. 14.33

```
P5 = POINT / 0.0, 3.75, 0.0
P6 = POINT / 3.0, 2.875, 0.0
L1 = LINE / P1, P2
L2 = LINE / P2, P3
C1 = CIRCLE / CENTRE, P6, RADIUS, 1.0
L3 = LINE / P3, LEFT, TANTO, CI
L4 = LINE / P4, RIGHT, TANTO, C1
L5 = LINE / P4, P5
L6 = LINE / P5, P1
PL1 = PLANE / P1, P2, P5
FROM / P0
RAPID
GO / TO, L1, TO, PL1, TO L6
GOTO / P1
GORGT / L1 / PAST, L2
GORFT / L2 / PAST, L3
GORFT / L3, TANTO, C1
GOFWD / C1, PAST, L4
GOFWD / L4 / PAST, L5
GOLFT / L5, PAST, L6
GOLFT / L6, PAST, L1
GOTO / P0
COOLNT / OFF
SPINDL / OFF
FINI
```

EXAMPLE 14.2 Write an APT program to cut the profile as shown in Fig. 14.34 with cutter speed and feed rate of 2000 rpm and 200 m/min respectively using a cutter of 11 dia., with an inside tolerance of 0.001 mm and 0.003 mm respectively.

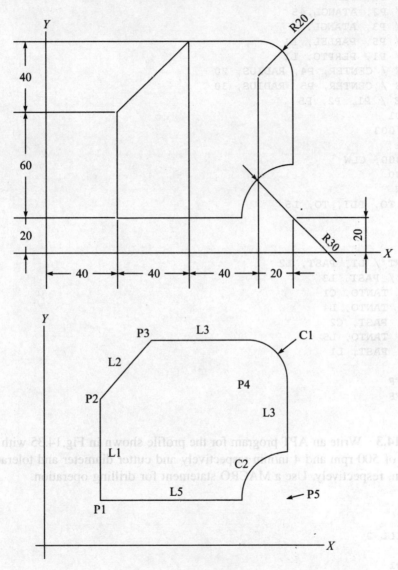

Fig. 14.34

Solution:

```
PARTNO/2
MACHIN / MILL 2
P0 = POINT / 0,0
P1 = POINT / 40, 20
P2 = POINT / 40, 80
```

```
P3 = POINT / 80, 120
P4 = POINT / 120, 100
P5 = POINT / 140, 20
L1 = LINE / P1, P2
L2 = LINE / P2, ATANGL, 45
L3 = LINE / P3, ATANGL, 0
L4 = LINE / P5, PARLEL, L1
L5 = LINE / P1, PERPTO, L1
C1 = CIRCLE / CENTER, P4, RADIUS, 20
C2 = CIRCLE / CENTER, P5, RADIUS, 30
PL1 = PLANE / P1, P2, P5
INTOL / 0.01
OUTTOL / 0.003
CUTTER / 11
SPINDL / 2000, CLW
FEDRAT / 200
COOLNT / ON
GO/TO, L1, TO, PLI, TO, L5
FROM / P0
RAPID
GOTO / P1
TLLFT, GOLFT / L1, PAST, L2
GORGT / L2 / PAST, L3
GORFT / L3, TANTO, C1
GOFWD / C1, TANTO, L4
GOFWD / L4, PAST, C2
GOFWD / C2, TANTO, L5
GORGT / L5, PAST, L1
GOTO / P0
COOLNT / OFF
SPINDL / OFF
FINI
```

EXAMPLE 14.3 Write an APT program for the profile shown in Fig.14.35 with cutter speed and feed rate of 500 rpm and 4 in/min respectively and cutter diameter and tolerance of 0.375 in. and 0.01 in. respectively. Use a MACRO statement for drilling operation.

Solution:

```
PARTNO/ 3
MACHIN / MILL 2
CLPRNT
INTOL / 0.01
OUTTOL / 0.01
CUTTER / 0.375
P0 = POINT / -1, -2, 0
P1 = POINT / 0, 0, 0
P2 = POINT / 9, 0, 0
P3 = POINT / 9, 4.5, 0
P4 = POINT / 0, 4.5, 0
P5 = POINT / 1.5, 0.9, 0
```

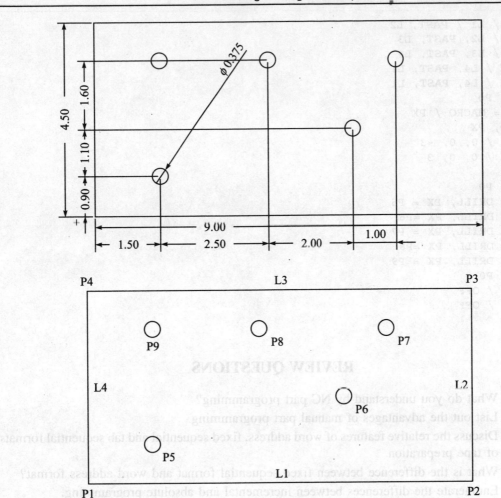

Fig. 14.35

```
P6 = POINT / 6, 2, 0
P7 = POINT / 7, 3.6, 0
P8 = POINT / 4, 3.6, 0
P9 = POINT / 1.5, 3.6, 0
L1 = LINE / P1, P2
L2 = LINE / P2, P3
L3 = LINE / P3, P4
L4 = LINE / P4, P1
PL1 = PLANE / P1, P2, P3
SPINDL / 500
FEDRAT / 4
COOLNT / ON
FROM / P0
GO/TO, L1, TO, PLI, TO, L4
GOTO / P1
```

```
GORGT / L1 / PAST, L2
GOFWD / L2, PAST, L3
GOLFT / L3, PAST, L4
GOBACK / L4, PAST, L4
GOBACK / L4, PAST, L1
GOTO / P0
DRILL = MACRO / PX
GO/ TO, PX
GODLTA / 0, 0, -3
GODLTA / 0, 0, 3
TERMAC
FROM / P0
CALL / DRILL, PX = P5
CALL / DRILL, PX =P6
CALL / DRILL, PX = P7
CALL / DRILL, PX =P8
CALL / DRILL, PX = P9
GOTO / P0
COOLNT / OFF
SPINDL / OFF
FINI
```

REVIEW QUESTIONS

14.1 What do you understand by NC part programming?

14.2 List out the advantages of manual part programming.

14.3 Discuss the relative features of word address, fixed sequential and tab sequential formats of tape preparation.

14.4 What is the difference between fixed sequential format and word address format?

14.5 Enumerate the differences between incremental and absolute programming.

14.6 What is the purpose of a part program? Explain the function of punched tape in an NC machine tool.

14.7 What do you mean by cutter radius compensation?

14.8 Discuss the structure of an APT program.

14.9 What are the steps involved in APT program?

14.10 Illustrate the procedure of APT programming using MACRO statement.

14.11 Define the following terms:

(a) Formats of tape preparation
(b) Parity check
(c) Part surface, drive surface and check surface
(d) Post-processor statement

14.12 What is the purpose of a tool presetter?

14.13 Discuss the steps involved in preparation of computer assisted part programming.

14.14 How computer assisted part programming is superior to manual part programming?

14.15 Discuss the advantages of computer assisted part programming.

14.16 Explain the important features of APT languages.

14.17 Discuss the importance of the various statements in APT languages.

14.18 List out the different types of computer aided part programming languages.

14.19 What is post-processor? Discuss the functions of the post-processor.

14.20 Discuss the various geometric statements used in the APT programming.

14.21 Define computer aided programming is preferred for NC-machine tools.

14.22 When would mirror images be used in a program?

14.23 What is offline programming?

14.24 Describe how an APT program generates an NC program.

14.25 What is a preparatory function?

14.26 What are miscellaneous functions?

14.27 What is a canned cycle?

14.28 What is the difference between a modal and non-modal command?

14.29 What is an indexer?

14.30 What is the difference between diameter and radius programming?

14.31 What is the role of a CNC part programmer?

14.32 How are macros written, stored and recalled into the main program?

PRACTICE PROBLEMS

14.1 Figures 14.36 to 14.46 are to be made using a CNC XL turn equipped with a FANUC controller. Prepare the part programs to completely machine the parts from rolled stock. The work material is aluminium. Clearly show the set point and axes on the sketch of the part. Prepare the planning sheet also.

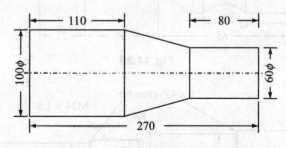

Fig. 14.36

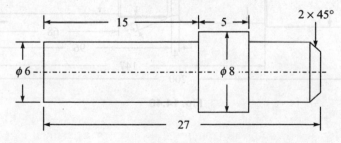

Fig. 14.37

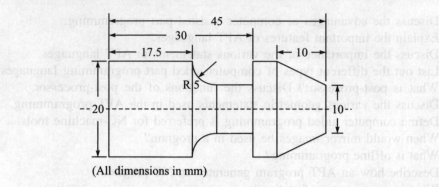

(All dimensions in mm)

Fig. 14.38

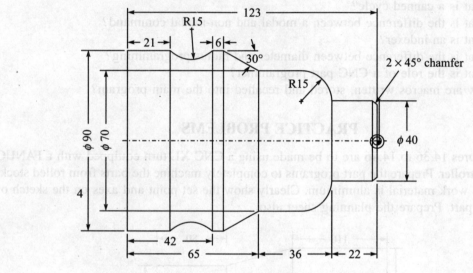

Fig. 14.39

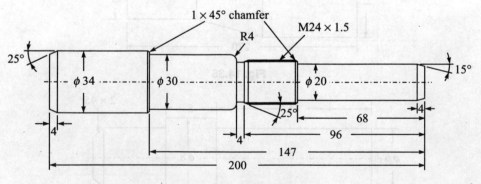

Fig. 14.40

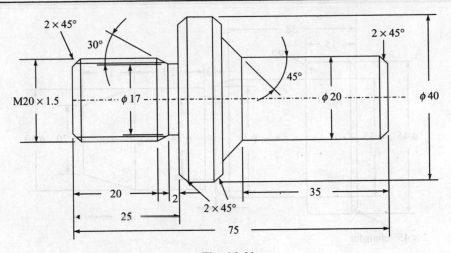

Fig. 14.41

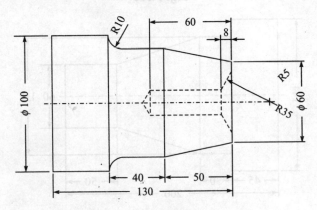

Fig. 14.42

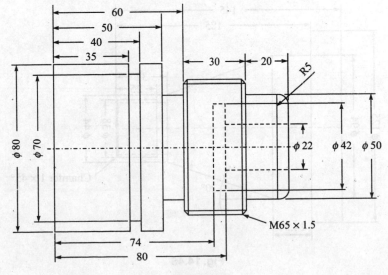

Fig. 14.43

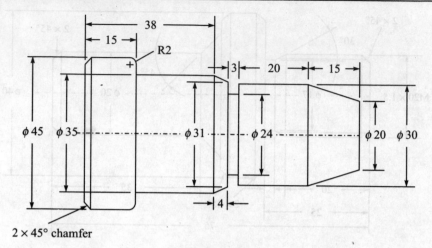

Fig. 14.44

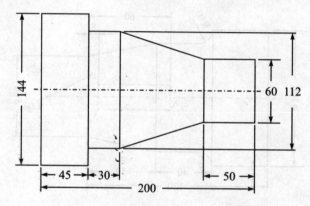

Fig. 14.45

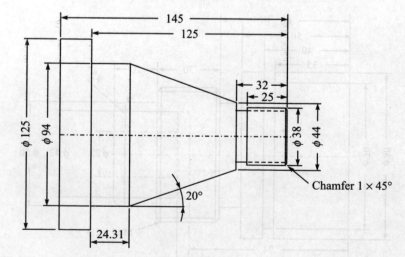

Fig. 14.46

14.2 Figures 14.47 to 14.56 are to be made using a CNC XL mill equipped with a FANUC controller. Prepare the part programs to completely machine the parts from rolled stock. The work material is aluminium. Clearly show the set point and axes on the sketch of the part. Prepare the planning sheet also.

Contouring

(a) Write a manual part program for machining the component shown in Fig. 14.47. Given Depth = 5 mm, Depth of cut for each pass = 0.5 mm, Number of parallel passes = 2.

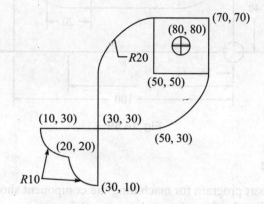

Fig. 14.47 Linear and circular interpolation.

Slotting

(b) Write a manual part program for machining the component shown in Fig. 14.48. Given Depth = 5 mm, Depth of cut for each pass = 0.5 mm.

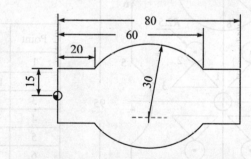

Fig. 14.48 Subprogram for Z depth.

Cutter Diameter Compensation

(c) Write a manual part program for machining the component shown in Fig. 14.49. Given Depth = 5 mm, Depth of çut for each pass = 0.5 mm.

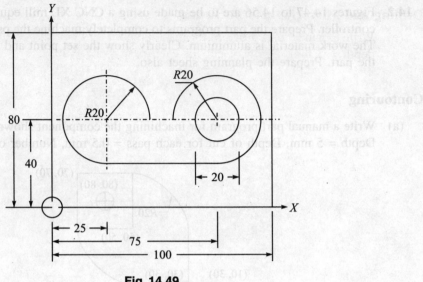

Fig. 14.49

Drilling and Mirroring

(d) Write a manual part program for machining the component shown in Fig. 14.50. Profile depth = 5 mm, Depth of cut for each pass = 0.5 mm.

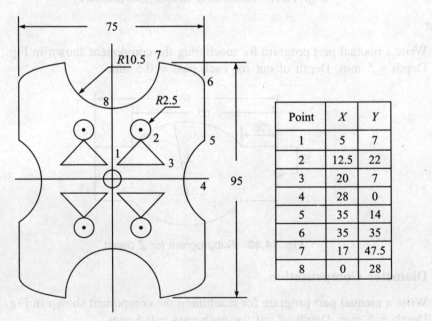

Point	X	Y
1	5	7
2	12.5	22
3	20	7
4	28	0
5	35	14
6	35	35
7	17	47.5
8	0	28

Fig. 14.50

Drilling and Mirroring and Pocketing

(e) Write a manual part program for machining the component shown in Fig. 14.51. Profile depth = 5 mm, Depth of cut for each pass = 0.5 mm.

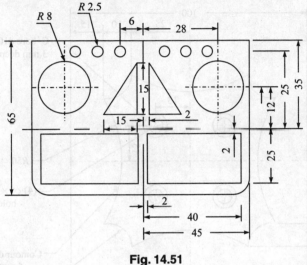

Fig. 14.51

Contouring Using Subprogram

(f) Write a manual part program for machining the component shown in Fig. 14.52. Profile depth = 5 mm, Depth of cut for each pass = 1 mm.

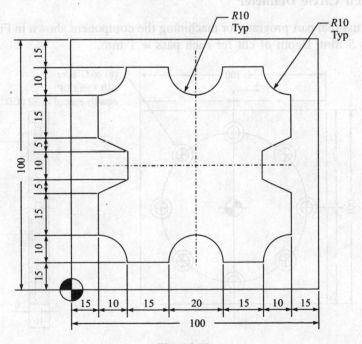

Fig. 14.52

Contouring Drilling and Contouring Boring

(g) Write a manual part program for machining the component shown in Fig. 14.53. Profile depth = 5 mm, Depth of cut for each pass = 1 mm.

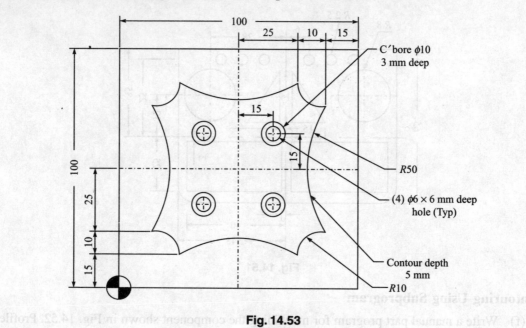

Fig. 14.53

Drilling on Pitch Circle Diameter

(h) Write a manual part program for machining the component shown in Fig. 14.54. Profile depth = 5 mm, Depth of cut for each pass = 1 mm.

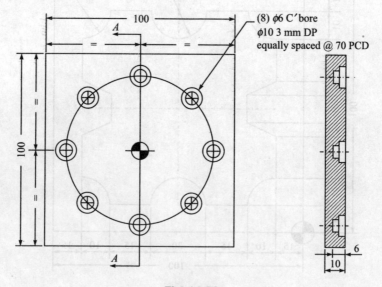

Fig. 14.54

Circular Pocketing

(i) Write a manual part program for machining the component shown in Fig. 14.55. Depth of cut for each pass = 1 mm.

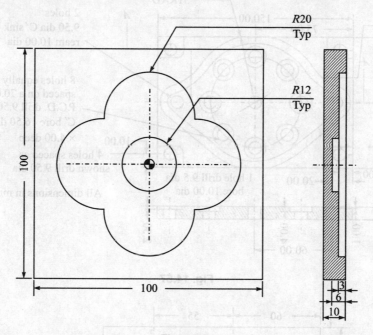

Fig. 14.55

Circular and Rectangular Pocketing

(j) Write a manual part program for machining the component shown in Fig. 14.56. Depth of cut for each pass = 1 mm.

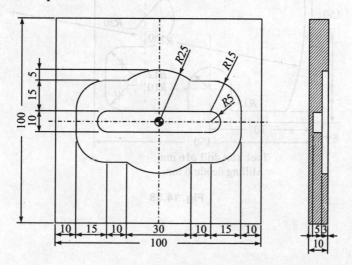

Fig. 14.56

14.3 For the following components (Figs. 14.57 to 14.67), develop the part programs using the APT language:

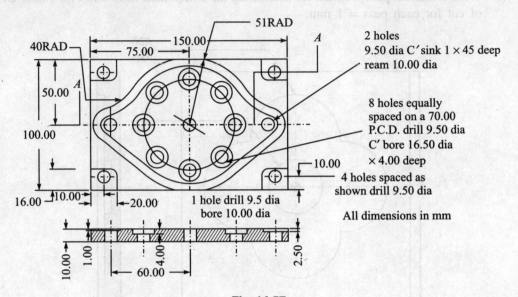

Fig. 14.57

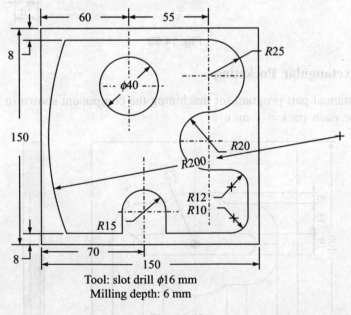

Tool: slot drill ϕ16 mm
Milling depth: 6 mm

Fig. 14.58

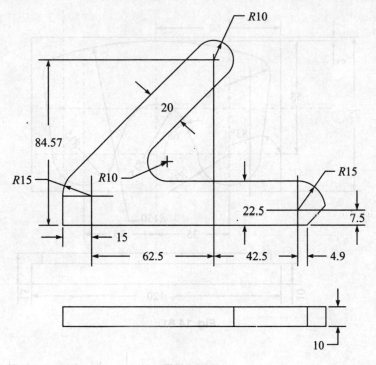

Fig. 14.59

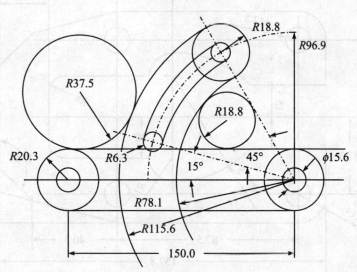

Fig. 14.60

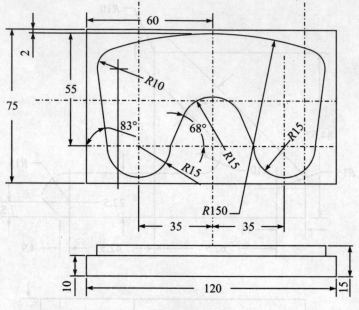

Fig. 14.61

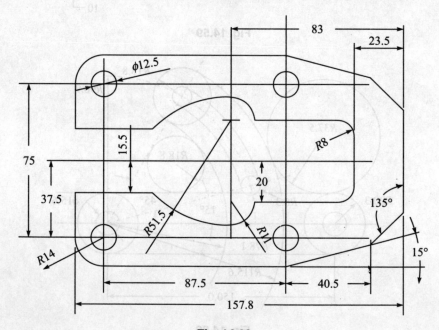

Fig. 14.62

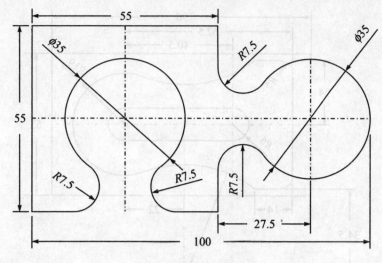

Fig. 14.63

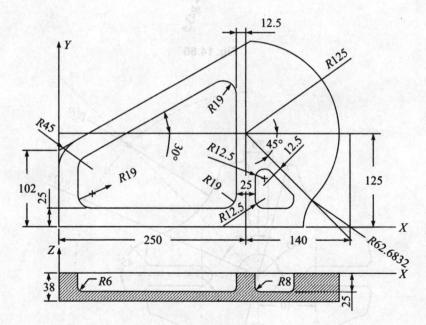

Fig. 14.64

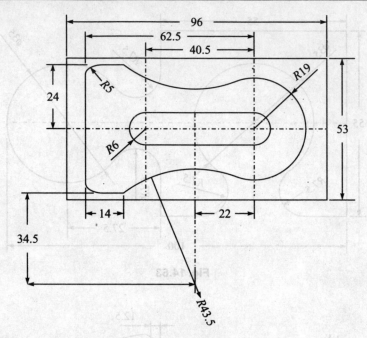

Fig. 14.65

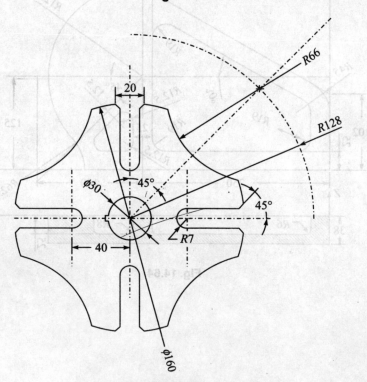

Fig. 14.66

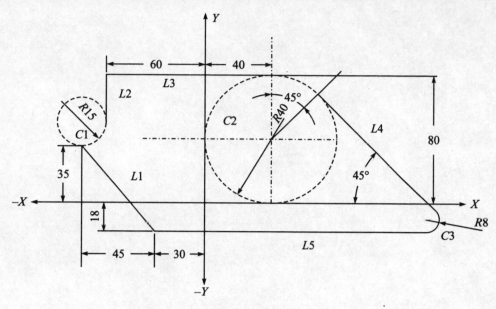

Fig. 14.67

Fig. 16.47

Part V

Information Systems in Manufacturing

GROUP TECHNOLOGY

Chapter

15

INTRODUCTION

When one deals with a variety of assignments, it has always been a common practice to group similar types of work together, with the intention of reducing the total work because of commonalities of approach and work procedures involved. This technology has been applied in process planning for small and medium batch manufacture.

Group technology (GT) deals with the identification of similar parts and group them for the purpose of manufacturing and design generally, whenever a product is undertaken for manufacture either its design is initiated from scratch or one examines existing drawings and makes modification in them. It is possible only if it is convenient to retrieve the needed information and, therefore, duplication would invariably takes place. So, an efficient design retrieval system is essential which would require some form of part identification and coding. For part identification and coding, the attributes of the component should be used namely of design and manufacturing, i.e, geometric shape, size and the process of manufacture.

This approach is different from the existing one and the working of the entire organization will be affected due to this approach. This technology streamlines procedure and is amenable to computerization and in the longer process, it forms the basis of integrated manufacture and thus would prove an important step in inducting automation in small and medium batch manufacture.

In various products, the components could be on the basis of similarity of their shapes and manufacturing procedures. If components are coded according to such characteristics their retrieval would be convenient which would not only help in avoiding duplication of efforts but also in enlarging the batch sizes of manufacture enabling reducing in time and cost involved.

15.1 PART FAMILIES

A part family is a collection of parts which are similar either because of geometry and size or because of similar processing steps required in their manufacture. The parts within a family are different, but their similarities are close enough to merit their identification as members of the part family. The major obstacle in changing over to group technology from a traditional production shop is the problem of grouping parts into families. There are three general methods to group the parts into family. They are:

(a) Visual inspection
(b) Production flow analysis (PFA)
(c) Parts classification and coding system

The visual inspection method is the least sophisticated and least expensive method. It involves the classification of parts into families by looking at either the physical parts or their photographs and arranging them into similar groupings.

The production flow analysis makes use of the information contained on route sheets rather than part drawings. Work parts with identical or similar routings are classified into part families. The disadvantage of PFA is that it accepts the validity of existing route sheets, with no consideration given to whether these process plans are logical or consistent.

The third method involves classifying the parts into families by examining the individual design and/or manufacturing attributes of each part. The classification results in a code number that uniquely identifies the parts attributes. This classification and coding may be carried out on the entire list of active parts of the firm or some sort of sampling procedure may be used to establish the part families. The parts shown in Fig. 15.1 belongs to different part families are very similar in terms of geometric designs, but quite different in terms of manufacturing attributes.

Fig. 15.1 Two parts have similar geometric features but different manufacturing attributes.

A family of parts with similar manufacturing process requirements but different design attributes. All parts are machined from cylindrical stock by turning, some parts are drilled and/or milling. The parts shown in Fig. 15.2 constitute a part family in manufacturing but their different geometries make them appear quite different from a design viewpoint.

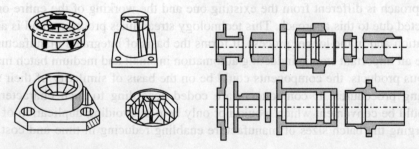

Fig. 15.2 A family of parts with similar manufacturing process requirements but different design attributes.

15.2 PARTS CLASSIFICATION AND CODING SYSTEMS

Classification of parts refers to grouping of parts based on the essential features of parts. The attributes used for classification of parts can be categorized into three types:

(a) Attributes with respect to part design (Design attributes)
(b) Attributes with respect to manufacturing of part (Manufacturing attributes)
(c) Combined attributes of design and manufacturing

TABLE 15.1 Some of the design and manufacturing attributes being considered in the industry.

Design attributes	Manufacturing attributes
Basic external shape	Major processes
Basic internal shape	Minor operations
Rotational or rectangular shape	Operation sequence
Length to diameter ratio	Major dimension
(Rotational parts)	Surface finish
Aspect ratio (rectangular parts)	Machine tool
Material type	Production cycle time
Part function	Batch size
Major dimensions	Annual production
Minor dimensions	Fixtures required
Tolerances	Cutting tools
Surface finish	

15.3 CODING METHOD

Coding refers to the process of assigning symbols to the parts. The symbols represent design attribute of parts, manufacturing features of parts or both. Classification refers to the process of categorization set of parts into part families. Coding methods are employed in classifying parts into part families.

15.4 CODES AND CODING SYSTEM STRUCTURE

A GT code is a string of characters capturing information about an item. A coding scheme is a vehicle for the efficient recording, sorting and retrieval of relevant information about objects. A large number of coding schemes have been developed. These coding schemes differ:

- In terms of the symbols they employ such as numeric, alphabetic or alphanumeric
- In the assignment of these symbols to generate codes

However, the variations in codes resulting from the way the symbols are assigned. They can be grouped into three distinct types of codes:

- Monocode or hierarchical code
- Polycode
- Mixed mode code

15.4.1 Monocode or Hierarchical Code

The structure of these codes is like a tree in which each symbol amplifies the information provided in the previous digit. Figrue 15.2 depicts the monocode generation scheme. The first digit (from zero to nine) divides the set of parts into major groups such as sheet metal parts, machined parts, purchased parts and components and so forth. The second and subsequent digits further partition the set into subgroups or each of these groups. For example, the second digit partitions the machined parts into rotational (0) and non-rotational (1) parts. Consider a code of 110 in Fig. 9.5. It represents a machined non-rotational part with a length to width ratio of less than one. The digits 1 in the first and second fields in code 110 have different meanings and contain different information. Therefore, the digits in a monocode cannot be interpreted independently; the interpretation depends on the information contained in the preceding symbol.

The major advantage of hierarchical code is that it captures a great deal information in a relatively short code. An example of monocode structure is depicted in Fig. 15.3. The applicability of these codes in manufacturing is rather limited as it is difficult to capture information on manufacturing sequences in a hierarchical manner. The disadvantage of this type of code is that it requires expertise to conceive such a coding system for a part spectrum. The polycode and monocode structures are often combined for use in practical manufacturing operations involving design information.

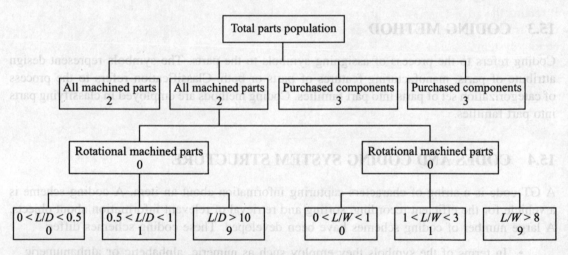

Fig. 15.3 Example of monocode.

15.4.2 Polycode

This is also known by many other names such as chain code, discrete code and fixed digit code. In polycode, the code symbols are independent of each other. Each digit in a specific location of the code describes a unique property of the work piece. It is easy to learn and useful in manufacturing situations in which the part functions or the manufacturing processes have to be described. An example of polycode is shown in Fig. 15.4. The length of a polycode may become excessive because of its unlimited combinational features. The numbers of mutually exclusive characteristics that can be inscribed in a 5-digit polycode are 50 (10 + 10 + 10 + 10 + 10). Thus, the polycode is simple to understand and interpret, and only the limited information can be given in the code.

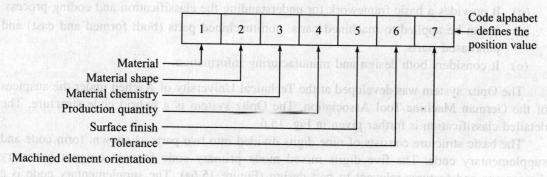

Fig. 15.4 Example of polycode.

15.4.3 Mixed Code

It retains the advantages of both mono and polycodes. Therefore, most coding systems use this code structure. For example, the Opitz classification system. A large number of classification and coding systems have been developed and a number of commercial codes are available. Some of the selected examples of worldwide classification and coding are given in Fig. 15.5.

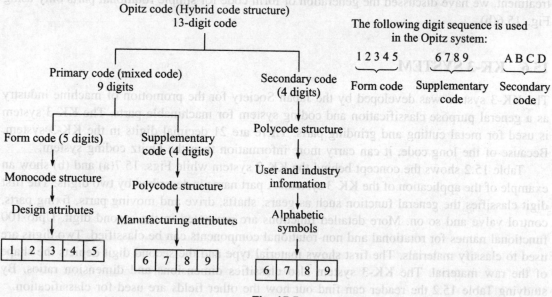

Fig. 15.5

Now, we will discuss few of the systems such as Opitz, KK-3 and MICLASS.

15.5 OPITZ CLASSIFICATION AND CODING SYSTEM

The advantages of this system are:

(a) It is non-proprietary.
(b) It is widely used.
(c) It provides a basic framework for understanding the classification and coding process.
(d) It can be applied to machined parts, non-machined parts (both formed and cast) and purchased parts.
(e) It considers both design and manufacturing information.

The Opitz system was developed at the Technical University of Aachen under the auspices of the German Machine Tool Association. The Opitz system is a hybrid code structure. The detailed classification is further given in Fig. 15.6.

The basic structure consists of nine digits divided into two parts as shown: form code and supplementary code. The five digits mixed mode primary code focuses on part geometry dimensions and features relevant to part design (Figure 15.6a). The supplementary code is a polycode consisting of four digits. This includes information relevant to manufacturing such as raw material, tolerances and surface roughness.

To help identify the production processes and production sequences, a secondary code consisting of four alphabetic symbols can be defined by the user. The user industry can give its own information in the secondary code, i.e., in the places of 4 digits A, B, C, D. The structure of secondary code is normally a polycode structure. Opitz classification and coding is available both in soft and hard copy forms. Though the original coding procedure is lengthy, for the simple treatment, we have discussed the generation of form code for simple rotational parts only using Fig. 15.6(b).

15.6 KK-3 SYSTEM

The KK-3 system was developed by the Japan Society for the promotion of machine industry as a general purpose classification and coding system for machinable parts. The KK-3 system is used for metal cutting and grinding parts. There are 21 decimal digits in the KK-3 system. Because of the long code, it can carry more information than the Opitz coding system.

Table 15.2 shows the concept behind the KK-3 system while Figs. 15.7(a) and (b) show an example of the application of the KK-3 system. A part name is described by two digits: The first digit classifies the general function such as gears, shafts, drive and moving parts, fixing parts, control valve and so on. More detailed functions are presented in the second digit. Upto 100 functional names for rotational and non-rotational components can be classified. Two digits are used to classify materials. The first shows material type and the second digit depicts the shape of the raw material. The KK-3 system also classifies dimensions and dimension ratios. By studying Table 15.2 the reader can find out how the other fields are used for classification.

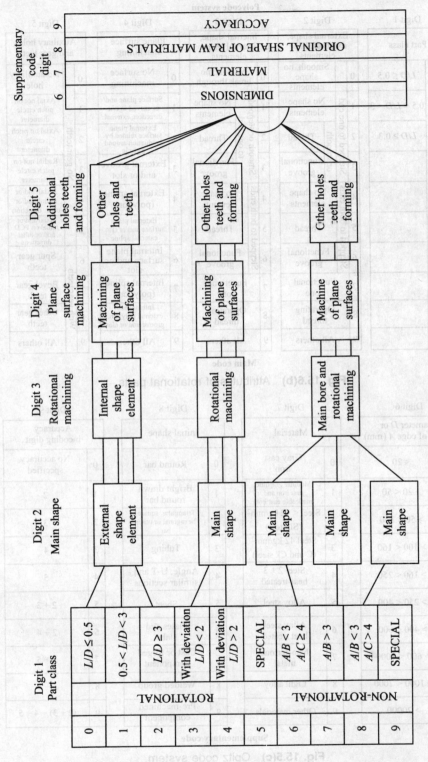

Fig. 15.6(a) Basic structure of the Opitz system of parts classification and coding.

Polycode system

	Digit 1			Digit 2			Digit 3			Digit 4		Digit 5
	Part class			**External shape, external shape elements**			**Internal shape, internal shape elements**			**Plane surface machining**		**Auxiliary holes and gear teeth**
0	$L/D \leq 0.5$		0		Smooth, no shape elements	0		No hole, no break through	0		No surface machining	0	No auxiliary hole
1	$0.5 < L/D < 3$	Rotational parts	1	Stepped to one end or smooth	No shape elements	1	Smooth or stepped to one end	No shape elements	1		Surface plane and/or curved in one direction, external	1	No gear teeth	Axial not on pitch circle diameter
2	$L/D \geq 0.3$		2		Thread	2		Thread	2		External plane surface related by graduation around a circle	2	Axial on pitch circle diameter
3			3		Functional groove	3		Functional groove	3		External groove and/or slot	3	Radial not on pitch circle diameter
4			4	Stepped to both ends	No shape elements	4	Stepped to both ends	No shape elements	4		External spline (polygon)	4	Axial and/or radial or other direction
5			5		Thread	5		Thread	5		External plane surface and/or slot, external spline	5	Axial and/or radial on PCD and/or other directions
6		Non-rotational parts	6		Functional groove	6		Functional groove	6		Internal plane surface and/or slot	6	With gear teeth	Spur gear teeth
7			7		Functional cone	7		Functional cone	7		Internal spline (polygon)	7	Bevel gear teeth
8			8		Operating thread	8		Operating thread	8		Internal and external polygon, groove and/or slot	8	Other gear teeth
9			9		All others	9		All others	9		All others	9	All others

Main code

Fig. 15.6(b) Attributes of rotational parts.

	Digit 6		Digit 7		Digit 8		Digit 9
	Diameter D or length of edge A (mm)		**Material**		**Initial shape**		**Accuracy incoding digit**
0	< 20	0	Gray cast iron	0	Round bar	0	No accuracy specified
1	$> 20 < 50$	1	Modular graphitic cast iron and malleable cast iron	1	Bright drawn round bar	1	2
2	$> 50 < 100$	2	Steel < 42 kg/mm^2 (SI-steel)	2	Triangular, square, hexagonal or other bar	2	3
3	$> 100 < 160$	3	Steel 42 kg/mm^2 (C and Ck steel)	3	Tubing	3	4
4	$> 160 < 250$	4	Steel 2 + 3 heat-treated	4	Angle, U-T and similar sections	4	5
5	$> 250 < 400$	5	Alloy steel	5	Sheet	5	2 + 3
6	$> 400 < 600$	6	Alloy steel heat treated	6	Plates and slabs	6	2 + 4
7	$> 600 < 1000$	7	Non-ferrous metal	7	Cast or forged component	7	2 + 5
8	$> 1000 < 2000$	8	Light alloy	8	Welded group	8	3 + 4
9	> 20000	9	Other materials	9	Pre-machined component	9	(2 + 3) + 4 + 5

Supplementary code

Fig. 15.5(c) Opitz code system.

TABLE 15.2   Structure of the KK-3 (Rotational Components)

Digit	Items		Rotational component
1	Parts name		General classification
2			Detail classification
3	Materials		General classification
4			Detail classification
5	Major dimensions		Length
6			Diameter
7	Primary shapes and ratio of major dimensions		
8	Shape details and kinds of processes	External surface	External surface and outer primary shape
9			Concentric screw threaded parts
10			Functional cut-off parts
11			Extraordinary shaped parts
12			Forming
13			Cylindrical surface
14		Internal surface	Internal primary shape
15			Internal curved surface
16			Internal flat surface and cylindrical surface
17			End surface
18		Non-concentric holes	Regularly located holes
19			Special holes
20			Non-cutting process
21			Accuracy

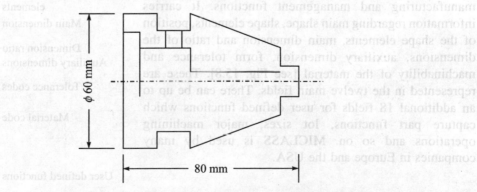

Fig. 15.7(a)   A typical part for coding under KK-3 system.

Code	Item	Component condition	Code
1	Name	Control valve	0
2	Name	Others (control valve)	9
3	Material	Copper bar	7
4	Material	Copper bar	
5	Dimension length	80 mm	2
6	Dimension diameter	60 mm	2
7	Primary shape and ratio of chief dimension	L/D 1.3	2
8	External surface	With functional tapered surface	3
9	Concentric screw	None	0
10	Functional cut-off	None	0
11	Extraordinary shaped	None	0
12	Forming	None	0
13	Cylindrical surface > 3	None	0
14	Internal primary	Piercing hole with diameter variation, no cut-off	2
15	Internal curved surface	None	0
16	Internal flat surface	None	0
17	End surface	Flat	0
18	Regularly located hole	Holes located on circumferential line	3
19	Spatial hole	None	0
20	Non-cutting process	None	0
21	Accuracy	Grinding process on external surface	4

Fig. 15.7(b) Example of KK-3 coding system.

15.7 THE MICLASS SYSTEM

MICLASS stands for Metal Institute Classification System. It was developed in the Netherlands and maintained in the USA by the Organization for Industrial Research (OIR). MICLASS was designed to standardize many design, manufacturing and management functions. It carries information regarding main shape, shape elements, position of the shape elements, main dimension and ratio of the dimensions, auxiliary dimension, form tolerance and machinability of the material [see Fig. 15.8]. These are represented in the twelve main fields. There can be up to an additional 18 fields for user defined functions which capture part functions, lot sizes, major machining operations and so on. MICLASS is used by many companies in Europe and the USA.

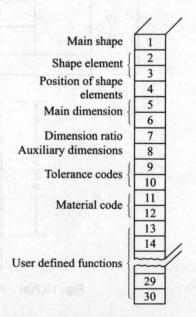

Fig. 15.8 MICLASS code structure.

15.8 MULTICLASS

It is a classification and coding system developed by the Organization for Industrial Research (OIR). The system is relatively flexible, allowing the user company to customize the classification and coding scheme to a large extent to fit its own products and applications. Multiclass can be used for a variety of different types of manufactured items, including machined and sheet metal parts, tooling, electronics, purchased parts, assemblies and subassemblies, machine tools and other elements. Upto nine different types of components can be included within a single multiclass software structure. Multiclass uses a hierarchical or decision tree coding structure in which the succeeding digits depend on values of the previous digits. In the application of the system, a series of menus, pick lists, tables and other interactive prompting routines are used to code the part. This helps to organize and provide the coding procedure. The coding structure consists of upto 30 digits. These are divided into two regions: one provided by OIR and the second designed by the user to meet specific needs and requirements. A prefix precedes the code number and is used to identify the type of part (e.g., a prefix value of 1 indicates machined and sheet metal parts). For a machined part, the coding for the first 18-digit positions (after the prefix) is summarized in Table 15.4.

TABLE 15.4   First 18 Digits of the Multiclass Classification and Coding System

Digit	Function
0	Code system prefix
1	Main shape category
2, 3	External and internal configurations
4	Machined secondary elements
5, 6	Functional descriptors
7–12	Dimensional data (length, diameter, etc.)
13	Tolerances
14, 15	Material chemistry
16	Raw material shape
17	Production quantity
18	Machined element orientation

15.9   HOW GT IS APPLIED

The formation of groups in small and medium batch manufacture is not an easy task. One of the ways of forming such groups could be visual. This requires some experience and can be applied when the number of products and their components is not very large. Grouping is done by looking similarities in shape, size and methods of manufacture. This practice would not be correct and as such cannot be recommended for adoption. However, it is cheap and less time consuming.

The other method is by forming a hypothetical composite part based on the features of all the components forming the group. Figure 15.9 shows a number of components which are manufactured on a lathe. The various operations to be done on them are setting the length, facing, turning, grooving, hole-making by drilling, boring, reaming, counterboring, chamfering, parting off, etc. Each job may not need all the operations to be performed on it. Now, an imaginary

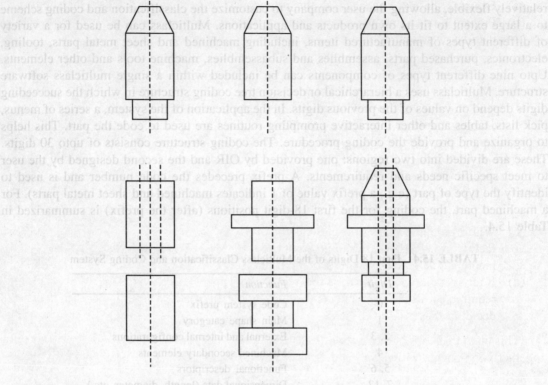

Fig. 15.9 Example for composite part analysis.

component is composed which involves all the operations needed for the group. The tool setting is done for such a component on a multitool set-up, say, a turret which can easily handle upto 11 tools. Depending on the particular component to be worked on, instructions are prepared to indicate the tools and the sequence in which they are to be used.

Such a methodology has resulted in the release of number of machine tools for use elsewhere which would otherwise have been poorly utilized. Whenever a new component is to be made its group is determined and then its sequence of operations and tooling is prepared.

15.10 PRODUCTION FLOW ANALYSIS (PFA)

In this analysis, grouping of parts is done in terms of the manufacturing sequence. For this, the route sheets of components are examined and the grouping is made on the basis of the facilities used and not on the sequence of operations. The defects of this analysis are: (a) It is based on the existing method of manufacture which may change after sometime. (b) It assumes that the route sheets have been correctly prepared which is often questionable and (c) Since this analysis is based on the existing practice within a company, it cannot be universally applied.

PFA is carried out in three stages, namely, factory flow analysis, group analysis and line analysis. These analyses help in arranging facilities and groups of components so as to minimize movement and improve the utilization of facilities. For example, group analysis is carried out as follows:

(a) Examine the route sheet of all the components to be manufactured in the shop.

(b) Prepare a matrix showing the operation numbers and the components number and fill it to show which component requires which operations (Fig. 15.10).

Operations	Components							
	1	2	3	4	5	6	7	8
1	X	X		X	X		X	
2	X	X	X		X	X	X	
3			X	X	X		X	
.								
.								
8		X						
9	X	X	X	X		X	X	

Fig. 15.10 Matrix: Operations–Components.

(c) While grouping parts any particular part should be included only in one group. For facility grouping, one machine type should be only in one group. Such operations as are incompatible should be in different groups.

(d) If an operation is required by only one or very few components or if some operation is required by all the components then these operations should not be taken note of while deciding the groups.

This matrix is rearranged (Fig. 15.11) by excluding operation numbers 2.8 and 9, and putting together components which need the same operations.

Operations	Components							
	1	2	4	5	8	3	6	7
1	X	X	X	X	(X)			
3						X		X
4						X	X	X
5	X		X	X				
6						X		X
7							X	X

Fig. 15.11 Rearranged matrix.

Components 1, 2, 4 and 5 are formed into a single group by excluding operation 1 on component 8. components 8, 3, 6 and 7 are formed into another group. Thus, the final grouping would be (Fig. 15.12).

Operations	Components							
	1	2	4	5	8	3	6	7
1	X	X	X	X				
2	X	X	X	X				
5	X	X	X	X				
9	X	X	X	X				
3						X		X
4						X	X	X
6						X		X
7						X	X	

Fig. 15.12 Grouping based on production flow analysis.

15.11 GROUP TECHNOLOGY MACHINE CELLS

Machine cell is defined as the groups of machines arranged to produce similar part families. Machine cell is a cellular arrangement of production equipment to achieve an efficient work flow within the cell, result in labour and machine specialization for the particular part families and thereby increase the productivity of the cell.

The three types of group technology machine cells are:

(a) Single machine cell
(b) Group machine layout
(c) Flow line design

The single machine approach can be used for work parts whose attributes allow them to be made on basically one type of process such as turning or milling.

The group machine layout is a cell design in which several machines are used together, with no provision for conveyorized parts movement between the machines. The cell contains the machines needed to produce a certain family of parts and the machines are organized with the proper fixtures, tools and operators to efficiently produce the parts family.

The flow line cell design is a group of machines connected by a conveyor system. Although this design approaches the efficiency of an automated transfer line, the limitation of the flow line layout is that all the parts in the family must be processed through the machines in the same sequence.

15.12 ORGANIZATION

GT being a new concept one has to be careful in its implementation. The team should be carefully chosen and the members should be answerable to the top management directly. People with authority should explain the concept of GT and its benefits to people at various levels of the organization by holding meetings. Since people may be prone to resist the change, one should be careful even about minor details during preparation and in the early stages of implementation. Frequent exchange of information should be encouraged amongst personnel involved. This may bring about good ideas, which would be helpful.

Every organization has its own way of working and the process of implementation cannot be copied from any other while forming the initial group of facilities for experimenting with GT, the selection of staff should be made responsible. They should be motivated for its success. As the activity is dependent on a group of people and in the initial stages production would slightly suffer, the people should be assured of suitably rewarding wages. Based on this the experience in a limited sphere, other cells can be created and improvements are carried out, e.g. standardization of designs, rationalization of material and tools.

15.13 CLASSIFICATION AND CODING SYSTEMS

The need for classification and coding systems for identification and retrieval of similar designs has been the backbone of GT. This would lead to variety reduction, standardization and design rationalization. Similarly, identification of components with similar production sequences and

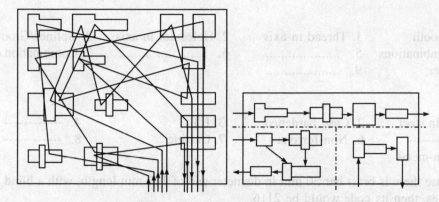

Fig. 15.13 Functional and group layouts.

methods of manufacture would enable the grouping of parts for process planning and production planning. The question arises whether separate systems should be used for design and manufacture. Since major identification is by shape, size and material which are permanent features, these could be defined in the primary code. The other information based on the method of manufacture (which may vary with the technology available and the scale of production) could form secondary code. The other parameters which could be considered for coding can be the functions of the components, the manufacturing operations and tooling.

15.14 CODE

Various structures of codes are possible. These could be hierarchical, polycode or mixed. The latter could be of fixed or variable length. A number of codes are of the hierarchical type where the first digit could be for the kind of work piece, the second could be for overall dimensions the third digit for some characteristic feature and the next one for material.

A typical example of such a hierarchical code system is VUOSO which was originated in Czechoslovakia.

Kind of workpiece

Without hole in the axis		Blind	Through	Geared	Flat	Box	Others
← Rotational →				← Non-rotational →			
1	2	3	4	5	6	7	8

Overall dimensions

Diameter	0–04 mm 0 1 2	40–80 mm 3 4 5	80–200 mm 9
L/D	< 1 1–5 >6	< 1 1–4	< 30

Features

0. Smooth	1. Thread in axis	2. Holes not in axis	3. Splines/Grooves
4. Combinations	5.	6.	7. Combination of 1, 2, 3
8 Taper	9.		

Material

1. Plain steel	2. Case hardened	3. HT	4. ------------
5. ----------	6. Non-ferrous	7. Casting	8. ------------
9. Non-metal			

Suppose there is brass bar 30 mm in diameter and of 100 mm length, with a blind threaded hole in axes, then its code would be 2116.

The hierarchical nature of this system indicates that all rotational workpieces having dimensions similar to this piece, would have the number 21 XX and so on.

15.15 COMPONENT CLASSIFICATION

For classification and subsequent coding, the components in the entire plant are studied.

(a) First these can be classified by the type of equipment used for their manufacture, e.g. capstan, milling machine, etc.

(b) These can be also classified by the geometric shape of the components, e.g. shaft type, bush type or body type, etc.

(c) These could be classified on the basis of the design and the operation of the components, e.g. stepped bush, tube, gear, etc.

(d) The component can also be grouped by the similarity of tooling used for them.

(e) Other features which could be used for classificatio are threaded features, the presence of some holes other than axial, slots, etc.

A secondary code can be assigned which would be more particular to the component but obviously cannot be used for primary identification. These again, in the hierarchical order can be material specifications, overall dimensions, tolerances and surface finish and be coded in digital form.

Secondary code

Materials: Various ferrous and non-ferrous metal and non-metal codes: 01–99

Tolerances: 1 to 9

Dimensions: Overall diameter or length and length/diameter ratio or length/width ratio codes: 01–99

Tolerances: 1 to 9

Surface finish: 1 to 9

Example: A code 335238 would mean material code 33, gross dimension code 52, tolerance code 3 and surface finish code 8.

Classification position of digit

1. Overall description of shape (Round/oblong)
2. Shape specific: shaft, bush, box type, etc.
3. Location of elements (Holes, threads, etc.)
4. Major dimensions (longer, length, diameter)
5. Proportion (*L/D*, *L/W* ratio)
6. Any other important dimension
7. Tolerances (dimensional)
8. Surface finish
9. Geometric tolerances
10. Material
11. Information related to manufacturing

15.16 SPECIFIC CLASSIFICATION SYSTEMS

A larger number of classification system are being used. The Opitz system is very comprehensive and consists of 5 digits for basic code form and 4 digits for supplementary code.

Form code

First digit.

0–5	rotating component
L/D < 0.5	represented by 0
L/D > 3.0	represented by 2
Any special shape	represented by 5
6–9	non-rotating component
e.g., *L/W* < 30	represented by 6
....	....
....	....
Any special shape	represented by 9

Second digit.

	Main Shape	External form
e.g	0	represents smooth surface
	1–3	represents one end stepped
	2	(2 for threaded)
	3	(3 for grooved)
	4–6	represents both ends stepped and so on.

Third digit.

	Internal shape	both etc.
e.g	0	for no hole
	1	for through hole

Fourth digit.

Plane surface matching, i.e., with or without slots, grooves, etc.

0	for no surface machining
3	for external groove or slot etc.

Fifth digit.

Other holes/gear teeth/forming etc.

0	represents no hole
2	represents P.C.D hole

The supplementary code is from 6th to 9th digit

Sixth digit.

Overall dimension, e.g. raw stock diameter

Seventh digit.

Material information, e.g. type of material heat treatment, etc.

Eighth digit.

Raw material shape

Ninth digit.

Accuracy tolerance requirement

Code system

This system has been developed and marketed by the manufacturing data system Inc. U.S.A. It is an eight-digit code where each digit can be represented by any figure from 0 to 9, A to F, i.e. hexadecimal. It has a hybrid structure because the first digit represents the basic shape of the component, e.g. 1 represents a concentric component other than profiles. The remaining digits are in a chain-like structure. For a concentric component, the value of the second digit would be based on the outside diameter, i.e.,

1. for single
2. for multiconcave cylinder
   ....
   ....
6. for single cone

The third digit refers to any central hole on the axis

1. no hole
2. through hole
3. one blind hole
4. through threaded hole
5. one threaded through hole
   ....
   ....

The fourth digit refers to any hole or holes other than the central one (e.g. 1 for longitudinal, 2 for radial, etc.). The fifth digit refers to grooves, threads (e.g. 1 for external groove, 2 for internal groove). The sixth digit gives miscellaneous information. The seventh digit refers to the maximum outside diameter or section (e.g. 1 for upto 2.54 mm, 3 for 4.06 to 6.86, etc.). The eighth digit indicates the maximum overall length (e.g. 1 for upto 25.4, 2 for 25.4 to 40.64 mm, etc.).

15.17 FACILITIES LAYOUT

Once the components have been classified, they can be put into categories on the basis of the facilities used. The facilities can be created depending upon the analysis, i.e., composite part analysis and production flow analysis. In the case of composite part analysis, a cell is formed based on a key machine which can manufacture all the components of that group. This results in a substantial saving in the setting time. In PFA, the cell is formed with a group of machines which cater to the components of that group. These machines are laid at one place which results in a considerable saving in throughput time when compared to if the components are made in a functional layout. (Fig. 15.13).

There is easier and more effective control of the production process. This also results in an enhanced utilization of the machine tools and therefore, need for lesser machine tools and subsequently requires lesser space and manpower. In a certain case, 28 machines were being used which involved at least 50 movements. When GT was applied in this batch manufacture, only 10 machines were needed involving only 11 movements. There is a large reduction in productive time like setting, waiting and transportation.

15.18 ADVANTAGES

1. Group technology allows similar designs to be easily modified from the existing designs from the database instead of starting from scratch.
2. Standard process plans can be developed for the groups. Greater efforts can be applied in optimizing the process plans.
3. Tooling and set-up standards can be developed for a part family and then standard set-up procedures can be used.
4. The use of GT allows faster production. Therefore, less inventory and work in process (WIP).
5. Material handling and movement is reduced.
6. Improvement in quality and reduction in scrap resulting in increase in production.
7. Improved utilization of machines and as a result reduced number of machines required. This increases the floor space available.
8. Increase in output per employee and hence increase in productivity.
9. Manufacturing lead time is reduced leading to reduction in overdue order.
10. Improved ability to respond to market changes increases job satisfaction and greater management—worker harmony.

15.19 LIMITATIONS

1. The cost of implementation is generally high. This is because an outside consultant is often required since inhouse expertise on GT is rarely available. It requires a long set-up time and painful debugging.
2. May not be suitable for a factory with a very large variety of products.
3. The entire production of the company cannot be put under the GT and hence GT will have to coexist with the conventional layouts.
4. There are too many GT codes in use and there is no one GT code that suits all applications.
5. It is often difficult to conceive all the operations for a group of components being taken care of in the cell created for it.
6. The range of product mix in a plant may be under the constant change in which case the GT cells may need constant revision, which is impractical.

15.20 GUIDELINES FOR IMPLEMENTING GT

1. Collect a complete variety of components being manufactured in the company.
2. Get an estimate of the quantity to be produced for each variety over a period of time. This period will depend on the policies of manufacture and stocking the requirements. Due to fluctuations in demand, it may not be very certain.
3. Obtain the process sheet for each component.
4. For implementation of GT, exclude special operations, e.g. heat treatment, painting, forging, etc.
5. Try to plan the production of all the parts of a family in one cell. This may mean deliberate utilization of secondary machines. This would be acceptable if these machines were inexpensive, otherwise such machines could be utilized by other cells too. This would enable flexibility in labour.
6. The layout should be such as to permit a redistribution of load amongst various cells whenever fluctuations in load arise.
7. Study the data on operations, their sequences for various components, the quantity required, the machine tool capacities, the set-up time and the machining times. Based on this, calculate the work load on each machine tool.
8. Collect the components which use (a) the same sequence of machine tools; (b) same machines; (c) similar machines.
9. Form cells on the basis of the above information. It is a good practice to adopt a cascade type of layout for cells. Such a layout enables an easy transfer of load amongst cells when the load varies. Inexpensive machines may be duplicated.
10. There are machines which are required for a larger number of components but their use in each of the cells is not much. It would be justified to keep them separate and made available to all the cells requiring them. This is particularly true for expensive machines. For example, forging machine and heat-treatment equipment can be formed into a single cell or two separate cells.

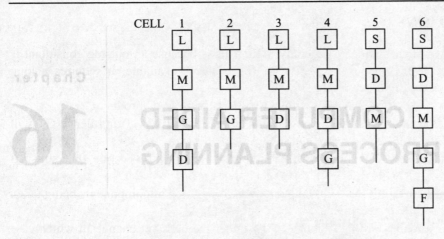

Fig. 15.14 Cascade layout.

REVIEW QUESTIONS

15.1 What do you understand by the term group technology?

15.2 Why is group technology more important in the present manufacturing scenario?

15.3 What is the basis for forming groups in technology?

15.4 Compare product layout and process layout.

15.5 Mention the advantages to be gained by the adoption of group technology.

15.6 What are the limitations of group technology?

15.7 Briefly explain the guidelines for implementing group technology.

15.8 Briefly explain about the Opitz coding system generally used in group technology.

15.9 Discuss advantages and disadvantages of Optiz code system.

15.10 Explain briefly MICLASS system with an example.

15.11 What is the method used for forming cells in group technology? Briefly explain with an example.

15.12 What is meant by the classification and coding approach? How is this different from the GT approach?

15.13 What factors must be considered in selecting a classification and coding system?

15.14 What is a production flow analysis? Discuss various steps involved in PFA.

15.15 What is composite part? Explain the concept of composite part with an example.

15.16 Discuss (a) Mono code (b) Poly code (c) Mixed code.

15.17 What are the main advantages of polycode over monocode?

15.18 What is a machine component incidence matrix?

15.19 What factors influence the grouping of a part machine matrix?

15.20 Discuss the impact of GT on machine utilization.

Chapter

COMPUTER AIDED PROCESS PLANNING

16

INTRODUCTION

This activity deals with determining the machine tools and the sequence of manufacturing operations involved in the manufacture of the components/products. Route sheets are used for the documentation of such activity. A process planner is expected to plan the process, determine the machining conditions and set the time standards for the component in question. He is to be familiar with the shop practice and equipment capabilities.

Every process planner would try to follow the best procedure based on his experience. Thus, it is possible that the routings would not be the same if different engineers are involved and it would also not be surprising if the same engineer may put up different routings for the same component if done at different occasions over a period of time. Such situations generally occur in a large variety manufacture when documentation is difficult to retrieve. These problems occur because no process standardization exists. This proves that conventional process planning can never be optimal. Because of their form, one is not sure whether the route sheets are updated or rewritten when the company acquires newer machining facilities.

For process planning to be done in a rational and consistent manner leading to optimal routing, the manual routine work needs to be eliminated, instead use should be made of modern computer software facilities, which on inputting specifications of the component will not only output the routings fast and correctly but can be stored efficiently and also communicated properly. This is *computer aided process planning*.

16.1 COMPUTER AIDED PROCESS PLANNING

Process planning can be considered under the two broad classes: retrieval and generative.

Retrieval or variant process planning is based on group technology approach. Standard process plans for part families are made and stored using the code system. When a process plan of a component is required, it can be speedily retrieved and edited, if necessary. In this system, a proper file organization is to be maintained and the coding used enables efficient retrieval.

The other type is the generative in which individual process plans are made for each component. In this, retrieval is not involved. As the component, features are input in the required way the system itself generates process plan. The logic used enables rationality and consistency in the process plans obtained. Figure 16.1 shows a typical computer aided process planning system.

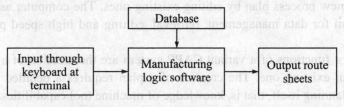

Fig.16.1 CAPP system.

16.2 RETRIEVAL TYPE CAPP SYSTEM

In this system separate files for part family, machine routing and operation sequence are created and stored. Algorithm is available which enables the user to identify the family to which the component belongs when he/she enters the part code number. On completion of the search, standard routing is retrieved from the machine routing file and then the standard operation sheet is retrieved from the operation reference file. These are edited to take care of any variation which a particular component may have from the part family. Other application software are also used to finally obtain the process plan in proper format with all the required details. Figure 16.2 shows the typical procedure.

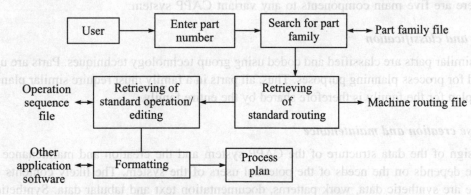

Fig. 16.2 Retrieval type CAPP system.

Retrieval type CAPP system is also known as *variant approach.*

The variant approach is the computerized approach to process planning and is only logical. With the computer speed and consistency, better and faster process plans can be generated. With the computerized process planning approach, the following benefits can be achieved: Clerical effort can be reduced, lead times can be cut, consistency of application of manufacturing data can be improved, the accuracy with which data is used can be improved, a standardized method of manufacturing can be introduced, records can be better maintained and manufacturing logic can be captured and used by anyone.

The variant approach (also called the **retrieval approach**), one of the two existing CAPP approaches to process planning, is a computer assisted extension of the manual approach. A process plan for a new part can be created by recalling, identifying and retrieving an existing plan for a similar part and making the necessary modifications required by the new part. The variant approach creates new process plan by editing existing ones. The computer assists by providing an efficient system for data management, retrieval, editing and high speed printing of process plans.

The two major functions of a variant CAPP system are the creation of a new plan and the modification of an existing one. The creation of a plan requires a detailed knowledge of the manual process planning itself, that is, knowledge of machine tool capabilities, the types of tools and the gage requirements. Also one should be knowledgeable of manufacturing processes, how parts are to be made, how they can be held, what should be done first and so on other data that is required is the manufacturing data (speeds, feeds, etc.) and how to use it to generate time and cost estimates.

The modification of an existing process plan is the most common function performed on a variant CAPP system. The variant system typically has a catalog or menu of standard based upon a coding scheme that codes parts into their families of parts. There is master process plan stored by the system, for each family of parts that can be retrieved and modified for any part that belongs to the family. When master plans are retrieved from the system, planners rely on their experience and manufacturing knowledge to either accept or edit these plans. A variant system also has the ability to examine the coding scheme of the part being planned in order to identify and retrieve any existing master plan that matches this part.

There are five main components to any variant CAPP system:

Coding and classification

In this similar parts are classified and coded using group technology techniques. Parts are usually grouped for process planning purposes. Thus, all parts in a family must require similar plans. The master plan for the family is therefore shared by the entire family.

Database creation and maintenance

The design of the data structure of the CAPP system and the creation and maintenance of its database depends on the needs of the potential users of the system. The likely contents of the database are synthetic data, work patterns, documentation text and tabular data. Synthetic data covers the elements of time and cost that a company uses up in building up its work values. Work patterns refer to the large building blocks in the database. These blocks define the relations between smaller data elements (blocks). Cross-referencing (via pointer) between the smaller and large blocks ensures the consistency of the data and information stored in the database. Documentation text is the text used in generating a process plan. This text forces all planners to use the same text and the same descriptions. The tabular data consists of the database records that describe the various manufacturing capabilities such as tooling, gaging, etc.

Logic processor

This is the heart of any CAPP system. Such processors can be divided into two categories: either low level or high level. A low level logic processor is one that stores the detailed method of manufacturing at the operation level. The high level logic processor is one that is capable of specifying the sequence of the process plan itself.

Documentation production

If the manufacturing sequence can be determined within the CAPP system, it is sensible that documents used at the shop floor level for controlling manufacturing should be produced by the computer. Documents that are produced from a CAPP system are method sheets, route sheets and tool kit sheets. Method sheets are typical manufacturing instructions at an operation level (low level) and give details of the exact method an operation must follow, including details such as speeds, feeds, tooling, etc. Routing sheets give details of the work through which the part passes and may include details of time value, tooling, etc. Tool kit sheets act as a listing from which tools can be ordered or designed prior to the beginning of production.

File maintenance

This covers the storage and control of records and the ability to retrieve and edit data as required.

Pointers

It gives the address of the value. By knowing the address of one element (block) the address of all the other elements in a (array) group can be identified.

16.3 GENERATIVE CAPP SYSTEM

This differs from the variant type in the same that the process plan is made from scratch. No standard plans exist and as such no retrieval is involved in this case. The software in the system is capable of taking technical and logical decisions (based on stored information pertaining to capabilities of machine tools available) when the user inputs description of the part in proper coded form and thus outputs the process plan. It builds up optimal process sequence based on part description provided and on the machining capabilities. This procedure will lead to rational, consistent and faster planning while the generative system can be used for any components. The variant type is limited to families of parts for which the data files have been created. Figure 16.3 shows the generative CAPP system.

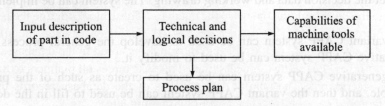

Fig. 16.3 Generative type CAPP system.

Generative Approach

The generative process planning approach is the automated approach to process planning. The generative approach does not require assistance from the user to generate a process plan. It usually accepts part geometrical and manufacturing data from the user and utilizes computerized searches and decision logics to develop the part process plan automatically. Generative process planning systems do not require or store predefined master process plans. The system automatically generates a unique plan for a part every time the part is ordered and released for manufacturing.

The generative approach generates a plan. In generating such a plan, an initial state of the part (stock) must be defined in order to reach the final state (finished part). The part taken between the two states defines the process plan or the sequence of processes. Two types of planning are available: forward and backward planning. In forward planning, we begin with the stock as the initial state and remove part features until the finished part the final state is obtained. In backward planning, we begin with the finished part as the initial state and fill it with part features until the stock is obtained.

A generative process planning system consists of two major components: coding scheme and decision logic. The coding scheme is geometry based and defines all the geometric features, feature sizes and locations and feature tolerances for all process related surfaces. The coding schemes must relate these features to their manufacturing requirements. It must relate them to the various manufacturing processes, the individual machine in each process, the available tooling and clamping. The coding scheme can be viewed as a universal language that relates all elements of manufacturing.

The decision logic involves the structuring of manufacturing planning logic and data into formats that will facilitate program coding and documentation. The decision logic results in determining the appropriate processing operations, selecting the machine for each operation, determining cut planning or other operation details subject to available tooling and fixturing and calculating the set-up and cycle time and cost for each operation.

16.4 HYBRID CAPP SYSTEM

This system is an interim system used when problems in building purely generative process planning systems occur. This system can be characterized as an advanced application of variant technology employing generative type features. The CAPP systems using this method should cooperate with process planner who possess technical knowledge. The process planner must be able to interpret the decision data and working drawing. The system can be implemented in three ways:

- The variant CAPP system can be used to develop the general process plan then the generative CAPP system can be used to modify it.
- The generative CAPP system can be used to create as such of the process plan as possible, and then the variant CAPP system can be used to fill in the details.
- Process planners can select either generative type for complicated part features or variant type for fast process plan generation.

16.5 PROCESS PLANNING SYSTEMS

Most of the process planning systems are of variant nature, some of them are CAM-I CAPP, MIPLAN, MITURN, MIAPP, UNIVATION, CINTURN, COMCAPPV, etc. There are some generative system, such as CPPP, AUTAP and APPAS.

16.5.1 CAM-I CAPP

The Computer aided manufacturing–international (CAM-I) automated process planning system (CAPP) is a database management system written in ANSI, standard FORTRAN. It provides a structure for a data base, retrieval logic and interactive editing capability. The coding scheme for part classification and the output format are added by the user.

16.5.2 MIPLAN and MULTICAPP

Both MIPLAN and MULTICAPP were developed in conjunction with Organization for Industrial Research (OIR). They are both variant systems that use the MICLASS coding system for part description. They are data retrieval systems which retrieve process plans based on part code, part number, family matrix and code range. By inputting a part code, parts with a similar code are retrieved. The process plan for each part is then displayed and edited by the user. Figure 16.4 shows multicapp system.

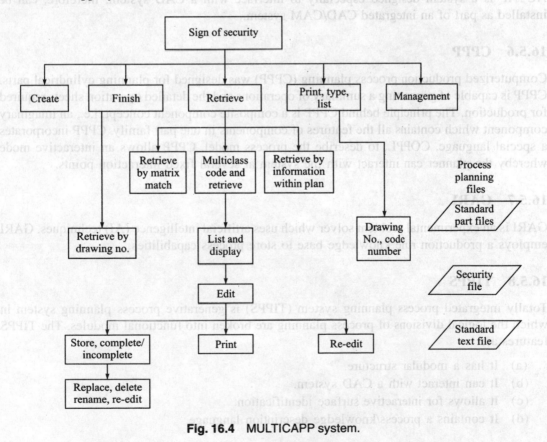

Fig. 16.4 MULTICAPP system.

16.5.3 APPAS and CAD/CAM

Automated Process Planning and Selection (APPAS) is a generative system for detailed process selection. APPAS describes the detailed technological information of each machined surface by means of a special code. CAD/CAM provides an interactive graphics interface to APPAS. Components can be modelled graphically and edited interactively.

16.5.4 AUTOPLAN

AUTOPLAN is generative, only in the detailing of the part. The four major modules of the system are:

(a) Group technology retrieval process plan retrieval
(b) Graphical planning aids—tooling layout, verification and work instruction and preparation
(c) Generative process planning
(d) Process optimization

16.5.5 AUTAP System

AUTAP uses primitives to construct a part similar to a constructive solid geometry (CSG). AUTAP is a system designed especially to interface with a CAD system, therefore, can be installed as part of an integrated CAD/CAM system.

16.5.6 CPPP

Computerized production process planning (CPPP) was designed for planning cylindrical parts. CPPP is capable of generating a summary of operations and the detailed operation sheets required for production. The principle behind CPPP is a composite component concept, i.e., an imaginary component which contains all the features of components in one part family. CPPP incorporates a special language, COPPL, to describe the process model. CPPP allows an interactive mode whereby the planner can interact with the system at several fixed interaction points.

16.5.7 GARI

GARI is an experimental problem solver which uses artificial intelligence (AI) techniques. GARI employs a production rule knowledge base to store process capabilities.

16.5.8 TIPPS

Totally integrated process planning system (TIPPS) is generative process planning system in which the logical divisions of process planning are broken into functional modules. The TIPPS features are:

(a) It has a modular structure.
(b) It can interact with a CAD system.
(c) It allows for interactive surface identification.
(d) It contains a process/knowledge description language.

Figure 16.5 shows flowchart of TIPPS.

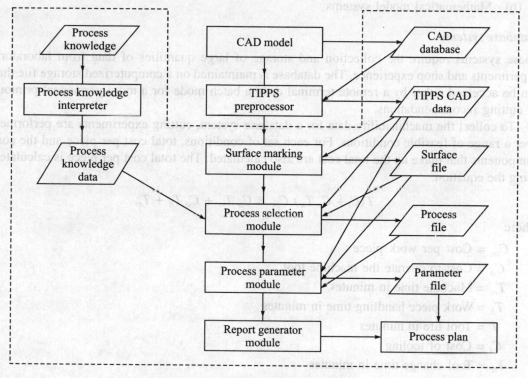

Fig. 16.5 Flowchart of TIPPS.

16.6 MACHINABILITY DATA SYSTEMS

In a machine shop, process planning includes selection of the cutting conditions like speed, feed and depth of cut w.r.t various machining operations. Depth of cut is determined based on the work piece geometry and sequence of operation. The objective of machinability data system is to select the cutting speed and feed rate, when the operation characteristics like type of machining operation, machine tool, cutting tool, work part and other operating parameters are defined.

The speed/feed selection problem can be solved by

(a) Experience and judgment of process planner or machine operator
(b) Handbook recommendations
(c) Computerized machinability data systems

The experience and judgement of process planner is the least systematic approach and carries the greatest risk.

Handbooks of machinability data are developed from a systematic analysis of large quantities of machining data. The cutting recommendations are based on laboratory experiments whose objective is to determine speeds and feeds.

Computerized machinability data systems have been classified into two types by Pressman and Williams as:

(a) Database systems

(b) Mathematical model systems

Database systems

These systems require the collection and storage of large quantities of data from laboratory experiments and shop experience. The database is maintained on a computerized storage file that can be accessed either by a remote terminal or in a batch mode for a more permanent printout of cutting recommendations.

To collect the machinability data for a database system, cutting experiments are performed over a range of feasible conditions. For each set of conditions, total cost per piece and the cost components that make up the total cost are also calculated. The total cost per piece is calculated using the equation:

$$T(C_t + C_o T_{tc}) \, C_{pc} = C_o T_m + C_o T_h + T_m$$

where

C_{pc} = Cost per work piece

C_o = Cost to operate the machine tool

T_m = Machine time in minutes

T_h = Work piece handling time in minutes

T = Tool life in minutes

C_t = Cost of tooling

T_{tc} = Tool change time in minutes

Mathematical model systems

These systems predict the optimum cutting conditions for an operation, i.e., determining optimum cutting speed for a given feed rate. Taylor's equation for tool life predicts optimum cutting speed as:

$$VT^n = C$$

where

V = Surface speed in m/s

T = Tool life

C and n = Constants

The equation for minimum cost cutting speed is:

$$V_{min} = \cfrac{C}{\left[\cfrac{1-n}{n} \cfrac{C_o T_{tc} + C_t}{C_o} \right]}$$

The cutting speed equation for maximum production rate is:

$$V_{max} = \cfrac{C}{\left[\cfrac{1-n}{n} T_{tc} \right]^n}$$

16.7 BENEFITS OF CAPP

Among the benefits derived from computer-automated process planning are the following:

Process rationalization

Manually prepared operation routings are not as consistent as that of computer automated operation routings. Automated process planning leads to more logical and consistent process plans. The same computer software is being used for all planners, which results in process rationalization.

Increased productivity of process planners

The systematic approach and the availability of standard process plans in the data files permit more work to be accomplished by the process planners. These benefits result in increased productivity of process planners. One system was reported to increase productivity by 600%.

Reduced turn around time

Working with the CAPP system, the process planner is able to prepare a route sheet for a new part in less time compared to manual preparation. This results in reduction of turn around time.

Improved legibility

Computer prepared documents are neater and easier to read than manually written route sheets.

Incorporation of other application programs

The process planning system can be designed to operate in conjunction with other software packages to automate many of the time consuming manufacturing support functions.

16.8 COMPUTER PROGRAMMING LANGUAGES FOR CAPP

Computer aided process planning involves almost all of the popular computer programming languages. BASIC is a simple interactive programming language, which has been used quiet widely at the beginning of CAPP development. FORTRAN is also an early computer programming language used and in maintaining its superior position in CAPP. PASCAL and 'C' language with the UNIX operating system because of their features such as clear control structure and powerful data structure are now important in CAPP.

Nowadays artificial intelligence computer programming languages such as LISP and PROLOG are being used increasingly. LISP is a functional programming language and offers flexibility in writing rules so that expert system builders can use logical programming language based on horn clauses. Programming in PROLOG involves writing logical formulae which indicate logic relations in problems. Both PROLOG and LISP are not very convenient sometimes for handling mathematical computations. This may lead to prolonging the execution time of the program and occupies more memory space. Nevertheless combining the advantages of both LISP and PROLOG will lead to a more efficient programming language for AI process planning.

REVIEW QUESTIONS

16.1   Briefly explain the need for computer-aided process planning.

16.2   What do you understand by process planning?

16.3   What are the various approaches available for computer aided process planning?

16.4   Give a brief description about the retrieval type of CAPP system.

16.5   Give a brief description about the generative type of CAPP system.

16.6   What are the differences between retrieval and generative type of computer-aided process planning?

16.7   Briefly explain the methodology to be followed for developing a retrieval type CAPP system.

16.8   Briefly explain the methodology to be followed for developing a generative type CAPP system.

16.9   Briefly explain the architecture of CAPP system with a neat sketch.

16.10 What is a composite part? Explain with an example the role of composite part concept in developing a CAPP system.

16.11 What are the various processes involved in set-up planning in the case of a rotational part?

16.12 Discuss the process of manual process planning taking a suitable example.

16.13 Compare manual and computer-aided process planning methods.

16.14 Give a brief description of hybrid CAPP system.

16.15 What are the benefits of CAPP?

16.16 What are the computer programming languages used for CAPP?

Part VI

Quality Control and Automated Inspection

Quality Control and Automated Inspection

COMPUTER-AIDED INSPECTION AND QUALITY CONTROL

Chapter

17

INTRODUCTION

In this age of intense competition, the survival of a product depends on its quality. The product should satisfy the requirements of the customer. The product should not only conform to specifications but also provide a reliable performance besides meeting the cost and delivery requirements.

The earlier practice of inspection was passive. It was akin to post-mortem, identifying good and bad components. With the introduction of mass production, interchangeability of the components was a must and so standardization at national and international levels for tolerance and gauge design was taken up. The inspection process failed to provide adequate feedback because of the time gap between the manufacture of the component and its inspection, and also due to the fact that the place of inspection and manufacture were away from each other. In the absence of feedback, quality control could not be controlled, resulting in higher costs. The onset of statistical quality control helped to some extent in bridging the gap between the inspector and the worker. The plots gave clue to the defects in the failures which proved helpful to the worker in initiating corrective measures. This system could only be used in mass production and where tolerances were not too low.

As the need for continuous monitoring arose due to stringent requirements of accuracy, in-process measurement was developed. For this sensors and comparators were designed which could provide the desired feedback continuously to enable manufacture of error free products. This was step towards integration with manufacture and it was possible to implement inspection devices on the transfer lines to achieve continuous production.

17.1 QUALITY ASSURANCE AND QUALITY CONTROL (QA & QC)

QA is carried out before the manufacturing of the product for maximizing the probability that the component would be manufactured according to the given specifications. The aspects to be considered are a proper selection of work material equipments, tooling and manufacturing processes. QC is carried out later and it enables detection of poor quality, suggests corrective action to be taken, the inspection procedures to be followed and specifications of gauges and instruments to be used. When these activities are computerized the procedure becomes automatic and output documentation is carried out in a systematic manner. However, this would be just an effort of automation to integrate it to the manufacturing system to save time and movements. As a development in this direction, non-contact sensors have been introduced which enable elimination of contact with component and this has led to high speed processing. The systems also permit feedback, say for tool wear adjustment, thereby leading to near perfect manufacture of the components.

17.2   INSPECTION AND TESTING

Inspection and testing constitute the operational part of quality control. They are performed before, during and after manufacturing to ensure that the quality level of the product is consistent with the accepted design standards.

17.2.1   QC Inspection

The term *inspection* refers to the activity of examining the product to determine if it meets specified design standards. The design standards are defined by the products. They relate to factors such as dimensions, surface finish and appearance. The objective of the inspection procedure is either

(a)   To take actual measurements of the values of the specified product characteristics or
(b)   Simply to check whether or not the specified characteristics meet the design standards.

In the first case, the factor (i.e., diameter of a turned part) is measured using an appropriate measuring instrument. For example, instruments for measuring diameter include micrometers, calipers or even a simple linear scale. In the second case, a device called a **gauge** is compared to the part to determine whether the quality characteristic of the product matches that of the gauge. For example, a GO/NOGO gauge might be used to check the part diameter. The gauge is designed so that if the part diameter is below a certain size, it will pass through the gauge opening. The gauge opening is, therefore, set at the upper acceptable size limit of the design specification.

The advantage of measuring the part characteristic is that the data can be collected about its actual value. The data might be recorded over time and used to make adjustments in the process so that future parts are produced with dimensions that are closer to the nominal design value. When a part dimension is simply checked with a GO/NOGO gauge, all that is known is whether the part is acceptable and whether it is too big or too small.

The advantage of gauging a part is that it can be done more quickly and at lower cost. Measuring the quality characteristic is a more involved procedure and takes more time. Whether measurement or gauging is done, inspection procedures are used at various stages during the manufacture of a product or part. Inspections are also performed on incoming raw materials to decide whether the materials should be accepted from the vendor.

17.2.2   QC Testing

Whereas inspection is used to assess the quality of the product relative to design specifications, *testing* is a term in quality control that is used relative to the functional aspects of the product.

QC testing is a procedure in which the item being tested is observed during actual operation or under conditions that might be present during operation. For example, a product might be tested by operating it for a certain period of time to determine whether it functions properly. If the product passes the test, it is approved for shipment to the customer. As another example, apart, or the material out of which the part is to be made, might be tested by subjecting it to a stress load that is equivalent to or greater than the load anticipated during normal operation.

Sometimes the testing procedure used on an item is damaging or destructive to the item. To ensure that the majority of the items are of satisfactory quality, a limited number of the items are sacrificed. The expense of destructive testing is significant enough that great efforts are made to devise methods that do not result in the destruction of the item. These methods are referred to as non-destructive testing (NDT) and non-destructive evaluation (NDE).

Another type of testing procedure involves not only testing the product to see that it functions properly, but also requires an adjustment or calibration of the product that depends on the outcome of the test. During the testing procedure one or more operating variables of the product are measured, and adjustments are made in certain inputs that influence the performance of the operating variables. For example, in testing certain appliances with heating elements, if the measured temperature is too high or too low after a specified time, adjustments can be made in the control circuitry (e.g., changes in potentiometer settings) to bring the temperature within the acceptable operating range.

17.3 COORDINATE MEASURING MACHINE

A coordinate measuring machine (CMM) consists of a contact probe and a means of positioning the probe in 3D space relative to the surfaces and features of a work part. The probe is not merely positioned relative to the part, its location can be accurately and precisely recorded to obtain dimensional data concerning the part geometry. These are controlled by computers or numerical control systems when a component is to be inspected for its profile and other features, the program or coordinate data is downloaded from the central computer. The systems are capable of transmitting data from the measuring machine back to the computer.

17.3.1 CMM Construction

In the construction of a coordinate measuring machine, the probe is fastened to some type of structure that allows movement of the probe relative to the part. The part is located on a worktable that is usually connected to the structure. There are several different physical configurations for achieving the motion of the probe, including the following common types:

Cantilever construction

In the cantilever configuration, the probe is attached to a vertical quill that moves in a z-axis direction relative to a horizontal arm that overhangs the worktable. The quill can also be moved along the length of the arm to achieve y-axis motion and the arm can be moved relative to the worktable to achieve x-axis motion. The advantages of this construction are the convenient access to the worktable and its relatively small floor space requirements. Its disadvantage is lower rigidity than some of the other CMM constructions.

Bridge construction

The bridge configuration is the most common type used in industry. Instead of a cantilevered arm to achieve the y-axis movement of the probe, the arm is supported on both ends like a bridge. This construction provides greater inherent rigidity and this makes the bridge construction more accurate than the cantilevered CMM.

Column construction

This construction is similar to the construction of a machine tool. Instead of achieving the relative motion exclusively by moving the probe, the column-type CMM obtains x-axis and y-axis relative motion by moving the worktable. The probe quill is moved vertically along a rigid column to obtain the z-axis motion.

Gantry construction

This construction is generally intended for inspecting large objects. The x-axis and y-axis motions are achieved by a construction similar to a gantry crane. The probe quill (z-axis) moves relative to the horizontal arm extending between the two rails of the gantry.

In all these constructions, special design features are used to build high accuracy and precision into the frame. These features include low-friction air bearings, installation mountings to isolate the CMM and reduce vibrations in the factory from being transmitted through the floor, and various schemes to counter balance the overhanging arm in the case of the cantilever construction.

One of the most important aspects in the design of a CMM is the probe. Most common are *touch-trigger* probes which use a highly sensitive electrical contact that emits a signal when the end of the probe is deflected from its neutral position in the slightest amount. Immediately upon contact the coordinate positions of the probe are recorded by the CMM controller. Any limited over travel of the probe quill due to momentum is neglected by the CMM. After the probe has been separated from the contact surface, it returns to the neutral position. Figure 17.1 shows a computer numerical control (CMM).

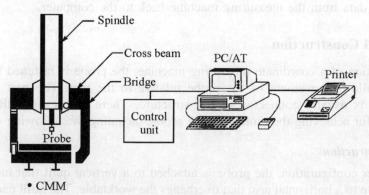

Fig. 17.1   CNC CMM.

During operation, the probe is brought into contact with the part surface to be measured and the three coordinate positions are indicated to a high level of accuracy. Typical accuracies of these machines are in the neighbourhood of +0.004 mm with a resolution of 0.001 mm. The measuring accuracy of a typical CMM is quoted $2.6 + L/300$ micrometers, where L is the measured length in mm.

Stationary granite measuring table

Granite table provides a stable reference plane for locating parts to be measured. It is provided with a grid of threaded holes defining clamping locations and facilitating part mounting. As the table has a high load carrying capacity and is accessible from three sides, it can be easily integrated into the material flow system of CIM.

Length measuring system

A three-axis CMM is provided with digital incremental length measuring system for each axis.

Air bearings

The bridge, cross beam and spindle of the CMM are supported on air bearings with high rigidity. They are designed insensitive to vibrations.

Control unit

The control unit allows manual measurement and self-teach programming in addition to CNC operation. The control unit is microprocessor controlled. Usually a joystick is provided to activate the drive for manual measurement.

CNC measuring centres are provided with dynamic probe heads and a probe changing system which can be operated manually or automatically.

Software

The CMM, the computer and the software together represent one system whose efficiency and cost effectiveness depend to a large extent on the software.

The features of a CMM software will include

  (a) Measurement of diameter, centre distances, lengths, geometrical and form errors in prismatic components, etc.

  (b) Online statistics for statistical information in a batch

  (c) Parameter programming to minimize CNC programming time of similar parts

  (d) Measurement of plane and spatial curves

  (e) Data communications

  (f) Digital input and output commands for process integration

  (g) Programs for the measurement of spur, helical, bevel and hypoid gears

  (h) Interface to CAD software

A typical software may also provide a generalized method for reverse engineering complex shaped objects. The component is digitized, taking a dense set of points, using a CNC CMM. The digitized data is then converted into a computer model which describes the true surface of the component, with allowance for the digitizing probe diameter. The model may then be expanded, offset or mirrored to an allowance for the manufacturing process.

Recent advances in CMM technology are based largely on greater intelligence features provided by the computer software. These advances include the capability for automatic work part alignment on the table, interactive programming of the CMM for inspection personnel who are not experienced in the use of computers. Besides this, the software has the capability to orient the coordinate system as required (between polar and cartesian coordinate systems). Similarly, translation of origin can be effected as desired.

Savings in inspection time by using CMM are significant. Typically between 5 and 10% of the time is required on a CMM compared to traditional manual inspection methods. Other advantages include consistency in the inspection process from one part to the next which cannot be matched by manual inspection, and reductions in production delays to get approval of the first work piece in a batch.

17.3.2 Principle of CMM

Figure 17.2 shows a schematic diagram of a typical coordinate measuring machine. CMM are of various designs namely, cantilever bridge, column, gantry type, and so on. Such machines need a deep foundation, carefully done, to isolate them from the vibrations which would otherwise affect their accuracy. They need to be located in an environmentally controlled space. The machine has a heavy base of granite and has a rigid structure. To explain the principle of CMM, a bridge type CMM can be described.

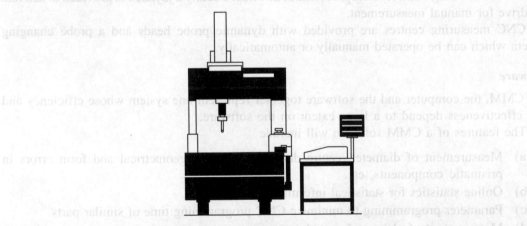

Fig. 17.2 Coordinate measuring machine.

The bridge can move in y-direction while the probe arm can move on the bridge in x-direction and the probe itself can be moved in z-direction. The weight of the moving elements is borne by the air bearings and thereby the bridge, the arm and the probe can be moved with very light force in their respective direction of movements. The measuring machine can be made to work in force axes by mounting a rotatory table on the bed. These movements can either be done manually or by switch or program control.

These machines have electromechanical probes of the touch trigger type and can be used for 3D information. Figure 17.3 shows an electromechanical probe for CMM. Figure 17.4 shows a multistyles probe for CMM are also available which are found quite useful while inspecting complex components. Standard blocks and spheres are supplied as accessories with the machines and are to be used for calibration and setting references.

The probes are generally of the ball ended type. However, conical probes which are more rigid, are convenient to use for determining the position of the centre of the hole. The touch trigger type has a number of different probes mounted on it enables measurement in difficult positions under servomotor control. The probe is carried on a light and flexible mechanism which communicates an electric signal when touched and the information is not affected by over-travel. The coordinates at that instant with respect to a reference are displayed on the indicating unit of the instrument.

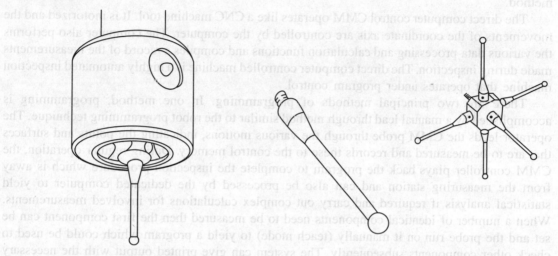

Fig. 17.3 Probe for CMM.   **Fig. 17.4** Multistylus probe for CMM.

17.4 OPERATION AND PROGRAMMING

The positioning of the probe relative to the part can be accomplished either manually or under computer control. The methods of operating and controlling a coordinate measuring machine can be classified as follows:

    (a) Manual control
    (b) Manual computer-assisted
    (c) Motorized computer-assisted
    (d) Direct computer control

In the manual control method, the human operator physically moves the probe along the machine's axes to make contact with the part and record the measurements. The probe is designed to be free-floating to permit easy movement along the coordinate axes. The measurements are proved on a digital readout, which the operator can record either manually or automatically. Any calculations on the data (e.g. calculating the centre and diameter of a hole) must be made by the operator.

The manual computer-assisted CMM provides some level of computer data processing and computational capability for performing these calculations. The type of data processing and computations range from simple conversions between US customary units and metric to more complicated geometry calculations such as determining the angle between two planes. The probe is still free-floating to permit the operator to bring it into contact with the desired part surfaces.

The motorized computer-assisted CMM uses a motor drive to power the probe along the machine axes under the operator's guidance. A joystick or similar device is typical as the means of controlling the motion. Features such as low-power stepping motors and friction clutches are utilized to reduce the effects of collisions between the probe and the part. The motor drive can be disengaged to permit the operator to physically move the probe as in the manual control method.

The direct computer control CMM operates like a CNC machine tool. It is motorized and the movements of the coordinate axis are controlled by the computer. The computer also performs the various data-processing and calculation functions and compiles a record of the measurements made during inspection. The direct computer controlled machine is a highly automated inspection machine that operates under program control.

There are two principal methods of programming. In one method, programming is accomplished by a manual lead through method similar to the robot programming technique. The operator leads the CMM probe through the various motions, indicating the points and surfaces that are to be measured and records these to the control memory. During regular operation, the CMM controller plays back the program to complete the inspection procedure which is away from the measuring station and can also be processed by the dedicated computer to yield statistical analysis if required and carry out complex calculations for involved measurements. When a number of identical components need to be measured then the first component can be set and the probe run on it manually (teach mode) to yield a program which could be used to check other components subsequently. The system can give printed output with the necessary identifications.

17.5 SPECIFICATIONS OF A CMM

Measuring range: Manual and motorized and along x, y and z axes in mm

Positioning accuracy: In microns (e.g. $3 + 5L/1000$, where L is length of the measuring stroke in mm)

Measuring accuracy: In microns (e.g. $13 + 20L/1000$)

Table clamping surface: $(x * y)$ in mm

Max weight of the component which can be placed in kg
Clearances under bridge and under probe in mm

A typical specification could also be the volume of the component it can handle, e.g., 600 mm cube. The measuring range is generally available upto one metre along each axis and the permissible weight of the component is upto 2 tonnes.

The gantry types of CMM are very useful for prototype or small batch work. Some machines have a magazine which carries a variety of probes. These can be called into action just as an ATC is used in a machining centre.

The other method of programming is accomplished in the manner of conventional NC part programming. The program is prepared offline and then downloaded to the CMM controller for execution. The programming statements for a computer controlled CMM include motion commands, measurement commands and report formatting commands. The motion commands are used to direct the probe to a desired inspection location in the same way a cutting tool is directed in a machining operation. The measurement statements are used to control the measuring and inspection functions of the machine calling the various data processing and calculation routines into play. Finally the formatting statements permit the specification of output reports to document the inspection.

17.6   ADVANTAGES OF CNC OPERATION OF CMM

CNC operation increases cost effectiveness through the following advantages:

    (a)   Shorter measuring times
    (b)   Higher throughput rates
    (c)   Better repeatability
    (d)   Economical even for small batches
    (e)   Simple operation
    (f)   Unmanned second and third shift inspection of parts if parts are loaded automatically

17.7   CMM BENEFITS AND TRENDS

The principal advantages of using coordinate measuring machines over manual inspection methods are:

Productivity

Because of the automated techniques included in the operation of a CMM inspection, procedures are speeded and labour productivity is improved.

Flexibility

A CMM is a general-purpose machine that can be used to inspect a variety of different part configurations with minimal change over time. In the case of the direct computer-controlled machine where the programming is performed offline the changeover time only involves making the physical set-up.

Reduced operator error

Automating the inspection procedure has the obvious effect of reducing human errors in measurements and set-ups.

Greater inherent accuracy and precision

A coordinate measuring machine is inherently more accurate and precise than the manual surface plate methods of inspection traditionally used.

    The technology of the coordinate measuring machine is evolving toward greater computer sophistication. The trends include increased functional capabilities of CMM systems. Examples are computer graphic display of the inspection results, online data analysis with feedback to the manufacturing process and three-dimensional contour analysis of surfaces such as cams and other complex shapes. Another trend is toward more user-friendly features, which make the CMM easier to program and easier to use. It is possible to program the CMM based on the part model contained in the CAD/CAM database in a manner similar to NC part programming with CAD/CAM.

Finally another future development is likely to be automated loading/unloading of the coordinate measuring machine.

17.8 LIMITATIONS OF CMM

The CMM's because of their sophistication, are located in a separate space away from the manufacturing area which gives rise to unnecessary transportation and handling. This also causes delay since the components need to be kept in a climatising room before being downloaded in the inspection machine room. This aspect has now been taken care of by mounting inspection probes on the CNC machine tools. Figure 17.5 shows a computer numerical control trigger probe. These can be brought into action by programs loaded on the CNC system and thus they serve quite the same function as a CMM. Figure 17.6 shows a computer numerical control probe with signal transmission device. The information picked by the probe is conveyed to the control system of the machine tools. The probe can be used for setting of the datum and recording of the tool offsets and thereby facilitates the manufacturing program by enabling the performance of job/tool setting inspection and feedback on a single setting on the CNC machine tool.

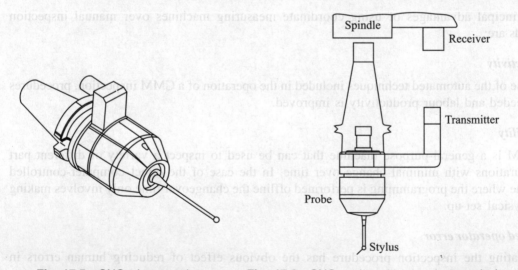

Fig. 17.5 CNC trigger probe.   **Fig. 17.6** CNC probe with signal transmission device.

17.9 OPTICAL INSPECTION METHODS

17.9.1 Machine Vision System

A vision system is defined as a system for automatic acquisition and analysis of images to obtain desired data for interpreting or controlling an activity. The system consists of image acquisition, image processing and interpretation. In image acquisition, using camera a digits representation of the image is acquired and stored. Processing involves manipulating the digits image to simplify and utilized for analytical routines used to interpret the data.

Using vision system, measurements can be carried out at any angle along all the three reference axes *X*, *Y* and *Z* without contacting the part. The measured values of the component parameters are then compared with the specified tolerances which are stored in the memory of the computer. The results are displayed on the VDU.

Reduction of tooling and fixture costs, elimination of robots and automated defect detection are the advantages of machine vision system.

The disadvantages of computer vision system are problem of dividing the picture into picture elements and to recognize the object in the viewing area.

17.9.2 Scanning Laser Beam Device

The scanning laser beam device is based on the principle of measurement of time rather than light. A laser is used to project a continuous thin beam of light. A rotating mirror deflects the beam so that it sweeps across the object to be measured. The light sensor is located at the focal point of the lens system to detect the interruption of the light beam as it is blocked by the object. The time lapse corresponding to the interruption of the light beam is measured to determine the desired dimension of the part. A microprocessor is programmed for conversion of the time lapse into a dimensional value and automatically reject the defective part. Figure 17.7 shows a laser based non-contact inspection system.

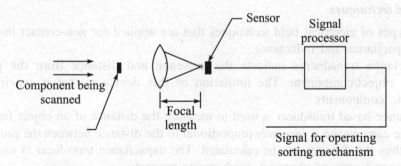

Fig. 17.7 Laser based, non-contact inspection system.

17.9.3 Photogrammetry

Photogrammetry involves the extraction of three-dimensional data from a pair of photographs taken at different angles. The two photographs are read by a device known as **monocomparator** to establish coordinates and positions of objects for inspection process. These data are then computer-analyzed to extricate the desired information. The drawback of photogrammetry technique is the need for photographs, which is a time-consuming and inconvenient step.

17.10 NON-CONTACT, NON-OPTICAL INSPECTION METHODS

There are three major types of non-contact non-optical inspection methods. They are electrical field techniques, radiation techniques and ultrasonicsis techniques. Figure 17.8 shows a scanning laser beam system.

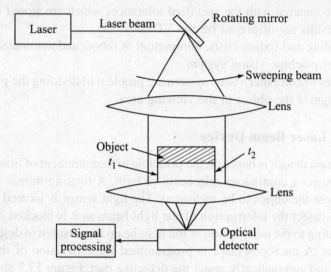

Fig. 17.8 Scanning laser beam system.

Electrical field techniques

The various types of electrical field techniques that are applied for non-contact inspection are reluctance, capacitance and inductance.

The reluctance transducers indicate the presence and distance from the probe of a ferromagnetic object/component. The limitation of this device is that it can inspect only electromagnetic components.

A capacitance-based transducer is used to measure the distance of an object from the face of a probe. The capacitance is *inversely* proportional to the distance between the probe face and the part, and thus the distance can be calculated. The capacitance transducer is used to detect various components/parts of electrical conductance property.

Inductance based systems operate by subjecting the object to an alternating magnetic field by means of an electromagnetic coil. This results in small circulating current, which in turn create their own magnetic field, which interacts with the primary field. This interaction affects the impedance of the coil, which can be measured and analyzed to determine the characteristics about the component/part. This technique can be utilized to determine the inner defects in metals.

Radiation techniques

The X-ray radiation technique is based on the principle that the amount of radiation absorbed by a material can be used to measure its thickness and other quality characteristics. X-ray radiation techniques are applied to the fields of metal working industry, welding, etc.

Ultrasonic techniques

Ultrasonic techniques are applied to the problem of determining dimensional features of work part by using high frequency (> 20,000 Hz) sound waves. In acoustical phase monitoring, sound waves are produced by an emitter and directed against the component. Assuming that all remains constant; the reflected sound pattern from the object should always be the same. During inspection, the sound pattern from the part is analyzed by a computer program and compared to the pattern of a standard quality part. If the pattern of the test part differs from the standard part, it is rejected.

The major advantage of non-contact inspection is that there is no need to reposition the work piece when it is being inspected. This results in faster processing. Since no probes are being used no problems of wear of probe or damage to component arises. These are machine vision systems which are microprocessor based TV (also called **computer vision**). The component stationary or moving from one station to another is scanned by the TV camera which is interfaced to digitize the information and input it to the computer. In the computer memory, the shape/size, i.e., data of the components is already stored. The input information is compared with the stored information and thereby inspection is carried out. The recognition of objects becomes difficult when there is overlap as may happen when being transferred by conveyor or system. The capability of the system to visualize the elements by clearly distinguishing between light and dark images is rather limited. Illumination of the objects plays an important part in machine vision. In vision inspection systems, part orientation and positioning are not important. The software provides the capability to determine the references and visualize the image. For gauging the component, the cursor is used to define the edges of the component and then the relevant boundaries are identified. The system applies a scale factor and the information gathered is compared with the stored values. The component identification is done by visualizing the shape, size and the arrangement of features on it. This is called **pattern recognition** and for this automatic edge detection is carried out. The edges and the holes are determined by the intensity of the image. The software enables translation of the digitized information of image into part shape and size. This technique can be used to check the dimension of the component and also to detect the presence or absence of the component at the work station. This could be integrated with the manufacturing system. A typical example being materials handling and robotics.

17.11   COMPUTER AIDED TESTING (CAT)

Testing is applied to assess the functional performance of a final product. There are different levels of automation in computer aided testing (CAT). The computer can be utilized to just monitor the test and analyze the results, but at this lower level the testing procedure is manually set up, initiated and controlled by a human operator. The computer receives the data from a data logger and prepares the test results report.

In sophisticated automation, computer-integrated test cells consist of series of testing stations interconnected by a materials handling system is utilized. During operation, a product is transferred by the handling system to an available test station. The test station automatically registers the product in the proper location and orientation, and attaches the required connecting apparatus to conduct the test. The testing starts with computer monitoring of the data and analyzing the results. If the product passes the test, it is automatically moved to next assembly operation. But if the product fails the test, the product is transferred to manual station for examination. The computer helps in diagnosing the problem and recommends the alternative repairs. Thus, CAT provides greater consistency in the test procedure.

REVIEW QUESTIONS

17.1 What is the significance of quality control in CIM?

17.2 What are the objectives of computer aided quality control?

17.3 Discuss the important benefits of computer aided quality control.

17.4 Differentiate between quality control and quality assurance.

17.5 Differentiate between QC inspection and QC testing.

17.6 What is the working principle of coordinate measuring machine?

17.7 State the applications of CMM.

17.8 Sketch and explain various types of CMMs.

17.9 Explain the method of part inspection using a CMM.

17.10 Discuss major non-contact inspection methods.

17.11 Describe the features of a flexible inspection system.

17.12 What are the various sampling methods of quality control?

17.13 What are the basic steps involved in designing a quality control system?

17.14 Explain the procedure for integration of CAD/CAM with inspection systems.

17.15 Explain computer aided quality management systems.

17.16 Explain computer inspection methods.

17.17 Explain about photogrammetry.

17.18 Explain the following terms:
   (a) Online inspection
   (b) Offline inspection
   (c) In-process inspection
   (d) Post-process inspection

17.19 What are the disadvantages and limitations of traditional quality control systems? Explain how they are overcome by computers and automation.

17.20 Explain how the inspection sensor technologies are classified.

17.21 Explain general types of non-contact inspection methods.

17.22 Explain the important benefits of applications of computer-aided quality control systems.

17.23 Explain the applications and advantages of integration of CAQC with CAD/CAM systems.

17.24 What are the important features available in a CMM software?

MACHINE VISION

Chapter

18

INTRODUCTION

Machine vision, also known as **computer vision** or **artificial vision,** is an important sensor technology finding application in many industrial applications. This is a process of sensing of vision data and its interpretation by computer. A typical machine vision system consists of a camera and digitizing hardware, a digital computer and hardware and software necessary to interface them. The interface hardware and software is referred to as pre-processor. The operating vision system consists of the following functions:

(a)   Sensing and digitizing image data
(b)   Image processing and analysis
(c)   Applications

The relationship between three functions are given in Table 18.1.

TABLE 18.1   Machine Vision System

Function	Sensing and digitizing	Image processing and analysis	Applications
Hardware	Lighting	Computer (processor)	Interface
	Camera	Monitor key	Actuators
	A/D	board	
	Frame grabber	Storage	
Techniques and	Signal conversion	Data reduction	Inspection
application	Sampling	Windowing	Identification
	Quantization	Digital	
	Encoding image	conversion	
	Storage/frame	segmentation	Visual servoing
	grabber lighting	Region growing	and navigation
	Structural	Education	
	Front/Back	feature	
	Beam splitter	extraction	
	Restroreflectors	Descriptors	
	Specular	object	
	Illumination	recognition	
		Tinplate matching	

18.1 IMAGING DEVICES

There are a variety of commercials imaging devices available. They are black and white videcon camera and solid state cameras. Solid-state cameras used for vision include charge coupled devices (CCD) and charge injectors devices (CID) and silicon bipolar sensor cameras. Vidicon camera and CCD are widely used.

18.1.1 Videcon Tube Camera

Figure 18.1 shows cross section of videcon tube and associated elements. In this system, the lens forms an image in the glass faceplate of the camera. The faceplate has an inner surface, which is created with two layers of material. The floor layer consists of a transparent signal electrode film deposited on the faceplate of the inner surface. The second layer is a thin photosensitive material deposited over the conducting film. The photosensitive layer consists of a high density of small areas. Each area generates a decreasing electrical resistance in response to increasing illumination. A charge is created on each small area on illumination. The charge accumulated for an area is a function by the intensity of impinging light over the specified time. This charge is scanned by an electron beam. For an accumulated positive charge, the electron beam deposits enough electrons to neutralize the charge. An equal number of electrons flow to cause current at video signal electrons. The current is a measure of light intensity for the amount of time with which the area is scanned.

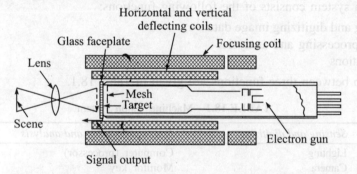

Fig. 18.1 Cross-section of vedicon tube and associated elements.

The current signal is converted into voltage signal, amplified, and converted into digital signal by a A/D converter and is stored. Each small element is called **one pixel** and is quantized to 6 to 8 bits by A/D converter. A digital signal with 6 bits will have 2 or 64 levels of signals called **gray levels**. This digitized quantity can be stored and used for further process.

In the charge coupled device (CCD), the image is projected by a video camera onto the CCD, which detects, stores and reads out accumulated charge generated by this light on each portion of the image. Light detection occurs through the absorption of light on a photoconductive substate like silicon. Charges accumulate under positive control electrode in isolated wells due to voltages applied to the central electrodes (see Fig. 18.2).

Each isolated well represents one pixel and can be transferred to output storage registers by varying the voltages on the metal control electrodes.

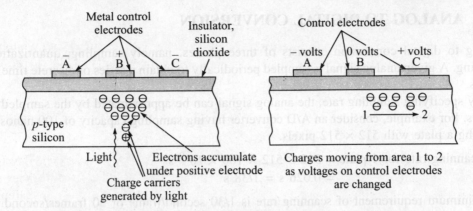

Fig. 18.2 Basic principle of CCD.

Figure 18.3 indicates one type of CCD register. In this device, charges are accumulated for the time. It takes to complete a single image after which they are transferred line-by-line to storage register. Register A accumulates the pixel charge produced by light image. Once accumulated for a single picture, the charges are transferred line by line to register B. The pixel charges are readout line-by-line through a horizontal register C to an output amplifier. During readout, register A is accumulating new pixel elements. The complete cycle is repeated nearly in 1/60 second.

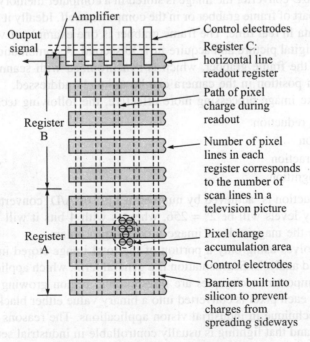

Fig. 18.3 One type of CCD imager.

18.2 ANALOG TO DIGITAL CONVERSION

Analog to digital conversion consists of three phases, namely sampling, quantization and encoding. A given analog signal is sampled periodically to obtain a series of discrete time analog signals.

By specifying sampling rate, the analog signal can be approximated by the sampled digital outputs. For example, consider an A/D converter having sampling capacity of 100 nanoseconds scanning a plate with 512×512 pixels.

$$\text{Scanning rate for 1 plate} = 512 \times 512 \times 100 \times 10^{-9} \text{ s}$$
$$= 0.026 \text{ s} = 1/38 \text{ s}$$

Minimum requirement of scanning rate is 1/30 second/frame or 30 frames/second.

Quantization refers to number of levels available in digitization. This depends on number of bits employed. For a 6 bit A/D converter, number of levels available is: $2^6 = 64$, i.e., 256 gray levels are available. Normally 4 bit is used which will give

$$2^4 = 16 \text{ levels}$$

Encoding is a process to distinguish between various amplitude levels. This is a function of spacing of each quantization level.

$$\text{Quantization level} = \text{Full scale rays}/2n$$

where n is number of bits in A/D converter.

Following the A/D converter the image is stored in a computer memory called **frame buffer**. The buffer may be part of frame grabber or in the computer itself. Ideally it is preferred to acquire a single frame of data in real time. The frame grabber is one example of video data acquisition, which will store a digital picture and acquire in 1/30 second. A combination of row and column counters is used in the frame grabber, which is synchronized with scanning of electron beam in the camera. Each position in the camera can be uniquely addressed.

In order to make image processing more effective, the following techniques are used:

(a) Image data reduction
(b) Segmentation
(c) Feature extraction
(d) Object recognition

Image data reduction is achieved by number of bits of A/D converter. For example, with 8 bit number of gray levels will be $2^8 = 256$, whereas with 4 bits it will be $2^4 = 16$. This will considerably reduce the magnitude of image processing problem.

Windowing involves using only a portion of the total image stored in the frame buffer for image processing and analysis. Segmentation is a general term, which applies to various methods of data reduction. Important techniques are thresholding, region growing and edge detection.

In thresholding, each pixel is converted into a binary value either black or white. This is the most widely used technique in industrial vision applications. The reasons are that it is fast and easily implemented and that lighting is usually controllable in industrial setting.

Region growing is a collection of segmentation techniques in which pixels are grouped in regions called **grid elements** based on attribute similarities.

Edge detection considers intensity change that occurs in the pixels at the boundary or edges of the past.

In machine vision, it is often necessary to distinguish one object from another. This is usually accomplished by necessary features that uniquely characterize the object. Some of the features that are used for feature extraction are:

(a) Gray level (maxi average or min)
(b) Area
(c) Perimeter length
(d) Diameter
(e) Min. enclosing rectangle
(f) Centre of gravity
(g) Eccentricity = Max. chord length B/Max. chord length A where chords A and B are perpendicular.

(h) Thickness area $= \dfrac{(\text{Perimeter})^2}{\text{Area}}$

$(\text{OR}) = \dfrac{\text{Diameter}}{\text{Area}}$

(i) No of holes

18.3 IMAGE PROCESSING AND ANALYSIS

The discussion in the preceding section described how images are obtained, digitized and stored in a computer. For use of the stored image in industrial applications, the computer must be programmed to operate on the digitally stored image. This is a substantial task considering the large amount of data that must be analyzed. Consider an industrial vision system having a pixel density of 350 pixels per line and 280 lines (a total of 98,000 picture elements), and a 6 bit register for each picture element to represent various gray levels; this would require a total of $98,000 \times 6 = 588,000$ bits of data for each $\dfrac{1}{30}$ s. This is a formidable amount of data to be processed in a short period of time and has led to various techniques to reduce the magnitude of the image-processing problem. These techniques include:

(a) Image data reduction
(b) Segmentation
(c) Feature extraction
(d) Object recognition

We will discuss these techniques of image data analysis in the following subsections.

18.3.1 Image Data Reduction

In image data reduction, the objective is to reduce the volume of data. As a preliminary step in the data analysis, the following two schemes have found:

(a) Digital conversion
(b) Windowing

The function of both schemes is to eliminate the bottleneck that can occur from the large volume of data in image processing.

Digital conversion reduces the number of gray levels used by the machine vision system. For example, an 8 bit register used for each pixel would have $2^8 = 256$ gray levels. Depending on the requirements of the application, digital conversion can be used to reduce the number of gray levels by using fewer bits to represent the pixel light intensity. Four bits would reduce the number of gray levels to 16. This kind of conversion would significantly reduce the magnitude of the image-processing problem.

18.3.2 Segmentation

Segmentation is a general term which applies to various methods of data reduction. In segmentation, the objective is to group areas of an image having similar characteristics or features into distinct entities representing parts of the image. For example, boundaries (edges) or regions (areas) represent two natural segments of an image. There are many ways to segment an image.

Three important techniques that we will discuss are:

(a) Thresholding
(b) Region growing
(c) Edge detection

In its simplest form, thresholding is a binary conversion technique in which each pixel is converted into a binary value, either black or white. This is accomplished by utilizing a frequency histogram of the image and establishing what intensity (gray level) is to be the border between black and white. This is illustrated for an image of an object shown in Fig. 18.4. Figure 18.4(a) shows a regular image with each pixel having a specific gray tone out of 256 possible gray levels. The histogram of Fig. 18.4(b) plots the frequency (number of pixels) versus the gray level for the image. For histograms that are bimodal in shape, each peak of the histogram represents either the object itself or the background upon which the object rests. Since we are trying to differentiate between the object and background, the procedure is to establish a threshold (typically between the two peaks) and assign, for example, a binary bit 1 for the object and 0 for the background. The outcome of this thresholding technique is illustrated in the binary-digitized image of Fig. 18.4(c). To improve the ability to differentiate, special lighting techniques must often be applied to generate a high contrast.

It should be pointed out that the above method of using a histogram to determine a threshold is only one of a large number of ways to threshold an image. It is, however, the method used by many of the commercially available robot vision systems today. Such a method is said to use a global threshold for the entire image. In some cases this is not possible and a local thresholding method as described below may be employed.

When it is not possible to find a single threshold for an entire image (for example, if many different objects occupy the same scene, each having different levels of intensity), one approach is to partition the total image into smaller rectangular areas and determine the threshold for each window being analyzed.

Thresholding is the most widely used technique for segmentation in industrial vision applications. The reasons are that it is fast and easily implemented and that the lighting is usually controllable in an industrial setting.

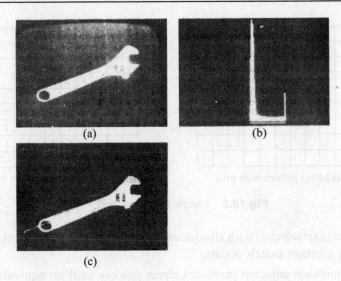

(a)

(b)

(c)

Fig. 18.4 Obtaining binary image by thresholding.

Once thresholding is established for a particular image, the next step is to identify particular areas associated with objects within the image. Such regions usually possess uniform pixel properties computed over the area. The pixel properties may be multidimensional, that is, there may be more than a single attribute that can be used to characterize the pixel (e.g., colour and light intensity). We will avoid this complication and confine our discussion to single pixel attributes (light intensity) of a region.

Region growing is a collection of segmentation techniques in which pixels are grouped in regions called **grid elements** based on attribute similarities. Defined regions can then be examined as to whether they are independent or can be merged to other regions by means of an analysis of the difference in their average properties and spatial connectiveness. For instance, consider an image as depicted in Fig. 18.5(a). To differentiate between the objects and the background, assign 1 for any grid element occupied by an object and 0 for background elements. It is common practice to use a square sampling grid with pixels spaced equally along each side of the grid. For the two-dimensional image of a key as shown, this would give the pattern indicated in Fig. 18.5(b). This technique of creating "runs" of 1s and 0s is often used as a first-class analysis to partition the image into identifiable segments or "blobs." Note that this simple procedure did not identify the hole in the key of Fig. 18.5(a). This could be resolved by decreasing the distance between grid points and increasing the accuracy with which the original image is represented.

For a simple image such as a dark blob on a light background, a runs technique can provide useful information. For more complex images, this technique may not provide an adequate partition of an image into a set of meaningful regions. Such regions might contain pixels that are connected to each other and have similar attributes, for example, gray level. A typical region-growing technique for complex images could have the following procedure:

(a) Select a pixel that meets a criterion for inclusion in a region. In the simplest case this could mean select white pixel and assign a value of 1.

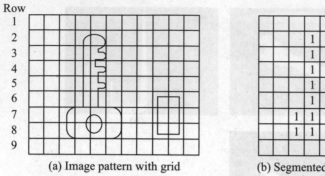

(a) Image pattern with grid   (b) Segmented image after runs test

Fig.18.5   Image segmentation.

(b)   Compare the pixel selected with all adjacent pixels. Assign an equivalent value to adjacent pixels if an attribute match occurs.

(c)   Go to an equivalent adjacent pixel and repeat process until no equivalent pixels can be added to the region.

This simple procedure of "growing" regions around a pixel would be repeated until no new regions can be added for the image.

The region growing segmentation technique described here is applicable when images are not distinguishable from each other by straight thresholding or edge detection techniques. This sometimes occurs when lighting of the scene cannot be adequately controlled. In industrial robot vision systems, it is common practice to consider only edge detection or simple thresholding. This is due to the fact that lighting can be a controllable factor in an industrial setting and hardware/computational implementation is simpler.

Edge detection considers the intensity change that occurs in the pixels at the boundary or edges of a part. Given that a region of similar attributes has been found but the boundary shape is unknown. The boundary can be determined by a simple edge following procedure. This can be illustrated by the schematic of a binary image as shown in Fig. 18.6. For the binary image, the procedure is to scan the image until a pixel within the region is encountered. For a pixel within the region, turn left and stop. Otherwise, turn right and stop. The procedure is stopped when the boundary is traversed and the path has returned to the starting pixel. The contour-following procedure described can be extended to gray level images.

18.3.3   Feature Extraction

In machine vision applications, it is often necessary to distinguish one object from another. This is usually accomplished by means of features that uniquely characterize the object. Some features of objects that can be used in machine vision include area, diameter and perimeter. A feature, in the context of vision systems, is a single parameter that permits ease of comparison and identification. A list of some of the features commonly used in vision applications is given in Table 18.2. The techniques available to extract feature values for two-dimensional cases can be roughly categorized as those that deal with boundary features and those that deal with area

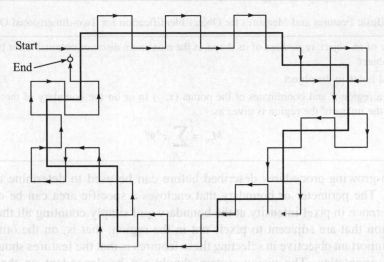

Fig. 18.6   Edge following procedure to detect the edge of a binary image.

features. The various features can be used to identify the object or part and determine the part location and/or orientation.

TABLE 18.2   Basic Features and Measures for Object Identification for Two-dimensional Objects

Gray level (maximum, average or minimum)

Area

Perimeter length

Diameter

Minimum enclosing rectangle

Centre of gravity: For all pixels (n) in a region where each pixel is specified by (x, y) coordinates, the x and y coordinates of the centre of gravity are defined as:

$$CG_x = \frac{1}{n} \sum_x x$$

$$CG_y = \frac{1}{n} \sum_y y$$

Eccentricity: A measure of *elongation*. Several measures exist of which the simplest is:

$$\text{Eccentricity} = \frac{\text{Maximum chord length A}}{\text{Maximum chord length B}}$$

where maximum chord length B is chosen perpendicular to A.

Aspect ratio: The length-to-width ratio of a boundary rectangle which encloses the object. One objective is to find the rectangle which gives the minimum aspect ratio.

Thinness: This is a measure of how thin an object is. Two definitions are in use:

$$\text{Thinness} = \frac{(\text{Perimeter})^2}{\text{Area}}$$

This is also referred to as *compactness*.

$$\text{Thinness} = \frac{\text{Diameter}}{\text{Area}}$$

TABLE 18.2 Basic Features and Measures for Object Identification for Two-dimensional Objects (*Contd.*)

The diameter of an object, regardless of its shape, is the maximum distance obtainable for two points on the boundary of an object.

Holes: Number of holes in the object.

Moments: Given a region R and coordinates of the points (x, y) in or on the boundary of the region, the *pq*th order moment of the image of the region is given as:

$$M_{pq} = \sum_{x,y} x^p y^q$$

The region-growing procedures described before can be used to determine the area of an object's image. The perimeter or boundary that encloses a specific area can be determined by noting the difference in pixel intensity at the boundary and simply counting all the pixels in the segmented region that are adjacent to pixels not in the region; that is, on the other side of the boundary. An important objective in selecting these features is that the features should not depend on position or orientation. The vision system should not be dependent on the object being presented in a known and fixed relationship to the camera.

The preceding measures provide some basic methods to analyze images in a two-dimensional plane. Various other measures exist for the three-dimensional case as well.

18.3.4 Object Recognition

The next step in image data processing is to identify the object the image represents. This identification problem is accomplished using the extracted feature information described in the previous subsection. The recognition algorithm must be powerful enough to uniquely identify the object. Object recognition techniques used in industry today may be classified into two major categories:

    (a) Template-matching techniques
    (b) Structural techniques

Template-matching techniques are a subset of the more general statistical pattern recognition techniques that serve to classify objects in an image into predetermined categories. The basic problem in template matching is to match the object with a stored pattern feature set defined as a model template. The model template is obtained during the training procedure in which the vision system is programmed for known prototype objects. These techniques are applicable if there is not a requirement for a large number of model templates. The procedure is based on the use of a sufficient number of features to minimize the frequency of errors in the classification process. The features of the object in the image (e.g., its area, diameter, aspect ratio, etc.) are compared to the corresponding stored values. These values constitute the stored template. When a match is found, allowing for certain statistical variations in the comparison process, then the object has been properly classified.

Structural techniques of pattern recognition consider relationships between features or edges of an object. For example, if the image of an object can be subdivided into four straight lines (the lines are called **primitives**) connected at their end points, and the connected lines are at right angles, then the object is a rectangle. This kind of technique, known as **syntactic pattern recognition**, is the most widely used structural technique. Structural techniques differ from decision-theoretic techniques in that the latter deals with a pattern on a quantitative basis and ignores for the most part interrelationships among object primitives. A detailed discussion of pattern recognition techniques is the subject of complete books and is beyond the scope of this text.

It can be computationally time consuming for complete pattern recognition. Accordingly, it is often more appropriate to search for simpler regions or edges within an image. These simpler regions can then be used to extract the required features. The majority of commercial robot vision systems make use of this approach to the recognition of two-dimensional objects. The recognition algorithms are used to identify each segmented object in an image and assign it to a classification (e.g., nut, bolt, flange, etc.).

18.4 APPLICATION

The machine vision can be used for inspection in the following ways:

- (a) Checking gross surface defects
- (b) Flaws in labelling
- (c) Verification of presence of components in assembly
- (d) Measuring for dimensional accuracy
- (e) Checking for presence of holes and their features in the component
- (f) Checking gray level of product
- (g) Other human inspection
- (h) Identification and sorting
- (i) Picking ports randomly oriented in a conveyor.
- (j) Visual servoing and navigation for part positioning, retrieving posts moving along conveyor, assembly, bin picking and seam tracking in arc welding.

REVIEW QUESTIONS

18.1 What is machine vision?

18.2 Name the components of machine vision system.

18.3 What are the hardware components of machine vision system?

18.4 Give specification of a machine vision system.

18.5 What is quantization level in machine vision?

18.6 Give application of object recognition in a machine vision.

18.7 Name the techniques employed for image processing and analysis.

18.8 Write short notes on

- (a) Image data reduction
- (b) Segmentation
- (c) Feature extraction
- (d) Object recognition

18.9 Discuss applications of machine vision system.

18.10 Explain how a machine vision system can be used for inspection and identification.

18.11 Discuss the application of machine vision for robotic guidance.

18.12 Name the feature that is used for characterization of an object.

18.13 Name the techniques of object recognition.

18.14 Discuss the principles of object recognition in machine vision system.

18.15 Explain the working principle of a charge coupled device (CCD) with a neat sketch.

18.16 Explain how an image is converted into digitized data and stored in a computer using videcon camera.

18.17 What is gray level of an image?

18.18 Name the different types of imaging devices.

18.19 What is a pixel?

18.20 What is scanning rate?

Part VII

Integration of Manufacturing Systems

Part VII

Integration of Manufacturing Systems

COMPUTER INTEGRATED PRODUCTION PLANNING SYSTEMS

Chapter

19

INTRODUCTION

The functions of a production planning and control system were accomplished by large staffs of clerical personnel. Members of the production planning and control department prepared the schedules, decided what materials and parts needed to be ordered, issued the individual orders to the work centres in the plant, and expedited the orders. For plants with large volumes of parts and complex products, the work involved in these activities was significant.

The types of problems commonly encountered in planning and managing production operations in a plant are the following:

Plant capacity problems

Production falls behind schedule due to lack of manpower and equipment resources. This results in excessive overtime, delays in meeting delivery schedules, customer complaints, backordering and so on.

Suboptimal production scheduling

The wrong jobs are scheduled because of a lack of clear order priorities, inefficient scheduling rules and the ever-changing status of jobs in the shop. As a consequence, production runs are interrupted by jobs whose priorities have suddenly increased, machine set-ups are increased, and jobs that are on schedule fall behind.

Long manufacturing lead times

In an attempt to compensate for problems 1 and 2, production planners allow extra time to produce the order. The shop becomes overloaded, order priorities become confused, and the result is excessively long manufacturing lead times.

Inefficient inventory control

At the same time that total inventories are too high for raw materials, work-in-progress and finished products, there are stock outs that occur on individual raw material items needed for production. High total inventories mean high carrying costs, while raw material stock outs mean delays in meeting production schedules.

Low work centre utilization

This problem results in part from poor scheduling (excessive product changeovers and job interruptions) and from other factors over which plant management has limited control (equipment breakdowns, strikes, reduced demand for products, etc.).

Process planning not followed

This is the situation in which the regular planned routing is superseded by an ad hoc process sequence. It occurs, for instance, because of bottlenecks at work centres in the planned sequence. The consequences are longer set-ups, improper tooling and loss efficient processes.

Errors in engineering and manufacturing records

Bills of materials are not current, route sheets are not up-to-date with respect to the latest engineering changes, inventory records are inaccurate and production piece counts are incorrect.

Quality problems

Quality defects are encountered in manufactured components and assembled products, thus causing delays in the shipping schedule. These problems give rise to the need for better systems to plan and control production operations. Today, industry has adopted the use of computer systems to perform much of the manual planning and clerical work that was previously done by human beings. In the ideal arrangement, the computerized systems are tied together within the company.

19.1 PRODUCTION PLANNING AND CONTROL

Production planning and control (PPC) is concerned with the logistics problems that are encountered in manufacturing, that is, managing the details of what and how many products to produce and when, and obtaining the raw materials, parts and resources to produce those products. PPC solves these logistics problems by managing information. The computer is essential for processing the tremendous amounts of data involved to define the products and the manufacturing resources to produce them and to reconcile these technical details with the desired production schedule. In a very real sense, PPC is the integrator in computer integrated manufacturing.

Planning and control in PPC must themselves be integrated functions. It is insufficient to plan production if there is no control of the factory resources to achieve the plan. And it is ineffective to control production if there is no plan against which to compare factory progress. Both planning and control must be accomplished, and they must be coordinated with each other and with other functions in the manufacturing firm such as process planning, concurrent engineering and advanced manufacturing planning. Now, having emphasized the integrated nature of PPC, let us nevertheless try to explain what is involved in each of the two functions, production planning and production control.

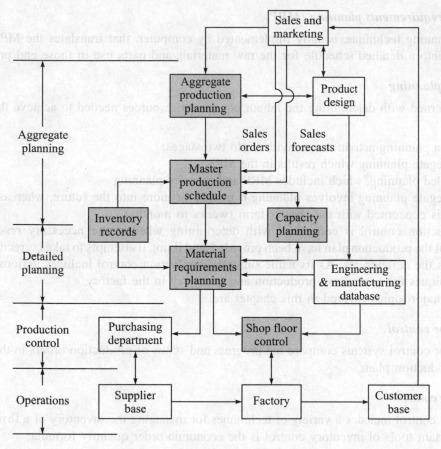

Fig. 19.1 Activities in a PPC system.

Production planning is concerned with:

(a) Deciding which products to make, how many of each, and when they should be completed
(b) Scheduling the delivery and/or production of the parts and products
(c) Planning the manpower and equipment resources needed to accomplish the production plan

Activities within the scope of production planning include:

Aggregate production planning

This involves planning the production output levels for major product lines produced by the firm. These plans must be coordinated among various functions in the firm, including product design, production, marketing and sales.

Master production planning

The aggregate production plan must be converted into a master production schedule (MPS) which is a specific plan of the quantities to be produced of individual models within each product line.

Material requirements planning (MRP)

It is a planning technique, usually implemented by computer, that translates the MPS of end products into a detailed schedule for the raw materials and parts use in those end products.

Capacity planning

It is concerned with determining the labour equipment resources needed to achieve the master schedule.

Production planning activities are divided into two stages:

Aggregate planning which results in the MPS

Detailed planning, which includes MRP and capacity planning

Aggregate planning involves planning 6 months or more into the future, whereas detailed planning is concerned with the shorter term (weeks to months).

Production control is concerned with determining whether the necessary resources to implement the production plan have been provided, and if not, it attempts to take corrective action to address the deficiencies. As its name suggests, production control includes various systems and techniques for controlling production and inventory in the factory.

The major topics covered in this chapter are:

Shop floor control

Shop floor control systems compare the progress and status of production orders in the factory to the production plans.

Inventory control

Inventory control includes a variety of techniques for managing the inventory of a firm. One of the important tools of inventory control is the economic order quantity formula.

Manufacturing resource planning

Also known as MRP II, manufacturing resource planning combines MRP and capacity planning as well as shop floor and other functions related to PPC.

Just-in-time production systems

The term *just-in-time* refers to scheduling discipline in which materials and parts are delivered to the next work cell or production line station just prior to their being used. This type of discipline tends to reduce inventory and other kinds of waste in manufacturing.

The activities in a modern PPC system and their interrelationships are depicted in Fig. 19.1. As the figure indicates, PPC ultimately extends to the company's supplier base and customer base. This expanded scope of PPC control is known as **supply chain management**.

19.1.1 Aid of Computers in PPC

The essential element in this exercise is preparing a database, the inputs for which would be the design and process planning. The data are component details, bill of material and process planning of various components; and these are entered in a master file which can be modified, appended to or deleted from periodically. The next exercise is material requirement planning, which would determine as per schedule, what, when and how much to order and if required, even modify the order priorities. Similarly, resources planning would determine requirements of tooling, machines and manpower. It is important that the production schedules match the capacity planned, i.e., manhours, machine hours, etc. Adjustments may have to be made in terms of overtime, additional shifts and subcontracting. In the longer run, it may mean acquiring more equipment and facilities or not. To effectively control the inventory and production, monitoring can be done with computers. With the database having information on various costing aspects of material, labour and equipment use, actual costs can be determined. On analysis, it can also determine why these costs are different from the ones expected. From this, the need of the common database for design, planning, manufacturing, inventory and costing should be clear.

19.2   MASTER PRODUCTION SCHEDULE

Master production schedule (MPS) represents a plan for manufacturing. It formalizes the production plan and converts it into specific material and capacity input requirements, i.e., it develops the quantities and dates to be exploded for generating per period requirements of production. The MPS is not a sales forecast of production. It also serves as a customer order backlog system. It considers changes in capacity or loads changes in finished goods inventory, fluctuations in demand. A detailed MPS also determines the economies of production by grouping various demands and making lot sizes. Thus, the MPS maintains the integrity of the total system backlogs, anticipated backlog and lower level component requirements.

Generally, the MPS consistent with the aggregate production planning from which it is derived. The aggregate production planning is an intermediate range planning in which strategies are devised to economically absorb demand fluctuations. The objective is to minimize future operating costs over particular planning horizons. Thus, as said earlier, MPS formalizes the production plan and converts it into specific material and capacity requirements. It is important to recognize that the MPS is not a control technique or a system, and also it is not a sales forecast of production but it is a feasible manufacturing plan rather. It is a logical representation of information for decision making.

19.2.1   Functions of Master Production Schedule

It has the following functions:

Translate aggregate plans
The aggregate plan sets a level of operations that roughly balances market demands with the material, labour and equipment of the firm. The MPS translates this plan into specific numbers of end items or modules to be produced in specific time periods.

Evaluate alternative schedules
MPS is a trial and error work and rework activity. Simulation capabilities enable planners to trial fit alternative master schedule. Detailed material and capacity requirements are then derived and

the planner can see exactly what lead times and delivery schedules would result with minimum total elapsed time.

Generate material requirements

MPS is used as the primary input to get required MRP system. Whenever the end items appear on the MPS, which signals that the MPS has to purchase or produce the necessary components in time.

Generate capacity requirements

The MPS is a prerequisite of capacity planning by reflecting an economical usage of labour and equipment capacities.

Facilitate information processing

The MPS determines when deliveries to be made for both make to stock and make to order by controlling the load on the plant. The MPS also coordinates the information sharing such as marketing capabilities, financial resources and personnel policies.

Maintain valid priorities

It is necessary to maintain the priorities of jobs whether it may be relative or absolute priorities. In either case, it should reflect true needs. Whenever the customers change the priorities, in such case, the priorities will be updated.

Effectively utilize capacity

To utilize capacity most effectively, the MPS may call for delaying some orders or building others ahead of demand.

19.2.2 Basic Inputs to MPS

Customer order
Product sales
Interplant requirements
Service parts requirements
Distribution warehouse requirements

The sales forecast provides the basis for extending the master schedule to generate the uncommitted or planned portion of inventories in anticipation of customer demand. This forecast leads directly into the order entry system. These inventories could fill seasonal peaks, promotional periods or new-product introductions, which create capacity overloads that could affect customer service. When the orders or bookings are temporarily below production capacity, inventories help prevent underutilization and delining productivity.

19.2.3 The Basic MPS Maintenance Step

(a) Load and level the capacity, using the most accurate MPS available, out through the combined lead-time. Use backlogs as necessary to satisfy the forecast.

(b) When new orders cannot be shipped from inventory, supply them from the first available planned lots.

(c) If planned quantity is not available for a new order, schedule the order at the end of the combined lead-time (freeze period) if capacity is available. If this is not feasible, extend the due date to the nearest future period that has capacity or reschedule a lower priority order in the earlier period to provide capacity.

(d) If items were not met from inventory and quantities were not planned, determine whether capacity is available earlier than the combined lead-time. If it is available, special handling will be required to coordinate long-lead items to meet an assembly date that is shorter than the freeze period.

(e) Finally, fill up any unused capacity with standard items to obtain a full and level load for all future periods.

The following are some of the rules which must be followed for MPS revisions:

Make changes as soon as the need is recognized

All major revisions should be done as soon as possible. This facilitates recognition of problems and issues to be dealt with such as long lead times, capacity adjustments and inventory adjustments in advance.

Newer reschedule a component

The component due date provided by the MPS is a firm requirement in manufacturing. If components are not available on time for some reason, such as equipment breakdown or strike, do not reschedule only the affected components. As a first priority, find other ways to get the component in time. Alternative scheduling may sometimes be feasible if safety time is built into the system or if low-priority items that use the same components can be scheduled. If nothing is feasible, always reschedule the highest level item, using the where-used list provided by pegging. Otherwise, you must be left with unnecessary in-process inventory in the system.

Maintain integrity

Never lie to your MPS. Do not close your eyes and hope that problems will go away! If you overstate your MPS without additional capacity, you will end up with many past due orders. If you understate, the efficiency of the system may be impaired. A good understanding between the master scheduler and the management is very important. If you (the master scheduler) budge for everyone in the system, you might be in the wrong job.

19.2.4 MPS Guidelines

(a) Work from an aggregate production plan
(b) Schedule common modules when possible
(c) Load facilities realistically
(d) Release orders on a timely basis
(e) Monitor inventory levels closely
(f) Reschedule as required

19.2.5 Determining Master Schedule Requirements

The general procedure for master production scheduling is to find all the gross requirements for the master-schedule item. This includes not only customer orders and forecasts of new demand but also any service parts or interplant demand. The next step is to "net out" the requirements (that is, subtract on-hand inventory). The net-requirements are then grouped (lot-sized) into planned orders to be released in a time-phased schedule. By working with planned orders, the schedulers can then check the planned workload against the time available at key work centres to ensure that capacity is adequate. If either the materials or capacity is insufficient then the master schedule must be revised and the scheduling process repeated.

19.3 MATERIAL REQUIREMENT PLANNING (MRP)

Material requirement planning is a computerized method for managing inventories and issuing orders for parts and materials. Figure 19.2 shows the structure of a material requirement planning system. It converts the master schedule for end products into a detailed schedule for the raw materials and components used in the end products. The detailed schedule identifies the quantities of each raw material and component item. It also tells when each item must be ordered and delivered so as to meet the master schedule for the final products. It is an effective tool for minimizing unnecessary inventory investment and is also useful in product on scheduling and purchasing of materials.

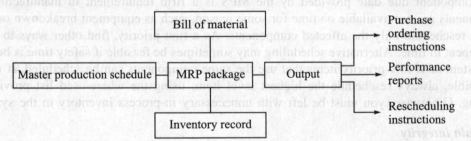

Fig. 19.2 Material requirement planning system.

19.3.1 Basic MRP Concepts

The basic MRP concepts are:

    (a)   Independent versus dependent demand

    (b)   Lumpy demand

    (c)   Lead time

    (d)   Common-use items

Independent demand

This means that demand for a product is unrelated to demand for other items, e.g., end products and spare parts. Independent demand patterns must usually be forecasted.

Dependent demand

In this, demand for the item is related directly to the demand for some other product, e.g., raw materials, component parts and subassemblies. MRP is the appropriate technique for determining quantities of dependent demand items.

Lumpy demand

It occurs in large increments. MRP is the appropriate approach for dealing with inventory situations characterized by lumpy demand.

Lead time

It is the time that must be allowed to complete the job from start to finish. In manufacturing, there are two kinds of lead times: ordering lead times and manufacturing lead times. An ordering lead time for an item is the time required from initiation of the purchase. Requisition to receipt of the item from the vendor. Manufacturing lead time is the time needed to process the part through the sequence of machines specified on the route sheet. It includes the operation time plus the non-productive time. In MRP, lead times are used to determine starting dates for assembling final products and subassemblies, for producing component parts, and for ordering raw materials.

Common-use items

In manufacturing, the basic raw materials are often used to produce more than one component type. Also, a given component may be used on more than one final product. MRP collects these common use items from different products to effect economics in ordering the raw materials and manufacturing the components.

19.3.2   Structure of a MRP System

MRP requires the following data:

- (a)   Master production plan
- (b)   On-hand inventories
- (c)   Product structure (bill of materials)
- (d)   Purchased or manufacturing order status by items
- (e)   Replenishment rules by item:
    - (i)   Lead time
    - (ii)   Order quality
    - (iii)   Scrap allowance
    - (iv)   Safety stock

Figure 19.3 represents a diagram showing the flow of data into the MRP processor and its conversion into output reports. Because so much data is required, a major MRP implementation problem is data accuracy.

The master production schedule represents everything the plans to ship: finished items, spares, etc. Manufacturing and procurement lead time must be considered; otherwise the product cannot be shipped on time. The master production schedule represents the independent demand, because it is the combination of orders on the books and forecasts. However, end items are usually composed of several components; consequently, end-item demand creates a demand for these components, called **dependent demand**.

Accurate inventory data is vital, with accuracies at least 95% being desirable. However, this level of accuracy is very difficult to attain. Online data entry systems are an aid. The entire inventory should be audited over a time cycle such as 6 or 12 months. However, only part of the inventory is counted at one time. Some companies use the Pareto approach (i.e., ABC analysis).

The bill of materials (BOM) designates what items and how many of each are used to make up a specific finished product. It is desirable to maintain a 99% level of accuracy for the BOM. The BOM is used to derive the dependent demands.

The status of an order may change after it is issued. Consequently, the status of previous orders must be considered each time the MRP analysis is performed, since the requirements calculated should represent the latest and most accurate information available.

Many types of replenishment rules are utilized within MRP. The selection of what rules to use depends on the inventory policies of the individual firm,

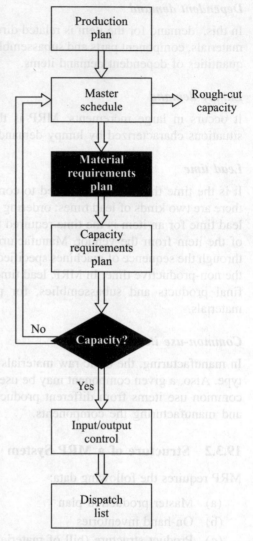

Fig. 19.3 Structure of a MRP system.

the value of the individual item, and management philosophy. Some of the replenishment rules are:

Fixed order quantity

In this case, inventory is replenished in a specified lot size. The exact quantity may be a function of equipment capacities, storage limitations or shelf life.

Economic order quantity (EOQ)

A mathematical expression for this rule is:

Formula

$$EOQ = \sqrt{\frac{2\,Cp \cdot A}{C_H}}$$

where

Cp = Order preparation cost

A = Annual demand for an item

C_H = Annual inventory carrying cost

Lot-for-lot order sizing

The lot size is the same as the requirement for the period covered. This lot-sizing method minimizes inventory carrying costs.

Fixed-period order quantity

The user specifies how many periods a planned order should cover. The sum of the requirements during this period becomes the order size.

19.3.3 Working of MRP

MRP logic tapes the master production schedule and breaks it down into dependent demands using the bill of materials. Some components may be used in different products or subassemblies; consequently, these demands must be consolidated. These consolidated demands are called **gross requirements**.

In the next step, inventory on hand is subtracted from gross requirements. At this time, any allowances for scrap and safety stock can be applied. Next, any purchased or manufacturing orders outstanding are subtracted, giving the **net requirements**.

Using the order quantity rules, the order quantities are computed and then offset from the requirements date to allow for procurement and manufacturing lead times.

Variations from previous plans must be considered before the current MRP cycle is completed. Because of these changes, orders may have to be expedited, delayed, increased, and/or decreased.

19.3.4 Computers in MRP

Fundamentally, MRP uses the computer in the following way:

(a) It compares a previously prepared master schedule with the bills of materials and determines which parts and materials are needed in what quantity to make the products planned.

(b) It then checks the existing inventory to see whether the necessary parts and materials are on hand or will be available when needed.

(c) If the parts and materials are not on hand or on order, it establishes an appropriate due date and then initiates or generates the necessary purchase or shop orders.

(d) If there is an open order for the parts or material, it re-evaluates the due date and, if necessary, calls for an adjustment.

In a nutshell, MRP helps to plan priorities, and that is what any reasonably good manual production-and inventory-control system should do.

19.3.5 MRP Systems

There are two basic styles of MRP systems, termed as the **regenerative approach** and the **net change approach**, respectively. These involve alternative approaches to the system-driven recalculation of an existing material plan based on changes in the input to that plan.

Regenerative MRP starts with the master production schedule and totally re-explodes it down through all the bills of materials to generate valid priorities. Net requirements and planned orders are completely regenerated at that time. The regenerative approach thus involves a complete re-analysis of each and every item identified in the master schedule, the explosion of all relevant BOM's, and the calculation of gross and net requirements for planned items. The entire process is carried out in a batch processing mode on the computer, and for all but the simplest of master schedules involves extensive data processing. Because of this, regenerative systems are typically operated in a weekly and occasionally monthly replanning cycle.

In the net change MRP approach, the materials requirements plan is continuously stored in the computer. Whenever there is an unplanned event such as a new order in the master schedule, an order being completed late or early, scrap or loss of inventory or indeed an engineering change to one of the BOM's, a partial explosion is initiated only for those parts affected by the change. If an event is planned, for instance, when an order is completed on time then the original material plan should still be valid. The system is updated to reflect the new status, but replanning is not initiated. Net change MRP can operate in two ways. One mode is to have an online net change system, by which the system reacts instantaneously to unplanned changes as they occur. In most cases, however, change transactions are batched (typically by day) and replanning happens over night.

19.3.6 MRP Database

A typical MRP database contains several related major sources of information, including:

    (a) The master parts information
    (b) Full inventory information
    (c) Bill of materials information
    (d) The manufacturing process or routing information for all manufactured and assembled items
    (e) Work centre information
    (f) Tooling information

19.3.7 MRP Output Reports

The MRP program generates a variety of outputs that can be used in the planning and management of plant operations. These outputs include:

Primary outputs

    (a) Order release notice, to place orders that have been planned by the MRP system
    (b) Reports obeying planned orders to be released in future periods
    (c) Rescheduling notices, indicating changes in due dates for open orders

(d) Cancellation notices, in directing cancellation of open orders because of changes in the master schedule

(e) Reports on inventory status

Secondary outputs

(a) Performance reports of various types, indicating costs, item usage, actual versus planned lead times, and other measures of performance

(b) Exception reports, showing deviations from schedule, orders that are overdue, scrap, and so on

(c) Inventory forecasts, indicating projected inventory levels in future periods

19.3.8 Benefits of MRP

The benefits of MRP system are:

(a) Improved level of customer service

(b) Better production scheduling

(c) Reduced inventory levels

(d) Reduced components shortages

(e) Reduced production lead times

(f) Reduced manufacturing costs

(g) Higher product quality

(h) Less scrap and rework

(i) Higher morale in production

(j) Improved communication

(k) Improved plant efficiency

(l) Improved competitive position

(m) Improved coordination with marketing and finance

The statistics touted for a well-managed MRP system are impressive: productivity up 5–30%, WIP inventories down 30–50%, late orders slashed 90%, labour requirements reduced 10%.

19.3.9 Classes of MRP Users

There are four classes of MRP users. The characteristics of the four classes are listed in Table 19.1. The lowest level is class D, in which the potential utility of material requirements planning is hardly realized at all. MRP is used basically as a data processing system with many of the traditional production control procedures still being used. At the top of the list is the class A MRP user. This is a company that uses material requirements planning together with capacity planning, shop floor control, and other components of a computer integrated production management system (CIPMS). The next step beyond the class A user is MRP II.

TABLE 19.1 The four classes of MRP users

Class of user	How company operates with MRP
A	Uses MRP as a model to run business (closed-loop system)
	Uses MRP as a game plan for sales, finance, manufacturing, purchasing and engineering
	Uses MRP schedules for foremen and purchasing people
	Needs no shortage list to override schedule
B	Uses MRP as a production-control system only
	Needs help from shortage list
	Has capacity planning and shop-floor-control system in place
	Has more inventory than necessary
C	Uses MRP as an inventory-ordering system only
	Needs shortage list for scheduling
	Uses overloaded master schedule
	Has more inventory than necessary
D	Uses MRP in data processing only
	Relies entirely on shortage lists and expediting
	Has mismanaged master schedule, if any
	Has poor inventory records

19.4 CAPACITY REQUIREMENT PLANNING (CRP)

The capacity requirement planning takes material requirements from the MRP system and converts them into standard hours of load and machine in various work centres. Thus, the CRP is concerned with determining what labour and equipment capacity is required to meet the current MRP as well as long term future production needs of the firm. It also serves to identify the limitations of the production resources so that an unrealistic master schedule is not planned. Figure 19.4 describes the relationship with PAC activities and MRP.

19.4.1 CRP Input

As depicted in Fig. 19.4, the major inputs for the CRP process are:

    (a) Planned orders and released from the MRP system

    (b) Loading information from the work centre status file

    (c) Routing information from the shop routing file

    (d) Changes, which modify capacity, give alternative routings or alter planned orders

    The released and planned orders from the MRP system are converted into standard hours of load by the CRP system. MRP systems assume that capacity is available when needed unless otherwise indicated. If the total load is greater than available capacity then either the capacity must be changed or the MPS revised. Thus, CRP is an iterative process that first simulates loads on work centres and relies upon planners to suggest changes if the plan cannot be met.

    When planning capacity, it is not necessary to plan every work centre in detail. Work centres that have known excess capacity will not be bottlenecks and so will not necessitate changes to the master schedule. (However, it may be desirable to plan these work centres for purposes of

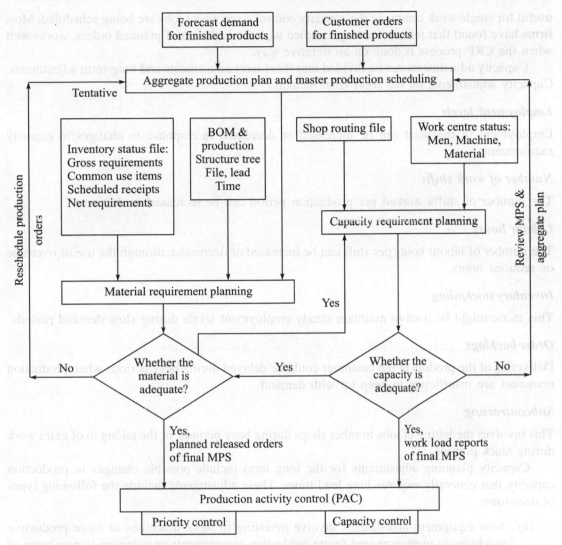

Fig. 19.4 Material requirement planning and capacity requirement planning.

control. The process of loading work centres with all the loads when they are required without regard to the actual capacity of the work centres is called **infinite loading**. Infinite loading gives planners a good grasp of the actual released order demand upon the system so that they can make decisions about using over time, using alternative routings, delaying selected orders, etc. CRP really goes beyond infinite loading, however, it includes planned orders as well as released orders and involves an iterative plan-re-plan process. There planning is continued until a realistic load is developed.

Finite loading can be done automatically by computerized loading systems; they simply limit the amount of load assigned to a work centre in each period. However, finite loading does not usually work well at the CRP stage because it forces "mechanical" changes back onto the master schedule that are not the best solutions to the scheduling problem. Finite loading tends to be more useful for single work centres in the capacity control stage where jobs are being scheduled. Most firms have found that infinite loading, coupled with the inclusion of planned orders, works well when the CRP process is done on an iterative way.

Capacity adjustments can be divided into short-term adjustments and long-term adjustments. Capacity adjustments for the short term include:

Employment levels

Employment in the plant can be increased or decreased in response to changes in capacity requirements.

Number of work shifts

The number of shifts worked per production period can be increased or decreased.

Labour hours

The number of labour hours per shift can be increased or decreased, through the use of overtime or reduced hours.

Inventory stockpiling

This tactic might be used to maintain steady employment levels during slow demand periods.

Order backlogs

Deliveries of the product to the customer could be delayed during busy periods when production resources are insufficient to keep up with demand.

Subcontracting

This involves the letting of jobs to other shops during busy periods, or the taking in of extra work during slack periods.

Capacity planning adjustments for the long term include possible changes in production capacity that generally requires long lead times. These adjustments include the following types of decisions:

(a) New equipment investments involve investing in more machines or more productive machines to meet increased future production requirements or investing in new types of machines to match future changes in product design

(b) New plant construction or purchase of existing plants from other companies

(c) Plant closings involve the closing of plants that will not be needed in the future

19.4.2 CRP Output

Aside from the rescheduling messages, which call for capacity modification or revision of the MPS, the major outputs of the CRP system are the verification of planned orders for the MRP system and the load reports. Good load reports have three characteristics:

(a) They are complete, that is, they have both planned and released orders.

(b) They are based upon valid priorities (which are up-to-date).

(c) They facilitate planning for the future.

19.5 INVENTORY CONTROL

On the one hand, it is desired that the activities in manufacturing should not be handicapped for want of material and at the same time over-stocking should be avoided to minimize on investment. An effective inventory control is one which achieves the best balance of these two contrary aspects. Right from the raw material and components to be purchased, the inventory needs to be considered for work in process, final products, cutting tools, jigs and fixtures and spare parts for equipment and services in the plant. One is well conversant with the 'economic order quantity' formula and based on the demand pattern, inventory holding and order costs, the quantity to order and the time entailed can be determined. Software is available to work these details out. Files are regularly updated and when the quantity in hand reaches the minimum level permissible, purchase orders for economically justified quantity are prepared for further action. The success of the computerized production management depends on how accurately, completely and timely, the transactions are updated.

The data for such exercise would basically consists of the component code number, its cost, order lot and lead time. Besides this, an estimate of overall requirement for a period of time, the calendar for receiving and issuing of items with their quantities and update position need to be listed. The file can also be used for storing revisions, purchasing information and remarks. The system works automatically but if any decision needs to be taken, necessary information and rules are built into the software to enable suitable action. In fact, the whole process is an exercise in detailing the quantity of each item, its ordering and delivery schedule to meet the requirement of the overall schedule over a time span. Its role can be well appreciated in purchasing materials and in scheduling production. Aggregation of some items and raw material common to many products may need to be done. Some of the items may be needed when some other items concerned with it is also available, i.e., subassemblies. Though production may be continuous, items could be obtained in batches to avail oneself of discounts offered by suppliers on bulk purchases. Taking into account the lead times, material requirement planning can help in scheduling, ordering, manufacturing and assembling activities.

19.6 COMPUTERIZED INVENTORY MANAGEMENT SYSTEM

Inventory management is concerned with getting optimum balance between two competing objectives. The objectives are to minimize investment in inventory and to maximize the service levels to the firm's customers and its operating departments. There are four types of inventory with which a manufacturing firm concerns are raw materials and purchased components in process inventory, finished product, maintenance, repair and tooling inventories.

To manage these four types of inventories, order point systems, material requirement planning are used.

Order point systems are concerned with determining when to order and how much to order. The problem of determining how much to order is based on economic order quantity. Computer software is written to perform this economic order quantity calculation for all the items stocked in inventory. The program keeps track of usage or demand rates for items in stock and computes annual demand for the item. Up-to-date files are maintained by the system on current stock levels so that when inventory falls below the reorder point, a new order for the optimum order quantity can be issued.

The inventory management module of computer integrated production management system is organized to obtain inventory accounting and inventory planning and control.

Inventory accounting is related with inventory transactions and inventory records. Inventory transactions include receipts, disbursements or issues, returns and loans. Inventory transactions allows for adjustments to be made in the records as a result of physical inventory count. The purpose of entering inventory transactions is to maintain accurate inventory records. Item master file is used to describe computerized inventory record file. The file contains item master data segment, inventory status segment, subsidiary data segment. Item master data segment gives item's identification and other data like time, cost and order quantity. Inventory status segment provides time phased record of inventory status. The subsidiary data segment contains information related to purchase orders, scrap or rejects, engineering change actions and so on. In addition to master record file, MRP module also helps in planning and control of inventories. Other functions of MRP are: determining economic lot sizes, determining safety stock levels, determining ordering prices and reorder points, ABC inventory analysis, automatic generation of requisitions for purchasing.

19.7 MANUFACTURING RESOURCE PLANNING (MRP II)

The initial versions of MRP in the early 1970s were limited to the planning of purchase orders and factory work orders and did not take into account such issues as capacity planning or feedback data from the factory for shop floor control. MRP was strictly a material and parts planning tool whose calculations were based on the MPS. It became evident that MRP should be tied to other software packages to create a more integrated PPC system. The PPC software packages that evolved from MRP became known as **manufacturing resource planning** or **MRP II**, to distinguish it from the original abbreviation and perhaps to suggest that it was second generation; that is, more than "just" MRP.

Manufacturing resource planning can be defined as a computer-based system for planning, scheduling, and controlling the materials, resources, and supporting activities needed to meet the MPS. MRP II is a closed-loop system that integrates and coordinates all of the major functions of the business to produce the right products at the right time. The term *closed-loop system* means that MRP II incorporates feedback of data on various aspects of operating performance so that the corrective action can be taken in a timely manner; that is, MRP II includes a shop floor control system. Figure 19.5 depicts a flowchart for MRP II system.

Application modules typically provided in a high-end MRP II system include the following:

Management planning

Functions included in this module are business strategy, aggregate production planning, master production scheduling, rough-cut capacity planning and budget planning.

Customer service

Typical components in the module are sales forecasting, order entry, sales analysis and finished goods inventory.

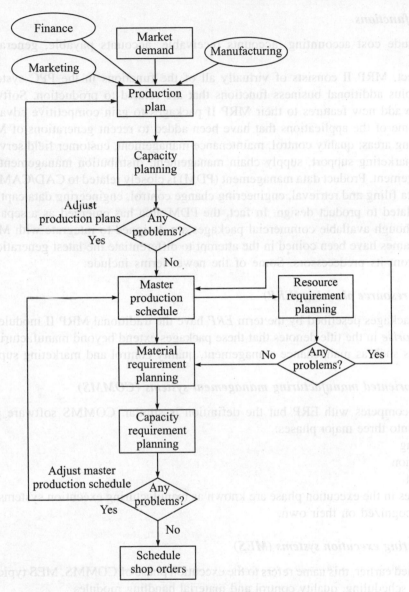

Fig. 19.5 Manufacturing resource planning (MRP II) system.

Operations planning

This is the MRP module, enhanced with capacity requirements planning. The output consists of purchase order and work order releases.

Operations execution

This includes purchasing, production scheduling and control, WIP inventory control, shop floor control and labour hour tracking.

Financial functions

These include cost accounting, accounts receivable, accounts payable, general ledger and payroll.

In effect, MRP II consists of virtually all of the functions in the PPC system shown in Fig. 19.1 plus additional business functions that are related to production. Software vendors continue to add new features to their MRP II packages to gain competitive advantages in the market. Some of the applications that have been added to recent generations of MRP II are in the following areas: quality control, maintenance management, customer field service, warranty tracking, marketing support, supply chain management, distribution management and product data management. Product data management (PDM) is closely related to CAD/CAM and includes product data filing and retrieval, engineering change control, engineering data capture and other features related to product design. In fact, the PDM area has emerged as a separate software market, although available commercial packages are designed to integrate with MPR II.

New names have been coined in the attempt to differentiate the latest generation of MRP II software from its predecessors. Some of the newer terms include:

Enterprise resource planning (ERP)

Software packages described by the term *ERP* have the traditional MRP II modules. Use of the word *enterprise* in the title denotes that these packages extend beyond manufacturing to include applications such as maintenance management, quality control and marketing support.

Customer-oriented manufacturing management systems (COMMS)

This term competes with ERP but the definition is similar. COMMS software packages are organized into three major phases:
Planning
Execution
Control
Modules in the execution phase are known as manufacturing execution systems, which have become recognized on their own.

Manufacturing execution systems (MES)

As mentioned earlier, this name refers to the execution phase of COMMS. MES typically includes production scheduling, quality control and material handling modules.

Customer-oriented management systems (COMS)

This term was coined by one of the originators of COMMS who started up his own commercial venture to market software and services for a more general clientele than only manufacturing. Hence, the word *manufacturing* was dropped from the title. What remained was customer-oriented management systems. Application modules in COMS are again similar to those in ERP and COMMS.

19.8 JUST IN TIME

As the developments in MRP and MRP II are taking place, Japanese have been quietly developing a number of innovations in manufacturing. The just in time or JIT is one such methodology that has revolutionized the manufacturing scene world over.

The APICS dictionary defines JIT as "JIT is a philosophy of manufacturing based on planned elimination of all wastes and continuous improvement of productivity".

The seven wastes that have been identified (Hall 1987) by the Japanese manufacturers, notably Toyota that lead to the continuous improvement in production processes are as follows:

Waste of overproduction

Eliminate by reducing set up times, synchronizing quantities and timing between processes, compacting layout, visibility and so forth. Make only what is needed now.

Waste of waiting

Eliminate through synchronizing workflow as much as possible and balance uneven loads by flexible workers and equipment.

Waste of transportation

Establish layouts and locations to make transport and handling unnecessary if possible. Then rationalize transport and material handling that cannot be eliminated.

Waste of processing itself

First question why this part or product should be made at all, then why each process is necessary. Extend thinking beyond economy of scale or speed.

Waste of stocks

Reduce by shortening set-up times and reducing lead times, by synchronizing work flows and improving work skills even by smoothing fluctuations in demand for the product. Reducing all the other wastes reduces the waste of stocks.

Waste of motion

Study motion for economy and consistency. Economy improves productivity and consistency improves quality. First improve the motions, then mechanism or automate. Otherwise there is danger of automating waste.

Waste of making defective products

Develop the production process to prevent defects from being made, so as to eliminate inspection. At each process, accept no defects and make no defects. Make processes safe to do this. From a quality process comes a quality product automatically.

Thus the goals for achieving the Just-In-Time manufacturing are the following:

(a) Zero defects
(b) Zero set-up time

(c) Zero inventories
(d) Zero handling
(e) Zero breakdowns
(f) Zero lead time
(g) Lot size of one

To achieve these goals, the key elements that one should follow in JIT approach are the following:

High quality

JIT systems eliminate all defects, which eliminates the scrap and rework that help to provide a smooth flow of materials through the plant. The quality is built into the process and is controlled by workers acting as their own quality inspectors.

Small lot sizes

In the conventional wisdom inventories are used as buffers for inefficiency in the production systems. In JIT, small lot sizes are practised which give the benefits of lower work in process inventory, lower lead time and uniform operating work load.

Set-up time reduction

Lower lot sizes calls for larger number of set-ups and hence it becomes necessary to reduce the set-up time.

Manufacturing methods

Use group technology with product families and flow based manufacturing that simplifies the material flow patterns. Use production flow analysis and develop flow based organization.

Product design

Design the products for ease of manufacturing and assembly. Increase product variety while reducing the process variety. Use modular design such that the same modules can be used in a number of products. This allows for large batch manufacture bringing in the economy of scale. As far as possible, use off-the-shelf components that provide high quality, at low cost and lower lead time. Design products with ease of automation such that mechanization is possible. Design customized products to a mass market at an economic cost.

The relationship with suppliers

In a JIT environment build long enduring relationships with suppliers. Treat suppliers as long term allies and give them access to information. Also try to support JIT suppliers with technical information and research.

19.9 SHOP FLOOR CONTROL

Shop floor control deals with managing the work-in-process. This consists of the release of production orders to the factory controlling the progress of the orders through the various work stations, and getting the current information of the status of the orders.

This can be shown in the form of a factory information system [Fig. 19.6]. The input to the shop floor control system is the collection of production plans. These can be in the form of master schedule, manufacturing capacity planning and MRP data. The factory production operations are the processes to be controlled.

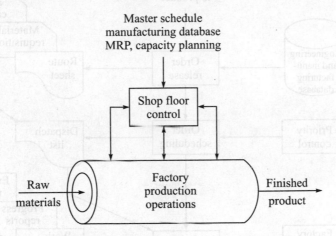

Fig. 19.6 Factory information system.

A typical shop floor control system consists of three phases. In a computer integrated manufacturing system these phases are managed by computer software. These three phases are connected with the production of shop floor control. These are executed by a combination of computers and human resources. The following sections describe the important activities connected with this task.

19.9.1 Order Release

The order release in shop floor control provides the documentation needed to process a production order. The documentation in the shop floor order may consist of the following documents:

(a) Route sheet
(b) Material requisition to draw necessary materials from the stores
(c) Job cards or other means to report direct labour time given to the order
(d) Instructions to material handling personnel to transport parts between the work centres in the factory
(e) Parts list for assembly, in the case of assembly operations

In a typical factory which works on manual processing of data, these documents move with the production order and are used to track the progress through the shop. In a CIM factory, more automated methods are used to track the progress of the production orders.

The order release is connected with two inputs. Authorization proceeds through the various planning functions (MRP, capacity planning). These provide timing and scheduling information. The engineering and manufacturing database provides the product structure and process planning

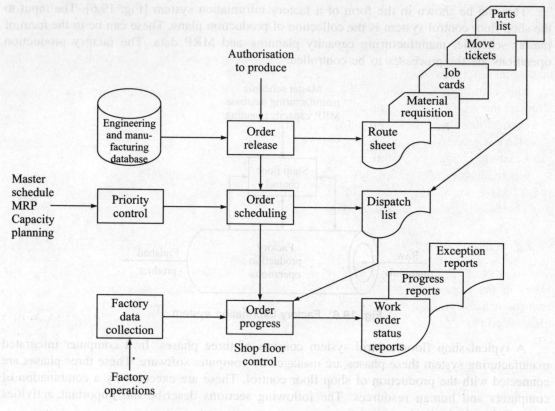

Fig. 19.7 Phases in shop floor control.

information needed to prepare the various documents that accompany the order through the shop.

19.9.2 Order Scheduling

This module assigns the production orders to various work centres, machine tools, welding stations, moulding machines, etc., in the plant. It follows directly from the order release module. Order scheduling executes the dispatch function in production planning and control. The order scheduling module prepares a dispatch list that indicates which production which production order should be accomplished at the various work centres. It provides the information on the relative priorities of the various jobs by showing the due dates for each job. By following the dispatch list in making work assignments and allocating resources to different jobs list in making schedule can be best achieved.

The order schedule module addresses to two important activities in shop floor production control.

    Machine loading

    Job sequencing

    Allocating the order to the work centres is termed as **machine loading** or **shop loading**, which refers to the loading of all machines in the plant. In most cases, each work centre will have a queue of orders waiting to be processed. This queue problem can be solved by job sequencing. This involves determining the order in which the jobs will be processed through a given work centre. To determine this sequence, priorities are given to jobs in the queue and the jobs are processed according to the priorities. Several queuing models are available in operations management to solve this problem.

This control of priorities is an important input to the order scheduling module. Rules to establish the priorities are:

Earliest due date: These are given high priority.

Shortest processing time: Shorter processing time orders are given high priority.

Least slack time: Orders with least time are given high priority.

Fluctuations in market demand, equipment breakdown, cancellation of the order by customer and defective raw material or delay in the receipt of order effect the priority. The priority control plan reviews the relative priorities of the order and adjusts the dispatch list accordingly.

19.9.3 Order Progress

The order progress module in the shop floor control system monitors the status of the various orders in the plant work-in-progress and other characteristics that indicate the progress and performance of production. The function of the order progress module is to provide the information that is useful in managing the factory based on the data collected from the factory. The order progress report includes:

Work order status reports

These indicate the status of the production orders. Typical information in the report includes the current work centre where each order is located, processing hours remaining before completion of each order, whether the job is on-time or behind schedule and priority level.

Progress report

A progress report records the performance of the shop during the period of master schedule and reports the number of operations completed and not completed during the time period.

Execution reports

These reports bring out the deviations from the production schedule (e.g. overdue jobs).

These reports are useful to production management in making the decisions about allocation of resources, authorization of the overtime hours and other capacity issues, and in identifying areas of problems in the plant that adversely affect the implementation of the master production schedule.

19.10 FUNCTIONS OF SHOP FLOOR CONTROL

Production managers are faced with the problem of acquiring up-to-date information on the progress of orders in the factory and making use of that information to control factory operations. This is the problem addressed by a shop floor control system. The functions of a shop floor control are classified by Raffish as follows:

Priority control and assignment of shop orders

Maintain information on work-in-process for MRP

Monitor shop order status information

Provide production output data for capacity control purposes

These functions are explained in the following subsections. The reader should refer Fig. 19.5 to recall the relationships and lines of communication between shop floor control and the other functions (MRP, capacity planning, etc.) in a computer-integrated production management system.

19.10.1 Priority Control and Assignment of Shop Orders

MRP and priority planning are concerned with the time-phased planning of materials, work-in-process and assembly of final product. Priority control can be considered as the execution phase of this planning process. Priority control is concerned with maintaining the appropriate priorities for work-in-process in response to changes in job order status. Suppose that the delivery date requirement for one batch of product was moved forward because demand for the product had increased such that current inventory levels were running low. Suppose also that the delivery date on another order had been pushed back due to low demand for that product. Priority control would be concerned with increasing the priority for the first job and decreasing the priority for the second job. The existence of priority control in the production management system recognizes that job priorities might change after the job order is originally issued to the shop.

The priorities for the jobs in the shop might be redetermined on a weekly or even daily basis. Once these priorities are established, the assignment of work to the work centres in the factory must be made. The assignment of shop orders is basically a problem in operation scheduling.

19.10.2 Maintain Information on Work-in-Process

Shop floor control is sometimes defined as a method of controlling the work-in-process in the factory. All of the functions and objectives of SFC boil down to this one goal of managing the parts and assemblies that are currently being processed in the shop. Information relating to quantities and completion dates for the various steps in the production sequence are compared against the plan generated in MRP. Any discrepancies due, for example, to parts scrapped in production, might require additional raw materials to be ordered and adjustments made in the priority plan for other components in that product.

19.10.3 Monitor Shop Order Status

This is similar to the work-in-process function and relies on the basic data. One of the principal reporting documents in SFC is the work order status report. As its name suggests, it provides information on the status of the orders in the shop.

This report should be updated several times per week, depending on the nature of the product and the processes in the shop. It might be sufficient to display the report on a CRT rather than in hard-copy form. However, an exception report should be printed periodically in a document form. This would indicate the orders that were behind schedule as well as other noteworthy exceptions that happened during the period (e.g., significant machine breakdowns).

The accuracy and correctness of the work order status report are independent on the correctness and timeliness of the basic data collected in the shop. These data deal with shop order transactions such as job completions, material movement, time turned in against an order and so on. The method of collecting these data is critically to the success and value of the shop floor control system. The design of this data collection (FDC—Factory Data Collection) system can take on various forms.

19.10.4   Production Output Data for Capacity Control

Capacity planning is the function of a computer-integrated production management system which determines the labour and equipment resources that will be required to accomplish the master production schedule. It is a planning function. On the other hand, capacity control is concerned with making adjustments in labour and equipment usage to meet the production schedule. To make these adjustments effectively, the capacity control function must have up-to-date information on production rates and order status from the factory data collection system.

19.11   COMPUTER PROCESS MONITORING

All the factory data collection systems described in the preceding section required some form of human participation. Computer process monitoring (also sometimes called **computer production monitoring**) is a data collection system in which the computer is connected directly to the workstation and associated equipment for the purpose of observing the operations. The monitoring function has no direct effect on the mode of operation except that the data provided by monitoring may result in improved supervision of the process. The industrial process is not regulated by commands from the computer. Any use that is made of the computer to improve process performance is indirect, with human operators acting on information from the computer to make changes in the plant operations. The flow of data between the process and the computer is in one direction only—from process to computer.

The components used to build a computer process monitoring system include transducers and sensors, analog-to-digital convertors (ADC), multiplexers, real-time clocks and other electronic devices. These components are assembled into various configurations for process monitoring. We discuss three such configurations:

Data logging systems
Data acquisition systems
Multilevel scanning

A particular computer process monitoring system is highly custom designed and may consists of a combination of these possible configurations. For example, a data acquisition system may include multilevel scanning.

19.11.1   Data Logging Systems

A data logger (DL) is a device that automatically collects and stores data for offline analysis. Strictly speaking, the data could be analyzed by a person without the aid of computer. Our interest here is in data logging systems that operate in conjunction with computers. Data loggers can be classified into three types:

1. Analog input/analog output
2. Analog input/analog and digital output
3. Analog and digital input/analog and digital output

Type 1 can be simple one-channel strip chart recording potentiometer for tracking temperature values using a thermocouple as the sensing device. Types 2 and 3 are more sophisticated instruments which have multiple input channels and make use of multiplexers and ADCs to collect process or experimental text data from several sources. The DL can be interfaced with tape punches, magnetic tape units, teletypes and printers, plotters and so on. They can also be interfaced with the computer for periodic transfer of data.

A programmable data logger (PDL) is a device that incorporates a microprocessor as part of the system. The microprocessor serves as a controller to the data logger and can be programmed by means of a keyboard. The programmable data logger can easily accommodate changes in rate or sequence of scanning the inputs. The PDL can also be programmed to perform such functions as data scaling, limit checking (making certain the input variables conform to prespecified upper and lower bounds), sounding alarms and formatting the data to be in a compatible and desirable format with the interface devices.

19.11.2 Data Acquisition Systems

The term *data acquisition system (DAS)* normally implies a system that collects data for direct communication to a central computer. It is, therefore, an online system, whereas the data logger is an offline system. However, the distinction between the DL and the DAS has become somewhat blurred as data loggers have become directly connected to computers.

Data acquisition systems gather data from the various production operations for processing by the central computer. The basic data can be analog or digital data which are collected automatically (transducers, ADC, multiplexers, etc.). It is a factory-wide system, as compared with data loggers, which are often used locally within the plant. The number of input channels in the DAS is, therefore, typically greater than in the DL system. For the data logger, the number of input channels might range between 1 and 100, while the data acquisition system might have as many as 1000 channels or more. The rate of data entry into the DL system might be 10 readings per second for multiple-channel applications. By contrast, the DAS would have to be capable of a data sampling rate of upto 1000 per second. Because of these differences, the data acquisition system would be typically more expensive than the data logger.

19.11.3 Multilevel Scanning

In this data acquisition system, it is possible for the total number of monitored variables to become quite large. Although it is technically feasible for all these variables to be monitored through multiplexing, some of the signals would not be needed under normal operating conditions. In such a situation, it is convenient to utilize a multilevel scan configuration, as illustrated schematically in Fig. 19.8. With the multilevel scanning, there would be two (or more) process scanning levels, a high-level scan and a low-level scan. When the process is running normally, only the key variables and status data would be monitored. This is the high-level scan. When abnormal operation is indicated by the incoming data, the computer switches to the low-level scan, which involves a more complete data logging and analysis to ascertain the source of the malfunction. The low-level scan would sample all the process data or perform an intensive sampling for a certain portion of the process that might be operating out of tolerance.

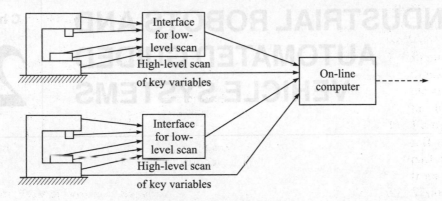

Fig. 19.8 Multilevel scanning.

REVIEW QUESTIONS

19.1 What are the typical problems encountered in production operations of a plant?

19.2 Define master production scheduling. Explain the factors that need to be taken into account while developing master production schedule.

19.3 What are the basic functions of master production schedule?

19.4 List out the inputs and output required for master production schedule.

19.5 Discuss the guidelines for master production schedule and explain how it is implemented.

19.6 What are the objectives of material requirements planning?

19.7 How material requirement planning is utilized in any organization? Explain.

19.8 Define MRP-II. Why do you call it as close loop?

19.9 What are the benefits of MRP-II?

19.10 What is the difference between MRP and MRP-II?

19.11 What is an inventory?

19.12 What is the information available in inventory file?

19.13 Define capacity requirement planning.

19.14 Differentiate between material requirement planning and capacity requirement planning.

19.15 Define JIT (Just-in-time).

19.16 Compare JIT and MRP and give suitable examples.

19.17 Does MRP lead to reduction in inventory costs? Justify your answer.

19.18 Discuss the functions of a computerized production planning and control system.

19.19 Write short notes on the following:

    (a) Computerized inventory management system

    (b) Use of computers in cost planning and control

19.20 Distinguish between manufacturing planning and manufacturing control as two broad categories of computer applications in manufacturing.

19.21 Discuss the possible computer applications in manufacturing planning activities.

INDUSTRIAL ROBOTS AND AUTOMATED GUIDED VEHICLE SYSTEMS

Chapter

20

INTRODUCTION

Webster defines a robot as an automatic apparatus or device that performs functions ordinarily ascribed to humans or operates with what appears to be almost human intelligence.

Definition of robot as per RIA (Robotics Institute of America)

A robot is a reprogrammable, multifunctional manipulator designed to move material, parts tools or specialized devices through variable programmed motions for the performance of a variety of tasks.

It is apparent that a robot must be able to operate automatically, which implies that it must have some sort of programmable memory. True to this definition of robot as an automatic machine, industrial robots are observed to perform following tasks (shown in the ascending order of technological complexity) in manufacturing:

Parts handling. This may involve tasks like

- (a) Recognizing, sorting/separating the parts
- (b) Picking and placing the parts at desired locations
- (c) Palletizing and depalletizing
- (d) Loading and unloading the parts on required machines

Parts processing. This may involve operations like routing, drilling, riveting, arc welding, grinding, flame cutting, deburring, spray painting, coating, sandblasting, dip coating, gluing, polishing, heat treatment.

Product building. This may involve assembly of typical products like

- (a) Electrical motors
- (b) Car bodies
- (c) Solenoids
- (d) Circuit boards and operations like bolting, riveting, spot welding, seam welding, inserting, nailing, fitting, adhesive bonding, inspection.

The automation of the above tasks greatly facilitates computer controlled manufacturing systems. Industrial robot has become an essential component of all flexible manufacturing systems, subsystems, cells and modules.

Definition

A robot is software-controllable mechanical device that uses sensors to guide one or more end-effectors through programmed motions in a workspace in order to manipulate physical objects.

20.1   STRUCTURE AND OPERATION OF ROBOTS

Although the mechanical, electrical and computational structure of robots can vary considerably, most of them have the following four major components in common:

Controller
Manipulator
Tooling
Sensory devices

20.1.1   Controller

Robot controllers generally perform three functions: They initiate and terminate the motion of the individual components of the manipulator in a desired sequence and at specified points. They store position and sequence data in their memory. They permit the robot to be interfaced to the "outside" world via sensors mounted in the area where work is being performed at the workstation.

To carry out these tasks, controllers must perform the necessary arithmetic computations for determining the correct manipulator path, speed and position. They must also send signals to the joint actuating devices (via interfaces) and utilize the information provided by the robot's sensors. Finally, they must permit communication between peripheral device and the manipulator. A typical standalone robot is shown in Fig. 20.1.

The machine language instructions (Program) to the robot to perform the desired tasks automatically are input through the keyboard of this unit. A teach pendant is usually provided for online non-textual commands. The mass storage media devices like floppy drive, magnetic cassette recorder and main frame disk drive can also be used to feed the programs to the controller. The programs can also be written in higher level languages like:

Algorithmic language (AL)
Versatile algorithmic language (VAL)
American robot BASIC (AR-BASIC)
Automatically programmed tools (APT)
Automatic parts assembly system (AUTO PASS)
Purdue algorithmic language (POL)
Rhino operating language (ROL)
Robot programming language (ROPL)
Robot programming systems (RPS)

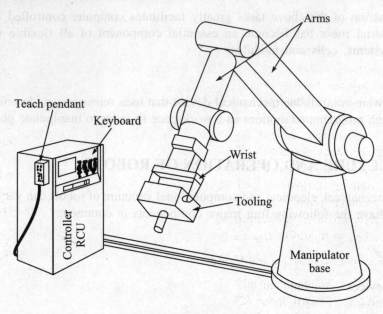

Fig. 20.1 A typical robot.

The higher level language programs are to be translated into machine language before these can be accepted by the robot control unit.

The controller converts the input program suitable signals which actuate the manipulation to perform the desired tasks. It can be equipped with process monitoring parameter, 'make decisions' on the basis of the sensed information and take suitable 'action'. The current technology relies upon computer and more specifically, on microprocessors to achieve the control function of the RCU. Some of the typical elements which go to make this unit are:

(a) Microprocessors

(b) Semiconductor memories like RAM and ROM containing software for operating the robot

(c) Input/output devices like teach pendant, keyboard terminal, floppy drive, remote safety push button, etc.

(d) A/D and D/A converters

(e) Power supplies including AC-DC converters

The RCU may not be a separate structure. It could be included within the manipulator itself.

20.1.2 Manipulator

This is a collection of mechanical linkages connected by joints to form an open-loop kinematic chain. Also included are gears, coupling devices and so on. The manipulator is capable of movement in various directions and is said to do "the work of the robot". Generally, joints of a manipulator fall into one of two classes. The first revolute, produces pure rotatory motion. The second prismatic, produces pure line or translational motion. Each of the joints axis about or along which the particular link either rotates or slides (translates). Every joint axis defines a degree of freedom (DOF) so that the total number of DOFs is equal to the number of joints. Many robots have six DOFs, three for positioning (in space) and three for orientation.

Regardless of its mechanical configuration, the manipulator defined by the joint like structure contains three main structural elements: the arm, the wrist and the hand or end effector. Besides the mechanical components, most manipulators also contain devices for producing the movement of the various mechanical members. These devices are referred to as actuators and may be pneumatic or electrical in nature. They are invariably coupled to the various mechanical links or joints (axes) of the arm either directly or indirectly. In this case, gears, belts, chains, harmonic drives or lead screws can be used.

It consists of base, arm and wrist of the robot joined in resemblance to a human arm. It also includes the power source, being either electric, hydraulic or pneumatic on receiving signals from RCU. This mechanical unit performs movement functioning desired degrees of freedom. The movement of the manipulator can be described in relation to its coordinate system which may be cartesian, cylindrical spherical or articulated. Depending on the controller, movement may be point to point motion or a continuous path motion. The actuation systems which are responsible for converting command signals into desired movements could be closed loop or open loop type. Some typical elements constituting the manipulator are: (a) Robot base, (b) Robot arm, (c) DC or AC servomotors, (d) Stepper motors, (e) Hydraulic actuators with hydraulic cylinder, (f) Pneumatic actuators with pneumatic cylinder, (g) Power supply, (h) Air supply, (i) Hydraulic supply, (j) Position feedback devices, (k) Velocity rate feedback devices and (l) Force feedback devices.

20.1.3 Tooling

It is the hand or the gripper of the robot, called the **end effector**. End effectors are divided into two classifications: Grippers (or hands) and end of arm tooling (EOAT). The gripper is used to lift parts or to transfer parts from one location to another. End of arm tooling makes changes in or operates on a part, e.g. arc weld, spotweld, deburr, grind, route, drill or drive screws.

The design of the gripper is essentially governed by the nature of the task to be performed by the robot. For instance, in part handling operations, nature, geometry, weight and orientation of the part and friction between part and gripper would influence the shape, size and material of the gripper. Some typical grippers for robots are shown in Fig. 20.2. In many applications, the task tool like spray painting gun, flame cutting torch, drilling spindle, etc. is directly fastened to the robot wrist and acts as the end effectors. Sensors, namely, approximation sensors, touch and slip sensors, compliance sensors, vision sensors and force, torque and velocity sensors may be provided as special purpose tooling as a part of end effectors. This imparts the desired versatility to the robot to perform its work similar to a human operator. For example, slight misalignments, while picking the parts, may be accommodated by providing compliance sensors. Similarly, vision sensors can greatly help the recognition and monitoring tasks.

Some typical operations performed by grippers are:

To move the fingers of the gripper with commanded speed
To grip the handled part with not more than the specified force
To open, move and close the fingers at specified positions
To measure dimensions of a part

Examples of grippers used as end effectors in industrial robot applications include the following:

Mechanical grippers, in which the part is held between mechanical fingers and the fingers are mechanically actuated.

Vacuum grippers, in which suction cups are used to hold flat objects.

Magnetized devices, for holding ferrous parts.

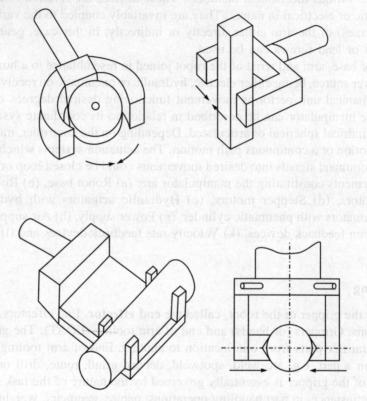

Fig. 20.2 Some typical grippers for robots.

20.1.4 Sensory Devices

Sensors used in industrial robotics can be classified into two categories although the same types of sensors might be used in both categories. The two classes are sensors that are internal to the robot and sensors that are external to the robot.

Sensors internal to the robot are those used for controlling position and velocity of the various joints. These sensors forms a feedback control loop with the robot controller. Typical sensors used to control the position of the robot arm include potentiometers and optical encoders. To control the speed of the robot arm, tachometers of various types are used. Sensors external to the robot are used to coordinate the operation of the robot with the other equipment in the cell. In many cases, these external sensors are relatively simple devices such as limit switches that determine whether a part has been positioned properly in a fixture, or to indicate that a part is ready to be picked up at a conveyor.

Tactile sensors

These sensors are used to determine whether contact is made between the sensor and another object. Tactile sensors can be divided into two types in robotics applications: touch sensors and force sensors.

Touch sensors are those that indicate simply that contact has been made with the object. Force sensors are used to indicate the magnitude of the force with the object. This might be useful in a gripper to determine the magnitude of the force being applied to grasp an object.

Turn is related to the angular displacement of the axis of the robot. One such angular displacement transducer which can be connected to a robot joint like an encoder. Linear potentiometric displacement transducers having similar principle of operation are also used.

Force and torque transducers

These sensors help the robot to perform the 'compliant' tasks. Compliance is the ability to modify an action, e.g. motion so that the task in hand is executed properly. It is with the help of feedback from force and torque transducers that the hands of robot would modify their motion so that the gripped object does not get crushed or reasonable amount of torquing effort is applied. These sensors can also be used as tactile sensors, i.e. whether a contact between the robot and another object has been made or not.

Strain gauges

Strain gauges are well known force transducers which transducer deformation due to a force (or torque) into change of resistance and therefore, current which in turn can be related to force. A closed loop system using strain gauges on the grippers of a robot is shown in Fig. 20.3. It shows a simplified version. In actual practice, several gauges may be used in a bridge configuration to amplify the signal. The mounted strain gauges feedback the force signal continuously to a signal comparator which in turn adjusts the command signal and the motor current to provide appropriate torque and therefore, force.

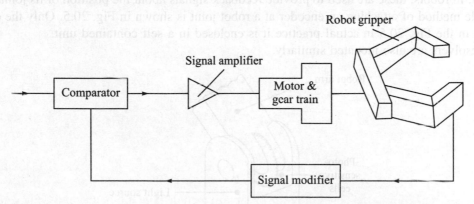

Fig. 20.3 Strain gauges as a force transducer in a robot.

Proximity sensors

These indicate when an object is close to the sensor. When this type of sensor is used to indicate the actual distance of the object, it is called a **range sensor.**

Machine vision and optical sensors

Vision and other optical sensors can be used for various purposes. A typical vision system is illustrated in Fig. 20.4. Optical sensors such as photocells and other photometric devices can be utilized for detection of the presence or absence of objects and are often used for proximity detection. Machine vision is used in robotics for inspection, parts identification, guidance and other uses.

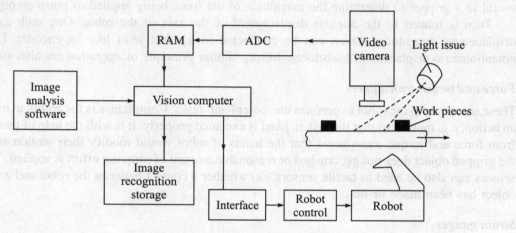

Fig. 20.4 A typical vision system.

Robot sensors

(a) position sensors

Encoders and resolvers

Encoders and resolvers can be used for transducer displacement of slides into suitable electrical signals. In robots, these are used to provide feedback signals about the position of its joints. One possible method of attaching an encoder at a robot joint is shown in Fig. 20.5. Only the disc is shown in the Fig. 20.5 in actual practice it is enclosed in a self contained unit.

Resolvers are also mounted similarly.

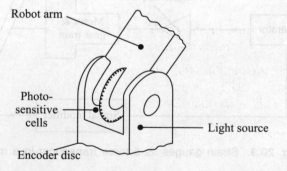

Fig. 20.5 Encoder mounted on a robot joint.

Precision potentiometers

These are relatively simple transducers in which a sliding contact wipes over a resistance element causing current changes in a circuit. The changes in the current are in turn related to angular displacement of the axis of the robot. One such angular displacement transducer which can be connected to a robot joint like an encoder is shown in Fig. 20.6. Linear potentiometric displacement transducers having similar principle of operation are also used.

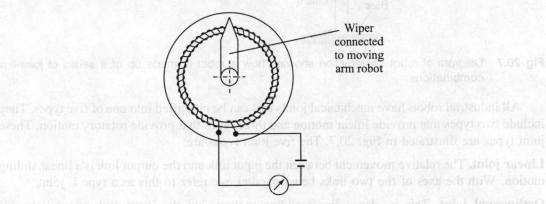

Wiper connected to moving arm robot

Fig. 20.6 Angular displacement potentiometer.

Optical encoder type torque transducers

The disc of incremental optical encoders can also be utilized to measure torque. The disc can be mechanically attached to a sensing shaft in the robot so that the shaft deflection due to torque causes a relative angular displacement between the two discs. The phase difference between the outputs of the two discs is then related to the torque.

Simple spring elements, diaphragms, proving rings, capacitive and piezoelectric type transducers can be used to sense force.

20.2   ROBOT ANATOMY

The manipulator of an industrial robot is constructed of a series of joints and links.

Joints and links

A joint of an industrial robot is similar to a joint in the human body. It provides relative motion between two parts of the body. Each joint provides the robot with a so-called degree of freedom (dof) of motion. Only 1 dof is associated with a joint. Robots are classified according to the total number of degrees of freedom they posses. Connected to each joint are two links, one which we call the **input link**, the other called the **output link**. Links are considered to be the rigid components of the robot. The purpose of the joint is to provide controlled relative movement between the input link and the output link.

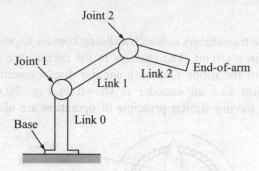

Fig. 20.7 Diagram of robot construction showing how a robot is made up of a series of joint-link combinations.

All industrial robots have mechanical joints that can be classified into one of five types. They include two types that provide linear motion and three types that provide rotatory motion. These joint types are illustrated in Fig. 20.7. The five joint types are:

Linear joint. The relative movement between the input link and the output link is a linear sliding motion. With the axes of the two links being parallel, we refer to this as a type L joint.

Orthogonal joint. This is also a linear sliding motion, but the input and output links are perpendicular to each other during the move. This is a type O joint.

Rotational joint. This type provides a rotation relative motion of the joints with the axis of rotation perpendicular to the axes of the input and output links. This is a type R joint.

Twisting joint. This joint also involves a rotatory motion but the axis of rotation is parallel to the axes of the two links. We call this a type T joint.

Revolving joint. In this joint, the axis of the input link is parallel to the axis of the output link which is perpendicular to the axis of rotation. We refer to this as a type V joint.

Each of these joint types has a range over which it can be moved. A typical range for a linear joint may be from several inches to several feet. The three types of joints which involve rotatory motion may have a range as small as a few degrees or as large as several complete turns.

Revolving joint—type 'V'

Most robots are mounted on a stationary base on the floor. We shall refer to that base and its connection to the first joint as link O. It is the input link to joint 1 the first in the series of joints used in the construction of the robot, the output link of joint 1. Link 1 is the input link to joint 2, whose output link is link 2 and so on. This joint-link numbering scheme is decipted in Fig. 20.8.

A typical robot manipulator can be divided into two sections: a body and arm assembly and a wrist assembly. There are usually 3 dof associated, with the body and arm and either 2 or 3 degrees of freedom usually associated with the wrist. At the end of the manipulator's wrist is an object that is related to the task that must be accomplished by the robot. For example, the object might be a work part that is to be loaded into a machine or a tool that is manipulated to perform some process. The body and arm of the robot is used to position the object and the robot's wrist is used to orient the object.

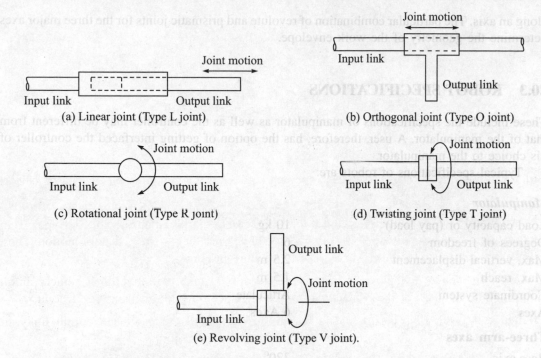

(a) Linear joint (Type L joint)

(b) Orthogonal joint (Type O joint)

(c) Rotational joint (Type R joint)

(d) Twisting joint (Type T joint)

(e) Revolving joint (Type V joint).

Fig. 20.8 Five types of joints commonly used in industrial robot construction.

To establish the position of the object the body-and-arm must be capable of moving the object in any of the following three directions:

Vertical motion (z-axis motion)

Radial (in-and-out or y-axis motion) motion

Right-to-left motion (x-axis motion)

Depending on the types of joints used to construct the body-and-arm, there are different ways to accomplish these motions. To establish the orientation of the object we can define three degrees of freedom for the robot's wrist.

Roll. This dof can be achieved by a 'T' type joint to rotate the object about the arm axis.

Pitch. This involves the up-and-down rotation of the object typically done by means of a type R joint.

Yaw. This involves right-to-left rotation of the object, also achieved typically using an R type joint.

Types of robot joints

Types	Notation	Symbol	Description
Revolute	R		Rotatory motion about an axis
Prismatic	P		Linear motion along an axis

Revolute joints R exhibit rotatory motion about an axis. They are the most common type of joint. The next most common type is a prismatic joint P, which exhibits sliding or linear motion along an axis. The particular combination of revolute and prismatic joints for the three major axes determine the geometry of the work envelope.

20.3 ROBOT SPECIFICATIONS

These include the specifications of manipulator as well as the controller may be different from that of the manipulator. A user, therefore, has the option of getting interfaced the controller of his choice to the manipulator.

Typical specifications of robots are:

Manipulator

Load capacity or (pay load)	10 kg
Degrees of freedom	6
Max. vertical displacement	2.5 m
Max. reach	1.5 m
Coordinate system	Articulate
Axes	6 Axes

Three-arm axes

Base axis	320°
Shoulder axis	180°
Elbow axis	270°

Three-wrist axes

Pitch	360°
Yaw	320°
Roll	360°

Speed

Max. pitch, yaw and roll	200/sec
Max. shoulder and base axis speed	45/sec
Max. elbow axis speed	120/sec

Drive system

All DC servo-driven axes

Positioning accuracy	+0.1 mm
Repeatability	+0.05 mm
Weight	980 kg
Power supply	440 V, 3 phase, 50 Hz

Tooling options
Vacuum cup gripper
Single action gripper
Parallel action gripper
Overgrip sensing
Special tooling can be designed and built

Controller

Programming method
Detachable teach pendant
Alphanumeric keyboard
Optional 12″ CRT for convenient editing of program cassette recorded for offline programming
Optional 3 ½″ floppy drive for offline programming
Programmable features
Point-to-point motion system
Continuous path motion system
Linear interpolation (4 axes)
Memory capacity
Type: solid state
Size: 64 k (extendable)
No. of addressable modules: 10
650 hours of battery back-up
Optional software features
Diagnostics
Max. operating temp: 50°C
Hardware features
Program selector for 8 programs
Manual speed over ride: + 10%
I/O-8, each rated at 115 V ac, 50 Hz, 3 amps
Upto five Rs 232C or Rs 422A optional, interfaces for vision system, force sensors, printer and CAD/CAM system
Optional electric limit switches for axes motions weight of RCU: 50 kg
Power supply: 220 V, single phase, 50 Hz

Load capacity

It is the highest load which can be carried by robot's arm or arms under the operating conditions. Heavy loads of the order of 1000 kg can be handled by commercially available robots. The rated capacity usually includes the weight of the gripper also and this should be duly considered for extra light load (below 1 kg) applications.

Degrees of freedom (dof)

The number of independent movements the wrist can perform in three-dimensional space relative to robot's base is called **dof of the robot**. By using three orthogonal translations and three rotations about the orthogonal axes, the state of an object, i.e., its location and orientation anywhere in the work volume of the robot can be completely defined. This suggests that a robot should have not more and not less than six dof. However, there are robots requiring fewer or more than six joints. Robots with fewer than six dof obviously have constrained movements.

Work space or work volume or working area

It is the three-dimensional space within which a robot's wrist is effective. Its size is determined by movement limits of the robot arm in linear directions and its rotations at joints. The shape of the work envelope depends on the coordinate system of movement of the robot.

Speed of movement

This is usually specified as the maximum value of speed of the end effector manifested at a specified location, under specified conditions such as no load maximum speed or rated load maximum speed, etc. For rotations, it is usually expressed as degrees/sec for the relevant joint. By taking into account the length of the fully extended arm, the maximum speed of the gripper may be computed. In most of the robots, the desired speed for a task is either programmable or manually adjustable at the RCU panel.

Accuracy and repeatability of movement

Accuracy relates to a robot's ability to move its end-effectors/gripped-tool to a commanded position within its work volume quantitatively. This is expressed as an error bringing out the difference between the target position and the actual position achieved by the robot.

Repeatability

It expresses the level of agreement between achieved positions for robot movement performed under the same method. Thus, resultant motion leads to several discordant displacements. As repeatability is a statistical term associated with several achieved positions for the same target, it can be expressed as the positional deviation from the average of these displacements. Figure 20.9 illustrates the difference between accuracy and repeatability.

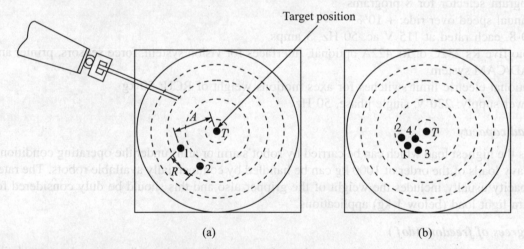

Fig. 20.9 Accuracy and repeatability.

Precision

The precision of a robotic manipulator is a measure of the spatial resolution with which the tool can be positioned within the work envelope.

Number of axes

Each robotic manipulator has a number of axes about which its links rotate or along which its links translate. The first three axes, or major axes, are used to establish the position of the wrist, while the remaining axes are used to establish the orientation of the tool or gripper. Since robotic manipulation is done in 3D space, a six axes robot is a general manipulator in the sense that it can move its tool or hand to both an arbitrary position and an arbitrary orientation within its workspace.

20.4 TYPES OF ROBOTS

The robots may be classified, on the basis of the end purpose, as follows:

Industrial robots

They have arms with grippers attached, which are finger like and can grip or pick up various objects. They are used to pick and place. These robots can be programmed and computerized sensory, welding and assembly robots usually have a self-contained micro or minicomputer.

Laboratory robots

They take many shapes and do many things. They have microcomputer brains, multijointed arms or advanced vision or tactile senses. Some of these may be mobile and others stationary.

Explorer robots

They are used to go where humans cannot go or fear to tread, e.g. outer space probes, to explore caves, dive far deeper under water and rescue people in sunken ships. They are sophisticated machines that have sensory systems and are remotely controlled.

Hobbyist robots

Most of the hobbyist robots are mobile and made to operate by rolling around on wheels propelled by small electric motors controlled by an on board microprocessor. Most hobbyist robots are equipped with speech-synthesis and speech recognition systems. They have an arm or arms and resemble a person in appearance.

Classroom robots

They are developed to assist the instructor in various aspects of the teaching-learning process.

Entertainment robots

They have the ability to speak and respond to the spoken word. They can be used to entertain the people at various events or operate as a roving advertisement.

20.5 ROBOT TECHNOLOGY LEVELS

Industrial robots have 3 levels of technology: low technology, medium technology and high technology. These different classes have unique characteristics and relate to different needs within the manufacturing areas.

Low-technology robots

They are used in the industrial environment for jobs like machine loading and unloading material handling press operation and very simple assembly operations.

Medium-technology robots

They are used primarily for picking and placing and for machine loading and unloading. They are a bit more sophisticated than low-technology robots.

High-technology robots

They can be used for material handling, press transferring, painting, sealing, spot welding, arc welding and a variety of different tasks found in the manufacturing environment.

20.6 COMMON ROBOT CONFIGURATIONS

The function of the arm is to position the pay load and that of the wrist is to orientate it. Robot arms are constructed members or links that move in relation to each other on rotational and/or linear joints. The arrangement and combination of these joints define the geometric configuration of the robot. The four classes of arm geometry within the coordinates systems are: cartesian, cylindrical, polar and revolute (or articulate or joint arm) [see Fig. 20.10]. The volume of the space created within the virtual surfaces swept by the robot arm at maximum and minimum reach is called the **robot work volume**. The shape of this volume is the work envelope.

Cartesian coordinates

A robot constructed around cartesian coordinate's configuration consists of three orthogonal slides. The three slides are parallel to the x, y and z axes of the cartesian coordinate system. By appropriate movement of these slides, the robot is capable of moving its arm to any point within its three-dimensional rectangular shaped workspace. Its work envelope is rectangular.

A robot with cartesian coordinate axes is basically used for point-to-point applications. So it is a useful system for robots that are loading and unloading material into machines. It is restricted to low-technology robotic system.

Cylindrical coordinates

The cylindrical coordinate system incorporates three degree of freedom or three axes: the rotational axis, up-and-down, z-axis and the reach or in-out axis R. Its work envelope traces out a cylinder. It is found mostly in pick and place arms, for parts feeding and assembly.

Polar (or spherical) coordinates

It has three axes: the rotational axis, the reach axis R and the bend axis. The work envelope traced out by the robot's manipulator is a sphere. This design is used mostly for machine loading, being well suited to a long straight reach into a press or moulding machine.

Revolute or jointed coordinate system

It is identified by three axes: the θ axis (wrist rotation), the β axis (shoulder rotation) and α axis (elbow rotation). This configuration is similar in appearance to the human arm. The robot arm is mounted to a base which can be rotated to provide the robot with the capacity to work with in a quasi-spherical space.

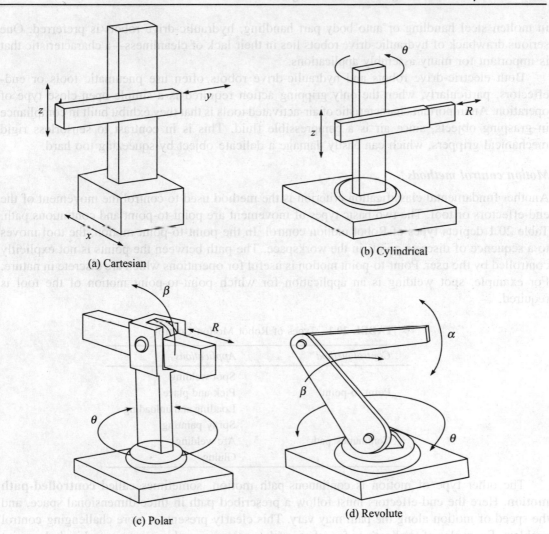

Fig. 20.10 Robot classification system.

20.7 CLASSIFICATION OF ROBOTS

In order to define the general notion of a robotic manipulator, it is helpful to classify manipulators according to various criteria such as drive technologies, motion control methods and work-envelope geometries.

Drive technologies

One of the most fundamental classification schemes is based upon the source of power used to drive the joints of the robot. The two most popular drive technologies are electric and hydraulic. Most robotic manipulators today use electric drives in the form of either DC servomotors or DC stepper motors. However, when high speed manipulation of substantial loads is required such as in molten steel handling or auto body part handling, hydraulic-drive robots is preferred. One serious drawback of hydraulic-drive robots lies in their lack of cleanliness—a characteristic that is important for many assembly applications.

Both electric-drive robots and hydraulic-drive robots often use pneumatic tools or end-effectors, particularly, when the only gripping action required is a simple open-close type of operation. An important characteristic of air-activated tools is that they exhibit built in compliance in grasping objects, since air is a compressible fluid. This is in contrast to sensorless rigid mechanical grippers, which can easily damage a delicate object by squeezing too hard.

Motion control methods

Another fundamental classification criterion is the method used to control the movement of the end-effectors or tool. The two basic types of movement are point-to-point and continuous path. Table 20.1 depicts types of Robot motion control. In the point-to-point motion, the tool moves to a sequence of discrete points in the workspace. The path between the points is not explicitly controlled by the user. Point-to-point motion is useful for operations which are discrete in nature. For example, spot welding is an application for which point-to-point motion of the tool is required.

TABEL 20.1   Types of Robot Motion Control

Control method	Applications
Point-to-point	Spot welding
	Pick and place
	Loading and unloading
Continuous path	Spray painting
	Arc welding
	Gluing

The other type of motion is continuous-path motion, sometimes called **controlled-path motion**. Here the end-effectors must follow a prescribed path in three-dimensional space, and the speed of motion along the path may vary. This clearly presents a more challenging control problem. Examples of applications for robots with continuous path motion control includes paint spraying, arc welding and application of glue or sealant.

Work-envelope geometries

The end-effectors or tool of a robotic manipulator is typically mounted on a flange or plate secured to the wrist of the robot.

The gross work envelope of a robot is defined as the locus of points in 3D space that can be reached by the wrist. The first three joints of a robot are referred to as the major axes. It is the major axes that are used to determine the position of the wrist. The axes of the remaining joints, the minor axes, are used to establish the orientation of the tool. The geometry of the work envelope is determined by the sequence of joints used for the first three axes. Six types of robot joints are possible. However, only two basic types are commonly used in industrial robots.

20.8   PROGRAMMING ROBOTS

The major advantage that robots have over fixed automation is their ability to be programmed. They can perform complex tasks under the control of stored programs which can be modified. They can move in response to real time inputs from sensors. Programming is an important feature of robots, and it is also a significant component in their design and application. The process of robot programming involves "teaching" it the tasks to be performed, storing the program, executing the program and debugging it. A number of different techniques and software approaches are used to program robots.

The kinematic analysis of a robot means the determination of position, velocity and acceleration of various links of the robot. Direct kinematics involves determination of position of the end effector for specified displacement of links. Inverse kinematics deals with the problem of the calculation of link movements to achieve the desired position of the end effector. The inverse kinematics problem is important in several robotic applications. The programming of the robot involves the coding of the data obtained from direct/inverse kinematic analysis so that end effector of the robot is located suitably for the required applications. The principal task of the robot program is to control the motions and actions of the manipulator. Due to the inaccuracies and uncertainties involved in the positions and movements of the robots, much of the programming involved deals with the detection and correction of errors.

20.9   PROGRAMMING METHODS

Manual method

This method is not really programming in the conventional sense of the word. It is more like setting up a machine rather than programming. It is the procedure used for the simple robots and involves setting mechanical stops, cams, switches or relays in the robot's control unit. For these low technology robots used for short work cycles (e.g. pick-and-place operations), the manual programming method is adequate.

Guiding

This is also known as the **walk through** or **playback method**. It involves counter balancing the manipulator arm so that it can be moved manually through the intended motions while its path is being recorded by the control system. This is the simple technique that does not require the operator to write any program code, but it is limited to relatively short and simple motions. With this method, all of the motion is under the control of the operator, who is physically back, at a different speed, if so desired, during the actual performance of the task. It is also possible to edit this program to optimize it.

Teach pendant

This is also known as the **lead through** method. It uses a control panel, called as **teach pendant**, which has buttons or switches that control the motion of the robot through a cable connected to the control system. The operator or programmer can lead the manipulator through the task one step at a time, recording each incremental move along the way. Only the major points in the path of the robot's movements are recorded, so the intermediate points must be interpolated or calculated by the control system. The control system then generates a program for the complete path of the robot. It is used for more complex and precise tasks.

Offline programming

A high-level language is used to write a control program which describes all the movements and actions of each of the actuators. The program can involve many steps, requiring a large number of lines of program code. In addition, a great deal of computation is normally associated with the translation and generation of path of motion. This method of control modifies its movements.

Online programming

This requires the availability of a robot, but it also gives the programmer the ability to see the robot actually executing the program as it is being developed. In most cases, this approach is used either as a substitute for a teach pendant to write relatively simple programs or to debug programs that have been written offline.

20.10 ROBOT PROGRAMMING FUNCTIONS

A robot programming system involves major functions to make the execution of the actual task possible.

World modelling

A significant part of specifying the robot's task is defining the positions of the objectives and features involved in the application. When the task environment is not known precisely, the position of the robot must be specified relative to the objects. Position data may be obtained from several sources such as robot sensors, geometric models or external sensing systems.

Path generation

The path of motion is normally specified by interpolating intermediate points from a sequence of motions and positions identified in the program. The position specifications must be translated from world coordinates to joint coordinates which the robot control system can understand. The type of path generated may not be the coordinates between the joints. Each joint may move independent of the others. This will normally result in a non-uniform, curved path. If straight line motion is desired, the motions of the joints must be coordinated.

Sensing

The use of sensors permits the robot to deal with uncertainties may include:
  The detection of errors
  The identification of objects or positions, using a vision system
  The initiation or termination of actions, using limit switches, proximity switches or feedback devices
  To use sensors, the robot control system must be able to connect and interpret sensory data from internal or external sources and generate control signals that will modify the actions of the

manipulator. The data acquisition, computation and output signal generation activities must be built into the control program. A sensor based robot must have a program that allows it to choose between alternative actions on the basis of data from the sensor and its model of the environment.

Programming support

Some of the functions that are necessary in a programming system for robots include:

   Editing tools for modifying programs
   Debugging tools for finding problems during program development
   Diagnostic tools for finding problems during operations
   Simulators for trying out robot programs without actually operating the robot
   Interfaces to controllers, peripheral equipment and other computers

20.11   BASIC TYPES OF ROBOT PROGRAMMING LANGUAGES

The robot programming languages can be classified according to the robot reference model, the type of control structures used the data, the type of motion spefication, the sensors, the interfaces to external machining and the peripheral used. The types of robot programming languages are following:

   Point-to-point motion languages
   Basic motion languages at the assembler level
   No structured high level programming languages
   Structured high level programming languages
   NC-type programming languages
   Object oriented languages
   Task oriented languages

The dependence of the robot programming languages on the robot control system makes necessary to standardize the interfaces, the test facilities and other software tools. A complete programming system supports a range of supporting tools, including the language, the operating system, the device controller and the debugging and application and programs. Most of the advanced software is written in high level languages. An operating system kernel of a task oriented architecture must include protocols for interprocessor communication, device drivers, interrupt handles and a runtime system. For peripheral and error deduction and recovery procedures, a hybrid control system have to be used. Robot programming must include the specification of motion.

Online and offline programming languages

Online and offline programming are used on different control levels of the robot cell. Offline programming supports the integration of robots into a CIM system. It allows the manipulation of CAD system data to reduce the programming time by eliminating the interaction with the physical devices.

An offline programming system needs a software development environment to specify the application requirements, to analyse the task and to decompose the global task to subtasks. An online programming system provides tools for debugging and testing the program. Most robot programs are written as *Robot-level programs,* i.e., they specify the task the robot is to perform in terms of the individual motions required of each of its actuators. Robot programming is done with high level languages that use a limited number of basic commands with motion and position specifications. High level interactive languages provide functions for data processing, computation and sensing as well as manipulation. The commands used in a programming language are subroutines that are provided by the system which define frequently used functions. They may be used to control the actions of the robot to perform computations or to process data. Some of the basic types of commands are:

> Motion and sensing functions (e.g., MOVE, MONITOR)
> Computation functions (e.g., ADD, SORT)
> Program-flow control functions (e.g., RETURN, BRANCH)

The most common command in most robot programs is to move the manipulator.

e.g. MOVE [<1, 3, 5><5, 10, 20>]

or in general terms

MOVE [<JOINTS><GOALS>]

The first command means that joints 1, 3 and 5 are to be moved by 5, 10 and 20 units, respectively.

There are also a number of frequently used functions that are not in the path itself (e.g., wait, continue, branch, O/P). Many robot languages have been developed. Each is unique in its design and capabilities. Some are more powerful and versatile than others. Some of the most widely used are:

VAL (Unimation)
AML (IBM)
HELP (General Electric)
RAIL (Auto Matrix)
AL (Standard University)
MCL (McDonnell Douglas Corp.,)
RPS (SRI)
JARS (Jet Propulsion Laboratory)

The VAL languages. VAL stands for Victor's assembly language. It is basically an offline language in which the program defining the motion sequence can be developed offline, but the various point location used in the work cycle are conveniently defined by lead through method.

VAL statements are divided into two categories: monitor commands and programming instructions.

The monitor commands are a set of administrative instructions that direct the operation of the robot system. The monitor commands would be used for functions such as:

Preparing the system for the user to write programs
Defining points in space
Commanding the PUMA to execute a program
Listing programs on CRT

The program instructions are a set of statements used to write robot programs. Programs in VAL direct the sequence of motions of the PUMA (robot). One statement usually corresponds to one movement of the robot's arm or wrist. Examples of program instructions include:

Move to a point
Move to a point in a straight line motion
Open gripper
Close gripper

The program instructions are entered into memory to form programs by first using the monitor command EDIT. This prepares the system to receive the program instruction statement in the proper order.

The MCL languages. The MCL stands for machine control language. The language is based on the APT NC language, but is designed to control a complete manufacturing cell, including a cell with robots. MCL is an enhancement of APT which possesses additional options and features needed to do offline programming of a robotic cell.

Additional vocabulary words were developed to provide the supplementary capabilities intended to be covered by the MCL language. These capabilities include vision, inspection and control of signals to and from the various devices that constitute the robotic work station. MCL also permits the user to define MACRO, like statements that would be convenient to use for specialized applications.

After the MCL program has been written, it is compiled to produce the CLFILE as output. The definition of the CLFILE has been extended to accommodate the new MCL features that go beyond the conventional cutter data in APT. The extensions include capabilities such as:

(a) The definition of the various devices within the work cell and the tasks which are performed by these devices
(b) Predefined frames of reference which are associated with the different machines or devices in the cell
(c) User defined frames of reference which could be used for defining the geometry of the work part
(d) The part identification and acquisition within the work cell

MCL represents a significant enhancement of APT which can be used to perform offline programming of complex robotic work cells.

20.12 ROBOT APPLICATIONS

About 90% of the robots in the world today are found in the industry. The need for industrial robots appeared primarily driven by the shortage of labour and the cost of labour. Table 20.2 shows the sale of robots in US, by application.

The robots are regarded as *steel collar workers*. Robots are extensively used in the manufacturing of automobiles, electronic goods and semiconductors. The automobile industry employs 20% of all the robots. Robots not only save costs but also allow shorter production costs. The characteristics of work which promote the application of robot are summarized as follows:

➢ Hazardous work environment.

TABLE 20.2 The sale of robots in U.S.A. by application

Applications	1985 unit sales	1990 unit sales	1995 unit sales
Machine tending	800	1650	2250
Material transfer	800	1650	2250
Spot welding	1300	1650	1500
Arc welding	500	1100	1350
Spray painting	500	1100	1050
Processing	250	770	1050
Electronic assembly	300	1320	2100
Assembly and inspection and others	550	1750	3400

➤ If a work is simple and highly repetitive, a robot will do the job more consistently and accurately.

➤ Handling of heavy parts.

➤ Manual operations requiring 2nd and 3rd shifts will be more economically performed by robots.

➤ If the dimensions and state of the workplace repeat without significant and unstructured variability, robots may be justified.

➤ A robot's capacity is mostly in situations where the part to be handled is in a known position and orientation.

The current day applications of robots can be categorized into two broad areas:

Industrial applications

Non-industrial applications

20.13 INDUSTRIAL APPLICATIONS

Robot applications in the industries today are primarily in four areas:

- Material handling
- Processing Applications
- Assembly
- Inspection

20.13.1 Material Handling

A material handling operation in which the primary objective is to move a part from one location to another without any complex constraints is the simplest operation for a robot. The application usually requires a relatively simple robot with few degrees of freedom and a simple controller.

These operations are commonly called **pick-and-place** (or pick-n-place) operations. The end-effector motion required for the work cycle is only point-to-point motion as the path traversed by end-effector is not important. In the simplest operation, the part is made available to the robot at a fixed known stationary location and orientation called **pickup** point. The robot end-effector approaches this known location (pickup point, point A), graphics the part in the end-effector (usually a gripper), moves away from this point to safe distance (point B), moves close to the position where the part is to be placed (point C) and places the part at the desired location (delivery point, point D).

In the pick-n-place operation, it is essential that the part is made available to the robot at the stationary pickup point in the specific orientation and position by some mechanical feeding or conveyor; the part placed at the delivery position (desired position) is moved away before the next part is delivered by the robot.

A typical pick-and-place operation is the loading of material or part and unloading it from a machine. The robot's task is to pick the part from a specific location (the pick point) and place it in desired position and orientation into the work holding device of the machine, which may be a chuck, vice, etc., once the part is loaded into the machines. Some manufacturing operations are performed on the part by the machine and then the part is unloaded from the machine. For unloading, the robot picks the part from the work holding device of the machine and places it at the delivery point or loads it into another machine.

In the machine loading and unloading operations, the timing of the robot and the machine must be coordinated. The robot's cycle time must match the machine's cycle time. For this coordination, the robot controller must establish communication with the machine or monitor the machine operation with the help of suitable sensors and controllers. An example of material handling is a loading and unloading in a machine cell.

A single-machine cell with a double-gripper robot

Consider a machine centre with input-output conveyors and a robot to load the parts onto the machine and unload the parts from the machine as shown in Fig. 20.11. A typical operation sequence consists of the following steps:

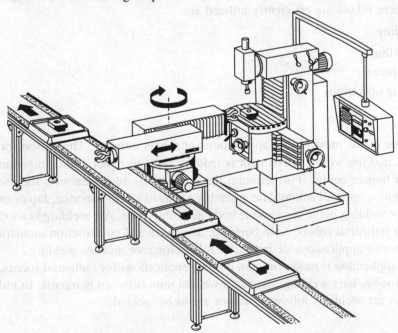

Fig. 20.11 A single-machine robotic cell.

(a) The incoming conveyor delivers the parts to a fixed position.

(b) The robot picks up a part from the conveyor and moves to the machine.

(c) The robot loads the part onto the machine.

(d) The part is processed on the machine.

(e) The robot unloads the part from the machine.

(f) The robot puts the part on the outgoing conveyor.

(g) The robot moves from the output conveyor to the input conveyor.

This operation sequence of the robotic cell is accomplished by a cell controller. Production rate is one of the important performance measures of such cells. We provide an example of determining the cycle time and production rate of a robotic cell.

Some common manufacturing operations where robots are successfully used are:

(a) Metal working operations like forging, etc.

(b) Plastic moulding

(c) Die casting

20.13.2 Processing Applications

A large number of robot applications are more than just material handling applications. A class of these applications where the end-effector is a tool instead of a gripper are classified as processing applications. With the tool the robot manipulator performs some manufacturing process function. The type of the tool depends upon the operation to be performed by the robot. Sometimes, the tool is permanently attached to the end of the tool chaining. Some of the processes where robots are efficiently utilized are:

Arc welding

Spot welding

Spray painting

Machining operations

Arc welding

Welding is one of the most visible applications of robots industry. The reason for using robots in welding is that the work environment is tedious, repetitive, hot and cramped and, therefore, unhealthy for human beings. Furthermore, the productivity increases with the use of robots in welding is high. Almost all automobile assembly plants in North America, Japan and Europe use robots for spot welding in the automobile body assembly lines. Arc welding is another important application of industrial robots. Ship building, aerospace and construction industries are among the many areas of applications of industrial robots for spot and arc welding.

Welding application is most common process applications for industrial robots. Arc welding requires continuous long welding to make a welded joint between two parts. In utilizing a robot for continuous arc welding, following issues are to be tackled:

(a) The arc welding process is usually in low quantity and changes often.

(b) The components to be welded have variations and may require the arc-welding path to change from part to part.

(c) The edges of parts to be welded are irregular.

(d) The path to be negotiated by the electrode is complex. For a robot negotiating, a straight-line motion is as difficult as negotiating a curved path.

(e) In the case of consumable electrode arc welding, the robot is required to move the electrode towards the parts being welded to maintain the arc, as the electrode is consumed.

(f) The quality of weld depends on the speed of electrode movement. If speed is low, more metal will deposit and if it is more, less metal will deposit and weld may be weak.

The arc-welding robot has to overcome all the above difficulties though it cannot possess the skill and judgment of the welder. To perform satisfactory welding operations, the manipulator should be capable of moving its tool-point (the end of electrode) along a trajectory in three-dimensional space.

It must be capable of continuous path movement, as a point-to-point robot cannot do continuous arc welding. In addition, it should have a feed mechanism for consumable electrode or filler metal wire. The work cell controller should coordinate the robot motion; electrode or wire feed, the spark gap, the welding current and other activities in the work cell.

The workspace of the welding robot should be large enough to accommodate size of the parts to be welded. A five degrees of freedom manipulator can weld parts in a plane while six degrees of freedom are required for welding complex contours. Robot programming for continuous arc welding requires a mechanism to feed the contour along which welding is to be performed as well as it may require interpolation algorithms for interpolating straight line or curved paths between two points.

Spot welding

Spot welding is a metal joining process in which two sheet metal parts fused together at localized points of contact. Two copper-based electrodes are used to squeeze the metal parts together and apply a large electrical current across the contact point to cause the fusion to occur. The electrodes, together with the mechanism that actuates them, constitute the welding gun in spot welding.

Because of its widespread use in the automobile industry for car body fabrication, spot welding represents one of the most important applications of industrial robots today. The end effecter is the spot welding gun used to pinch the car panels together and perform the resistance welding process. The welding gun used for automobile spot welding is typically very heavy. Prior to the use of robots in this application, human workers performed the operation and the heavy welding tools were difficult for the workers to manipulate accurately. As a consequence, there were many instances of missed welds, poorly located welds, and other defects, resulting in overall low quality of the finished product. The use of industrial robots in this application has dramatically improved the consistency with which the welds are made.

The types of robots for spot welding are usually large, with sufficient payload capacity to wield the heavy welding gun. Five or six axes are generally needed to achieve the positioning and orientation required. Playback with point-to-point control are used, and the programming is accomplished using the powered lead through method jointed arm and polar coordinate robots are the most common anatomies in the automobile spot welding lines. In-line robot cell layouts with 20 to 30 robots are typical.

Spray painting

Besides welding, spray painting is an important application of robots in automobile and other manufacturing industries. The robots with spray guns painting bodies and thousands of body parts are now commonly seen in automobile factories. The primary reason for the increased use of robots in painting is their high level of consistency, which is difficulty to achieve with human operators. Furthermore, spray painting poses health risks and is environmentally undesirable for human operators.

Machining operations

Machining is very important operation and practically all the industrial parts would have some amount of machining operation being carried at some stage. The machining processes that are most common are drilling, milling, turning and grinding. In addition to these processes the specialized processes such as gear cutting are also carried out. Robot is generally used to load a raw casting or forging on to the machine to be properly located and machined. After the completion of the machining operation the part will be unloaded from the machine. Since this being a major application of robotics is separately discussed

20.13.3 Assembly Applications

Assembly is the final stage of manufacturing and it is manual labour intensive. It accounts for as much as 40 to 50 percent of human labour required to get the finished product. Given the high rate of occurrence of assembly operations in manufacturing industry, automated assembly system have been traditionally applied to high-volume products requiring large investment in custom-designed equipment to perform the specific assembly. However, majority of products requiring assembly operations have low or medium production volumes and large investments in specialized assembly equipment cannot be economically justified. Robotics or other programmable and flexible automation systems are ideal for these applications.

Assembly of components is yet another but difficult area of application for robots. The difficulty arises because of the complexity of the assembly process, which involves orientation, alignment and joining operations such as welding, riveting and press fitting. Because the assembly process is mostly manual, assembly tasks sometimes take more time than is spent on producing parts. Therefore, there is a need to make concerted efforts to automate the assembly process. However, not every final product may quality for automated assembly. Boothroyd et al. (1982) provided a number of guidelines for evaluating the assembility of objects. Redford (1986) considered the following requirements necessary for improved, more efficient robotic assembly:

  (a) Faster robots
  (b) Limited-capability, cheap robots
  (c) Versatile, inexpensive grippers

(d) Identification of assembly families

(e) Improved assembly efficiency

(f) Low-cost feeders

The design for robotic assembly should evolve from the design of the product. For example, establishing assembly sequences requires identification of potential jigging and gripping surfaces, grip and assembly forces, clearances and tolerances, and other issues that must be accounted for in the component design. For machine assembly, the tolerances on grip and jig surfaces must be adequate with respect to mating surfaces. This aspect may not be critical in the case of manual assembly.

20.13.4 Inspection Applications

Applications of robots in inspection is a growing area. Quality control in industrial manufacturing operations is essential in the competitive world. A manual inspection of the finished product as well as the in-process parts is a tedious and monotonous task for a human operator. In large quantity production, 100 percent inspection is impossible, and sampling and procedures of statistical quality control are restored to. Robotics technology is expected to play a significant role in making 100 percent inspection possible.

Inspection function is required in every stage of manufacturing from raw materials to finished products. Robots can be used to inspect physical dimensions, surface finish and other characteristics of the raw materials, intermediate stages of part, finished parts, subassemblies or finished products. To perform the inspection tasks, robot requires various sensors or vision systems.

Sensor based inspection

Specific physical dimensions can be determined by the robot with the help of sensors placed on the gripper fingers of a material-handling robot. In material each part is grasped by the gripper and moved from one place to other. The specific physical size information can be obtained by the sensors fitted on the fingers of the gripper and the robot can determine whether the size is within tolerance limits or not. If the size is correct, part is placed on the desired place and if the size is incorrect, the part may be dropped into a waste bin.

Vision based inspection

It was learned there that from the image of an object several features of the object can be extracted. These features provide valuable information required in inspection. A typical robotic vision system is capable of analyzing a two-dimensional scene. The two-dimensional view of the scene is obtained from a single camera. If more features of the object are required to be inspected, the camera is manoeuvred into different positions by the robot manipulator; alternately, the robot manipulator can be used to present the part to a stationary vision system (camera), in different orientations. Thus, robot's task is a sort of material handling task. In the design of a robotic vision inspection system, several factors must be considered. These factors are:

(a) Proper lighting of the workspace.

(b) Camera with required resolution to extract dimensions and other features to the desired accuracy.

(c) The field of view should be large enough to accommodate the part.

(d) Robot should have sufficient degrees of freedom to manipulate the camera or the part, as the case may be.

20.14 NON-INDUSTRIAL APPLICATIONS

The need for industrial robots (for manufacturing automation) appears to be primarily driven by the shortage of labour and the cost of labour. While only Japan has embraced robotics in a big way, it would appear that it is only a matter of time before other industrialized nations follow suit. However, there are applications in hostile environments in which it is necessary to use robots (for example, in space, nuclear plants) or it is too dangerous to use humans (for example, military operations). There are others where the physical task demands skills that humans simply do not have (for example, surgery). Some of these are briefly described hereinafter.

Space robotics

Space exploration needs human intelligence but does not need the physical presence of human bodies. In principle, human operators on earth can control partially autonomous vehicle and manipulators on the moon or on distant planets.

Hazardous environments

DOE uses robotics technology for automating the manufacturing of explosive components and for dismantling radioactive or toxic weaponry. The US Navy is trying to use robotics technology for detecting and defusing mines in shallow water. A remotely controlled underwater submersible was used when the Titanic was salvaged several years ago.

Virtual reality

Virtual reality systems (simulators) can be used for training and educating people. An important component of these systems is the haptic interface, that allows the user/operator to feel the virtual environment and exert forces on it. Thus, a virtual reality system is a robot plus high-resolution displays.

Highways

Cars are being equipped with increasingly sophisticated sensors, navigation systems and controllers. The IVHS project is aimed at building an intelligent highway system in which operations such as merge, change-lane and exit can be automated so that the human driver acts only in a supervisory mode. Highway maintenance and construction are areas where robotic systems can be used for automation.

In addition, there are many areas in the service industry where robotics can be expected to play a major role and according to J. Englelberger et al (1986), "service robotics will surely outstrip industrial robotics". Some possible application areas in the service industry are:

Medical robotics

The US pioneered research in this area. In robot-assisted surgery, the surgeon directs the robot to make controlled, high-precision incisions with accuracy far better than a human surgeon can.

The latest advance in laparoscopic surgery involves inserting a micro-robot through a small incision in the body and teleoperate it to perform surgery, suturing, etc. Now, Japan and Europe have active research programs in this area.

Personal care for disabled people

There are many assistive devices for people with disabilities. Robots can be vocational assistants by operating as arms for paraplegics. They can be used to fetch papers or pick up the phone. In a home, they can be used to push open doors, get water from a faucet, and pick up trash from the floor. Since a human user controls the personal robot, the robot need only have very limited intelligence.

Entertainment robots

Entertainment robots is a fast growing market that is fuelled by growth in theme parks. In Disney's theme parks, robots are used to create animated figures. Ford used a robot to advertise its new 1996 models. Virtual reality systems are also ready to take off.

Custodian robots

Cleaning public restrooms is a tedious and dirty job and best left to a robot. Since restrooms are fairly structured (most toilets, urinals and sinks look similar), the cleaning operation can be used to clean carpets.

Robot attendant at gas station

A robot system can be used to fill up the gas tank without getting out of the car. This would be a great benefit if it is very hot or cold outside or if it is very late at night.

20.15   AUTOMATED GUIDED VEHICLE SYSTEMS

An automated or automatic guided vehicle system (AGV) is a materials handling system that uses independently operated, self-propelled vehicle that are guided along defined pathways in the floor.

The vehicles are powered by means of on-board batteries that allow operation for several hours (8 to 16 hours) between recharging. The definition of the pathways is generally accomplished using wires embedded in the floor or reflective paint on the floor surface.

Guidance is achieved by sensors on the vehicles that can follow the guide wires or paint strip.

20.16   TYPES OF AGVs

There are a number of different types of AGVs. Figure 20.12 illustrates three types of automated guided vehicles. The types can be classified as follows:

20.16.1   Driverless Trains

This type consists of a towing vehicle that pulls one or more trailers to form a train. It was the first type of AGVs to be introduced. It is useful in applications where heavy pay loads must be

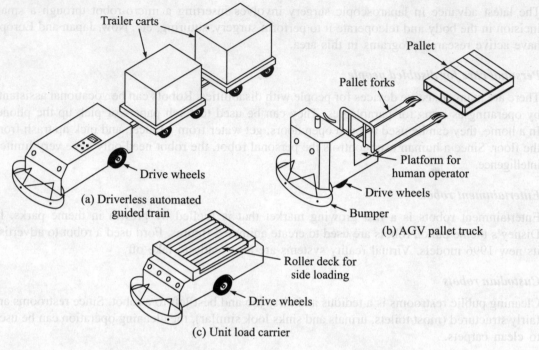

Fig. 20.12 Three types of automated guided vehicles.

moved large distances in warehouses or factories with intermediate pickup and drop-off points along the route.

20.16.2 AGVs Pallet Trucks

Automated guided pallet trucks are used to move palletized loads along predetermined routes. In the typical application, the vehicle is backed into the loaded pallet by a human worker who steers the truck and uses its forks to elevate the load slightly. Then the worker drives the pallet truck to the guidepath, programs its destination and the vehicle proceeds automatically to the destination for unloading. The capacity of an AGVs pallet truck ranges upto 6000 lb and some trucks are capable of handling two pallets rather than one. The recent modification related to the pallet truck is the forklift AGV. This vehicle can achieve significant vertical movement of its forks to reach loads on shelves.

20.16.3 AGVs Unit Load Carriers

This type of AGVs is used to move unit loads from one station to another station. They are equipped for automatic loading and unloading by means of powered rollers, moving belts, mechanized lift platforms or other devices. Variations of unit load carrier include light-load AGVs and assembly line AGVs.

The light-load AGV is a relatively small vehicle with a corresponding light load capacity (500 lb or less). It does not require the same large aisle width as the conventional AGV. Light-load guided vehicles are designed to move small loads (single parts, small baskets or totes pans) through plants of limited size engaged in light manufacturing. The assembly line AGVs is designed to carry a partially completed subassembly through a sequence of assembly workstations to build the product.

AGV's technology is far from mature, and the industry is continually working to develop new systems in response to new application requirements. An example of a new and evolving AGV's design involves the placement of a robotic manipulator on an AGV to provide a mobile robot for performing complex handling tasks at various locations in a plant.

20.17 APPLICATIONS

Automated guided vehicle systems are used in a growing number and variety of applications. The applications can be grouped into five categories:

Driverless train operations

These applications involve the movement of large quantities of materials over relatively large distances. For example, the moves are within a large warehouse or factory building or between buildings in a large storage depot. For the movement of trains consisting of 5 to 10 trailers, this becomes an efficient handling method.

Storage/distribution systems

Unit load carriers and pallet trucks are typically used in these applications. These storage and distribution operations involve the movement of materials in unit loads from or to specific locations. The applications often interface the AGVs with some other automated handling or storage system such as an automated storage/retrieval (AS/RS) in a distribution centre. The AGVs deliver incoming items or unit loads from the receiving dock to the AS/RS, which places the items in storage and the AS/RS retrieves individual pallet loads or items from storage and transfers them to vehicles for delivery to the shipping dock.

When the rates of incoming loads and the outgoing loads are in balance, this mode of operation permits loads to be carried in both directions by the AGVs vehicles, thereby increasing the handling system efficiency.

This type of storage/distribution operation can also be applied in light manufacturing and assembly operations in which work-in-progress is stored in a central storage area and distributed to individual workstations for assembly or processing. Electronics assembly is an example of these types of applications.

Assembly-line operations

AGVs are being used in a growing number of assembly-line applications, based on a trend that began in Europe. In these applications, the production rate is relatively low and there are a variety of models made on the production line. Between the workstations, components are kitted and placed on the vehicle for the assembly operations that are to be performed on the partially completed product at the next station. The workstations are generally arranged in parallel configurations to add to the flexibility of the line. Unit load cariers and light-load guided vehicles are the type of AGVs used in these assembly lines.

Flexible manufacturing systems

In this application, the guided vehicles are used as the materials handling system in the FMS. The vehicles deliver work from the staging area to the individual workstations in the system. The vehicles also move work between stations in the manufacturing system. At a workstation, the work is transferred from the vehicle platform into the work area of the station for processing. After the completion of processing at that station, a vehicle returns to pick up the work and transport it to the next area. AGV systems provide a versatile material handling system to complement the flexibility of the FMS operation.

Miscellaneous applications

Other applications of automated guided vehicle systems include non-manufacturing and non-warehousing applications such as mail delivery in office buildings and hospital material handling operations. Hospital guided vehicles transport meal trays, linen, medical and laboratory supplies and other materials between various departments in the building. These applications typically require movement of the vehicles between different floors of the hospital and hospital AGVs have the capability to summon and use elevators for this purpose.

20.18 FUNCTIONS OF AGVs

There are several functions that must be performed to operate any automated guided vehicle system successfully. These are:

Vehicle guidance and routing
Traffic control and safety
System management

20.18.1 Vehicle Guidance and Routing

The term *guidance system* refers to the method by which the AGVs pathways are defined and the vehicle control systems that follow the pathways.

There are two principal methods to define the pathways along the floor embedded guide wires and paint strips. The guide wire system is commonly used in warehouse and factory applications.

In the guide wire method, the wires are usually embedded in a small channel, i.e., about 1/8 in wide and ½ in deep. After the guide wires are installed, the channel slot is filled so as to eliminate the discontinuity in the floor surface. A frequency generator provides the guidance signal carried in the wire. The signal is of relatively low voltage, low current and has a frequency in the range 1 to 15 kHz. This signal level creates a magnetic field along the pathway that is followed by sensors on-board each vehicle. Figure 20.13 illustrates the operation of a typical sensor system.

Two sensors (coils) are mounted on the vehicle on either side of the guide wire. When the vehicle is moving along a course such that the guide wire is directly between the two coils, the intensity of the magnetic field measured by each coil will be equal. If the vehicle strays to one side or the other, or if the guide wire path curves, the magnetic field intensity at the two sensors will be different.

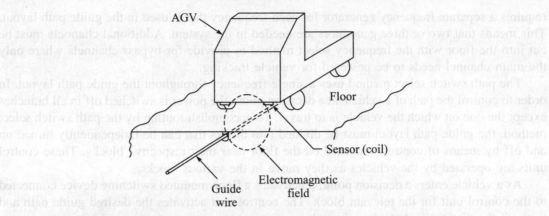

Fig. 20.13 Operation of the on-board sensor system that uses two coils to track the magnetic field in the guide wire.

This difference is used to control the steering which makes the required changes in vehicle direction to equalize the two sensor signals, thereby tracking the defined pathway.

When paint strips are used to define the vehicle pathways, the vehicle possesses an optical sensor system that is capable of tracking the paint. The strips can be taped, sprayed or painted on the floor. One system uses a 1-in wide paint strip containing fluorescent particles that reflect an ultraviolet (UV) light source on the vehicle. A non-board sensor detects the reflected light in the strip and controls the steering mechanism to follow it.

The paint guidance system is useful in environments where electrical noise would render the guide wire system unreliable or when the installation of guide wires in the floor surface would not be appropriate. One problem with the paint strip guidance method is that the paint strip must be maintained.

Routing

Routing in AGVs is concerned with the problem of selecting among alternative pathways available to a vehicle in its travel to a defined destination point in the system. A typical layout that exploits the capabilities of modern AGV's technology, contains feature such as multiple loops, branches, side tracks and spurs in addition to the required pickup and drop-off stations. Vehicles in the system must decide which path to take to reach the defined destination point.

When a vehicle approaches a branching point in which the guide path splits into two (or more) directions, a decision must be made as to which path the vehicle should take. This is referred to as a *decision point* for the vehicle.

There are two methods used in commercial AGVs to permit the vehicle to decide which path to take:

(a) Frequency select method

(b) Path switch select method

In the frequency select method, the guide wires leading into the two separate paths at the branch have different frequencies. As the vehicle enters the decision point, it reads an identification code on the floor to identify its location. Depending on its programmed destination the vehicle selects one of the guide paths by deciding which frequency to track. This method requires a separate frequency generator for each frequency that is used in the guide path layout. This means that two or three generators are needed in the system. Additional channels must be cut into the floor with the frequency select method to provide for bypass channels where only the main channel needs to be powered for vehicle tracking.

The path switch select method uses a single frequency throughout the guide path layout. In order to control the path of a vehicle at a decision point, the power is switched off in all branches except the one on which the vehicle is to travel. To accomplish routing by the path switch select method, the guide path layout must be divided into blocks that can be independently turned on and off by means of controls mounted on the floor near their respective blocks. These control units are operated by the vehicles as they move in the various blocks.

As a vehicle enters a decision point, it activates a floor-mounted switching device connected to the control unit for the relevant block. The control unit activates the desired guide path and turn off the alternative branch or branches.

20.18.2 Traffic Control and Safety

The purpose of traffic control for an AGV is to prevent collisions between vehicles travelling along the same guide path in the layout. This purpose is usually achieved by means of a control system called the **blocking system**. The term *blocking* suggests that a vehicle travelling along a given guide path is in some way prevented from hitting any vehicle ahead of it.

There are several means used in commercial AGV systems to accomplish blocking. They are:

(a) On-board vehicle sensing
(b) Zone blocking

On-board vehicle sensing and zone blocking are often used in combination to implement a comprehensive blocking system.

On-board vehicle sensing (forward sensing) involves the use of some form of sensor system to detect the presence of vehicles and carts ahead on the same guide wire. The sensors used on commercial guided vehicles include optical sensors and ultrasonic systems.

When the on-board sensor detects an obstacle (another guided vehicle) in front of it, the vehicle stops. When the obstacle is removed, the vehicle proceeds. Assuming that the sensor system is 100% effective, collisions between vehicles are avoided and traffic is controlled. The effectiveness of forward sensing is limited by the capability of the sensor system to detect vehicles in front of it on the guide path since the sensors themselves are most effective in detecting obstacles directly ahead of the vehicle. These systems are most appropriate on layouts that contain long stretches of straight pathways. They are less effective at turns and convergence points where forward vehicles may not be directly in front of the sensor.

The concept of zone control is simple. The AGV's layout is divided into separate zones, and the operating rule is that no vehicle is permitted to enter a zone if that zone is already occupied by another vehicle.

The length of a zone is sufficient to hold one vehicle (or a train in driverless train systems) plus an allowance for safety and other considerations. These other considerations include the number of vehicles in the system, the size and complexity of the layout and the objective of minimizing the number of separate zone controls. When one vehicle occupies a given zone, any trailing vehicle is not allowed into that zone.

The leading vehicle must proceed into the next zone before the trailing vehicle can occupy the given zone. By controlling the forward movement of vehicles in the separate zones, collisions are prevented and traffic in the overall system is controlled. Figure 20.14 illustrates the concept of zone control to implement blocking system.

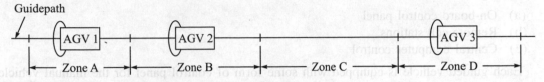

Fig. 20.14 Zone control to implement blocking system.

One means of implementing zone control is to use separate control units for each zone. These controls are mounted along the guide path and are actuated by the vehicle in the zone. When a vehicle enters a given zone, it activates the block in the previous (upstream) zone to block any trailing vehicle from moving forward and colliding with the present vehicle. When it moves into the next (downstream) zone, it activates the block in that zone and deactivates the block in the previous zone. In effect, zones are turned off and on to control vehicle movement by the blocking system.

In addition to avoiding collisions between vehicles, a related objective is the safety of human beings who might be located along the route of the vehicles travelling in the system. There are several devices that are usually included on AGV to achieve this safety objective. One of the safety devices is an obstacle-detection sensor located at the front of each vehicle. This is the same on-board sensor as that used in the blocking system to detect the presence of other vehicles located in front of the sensor. The sensor can detect not only other vehicles, but also people and obstacles in the path of the vehicle. These obstacle-detection systems are based on optical, infrared or ultrasonic sensors. The vehicles are programmed either to stop when an obstacle is sensed ahead of it or to slow down. The reason for slowing down is that the sensed object may be located off to the side of the vehicle path, or directly ahead of the vehicle beyond a turn in the guide path. In either of these cases, the vehicle should be permitted to proceed at a slower speed until it has passed the object or rounded the turn.

Another safety device included on virtually all commercial AGVs is an emergency bumper. The bumper surrounds the front of the vehicle and protrudes ahead of it by a distance which can be a foot or more. When the bumper makes contact with an object, the vehicle is programmed to brake immediately. Depending on the speed of the vehicle, its load and other conditions, the braking distance will vary from several inches to several feet. Most vehicles are programmed to require manual restarting after an obstacle encounter has occurred with the emergency bumper.

Other safety devices on the vehicles include warning lights (blinking or rotating lights) and/or warning bells. These devices alert people that the vehicle is present. If the vehicle strays by more than a few inches from the defined path, the vehicle is programmed to stop.

20.18.3 System Management

Managing the operations of an AGV deals principally with the problem of dispatching vehicles to the points in the system where they are needed (e.g., to perform pickups and deliveries) in a timely and efficient manner. The system management function depends on reliable operation of the other system functions such as guidance, routing and traffic control). There are a number of methods used in commercial AGV systems for dispatching vehicles. These methods are used in combination to maximize responsiveness and effectiveness of the overall system. The dispatching methods include:

(a) On-board control panel
(b) Remote call stations
(c) Central computer control

Each guided vehicle is equipped with some form of control panel for the manual vehicle control, vehicle programming and other functions. Most commercial vehicles have the capacity to be dispatched by means of this control panel to given station in the AGV's layout.

Dispatching with an on-board control panel represents the lowest level of sophistication among the possible methods. Its advantage is that it provides the AGVs with flexibility and responsiveness to changing demands on the handling system. Its disadvantage is that it requires manual attention.

The use of remote call stations is another method that allows the AGVs to respond to changing demand patterns in the system. The simplest from of call station is a press button mounted near the load/unload station. This provides a signal to any passing vehicle to stop at the station in order to accomplish a load transfer operation.

The vehicle might then be dispatched to the desired location by means of the on-board control panel.

More sophisticated call stations consist of control panels mounted near the various stations along the layout. This method permits a vehicle to be stopped at a given station, and its next destination to be programmed from the remote call panel.

This represents a more automated approach to the dispatching function and is useful in AGVs that are capable of automatic loading and unloading operations.

Both of the call station methods involve a human interface with the AGVs at the load/unload station. It is also possible to automate the call function at an automatic load/unload station. One example is an automated production workstation that receives raw materials and sends completed parts by means of the AGVs. The workstation is interfaced with the AGVs to call for vehicles as needed to perform the loading and unloading procedures.

In large factory and warehouse systems involving a high level of automation, the AGVs servicing the factory or warehouse must also be highly automated to achieve efficient operation of the entire production-storage-handling system. Central computer control is used to accomplish automatic distpatching of vehicles according to a preplanned schedule of pickups and delivery in the layout and/or in response to calls from the various load/unload stations in the system. In this dispatching method, the central computer issues commands to the vehicles in the system concerning their destinations and operations to perform. To accomplish the dispatching function, the central computer must possess real-time information about the location of each vehicle concerning which vehicles to dispatch to what locations. Hence, the vehicles must continually communicate their whereabouts to the central controller.

There are differences in the way these central computer dispatching systems operate. One of the differences involves the distribution of the decision-making responsibilities between the central controller and the individual vehicles. At one extreme, the central computer makes nearly all the decisions about routing of vehicles and other functions. The central computer plans out the routes for each vehicle and controls the operation of the guide path zones and other functions.

At the opposite extreme, each individual vehicle possesses a substantial decision-making capability to make its own routing selections and to control its own operations.

The central computer is still needed to control the overall scheduling and determine which vehicles should go to the various demand points in the system. However, the vehicles themselves decide which routes to take and control their own load transfer operations. Vehicles in this second category are reffered to as *smart* vehicles.

To accomplish the system management function, it is helpful to monitor the overall operation of the AGVs by means of some form of graphics display. Even with central computer control it is still desirable for human manager to be able to see the overall system operations, in order to monitor its general status and to spot problems (e.g., traffic jams, breakdowns, etc.). A CRT colour graphics display is used for these purposes in modern guided vehicle systems.

Another useful tool in carrying out the systems management function is a system performance report for each shift of AGV's operation. These periodic reports of system performance provide summary information about proportion uptime, downtime, number of transactions (deliveries) made during a shift and more detailed data about each station and each vehicle in the system. Hard-copy reports containing this type of information permit the system managers to compare operations from shift to shift and month to month to maintain a high level of overall system performance.

REVIEW QUESTIONS

20.1 Define a robot and describe the elements of a robotic system.

20.2 Explain the applications for which robots are generally used.

20.3 Briefly explain the classification of robots.

20.4 What are the various types of motion control possible in robots?

20.5 What are the different types of drives used in robots?

20.6 Explain the various types of end effectors used in robot arm.

20.7 Define the following terms:
   (a) Spatial resolution
   (b) Accuracy
   (c) Repeatability

20.8 Discuss the various types of robotic joints and state their applications.

20.9 Discuss the various robot configurations and explain their applications.

20.10 Discuss the differences between polar arm and articulated arm configuration.

20.11 How do you specify a robot?

20.12 Discuss the features of SCARA robot.

20.13 Sketch and explain different types of grippers commonly used in robots.

20.14 What are the methods used for programming a robot?

20.15 State the differences between lead through programming and teach pendant programming.

20.16 Discuss the basic types of robot programming language.

20.17 What are the limitations of first generation robot programming language?

20.18 Define work volume of a robot.

20.19 Explain the anatomy of a robot.

20.20 Discuss the industrial applications of robots.

20.21 Explain the following terms:

(a) Degrees of freedom of robots

(b) Drives and controls

(c) Manipulator configurations and their relative features

20.22 Describe various functions performed by the controller and manipulator of an industrial robot.

20.23 What are the essential features of high level robot languages?

20.24 Discuss the various sensory devices used in robots.

20.25 Discuss contact and non-contact types sensors used in robot end-effectors.

20.26 What is path control of a robot? How path of a robot can be controlled?

20.27 What is automated guided vehicle?

20.28 What are the various types of AGV's that are used in automation manufacturing?

20.29 Describe the advantages to be gained by the use of automated guided vehicle in manufacturing shop.

20.30 What are the various guidance methods available for automated guided vehicle?

20.31 Describe briefly about the wire guidance used in automated guided vehicle movement.

20.32 Describe the traffic control patterns used in AGV's traffic management.

Chapter 21

FLEXIBLE MANUFACTURING SYSTEMS

INTRODUCTION

A flexible manufacturing system (FMS) is an individual machine or group of machines served by an automated materials handling system that is computer controlled and has a tool handling capability.

Flexibility means that a manufacturing system is versatile and adaptable which is also capable of handling relatively high production runs. Versatile means that it can produce a variety of parts and adaptable in that it can be quickly modified to produce a completely different line of parts.

The key elements necessary for a manufacturing system to qualify as an FMS are as follows:

(a) Computer control
(b) Automated materials handling capability
(c) Tool handling capability

FMS is the result of integrating CNC, DNC, GT and automated materials handling systems into one system. Human tasks include the following:

(a) Equipment trouble shooting, maintenance and repair
(b) Tool changing and setup
(c) Loading and unloading the system
(d) Data input
(e) Changing of part programs
(f) Development of programs

The evolution of manufacturing starts from the manual operation and ends up with computer integrated manufacturing. The evolution of manufacturing is depicted in Fig. 21.1. The transfer lines are capable of producing large volume of parts at high production rates but have low flexibility. CNC machines are used to produce small volume parts at low production rates but have high flexibility. An FMS can produce medium volume of parts at medium production rates with medium flexibility. The manufacturing concepts based in volume and variety mix is shown in Fig. 21.2.

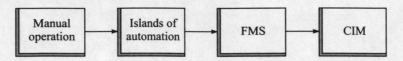

Fig. 21.1 The evolution of manufacturing.

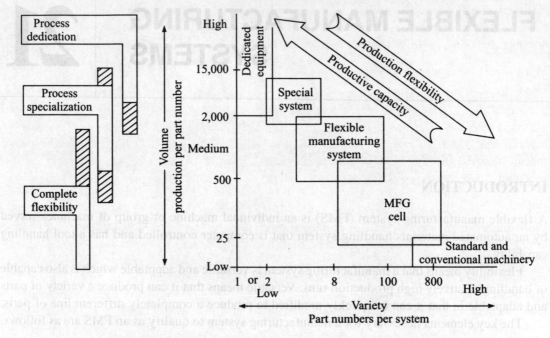

Fig. 21.2 Manufacturing concepts based on volume and variety mix.

A FMS integrates the following technologies:

(a) Flexible automation
(b) Group technology
(c) CNC machine tools
(d) Automated material handling
(e) Computer control of machines

21.1 COMPONENTS OF FMS

The components of a FMS are:

Workstations

The workstations vary according to the type of part being produced. In metal cutting systems, the machines are usually computer numerically controlled (CNC) horizontal spindle machining centres, if prismatic work pieces are to be produced, or turning centres if rotational work pieces. Some systems consist of both types of machines, when work pieces involving both types of operation are required. Other systems include single-purpose machines, as opposed to machining centres which are designed to perform a range of processes. In addition to metal working machines, there may also be gauging machines or other types of inspection machines. There are systems for sheet-metal operations, P.C.B manufacture and assembly operations.

Load and unload stations

Parts have to be introduced into the system at some point and there are usually load-unload stations, where parts are placed on pallets, usually by human operators. In some cases, parts may be supplied by an orienting device and loaded by robot. Unloading is usually done at the same stations, but there may be separated unload stations.

Work pieces transport equipment

Work pieces must be transported from the load positions to the production equipment and back for unloading. Three types of equipment are in common use, namely conveyors, vehicles and robots. There are loop-lines in conveyor systems for each workstation. Conveyor systems are less popular. There are several types of vehicles: Railcars and AGVs. There are several types of mobile floor-mounted robots, which can be used for work piece transport. Overhead gantry-mounted robots are popular for both work pieces and tool handling.

Pallets

Work pieces are normally held in pallets of some sort for transport and locating on machine tables. Two types are common: one type of pallet serves just as a carrier for a batch of small parts, to facilitate and reduce the frequency of movements, perhaps by a robot. This type is common in systems which use conveyors and gantry robots, but are also used in AGV systems. The other type of pallet is one on which one or more parts are accurately located and which is itself moved onto the machine table and held in position while machining operations are performed on the parts.

Fixtures

Fixtures are used to locate parts precisely on pallets. They are usually specific to one type of part so that each part requires a different fixture. In some cases, however, several types of part may be sufficiently similar to make use of the same fixture. The fixtures may be permanently bolted on the pallets, or they may be removed from the pallet when a part requiring a deferent fixture is to be introduced into the system and placed on the pallet.

Tools

Most operations require some form of tooling specific to the particular operation being performed typically cutting tools in machining centres. Machining centres have tool magazines in which a set of tools can be held so that any operation on a range of work pieces can be performed. Tools have to be changed, because of their tool life or because the part to be worked requires tools which are not currently in the tool magazine.

21.2 TYPES OF FMS SYSTEMS

There are various ways to classify flexible manufacturing systems. One classification that is sometimes made in FMS terminology is the difference between a flexible manufacturing system and a manufacturing cell.

The term *cell* can be used to refer to a machine grouping that consists of either manually operated or automated machines or combinations of the two. The cell may or may not include automated material handling and it may or may not be computer controlled.

The term *flexible manufacturing system* generally means a fully automated system consisting of automated workstations, automated materials handling and computer control.

The term *manufacturing cell* is used largely in connection with group technology but both cells and FMS rely on a GT approach in their design. A distinction between a FMC and FMS is in the number of machines in the grouping. A grouping of four or more machines in a system and three or fewer machines constitute a cell. For example, a grouping of several machines served by a robot and capable of processing a family of parts is commonly called a **flexible manufacturing cell**.

A flexible manufacturing system can be described as being either a dedicated FMS or a random order FMS. A dedicated FMS is used to produce a much more limited variety of part configurations. The geometry differences are minor and the product design is considered stable. Therefore, the machine sequence is identical or identical for all parts processed on the system. This means that a flow line configuration is generally most appropriate and that the system can be designed with a certain amount of process specialization to make the operations more efficient. Instead of using general-purpose machines, the machines can be designed for the specific processes required to make the limited part family.

The random-order FMS is the more appropriate type under the following conditions: (a) The part family is large. (b) There are substantial variations in the part configurations. (c) There will be new part designs produced on the system and engineering changes in parts currently made on the system and the production schedule is subject to change from day to day. To accommodate these variations the random-order FMS must be more flexible than the dedicated FMS.

It is equipped with general-purpose machines to deal with the variations in product and is capable of processing parts in various sequences (random order).

A classification in flexible machining systems is based on the part geometry being processed. Machined parts can usually be divided into either two categories: prismatic parts are cube like and require milling and related machining operations to shape them. Round parts are cylindrical or disk shaped and require turning and related rotational operations.

Flexible manufacturing cells (FMC)

Flexible manufacturing cells consist of one or more CNC machine tools, general purpose or of special design interfaced with automated material handling and tool changers. FMCs are capable of automatically machine a wide range of different work pieces.

A turning centre fitted with a gantry loading and unloading system and pallets for storing work pieces and finished parts is a typical flexible turning cell. Automatic tool changers, tool magazines, block tooling, automatic tool offset measurement, automatic chuck change and chuck jaw change make the cell more productive. One or two horizontal machining centres with modular fixturing, multiple pallets, advanced tool management system, robots or other material handling systems to facilitate access of the jobs to the machine is a flexible machining cell.

Flexible transfer lines (FTL)

Flexible transfer lines are designed for high volume production wherein a part undergoes different types of operations. As each operation is assigned to and performed on only one machine, there is a fixed route for each part through the system. The material handling system is usually a pallet or conveyor. It also consists of SPMs and robots. In FTL, a number of different work pieces are manufactured as the scheduling is easier and the resetting procedure is automatic.

Flexible machining systems

Flexible machining system consists of several flexible automated machine tools of universal or special type which are flexible interlinked by an automatic work piece flow system so that the different work pieces can be machined at the same time. Different machining times at the individual stations are compensated for by central or decentralized work piece buffer stores. Flexibility is applied by usage of CNC control, flexible transport system and by adapting to changes in the volumes in the product mix, machining process and sequences.

21.3 FLEXIBILITY IN FMS

Flexibility is one of the key concepts used in the design of modern automated manufacturing system such as flexible manufacturing systems.

Flexibility can be defined as the collection of properties of a manufacturing system that support changes in production activities or capabilities. Normally, the changes are due to both internal and external factors. Internal changes could be due to equipment breakdowns, software failures, worker absenteeism, variability in processing time and so forth. Some degree of redundancy in the system is required to cope with internal changes. This, however, adds more capacity. Changes in product design, demand and product mix typically represent external changes. To absorb uncertainties due to product design changes, the manufacturing system must be versatile and able to produce the intended variety of part types with minimal cost and lead-time. Some degree of redundancy may, however, be required to cope with changes in the demand and product mix. We can, therefore, say that the flexibility refers to the ability of the manufacturing system to respond effectively to both internal and external changes by having built-in redundancy of versatile equipment. A number of types of flexibility have been listed follows:

Machine flexibility

Machine flexibility refers to the capability of a machine to perform a variety of operations on a variety of part types and sizes.

The other measure of machine flexibility is the universe of part types that a machine can produce. It implies the ease with which the parts are changed over from one part type to another on a machine. The changeover time, which includes setup, tool changing, part program transfer and part move times, is an important measure of machine flexibility.

Routing flexibility

Routing flexibility means that a part(s) can be manufactured or assembled along alternative routes. The alternative routes are possible if manufacturing or assembly operations can be performed on alternative machines, in alternative sequences, or with alternative resources. Because a wide variety of operations can be performed on flexible machines, they provide routing flexibility.

Routing flexibility is used primarily to manage internal changes resulting from equipment breakdowns, tool breakages, controller failures and so on. Of course, the internal changes can be minimized by such measures as online monitoring of failures and preventing maintenance programs.

Process flexibility

Process flexibility, also known as **mix flexibility**, refers to the ability to absorb changes in the product mix by performing similar operations or producing similar products or parts on multipurpose, adaptable, CNC machining centres. Mix flexibility provides protection against market variability by accommodating changes in product mix due to the use of shared resources. However, extreme mix variations would result in requirements for a greater number of tools, fixtures and another resource.

Product flexibility

Product flexibility, also known as **mix change flexibility**, refers to the ability to change over to a new set of products economically and quickly in response to markets or engineering changes to operate on a make-to-order basis. The changeover time includes time in all activities of new product manufacture, such as design, process planning, tooling and fixturing. This time can be minimized by using the principles of group technology in design and manufacturing as well as using flexible machines and software.

Production flexibility

Production flexibility refers to the ability to produce a range of productions without adding major capital equipment, even though new tooling or other resources may be required. The product envelope, that is, the range of products that can be produced by a manufacturing system at moderate cost and time, is determined by the process envelope.

Expansion flexibility

Expansion flexibility refers to the ability to change a manufacturing system with a view to accommodating a change product envelope. In the case of production flexibility, there is no change in the major capital equipment. In the case of expansion flexibility, there are additions as well as replacement of equipment, but these changes are easy to make because such provisions are made in the original manufacturing system design.

21.4 BUILDING BLOCKS OF FMS

A flexible manufacturing system consists of more than one complementary or supplementary machine, many of which are of the CNC type, served by compatible common and preferably automatic tool and work piece supply systems under the supervision of an integrated computer control.

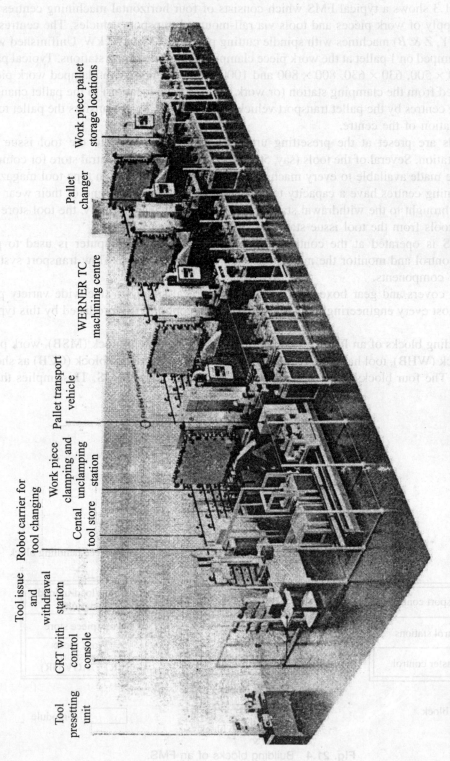

Fig. 21.3 An FMS with four horizontal machining centres and automatic supply of work pieces and tools via rail-mounted transport vehicles.

Work piece pallet storage locations

Pallet changer

WERNER TC machining centre

Pallet transport vehicle

Work piece clamping and unclamping station

Cental unclamping tool store

Robot carrier for tool changing

Tool issue and withdrawal station

CRT with control console

Tool presetting unit

Figure 21.3 shows a typical FMS which consists of four horizontal machining centres and automatic supply of work pieces and tools via rail-mounted transport vehicles. The centres are four axis (X, Y, Z & B) machines with spindle cutting powers of 15 to 32 kW. Unifinished work pieces are clamped on 1 pallet at the work piece clamping and unclamping stations. Typical pallet sizes are 500×500, 630×630, 800×800 and 1000×1000 Pallets with clamped work pieces are transported from the clamping station (or work piece storage location) to the pallet changers of machining centres by the pallet transport vehicle. The pallet changer transfers the pallet to the machining station of the centre.

The tools are preset at the presetting unit and manually carried to the tool issue and withdrawal station. Several of the tools (say, 500) always remain in the central store for common usage and are made available to every machine by robot carrier. The chain type tool magazines of the machining centres have a capacity to carry 40 to 72 tools. At the end of their wear life, the tools are brought to the withdrawal station by the robot carrier. Similarly, the tool store can be supplied tools from the tool issue station by the robot carrier.

The FMS is operated at the control console and the system computer is used to plan coordinate, control and monitor the machines, tooling and work pieces flow transport systems and its other components.

Cylinder covers and gear boxes are some of the typical small lot size wide variety parts covering almost every engineering industry which can be continuously produced by this typical FMS.

The building blocks of an FMS system are the machining system block (MSB), work piece handling block (WHB), tool handling block (THB) and computer control block (CCB) as shown in Fig. 21.4. The four blocks have to be compatible for building an FMS. This implies that it

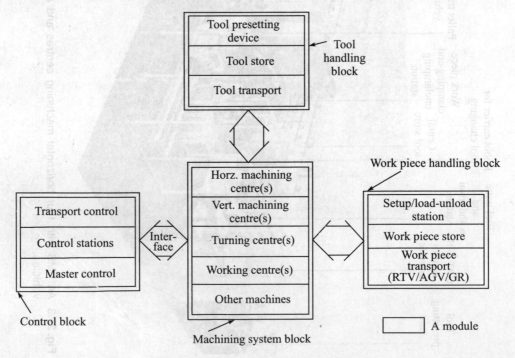

Fig. 21.4 Building blocks of an FMS.

should be possible to interface all the blocks for an integrated computer control of all the automated functions.

21.5 MACHINING SYSTEM OF FMS

The machining system of an FMS is capable of performing several of the operations on the work piece automatically. Those CNC machining centres, turning centres, working centres and conventional machines which may supplement each other for completion of the operations on the work piece are chosen to build the machining system of the FMS. Several complementary machines of the same kind are also employed. The latter combination is sometimes identified as an FMC. The essential requirement in all cases of building FMS is that it should be possible to interface the CNC machines with the FMS-computer. The controller of these machines should have the provision for DNC/LAN/MAP-communication.

21.5.1 Horizontal Machining Centres

The horizontal machining centre having a chain-type tool magazine and a pallet changer is emerging as one of the standard modules for the FMS systems handling prismatic work pieces. Figure 21.5 shows a machining centre with pallet changer. Some of the features incorporated in these centres and their capabilities are following:

(a) Several types of machining operations can be automatically performed in the desired sequence.
(b) Complete machining of work piece with a minimum number of setups is possible. Fully guarded enclosures permit closed working space, good accessibility and reliable protection from flying chips and splashes.
(c) The tool magazine can be expanded to contain tools upto about 150 when one chain capacity is inadequate. Even two chain magazines can be provided. Enclosed tool changing equipment provides protection from chips and coolant ensuring trouble-free smooth operation.
(d) Stepless speed control of the main spindle motor whether DC or AC can be provided.
(e) Work piece-carrying pallets which can be changed are provided to increase the productivity of the machines. The work piece handling system can be standardized to feed in and out the pallets from pallet changers and transport-devices. The pallet changers can also be fed with linear or rotatory shuttles having fixed number of pallets. Alternately, the pallet changers may be fed from stationary expandable pallet stores. Several pallets provided by any of the methods mentioned earlier can provide buffer work piece storage for unmanned shifts during night.
(f) Hydraulic or pneumatic connections can be provided to the pallets for automatic clamping and unclamping of the work pieces.
(g) Probes for inprocess gauging can be easily added to these machines.
(h) A program controlled washing device for the pallets carrying machined work pieces can be incorporated within the working envelope of the machining centre.
(i) Five-axis machining is easily achievable by using special heads or by adding standalone axis A to the four axes (X, Y, Z and B) usually provided on these machines.

(j) To meet the requirements of several types of FMS, the machining centre is currently made from basic modules which can be varied to build a machining centre of the desired specifications. Figure 21.6 lists some basic modules of such machining centres.

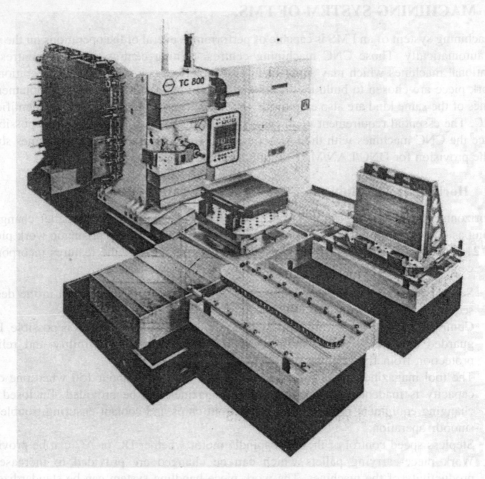

Fig. 21.5 A machining centre with pallet changer.

21.6 TOOL MANAGEMENT SYSTEMS

In an automated manufacturing system, though the automation of work flow systems as well as the process control and monitoring functions have been very well put into practice, the cutting tool supply is still a weak point. Managing the cutting tools in any manufacturing system is extremely important, since they contribute towards the quality of the product produced. The success of manufacturing depends to a great extent on making available the right tool in the right condition at the right time. Any deviation from this rule would make the system performance suffer.

There are five basic modules which supports the tool management system of the FMS:

(a) Tool coding systems
(b) Tool supply systems
(c) Work and tool probing
(d) Tool monitoring systems
(e) Tool management systems

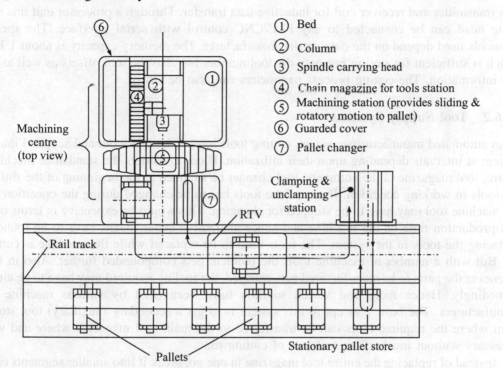

① Bed
② Column
③ Spindle carrying head
④ Chain magazine for tools station
⑤ Machining station (provides sliding & rotatory motion to pallet)
⑥ Guarded cover
⑦ Pallet changer

Fig. 21.6 A machining centre built from its seven basic modules and fed from an RTV and stationary pallet store.

21.6.1 Tool Coding Systems

In earlier control systems, the cutting tools were identified by means of their position in the cutting tools magazine. This does not allow for efficient transfer of cutting tools in a manufacturing system. Hence systems are developed for coding individual tools so that their designations can be directly read from the tools themselves instead of relying on the physical position. This would allow for random location of tools anywhere in the system.

One simple way of coding is the use of the bar code on the tool. When the tool enters the system the tool number can be read which would be used for getting the relevant information from the database. The bar code strip is pasted on the adopter of the tool.

Another more preferred method is the use of data carriers which are basically small chips (IC) located in the tool adopter (pull stud) or any suitable location on the tool adopter. This data carrier can be written to and read. These are generally electrically erasable and programmable read only memories (EEPROM) which can retain data for about 10 years and do not require batteries. These are very small in size about 12 mm in diameter and 5 mm in thickness. These are sealed in a water proof stainless steel housing. Further the sensing face of the carrier is coated with abrasion resistant oxide ceramic to protect it from hot metal chips.

The total system consists of data carrier, read/write head and a processor unit. The reading and writing processes are carried out by non-contact inductive means. This read/write head can communicate with the data carrier from a distance of about 4 mm. The read/write head consists of a transmitter and receiver coil for inductive data transfer. Through a processor unit this read/write head can be connected to any PLC/CNC control with serial interface. The specific protocols used depend on the data carrier manufacturer. The memory capacity is about 1 K bit which is sufficient for storing not only the tool number but also the tool offsets as well as tool life information. The cutting process parameters can also be stored.

21.6.2   Tool Supply Systems

In an automated manufacturing system, cutting tools have to be taken out and supplied into the system at intervals depending upon their utilization. If one considers the standalone machining centre, tool magazine of the automatic tool changer be supplied at the beginning of the shift, all the tools in working condition. If any of the tools has to be changed during the operation then the machine tool may have to be stopped for changing. This would be expensive in terms of the lost production time on the machine and hence alternative better means have to be found for replacing the tools in the system. The tools need to be replaced while the machine is cutting.

But with a number of machine tools the problem gets compounded further. Also in FMS whenever the parts to be manufactured gets changed, the tooling required may have to be altered accordingly. Hence more and varied solution have been tried by various machine tool manufacturers. The basic concept in this system is to get a secondary (auxiliary) tool storage from where the required tools can be transferred to the main tool magazine where and when necessary without much effort and loss of cutting time.

Instead of replacing the entire tool magazine in one go, break it into smaller segments called **cassettes**, so that each individual cassette is replaced at a time in place of the complete magazine. This would mean a small tool change time and comparatively straightforward way of replacing the tools. The number of tools in the system will be 120 or so.

There are a number of ways in which this method is adopted. In the quick tool change (QTC), the cassette is in the form of a linear array capable of storing eight tools. A simple transfer machine removes old tools with new tools by a comb cassette. The chain tool magazine used in this case is divided into a variably overlapping storage and changing area as well as fixed areas for standard tools and worn tools. The machine tool automatically deposits the tools in the respective areas after their use in spindle. The movement into the respective areas of the magazine chain takes parallel to the machining time. The tools which are worn out are first transferred to the tool positions meant for transfer. Then the tool chain indexes to bring all the tools to the transfer position. The empty comb cassette would then transfer the tools from the chain. The cassette is withdrawn while a new cassette with fresh tool is brought to the tool change position. These tools then get transferred to the tool chain. This system can work in manual, semi-automatic or fully automatic mode depending upon the desired application.

TABLE 21.1 Tool Supply Concepts in Manufacturing Systems

Tool storage and transport	Storage capacity per machine tool	Time fraction per tool tool changeover	Bound capital for tools in the manufacturing system
One-machine integrated circulating magazine	40–100 tools	Large	Large
Two-machine integrated circulating magazine	60–120 tools	Small	Large
Interchangeable tool magazines	20–60 tools	Medium	Medium
Successive interchange of a segment of tools from the magazine	40–100 tools	Small	Medium
Successive interchange of single tools from a stationary auxiliary magazine	20–40 tools	None	Medium-small
Separate tool highway	20–40 tools	None	Small

21.6.3 Work and Tool Probing

The touch trigger probe is basically a tool used for measurement. It consists of a spherical sapphire unit connected elastically to a rigid measuring unit. Figure 21.7 depicts a touch trigger probe system. When the probe moves along a direction and touches a surface, the elastic connection deflects and triggers the measuring system. The trigger is basically an ommi-directional switch capable of detecting deflections in any direction. The deflection of the probe is in any direction. The deflection of the probe is measured and from that the physical position of the surface where the probe contacts gets recorded. Since the probe tip is spherical, the contact between the probe and the measured surface is a point contact. These are the basic measuring units used in coordinate measuring machines (CMM).

It is possible to use these probes in CNC machine tools for the purpose of measurement as well as tool setting. The CNC controller has a tool probing instruction or cycle through which the measured position can be stored into parametric variables. These parameters may be computed to evaluate the necessary inspection function. This system not only enhances the utility of the CNC machine tool, but also improves the productivity by reducing the setting time for complex jobs, by measuring the tool offsets on the machine, etc. It improves the quality of the jobs produced by measuring the exact offsets just before the machining to start for each of the parts in a batch. Probing could also be used for tool breakage monitoring.

The various types of transmission methods available for the measured value from the probe to the CNC controller are.

Direct hard wired connection

This may be used only for the tool setting probes. Since all the elements concerned with the measurement are stationary and fixed in space.

Inductive transmission

This system is also used for measuring probes. A probe is treated like any other tool and is placed in the spindle. During the gauging cycle, the probe signals are transmitted across inductive modules as shown in Fig. 21.8.

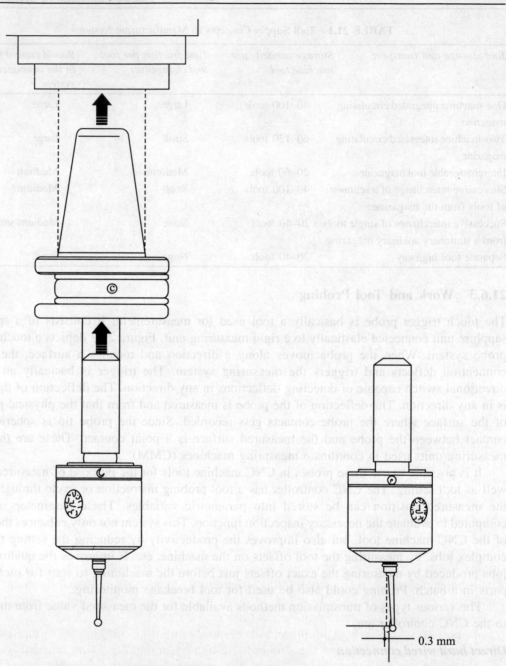

0.3 mm

Fig. 21.7 Touch trigger probe system.

Optical transmission

This system is also used for measuring probes. The optical system communicates by invisible infrared coded messages which convey probe and control signals between the optical transmitter and receiver as shown in Fig. 21.8. These are the most convenient form since the optical receiver can be located outside the machining area and no specific location is required and are therefore widely used. They also used batteries, and hence required minimum wiring.

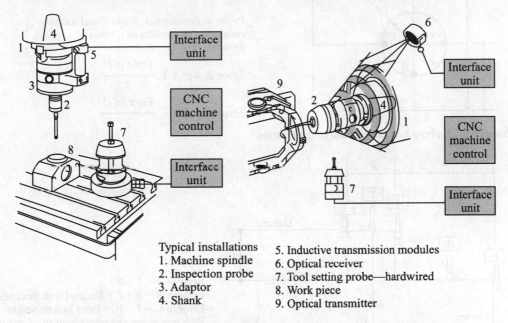

Typical installations
1. Machine spindle
2. Inspection probe
3. Adaptor
4. Shank
5. Inductive transmission modules
6. Optical receiver
7. Tool setting probe—hardwired
8. Work piece
9. Optical transmitter

Fig. 21.8 Type of transmission systems used in machining centres.

Work probing

Similar arrangements may be made for connecting the probe systems in turning centres also. Once a component is fully machined or after a critical feature is machined, probing may be carried out to check that the generated features are accurate. By making use of the machine tool axis transducers, it is possible to measure any feature on the component. Figure 21.9 explains two examples of the kind of inspection that may be carried out.

The inspection can be in-cycle or post-process. In-cycle gauging refers to the technique of measuring the component using probes in between cuts. If the probing done between the cuts, the last cut may or may not be carried out, depending upon the required dimension measured. The CNC part program is suitably written with probing cycles and skipping the blocks based on the results of measurement. Therefore, in-cycle probing would allow for all the jobs to be controlled to closer tolerances required in modern industry.

In post-process gauging, the inspection is carried out after completing all the machining, without removing the job from the workholding fixture. In this case, it is established whether or not the part is within the tolerance limits. This would allow for the remedial cuts to be taken to bring the job within the tolerance limits required for the critical dimensions. In this case, automatic correction of size is not possible since the system can only give the results of inspection and then the remedial action is to be manual.

Inspection probes can also be used for manual datuming of the components in case of machining centres. Before the component is clamped, the probe be brought near the surface to

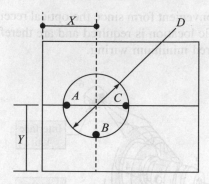

Probe at positions A, B and C and use the resultant error values as follows:

Error in $D = -A - C$

$$\text{Error in } X = A + \frac{\text{Error in } D}{2}$$

$$\text{Error in } Y = B + \frac{\text{Error in } D}{2}$$

(a) Inspection of a bore for diameter and centre position

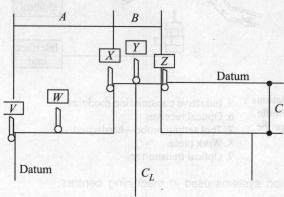

Error in $A = V - X$
Error in $B = X + Z = $ Error of web thickness
Error in $C = Y - W = $ Error in web height
Web centre line out of position from datum

$$= \text{Error in } A + \frac{\text{Error in } B}{2}$$

(b) Inspection of a web relative to a datum face

Fig. 21.9 Probing examples.

be used for datum (set point/surface) and accordingly the set point values entered in the controller registers.

Tool probing

The types of probes required for tool probing are different from the ones used for inspection. The probe head is flat faced with a square cross-section as shown in Fig. 21.10. The tool can be brought in contact with the probe tip in both the directions to obtain the tool offsets. The typical tool setting times are of the order of 20 seconds including the automatic entry offset values into the controller memory. The same probe would also be used for monitoring the tool breakages in between the cuts.

21.6.4 Tool Monitoring Systems

In the case of tool monitoring systems, the tool has to be continuously monitored while it is cutting. This would allow for continuously looking for tool wear as well as the times when the tool breaks because of unforeseen conditions in the machining system. The various methods adopted for these two functions are slightly different.

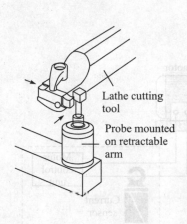

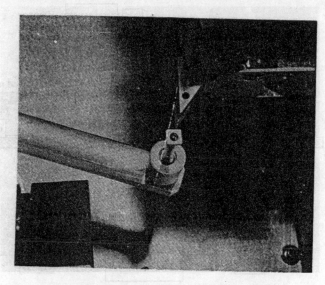

Fig. 21.10 Probe type used for tool offset determination in turning centres.

Tool wear monitoring

Tool wear is a phenomenon whose behaviour can be explained qualitatively but not quantitatively. Though some tool life equations do exist, their universal adaptability for all parameter ranges are doubtful. Direct in-process measurement of tool wear is difficult in view of the location of the wear and the measurement techniques employed.

As a result tool wear measurement has to be indirect. Some parameters used for measuring tool wear are:

Cutting power

Cutting forces

Vibrations

Acoustic emission

Tool temperatures

Of these variables mainly the cutting forces and power based tool monitoring systems are commercially and widely available, whereas the others are still not widely used in practice.

The power consumed during a machining process is a function of the forces acting. The cutting forces depend upon the quality and condition of the cutting edge. As time progresses, the power consumed by the tool for the same material removal increase with increase in tool wear. Thus, power measurement is an indirect way of monitoring the life of the tool.

Power may be measured by a power meter (watt hour meter) installed in the spindle motor circuit.

A more effective method is to determine the resistance offered to the tool motion which can be measured in terms of the power consumed by the axes motors. In applications like drilling, measuring the current consumed by the feed motor in the spindle (Z-axis) direction would be a good indicator for tool condition. An arrangement for such tool condition monitoring using a current sensor is shown in Fig. 21.11.

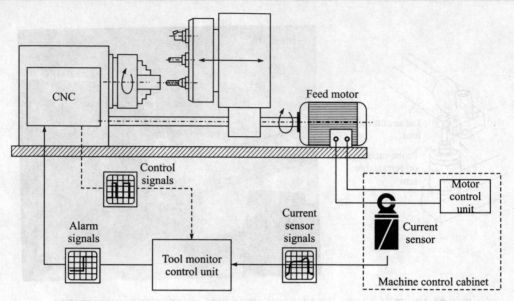

Fig. 21.11 Schematic of CNC turning centre equipped with a tool monitoring system based on current sensor.

Another system of tool condition monitoring is by the measurement of the torque on the main spindle. Figure 21.12 shows the schematic of a tool condition monitoring system based on the measurement of torque of the main spindle. In this system, the spindle torque is measured in terms of the differential twist separated by a small distance.

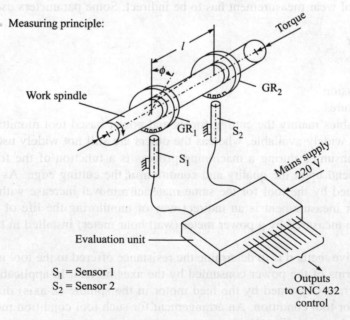

Fig. 21.12 Schematic of a tool condition monitoring system based on the measurement of torque of the main spindle.

Direct measurement of cutting force is a better method for tool condition monitoring rather than power. Hence a number of systems based on force measurement are available commercially. The problem encountered often is that the force sensor should be located close to the source of power, i.e., the tool. However, with the ability to change tools during the machining, it is necessary to have a force sensor located in the tool holding structure rather than with the tool itself. Figure 21.13 shows a set up for tool condition monitoring using, plate sensor under the tool turret in a turning centre.

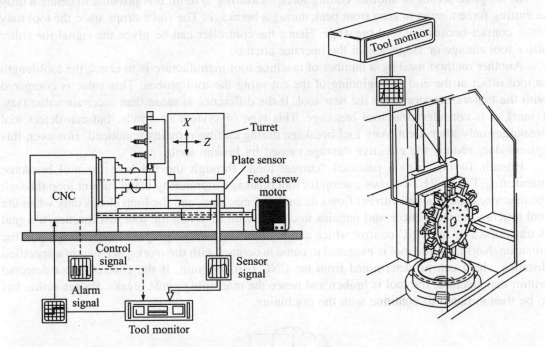

Fig. 21.13 Schematic of a setup for tool condition monitoring using a plate sensor under the tool turret in turning centre.

The plate sensor consists of a structure which is simple and has strain gauges to directly measure the feed force. The simplicity of the construction of the plate sensor helps in its easy adaptability to any machine tool. The threshold signal for limiting tool wear from the plate sensor is linked to the CNC controller which can automatically call up a sister tool for replacement of the worn-out tool before a catastrophic failure. This also allows for uninterrupted cutting and reduces the costly breakdowns of the CNC machine tool.

Another possibility of tool condition monitoring is through the measurement of vibrations of the cutting tool. The vibration signature of a cutting tool is a good indicator of the quality of the cutting edge. The vibration spectra at the beginning when the tool is sharp, can be compared with those at each time and the shift taking place in amplitudes and dominant frequencies can be measured. These are useful for identifying the tool failure criterion. The failure criterion used is the power spectral density of the cutting tool vibration which is a good indication of the energy consumed for cutting by the cutting tool.

Tool breakage monitoring

Another problem faced is the breakage of tool during cutting which if not detected in time may lead to various problems associated with spoiled jobs, particularly in unmanned machining shifts. Hence it is necessary to have systems which can detect the breakage of tools through some means and give an alarm to the operator, or automatically replace the tool by a sister tool from the secondary tool storage.

In the plate sensor or another cutting force measuring system, it is possible to detect a drop in cutting force almost to zero from peak during a breakage. The force drops since the tool may loose contact because of tool breakage. Hence the controller can be given the signal for either sister tool change or stoppage till the operator attends.

Another method used by a number of machine tool manufacture is to check the tool length or tool offset at the end or beginning of the cut using the tool probes. This value is compared with the tool values stored for the new tool. If the difference is more than a certain value (say, 1 mm), it is considered as tool breakage. This type of system is simple, but can detect tool breakages only after a cut. Any tool breakage during cutting remains unnoticed. However, this system does reduce the effective damage caused by broken tools.

Ernault Toyoda uses a patented "current loop" through the machine for tool breakage monitoring. Figure 21.14 shows a setup for tool breakage monitoring using current loop through the machine tool. A small current flows through the machine, and the loop closes only when the tool touches the work piece and remains in contact with it. When the tool just touches, a signal is transmitted to the CNC control which can be used for various purposes depending upon the situation. Normally, the tool is expected to come in contact with the workspace within a specified distance which can be determined from the CNC part program. If the contact is not detected within that distance, the tool is broken and hence the machining cycle breaks. Proper action has to be then initiated to continue with the machining.

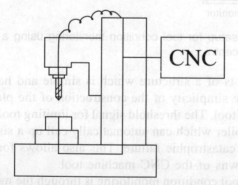

Fig. 21.14 Schematic of a setup for tool breakage monitoring using current loop through the machine tool.

They used this novel facility for reducing the tool idle time by accelerating the tool movement when the contact is not established between the tool and work piece by doubling the programmed feed rate.

Another approach employed by Deckel is the use of infrared beam to detect the tool breakage in their range of machining centres DC 30 and DC 40. Figure 21.15 depicts a typical system for tool breakage monitoring using infrared fixed position probe in a machining centre. At the end of each machining sequence, the spindle moves the tool automatically to the measuring location where the tool is expected to break infrared beam. If the beam could not be broken, the tool length is small and hence indicates tool breakage.

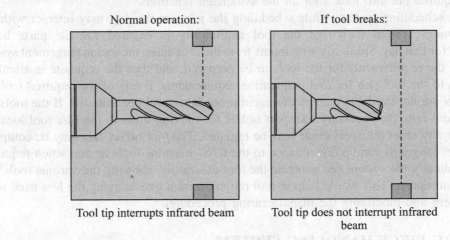

Normal operation: If tool breaks:

Tool tip interrupts infrared beam   Tool tip does not interrupt infrared beam

Fig. 21.15 Typical system for tool breakage monitoring using infrared fixed position probe in a machining centre.

21.6.5 Tool Management Systems

In this system, software systems can be presented to track the tools in the manufacturing systems.

Tool life monitoring

There are various systems available as part of modern CNC controllers which are termed as **tool management systems**. These systems are able to keep track of the actual time for which each of the tool is in use. When the tools are entering the system (loading in the tool magazine), the expected tool life is entered in the controller memory along with all other tool related information such as offsets. This value is generally a conservative estimate of the tool life expected based on the work materials and the process parameters used. This value can be generally obtained form machinability data banks used on the shop floor or from previous experience.

As machining proceeds with various components, the controller records the actual time for which each tool is used. This time used when subtracted from the expected tool life gives the left-over tool life for each tool. When the left-over tool life is less than a certain nominal value, the controller initiates the action for replacing it by a sister tool. Controllers manufactured by Yamazaki, Fanuc, Sinumeric and others have this type of tool management option available with them.

It is also possible to have tool life monitored outside the controllers. For example, the tool life of all the tools present in the manufacturing system consisting of more than one machine tool may be monitored in an IBM PC compatible.

Tool management system

A true tool management system is much more than a simple tool life monitoring system and interfaces with the factory information system. This system would not only see that the tools be kept as required but also look after all the associated functions.

The shop scheduling system, while scheduling the jobs for the day, may interact with the tool management system such that the tool availability is ensured for the parts being manufactured for that day. Similarly with inputs from the scheduler, the tool management system may generate the requirements for the tools to be prepared, and thus the requisite instructions may be given to the tool crib for tool preparation requirements. If any of the required tools or some elements are not available then procurement procedures could be initiated. If the tools are ready at the tool crib, the necessary transport orders for moving them to the M/c tool location using AGV or any other transport mode may be initiated. The tool offset data may be compiled and tool offset programs can be downloaded to the CNC machine tools as and when required. Finally, periodically the system can generate the tool utilization, showing the various tools and their utilization reports. This would help in tool rationalization by removing the less used tools from the system and modifying the manufacturing procedures.

21.7 WORK PIECE HANDLING SYSTEM

This system has the function of timely supply of unmachined work piece from the storage to the machining centres and transporting the machined parts from the centres to the locations. This block generally consists of the following modules:

Work piece setup or loading and unloading station
Work piece store and
Work piece transport

Work piece setup or loading and unloading station

Prismatic components meant for machining centres are generally setup on pallets. The pallets are now-a-days standardized to have compatible mechanical interface with the storage locations, transport systems, pallet changer and the machining centres. The palletization may be done manually or with the help of a robot. Fixtures are used to locate the parts precisely on the pallets. The fixtures should be able to accommodate several parts belonging to a family. Principles of group technology (GT) are generally employed in this connection. The fixtures may even be permanently fixed to the pallets which in turn may also have the provision for automatic clamping actuated hydraulically or pneumatically. Several setup stations may be combined to constitute a central clamping and unclamping station. When the work pieces have relatively low weight and size, they may be directly loaded onto the machine by a robot or a manipulator. In such cases, the work pieces may be loaded manually or automatically on tray like pallets. A transport pallet of this type may be used to accommodate several work pieces in proper position. Rotational work pieces may also be loaded into turning centres in a similar fashion.

Work piece holding

Work piece holding with the help of fixtures and clamps is an important process detail which greatly influences the machine tool utilization rate and round-the-clock operations of FMS. In addition to the conventional fixture design requirement of accurate positioning, location and clamping, FMS fixtures have to accommodate part families and minimize changeover times. Modular fixturing which employs reusable and standardized interchangeable base plates, sub-plates, tombstone type bases help meet the objectives of FMS and are finding favour with FMS users.

Several work pieces are clamped into tombstone type of fixtures. The base plates including the tombstone type also can be mounted directly on the pallets which could be transported to the pallet changers. This results in long enough pallet marching time, decreases changeover times and improves machine time utilization. Mechanical and hydraulic clamping reduce pallet preparation times further. Due consideration has of course to be given to the clamping pressures otherwise the parts measured in clamped position on a coordinate measuring machine (within the FMS cycle) would spring out of tolerance when unclamped.

The pallet fixturing instructions are provided at the clamping and unclamping stations. This involves details of the pallet, fixture components, work piece, clamps, retrieval of components, assembly and disassembly instructions, and storage of disassembled components. Once the pallet assembly is ready, it is labelled and identification data is input into the system computer of the FMS.

Pallet/work piece identification

The work pieces carried on pallets are generally indirectly identified by their pallets. The pallets may be recognized from their magazine location or pallet changer position in stand-alone machines. A typical method of identifying the pallets and the relevant work pieces is illustrated in Fig. 21.16. In this case, the machine operator specifies the part program that can be used for machining a newly built-up pallet. The sequences, in which part programs are to be processed as well as the number of parts to be machined with each program, are also specified. Before each pallet change, a control program module checks if any pallet magazine contains parts with current or higher priority, and accordingly moves the pallet to the machining station.

This permits random pallet positioning. Again in standalone machines, the pallet changer position can be associated with certain part program numbers to recognize the work pieces/pallets. Further pallet positions can also be recognized. For example, four different types of work pieces on tombstone type of fixture per pallet can be identified by associating them with different indexed positions of the pallet.

In FMS, where work pieces clamped to fixture are delivered to machining centres on pallets, methods similar to tool coding can be employed to identified the pallets. Further, each pallet may be identified by using binary coded pins, optical character recognition devices and even sophisticated vision systems. As mentioned earlier, the pallets generally carry prismatic components. The rotational components on the other hand are loaded into turning centres directly without using any pallets. In this case, work piece identification is relatively difficult and is carried out by using either touch trigger probing or vision based systems. A machine vision system can recognize work pieces by comparing their configurations images with the ones already stored in the memory of the system. On the other hand, a probe recognize a work piece by some of its unique dimensions. Once a component is identified, the system computer can transmit the relevant part program to the machine. This can be preceded by automatic setup of the work piece also, which can be achieved by probing.

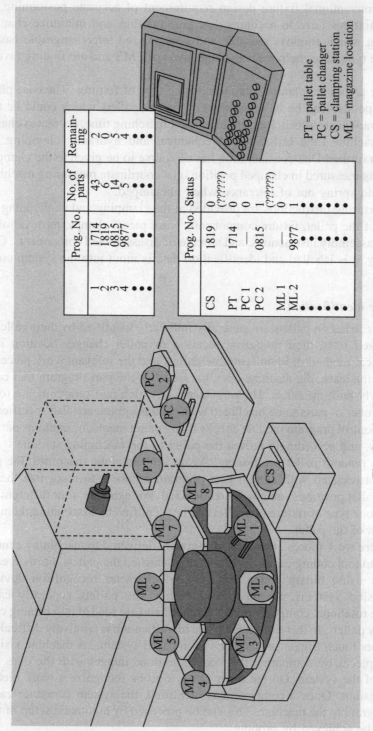

PT = pallet table
PC = pallet changer
CS = clamping station
ML = magazine location

Fig. 21.16 Random pallet positioning.

The measurement probes determine surfaces or datum's points on the work piece and accordingly enter zeroes for the axes at those locations. In addition to the cycle which defines work piece zero, several other measuring cycles can be incorporated in the probe software modules.

Work piece store

Pallets with work pieces on them may be stored in expandable work piece stores. The storage position for each of the pallets may be fixed or alternately identified as explained in pallet workpiece identification. Such stores may be used as buffer storage points for an unmanned shift.

Work piece storage and retrieval can also be automated in FMS. An FMS is generally a small-lot but great-variety manufacturing system involving frequent modifications of parts and work piece changeovers on the pallets.

At the same time minimization of setup time is one of the prime objectives in FMS. In this respect, AS/RS is gaining wider acceptance for becoming a centrally controlled pallet/work piece store working unmanned around the clock. The number of pallets which can be stored may vary between 400 to 10,000 for small scale to large scale AS/RS. The system essentially consists of storage racks, a PLC based or computer controlled stacker crane, the location and accuracy of which is comparable to that of a robot. Employment of servomotors in the cranes help achieve positioning accuracy and reliability. Elevating speeds of the stacker could be as high as 180 m/min. The AS/RS computer is also networked in the LAN of the FMS for necessary data processing and integrated operations. Several software modules continue to be developed to enhance AS/RS functioning so as to include sorting of several types of work pieces, setting up pallets, rearrangements and real time changes and positioning and allocation of cranes, and optimum utilization of the AS/RS.

Work piece transport

The following devices are used for transporting the work pieces from the store to the machine.

    (a)   Rail-mounted transport vehicles (RTV)
    (b)   Automatically guided vehicles (AGV)
    (c)   Gantry robots (GR)

An RTV can carry work pieces from the store or set up stations to all those machines which can be interlinked with rails. This may sometimes impair access to the machines but offers a reliable transport travelling at high speed on linear routes.

AGVs, on the other hand, do not require any rails and transport even over non-linear routes and are widely employed. Most of these unmanned industrial trucks are inductively controlled and consequently move at lower speeds than RTVs. Several AGVs may be used in an FMS to cover all the routes.

Gantry robots which are essentially mobile may be used to load the work pieces directly on to the machines from the pallets which store work pieces. This, of course, requires robot grippers which suit the work piece geometry. Further the components should not have large dimensions and their weight should be small. It may be noted that, in this case, several of the expensive fixtures, which would otherwise be used to mount the work pieces on the pallets, are not required. Alternatively a gantry robot moving along a linear overhead gantry may be used to transport work piece-set pallets from the store or setup stations to the machines.

21.8 FMS CONTROL

All the elements present in an FMS such as the CNC machine tools, material handling units (AGVs), work pieces and tools are to be controlled in real time. This means that the FMS control software ensures for timely supply of tools, work pieces and programs to connected machine tools. This should have modules for:

(a) Production scheduling to schedule various production operations of the FMS based on the parts to be made entered at the remote job entry (RJE) terminal.
(b) Transport management to take care of the work piece and tool movements in FMS under the direct control of the FMS supervisor.
(c) Tool management to arrange the availability of the right tool in the right condition at the right time in the right place. This provides the necessary tool offsets required by the various programmers.
(d) Simulations a powerful tool, which can be used for the design purpose as well as for knowing the condition of the present status of the production operations within FMS.
(e) Production control coordinating various production operations of the FMS modules by direct communications with their controllers (CNC, AGV, etc.).
(f) Machine diagnostics to obtain any malfunctions of the FMS modules.
(g) Managing part programs, data files of tools and work pieces and their storage positions.
(h) Maintenance planning based on the feedback on the health of the FMS components to properly plan the maintenance schedules.

Some of the operations shown are time critical and hence need to be done on real time basis, while many functions are not time critical and hence can be done when the processing time is available. For this purpose, a distributed control is generally preferred with the real time control done with a slave computer dedicated for the application.

The typical operating procedure of an FMS is as follows:

1. In the beginning of the day's job orders will be entered, either directly or downloaded from the plant computer.
2. The system checks that the route sheet and process plans for all the jobs to be done are available. Else action has to be taken to initiate to obtain this information. In the mean time those jobs are to be removed from the job list that can re-enter when the process plans are available.
3. Check for the availability of the machine tools, raw material and cutting tools. It is necessary to ensure that the required tools and work pieces in a form suitable for immediate use be made available before the start of the day. For this purpose, it is possible to have a look ahead capability in the software to make a trail schedule run a little earlier to provide instructions to the tool crib and work preparation areas.
4. Once having all the equipments in place, schedule the parts based on the priorities assigned. Make a simulation run of the schedule to see that everything is in order. Based on the prepared or feasible schedule create the job list sequence and create the various tasks in chronological order to be executed by the FMS controller.
5. Send tool orders to the tool crib and fixture and blank orders to the work preparation areas.

6. Start executing the created sequence of tasks.

7. It is possible to enter any new orders, which may change the created task chronological order. Also some priorities for the jobs may be changed in between, which needs rescheduling of the remaining tasks and provision may need to be provided in the software for this function as well.

8. The current status of jobs can be seen on the FMS terminal from where manual intervention by the supervisor can be initiated when necessary. In addition, a number of reports on the various elements of FMS can always be made available. This will help in the initiation and implementation of optimal decisions by the supervisor when necessary.

FMS controller functions

Typical FMS controller software functions and their interface with the hardware are listed as follows:

Capacity planning module checks the FMS for the early due date of the parts released for the FMS. This will be able to specify as a rough guide the following information, which needs to be accepted by the user:

(a) Machine loading

(b) Pallet/fixture requirement

(c) Due dates for completion of the parts

Selection of the planned order depends upon:

(a) Tool mix per machine tool magazine

(b) Transport units required

(c) Machine capacity/loading required

(d) Part programs required

Fine planning module is meant to tweak the created schedule by the operator for any specific requirement. In order to carry out the software functions, the data related to the FMS should be properly presented. Master data have to be setup only once when the FMC is put into operation. This consists of the system specific data relating to the architecture of the system in terms of the machine tools and system specific data relating to their architecture of the system in terms of the machine tools and material handling equipment present, their relative positioning and other information of the layout .

Resources data in terms of the resource required for the manufacturing of the specific part mix are to be entered. These are tool master data and work piece carrier master data (data relating to work piece carrier and clamping fixtures).

Control data (product specific data) are the actual resources required to work on the specific part mix in question such as NC programs, tool layouts and work schedules (technical control data) and manufacturing orders (organizational control data).

Manufacturing control

This is the most complex of all the modules in the FMS software. It has to execute the schedule created in real time and make corrections to it continuously depending upon the exigencies in the system operation. It has to carry out the real time control of the work stations, load/unload stations, pallets, human operators, tools and material handling units. This has to be continuously done with the aim of optimizing the utilization of all the resources of the FMS.

Monitoring

It continuously monitors the various functional units within FMS in order to achieve the unattended operation. Some of the functions monitors are:

(a) CNC machine tool status
(b) Tool life
(c) Cutting abnormalities
(d) Tool breakages
(e) Error diagnostics

Tool crib

In the tool crib, which receives orders from planning for the assembly of tools required for the specific parts, after the tools are assembled, they are to be set and tool offset is measured in the setting equipment. The tool offset files will be entered into the FMS computer and also the tools will be kept ready for transportation. The necessary information may be sent to the transport module to arrange for the transport service.

Work preparation

The instructions in terms of which parts need to be setup or unloaded from the pallet will be issued. Based on which the operation will be carried out normally by the human operator and then communicate the data related to the pallet offsets to the machine tool. Further information need to be send to the transport for getting the AGV.

Transport

This module will have to administer the use of the AGVs for various requirements generated within FMS. Typical transport orders could be

(a) Pallet from load area to machine tool or buffer
(b) Pallet from machine tool to buffer or unload area
(c) Tools from tools store to the tool magazine
(d) Tools from machine tool to tool crib

Machine tools

All the information required for the running of the work stations is coordinated by this module.

The following data will be generated during the course of the functioning of the FMS: Status data to describe the current situation with regard to resources in FMS such as plant status data, resource data in terms of work piece carrier data and tool data. Log data is operational data and machine data required for later analysis (diagnostics) are recorded, evaluated by the software function modules or filed with details of data and time of day (logged); such data include:

Status and messages

Machine specific messages from the various work stations present in FMS such as those from CNC, PLC, handling devices and transport system

Status and operational messages such as NC start, NC end and NC program run time.
Alarms such as machine fault are breakdown
Tool specific messages such as tool breakage or end of tool life
NC messages such as load NC program

21.9 FMS LAYOUT CONFIGURATIONS

The types of layout configuration for FMS are:

(a) In-line
(b) Loop
(c) Ladder
(d) Open field
(e) Robot centred cell

In-line configuration

The in-line configuration is shown in Fig. 21.17. It is most appropriate for systems in which part progress one workstation to the next in a well defined sequence with no back flow. But depending on the flexibility and storage features of the handling system, it is possible to back flow in primary handling system. The back flow is made convenient with the arrangement of the shuttle flow at workstation as shown in Fig. 21.18.

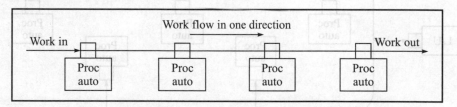

Fig. 21.17 Basic in-line configurations.

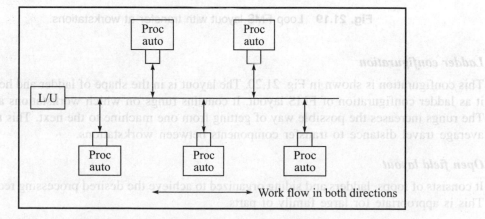

Fig. 21.18 Line with transfer at workstations.

Loop configuration

The loop configuration is shown in Fig. 21.19. Parts usually flow in one direction around the loop with the capability to stop at any station. The load/unload stations are located at one end of the loop.

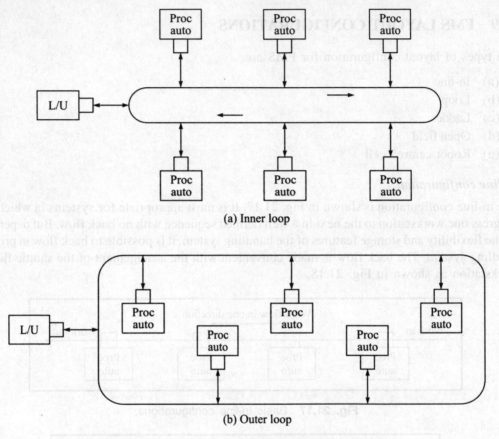

(a) Inner loop

(b) Outer loop

Fig. 21.19 Loop FMS layout with transfer at workstations.

Ladder configuration

This configuration is shown in Fig. 21.20. The layout is in the shape of ladder and hence named it as ladder configuration of FMS layout. It contains rungs on which workstations are located. The rungs increases the possible way of getting from one machine to the next. This reduces the average travel distance to transfer components between workstations.

Open field layout

It consists of loops, ladders and siding organized to achieve the desired processing requirements. This is appropriate for large family of parts.

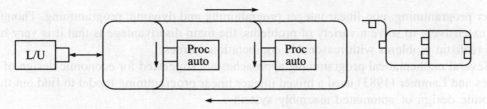

Fig. 21.20 Ladder type FMS layout.

Robot centred cell

It is a relatively new form of flexible system in which one or more robots are used as the material handling system. Industrial robots can be equipped with grippers that make them well suited for the handling parts. Robot centred cell is shown in Fig. 21.21.

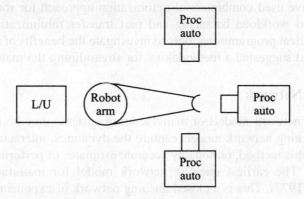

Fig. 21.21 Robotic centred cell.

21.10 FMS MODELLING AND ANALYSIS

Flexible manufacturing systems are complex for study and analysis. Researches have used a number of mathematical and simulation tools, techniques to model and solve these problems. These basic tools and techniques are:

- (a) Mathematical programming
- (b) Queuing networks
- (c) Markov models
- (d) Perturbation analysis
- (e) Simulation
- (g) Petri nets

21.10.1 Mathematical Programming

Mathematical programming has been used to formulate and solve some of the design, planning and scheduling problems. Different formulations that are used include linear programming, linear integer programming, non-linear integer programming and dynamic programming. Though it is used extensively to solve a variety of problems, the main disadvantage is that it is very hard to solve realistic problems with reasonable computational times.

Several mathematical programming approaches are suggested for economic design of FMS. Graves and Lammer (1983) used a mixed integer linear programming model to find out the best economic design of automated assembly systems.

Kusiak A (1984) discussed various clustering algorithms and used integer programming to address issues of scheduling, planning, etc. Some of the researchers have tried to solve the complex operational level problems with mathematical programming. Wilson (1987) suggested a methodology to formulate and solve a set of sequencing problems using integer programming. Chang and Sullivan (1984) used an extremely large mixed integer programming model for solving realistic problems in a system. Chen and Chung (1991) have presented three loading and routing models which aim at maximizing the operation assignments and the inherent FMS flexibility. Arbib et al (1990) have used combinatorial optimization approach for routing and scheduling. They have considered workload balancing and part transfer minimization as objectives. Co (1992) used mathematical programming tools to investigate the benefits of restricting production flexibility of FMS and suggested a methodology for streamlining the material flow.

21.10.2 Queuing Network

A survey of queuing network models of manufacturing systems may be found in Suri (1984), Buzacott (1986). Queuing network models capture the dynamics, interactions and uncertainties in the system. Using this method, reasonably accurate estimates of performance can be obtained in an efficient way. The earliest queuing network model for manufacturing systems was CAM-Q by Solberg (1977). This is a closed queuing network of exponential serves with single customer class, computationally simple, qualitatively accurate but highly aggregated.

Queuing network models are useful in the preliminary design of manufacturing systems but are not accurate enough at the detailed design/operations stage. Also, when non-product form feature such as blocking, priority queuing disciplines, synchronization and multiple resource holding are manifested in the given system, approximate techniques will have to be used leading to less accuracy and decreased efficiency.

21.10.3 Markov Model

In this approach, the given system is modelled as a discrete parameter or continuous parameter Markov chain or as a semi-Markov process with discrete state space. It assumes existence of a steady state and the performance estimates are obtained as functions of the steady state probabilities. Two major problems in this approach are state space explosion and complexity of state space generation. The latter arises due to the large number and non-trivial nature of interactions in FMS.

21.10.4 Perturbation Analysis

This is a recently developed technique by Ho (1985) which enables parameter sensitivities to be computed on line, in real time. A typical application of perturbation analysis to compute sensitivities of performance measure to input parameter variations is reported in Suri (1985). Perturbation analysis works by observing a single experiment called the **nominal experiment**, which could be on the actual system or based on a simulation. By performing simple calculations, answers are obtained to questions such as what would be the performance if machines were faster or if there were more fixtures of a certain type. Perturbation analysis can handle detailed features of the systems but cannot predict accurately the effects of large changes in decisions.

21.10.5 FMS Simulation

Simulation is a useful computer technology in FMS modelling, design and operation. Simulation modelling allows one to describe real world objects in FMS, such as moving of work piece from one place to another. There are three approaches to simulation modelling for FMS. The first one network or graphical models, where some objects (such as machines) may be represented by graphical symbols placed in the same physical relationship to each other as the corresponding machines are in the real world. The graphical aspects of this kind of models are relatively easy to specify and once completed, they also provide a communication vehicle for the system design which can be readily understood by a variety of people. SLAM and SIMAN are two widely used network modelling tools for FMS.

The second approach to FMS modelling is called **data driven simulation**. The model consists of only (or mainly) numerical data. That information usually represents, for example, a simple count of machine in a system, or a table of operation times for each process on the route of a given part type. The nature of this information such that, if it were collected in the factory information system, it would only be necessary to access it and place it in proper format in order to run a simulation of the corresponding real world system. This concept is quite close to automated simulation. It has the ultimate easy for use. The first such programs for FMS was developed at Purdue. It was called the **general computerized manufacturing system (GCMS) simulator**.

The third approach for FMS modelling is using a base programming language, such as SIMULA and SIMSCRIPT. These base programming languages provide more model specific constructs that can be used to build a simulation model. So, it has a much stronger modelling capability. Another method for DEDS simulation called **activity cycle diagram (ACD)** can also be used in FMS simulation. This is a diagram used in defining the logic of a simulation model. It is equivalent to a flow chart of a general purpose computer program. The ACD shows the cycle for every entity in the model. Conventions for drawing ACD are as follows:

(a) Each type of entity has an activity cycle.
(b) The cycle consists of activities and queues.
(c) Activities and queues alternate in the cycle.
(d) The cycle is closed.
(e) Activities are depicted by rectangle and queues by circles or ellipses.

Figure 21.22 presents an ACD (activity cycle diagram) for a machine shop. Jobs are arriving from the outside environment. Jobs are waiting in a queue for the machine. As soon as the machine is available, a job goes to the machine for processing. Once processing is over, the job again joins a queue waiting to be dispatched.

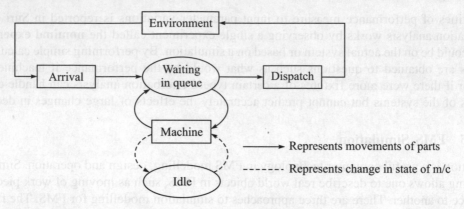

Fig. 21.22 Activity cycle diagram.

The ACD gives better understanding about the FMS to be simulated. So it is widely used for FMS simulation.

21.10.6 Petri Nets and its Applictions in FMS Modelling

A Petri Net (PN) may be identified as a particular kind of bipartite directed graphs populated by three types of objects. These objects are places, transitions and directed arcs connecting places to transitions and transitions to places. Pictorially, places are depicted by circles, transitions by bars or boxes. A place is an input place to a transition if there exists a directed arc connecting this place to the transition. A place is output places of a transition if there exists a directed arc connecting the transition to the place. Figure 21.23 represents a simple PN, where places p_1 and p_2 are input places to transition t_1, place p_3 is the output place of t_1.

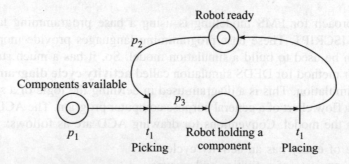

Fig. 21.23 Simple pertinent example.

Formally, a PN can be defined as five tuple PN = $(P, T\ I, O, m_0)$, where

1. $P = \{p_1, p_2, p \dots p_n\}$ is a finite set of places
2. $T = \{t_1, t_2 \dots t_m\}$ is a finite set of transitions, $P \cup T \neq \Phi$, $P \cap T = \Phi$
3. $I : (P \times T)\ |\rightarrow N$ is an input function that defines directed arcs from places to transitions, where N is a set of non-negative integer.
4. $(P \times T) \rightarrow N$ is an output function that defines directed arcs from transitions to places
5. $m_0 : P\ |\rightarrow N$ is the initial marking

The state of the modelled systems is represented by the tokens (small dots within the places) in every place. For example, in the Fig. 21.23 a small dot in place p_i means components available. The change of the states represents the system evolution. State changing is brought by firing a transition. The result of firing a transition is that for every place connected with the transition, after firing of the transition, a token will be removed from its input place and a token will be added to its output place. As an example of figure the firing of transition t_1 will cause the tokens in place p_1, p_2 disappear and a token will be added to place p_3.

Due to the advantage of its formal theory background, natural link with DEDS, and mature simulation tool, PN is well suited in FMS modelling. We give a two-machine production line [see Fig. 21.24] to demonstrate the modelling of FMS using PN.

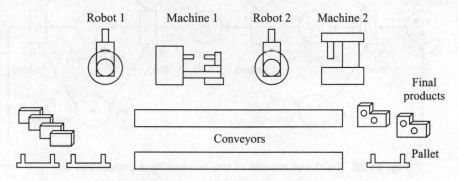

Fig. 21.24 A two-machine production line.

The production line consists of two machines (M_1 and M_2), two robots (R_1 and R_2) and two conveyors. Each machine is serviced by a dedicated robot that performs load/unload task. One conveyor is used to transport work piece, a maximum low at one time. The other conveyor is used to transport empty pallets. There are three pallets available in the system. Each work piece is machined on M_1 and then on M_2. The machining time is 10 time units on M_1 and 16 units on M_2. The load and unload tasks takes 1 time unit.

As the same with modelling general FMS or other systems, the modelling of this system using PN takes several steps.

Major activities are identified. In this example, they are R_1 loading; M_1 processing, R_1 unloading, R_2 loading, R_2 processing, R_2 unloading. The resources are raw materials with pallets, conveyors, M_1, M_2, R_1 and R_2. The relationships between the four major activities form a sequential order.

A partial PN model is defined to describe the four activities and their relations are shown in the Fig. 21.25(a). Here four transitions are used to represent four short operations, i.e., R_1 is loading, R_1 unloading, R_2 loading, R_2 unloading. Two places are used to represent two long operations, i.e., M_1 and M_2 processing.

Through stepwise process, gradually adding resources, constraints and links into the partial PN model will finally form the refined model as shown in Fig. 21.25(b).

The model is checked to use whether it satisfies the specification. The PN simulation tool can also be used in this phase to check the model. If some problems are found, the model will be modified.

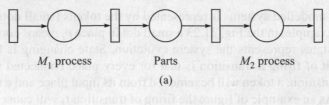

M_1 process   Parts M_2 process

(a)

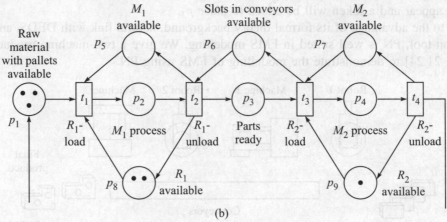

(b)

Fig. 21.25 Petri net models of the two-machine production line.

21.11 FMS BENEFITS

A number of benefits can be expected in successful FMS applications. The principal benefits are the following:

Increased machine utilization

FMSs achieve a higher average utilization than machines in a conventional batch production machine shop. Reasons for this include:

(a) 24 hr/day operation
(b) Automatic tool changing at machine tools
(c) Automatic pallet changing at workstations
(d) Queues of parts at stations
(e) Dynamic scheduling of production that takes into account irregularities from normal operations.

It should be possible to approach 80–90% asset utilization by implementing FMS technology.

Fewer machines required

Because of higher machine utilization, fewer machines are required.

Reduction in factory floor space required

Compared with a job shop of equivalent capacity, an FMS generally requires less floor area. Reductions in floor space requirements are estimated to be 40–50%.

Greater responsiveness to change

An FMS improves response capability to part design changes, introduction of new parts, changes in production schedule and product mix, machine breakdowns and cutting tool failures. Adjustments can be made in the production schedule from one day to the next to respond to rush orders and special customer requests.

Reduced inventory requirements

Because different parts are processed together rather than separately in batches, work-in-process (WIP) is less than in a batch production mode. The inventory of starting and finished parts can be reduced as well. Inventory reductions of 60–80% are estimated.

Lower manufacturing lead times

Closely correlated with reduced WIP (work-in-process) is the time spent in process by the parts. This means faster customer deliveries.

Reduced direct labour requirements and higher labour productivity

Higher production rates and lower reliance on direct labour translate to greater productivity per labour hour with an FMS than with conventional production methods. Labour savings of 30–50% are estimated.

Opportunity for unattended production

The high level of automation in an FMS allows it to operate for extended periods of time without human attention. In the most optimistic scenario, parts and tools are loaded into the system at the end of the day shift, and the FMS continues to operate throughout the night so that the finished parts can be unloaded the next morning.

21.12   FMS PLANNING AND IMPLEMENTATION ISSUES

Implementation of an FMS represents a major investment and commitment by the user company. It is important that the installation of the system be preceded by thorough planning and design and that its operation be characterized by good management of all resources: machines, tools, pallets, parts and people. Our discussion of these issues is organized along these lines:

- FMS planning and design issues
- FMS operational issues

21.12.1   FMS Planning and Design Issues

The initial phase of FMS planning must consider the parts that will be produced by the system. The issues are similar to those in GT machine cell planning. They include:

Part family considerations

Any FMS must be designed to process a limited range of part (or product) styles. The boundaries of the range must be decided. In effect, the part family that will be processed on the FMS must be defined. The definition of part families to be processed on the FMS can be based on the product commonality as well as on part similarity. The term *product commonality* refers to different components used on the same product. Many successful FMS installations are designed to accommodate part families defined by this criterion. This allows all of the components required to assemble a given product unit to be completed just prior to beginning of assembly.

Processing requirements

The types of parts and their processing requirements determine the types of processing equipment that will be used in the system. In machining applications, non-rotational parts are produced by machining centres, milling machines and like machine tools; rotational parts are machined by turning centres and similar equipment.

Physical characteristic of the work parts

The size and weight of the parts determine the size of the machines at the work stations and the size of the material handling system that must be used.

Production volume

Quantities to be produced by the system determine how many machines will be required. Production volume is also a factor in selecting the most appropriate type of material handling equipment for the system.

After the part family, production volumes, and similar part issues have been decided, design of the system can proceed. Important factors that must be specified in FMS design include:

Types of workstations. The types of machines are determined by part processing requirements. Consideration of workstations must also include the load/unload station(s).

Variations in process routings and FMS layout. If variations in process sequence are minimal then an inline flow is most appropriate. As product variety increases, a loop is more suitable. If there is significant variation in the processing, a ladder layout or open field layout are the most appropriate.

Material handling system. Selection of the material handling equipment and layout are closely related, since the type of handling system limits the layout selection to some extent. The material handling system includes both primary and secondary handling systems.

Work-in-process and storage capacity. The level of WIP allowed in the FMS is an important variable in determining utilization and efficiency of the FMS. If the WIP level is too low, then stations may becomes starved for work, causing reduced utilization. If the WIP level is too high, then congestion may result. The WIP level should be planned, not just allowed to happen. Storage capacity in the FMS must be compatible with WIP level.

Tooling. Tooling decisions include types and numbers of tools at each station. Consideration should also be given to the degree of duplication of tooling at the different stations. Tool duplication tends to increase routing flexibility.

Pallet fixtures. In machining systems for non-rotational parts, the number of pallet fixtures required in the system must be decided. Factors influencing the decision include: levels of WIP allowed in the system and differences in part style and size, parts that differ too much in configuration and size require fixturing.

21.12.2 FMS Operational Issues

Once the FMS is installed, then the existing resources of the FMS must be optimized to meet production requirements and achieve operational objectives related to profit, quality and customer satisfaction. The operational problems that must be solved include:

Scheduling and dispatching

Scheduling of production in the FMS is dictated by the master production schedule. Dispatching is concerned with launching of parts into the system at the appropriate times. Several of the problem areas below are related to the scheduling issue.

Machine loading

This problem is connected with allocating the operations and tooling resources among the machines in the system to accomplish the required production schedule.

Part routing

Routing decisions involve selecting the routes that should be followed by each part in the production mix to maximize use of work station resources.

Part grouping

This is concerned with the selection of groups of part types for simultaneous production, given limitations on available tooling and other resources at workstations.

Tool management

Managing the available tools includes decisions on when to change tools, allocation of tooling to workstations in the system and similar issues.

Pallet and fixture allocation

This problem is concerned with the allocation of pallets and fixtures to the parts being produced in the system.

21.13 APPLICATIONS OF FMS

Some of the applications of FMS are in the following areas:

    (a) Machining
    (b) Assembly
    (c) Sheet metal press working
    (d) Forging
    (e) Plastic injections
    (f) Welding

(g) Textile machinery manufacture

(h) Semiconductor component manufacture

21.14 AUTOMATED STORAGE AND RETRIEVAL SYSTEMS (AS/RS)

An automated storage and retrieval system is defined by the Materials Handling Institute as "A combination of equipment and controls which handles, stores and retriever materials with precision, accuracy and speed under a define degree of automation. In general, an AS/RS performs a basic set of operations without human intervention, regardless of the specific type of system that is employed".

The operations performed by AS/RS are

(a) Automatic removal of an item from a storage location

(b) Transportation of an item to a specific processing or interface point

(c) Automatic storage of an item in a predetermined location, having received an item from a processing or interface point.

21.14.1 Types of AS/RS

Unit load AS/RS

The unit load AS/RS is used to store and retrieve loads that are palletized or stored in standard size containers. The loads are generally over 500 lb per unit. In general, a unit load system is computer controlled, having automated S/R machines designed to handle unit load containers. Each S/R machine is guided by rails in the floor. On the frame of the S/R machine, it is a shuttle, which is the load supporting mechanism that moves loads to and from storage locations and the pick up and deposit stations. Usually, a mechanical clamped mechanism on the S/R machine handles the load. However, other mechanisms can be used such as a vacuum or a magnet based mechanism for handling sheet metal. A unit load As/Rs is pictured in Fig. 21.26.

Mini load AS/RS

A mini-load system is designed to handle small loads such as individual parts, tools and supplies. The system is suitable for use where there is a limit on the amount of space that can be utilized and where the volume is too low for a full scale unit load system and too high for a manual system. A smaller investment and flexibility of handling small items make it popular choice in industry.

Person-on-board AS/RS

It allows storage of items in less than unit load quantities. A person rides on a platform with the S/R machine to pick up individual items from a bin or drawer. This provides order picking ability, which can reduce the time it takes to fill an order. The operator can select the items and place them in a module, which is then carried by the S/R machine to the end of a conveyor to reach its destination. The platform may contain additional devices, some automatic to facilitate lifting heavy items.

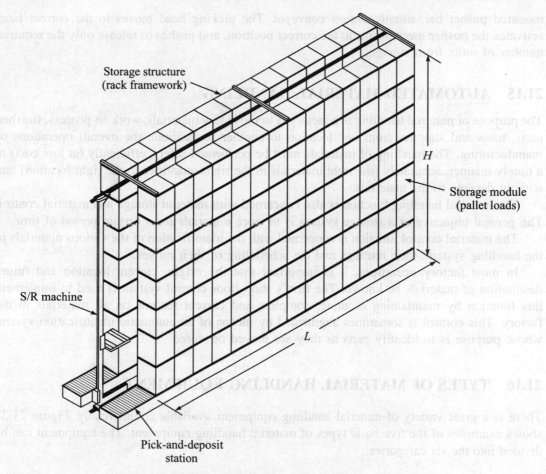

Storage structure
(rack framework)

Storage module
(pallet loads)

S/R machine

H

L

Pick-and-deposit
station

Fig. 21.26 A unit load automated storage/retrieval system.

Deep lane AS/RS

It is another type of unit load system. The items are stored in multi deep storage with up to 10
items per row rather than single or double deep. This leads to a high density of stored items,
permitting high usage of the unit. Each rack permits flow through of items, that is, an item is
deposited on one side of the storage rack and removed from the other side. The S/R vehicle
operates in the aisle and delivers loads to a rack entry vehicle. The rack entry vehicle is typically
a moving platform that carriers the load into the storage rack, deposits it there, and returns to
the S/R machine for the next load. The S/R machine is similar to the unit load S/R machine,
except that S/R machines have specialized functions such as controlling rack entry vehicles.

Automated item retrieval system

This system is designed for automatic retrieval of the individual items or cases for storage. The
storage system consists of items stored individually in a flow through system that can be
automatically release from storage and automatically brought to a required point. The items are
stored from the rear, as in the deep lane system, and are retrieved from the front by use of rear

mounted pusher bar, usually onto a conveyor. The picking head moves to the correct lane, activates the pusher mechanisms to the correct position, and pushes to release only the required number of units from storage.

21.15   AUTOMATED MATERIAL HANDLING

The purpose of material handling in a factory is to move raw materials, work-in-process, finished parts, tools and supplies from one location to another to facilitate the overall operations of manufacturing. The handling of materials must be performed safely, efficiently (at low cost) in a timely manner, accurately (the right materials in the right quantities to the right location) and without damage to the materials.

The material handling function is also concerned with material storage and material control. The general objective of a storage system is to store materials for a certain period of time.

The material control function is concerned with the identification of the various materials in the handling system, their routings and the scheduling of their moves.

In most factory operations, it is important that the origin, current location and future destination of materials be known. The firm's shop floor control system is used to implement this function by maintaining accurate, complete and current records on all materials in the factory. This control is sometimes augmented by means of an automatic identification system whose purpose is to identify parts as they are moved or stored.

21.16   TYPES OF MATERIAL HANDLING EQUIPMENT

There is a great variety of material handling equipment available commercially. Figure 21.27 shows examples of the five basic types of material handling equipment. The equipment can be divided into the six categories:

Hand trucks

Platforms with wheels for manual movement of items, unit loads and bulk materials. Examples include wheelbarrows, dollies, two-wheeled trucks, four-wheeled trucks, hand-lift or manually operated fork-lift trucks.

Powered trucks

Powered vehicle platform for mechanized movement item, unit loads and bulk materials. Driven by human, powered by battery gasoline or propane gas. Examples include walkie trucks, riding trucks, fork-lift truck side loaders, tractor-trailer trains and industrial crane trucks.

Cranes, monorails and hoists

Handling devices usually manually operated designed for lifting, lowering and transporting heavy objects. Examples include bridge cranes, gantry cranes, jib cranes, overhead monorails, hand and powered hoists.

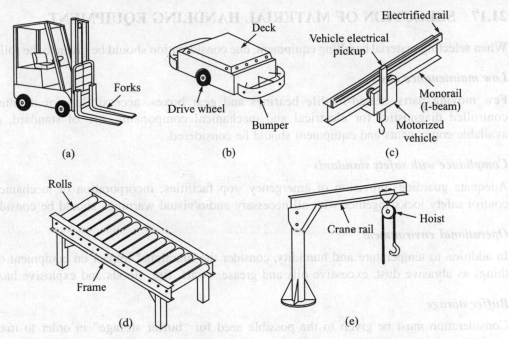

Fig. 21.27 Examples of the five basic types of material handling equipment.

Conveyors

Large family of handling devices, often mechanized, sometimes automated, designed to move material between specific locations over a fixed path, generally in large quantities or volumes. Examples include gravity conveyors, powered conveyors (rollers, belt, chain, overhead, in-floor two and cart-on-track).

Automated guided vehicle systems (AGVs)

They are battery-powered, automatically steered vehicle designed to follow defined pathways. Some are capable of automatically loading and unloading unit loads usually interfaced with other automated systems to achieve full benefits of integrated automation. Examples include driverless trains, pallet trucks and unit load carriers.

Other handling equipment

Industrial robots and other parts manipulators
Transfer mechanisms used in automated flow lines
Dial indexing tables
Elevators
Pipelines
Containers
Highway tractor-trailers
Railway trains
Cargo aircraft
Ships, barges and other marine vessels

21.17 SELECTION OF MATERIAL HANDLING EQUIPMENT

When selecting material handling equipment, due consideration should be given to the following:

Low maintenance costs

Few moving parts, sealed-for-life bearings and gear boxes, accessibility for maintenance, controlled diagnostics for electrical and mechanical components, use of standard, readily available components and equipment should be considered.

Compliance with safety standards

Adequate guarding, provision of emergency stop facilities, incorporation of mechanical and control safety locks together with all necessary audio/visual warnings should be considered.

Operational environment

In addition to temperature and humidity, consider the functional impact on equipment of such things as abrasive dust, excessive oils and grease, swarf, fire, floods and explosive hazards.

Buffer storage

Consideration must be given to the possible need for "buffer storage" in order to maximize production equipment utilization and ensure continuous operation.

Effective use of operating time

The maximization of operating time of high cost dedicated units, e.g. robots is an important factor in the development of an economic manufacturing system.

21.18 PRINCIPLES OF MATERIAL HANDLING (CIC MHE)

The principles for material handling are:

Principle 1   Unit load principle. Unit loads shall be appropriately sized and configured in a way which achieves the material flow. A unit load is one that can be stored or moved as a single entity at one time, such as a pallet, container or tote, regardless of the number of individual items that make up the load.

Less effort and work are required to collect and move many individual items as a single load than to move many items one at a time. Large unit loads are common both pre- and post-manufacturing in the form of raw materials and finished goods.

Principle 2   System principle. Material movement and storage activities should be fully integrated to form a coordinated, operational system that spans receiving, inspection, storage, production, assembly, packaging, unitizing, order selection, shipping, transportation and the handling of returns.

Principle 3   Systems flow principle. The flow of materials should be integrated with the flow of information in handling and storage systems. The information for each item moved should include identification, origination (pickup) point and destination point.

Principle 4   Automation principle. Material handling operations should be mechanized and/or automated where feasible to improve operational efficiency, increase responsiveness, improve consistency and predictability, decrease operating costs and eliminate repetitive or potentially unsafe manual labour.

Principle 5   Space utilization principle. Effective and efficient use must be made of all available space. When transporting loads within a facility, the use of overhead space should be considered as an option. Use of overhead material handling systems saves valuable floor space for productive purposes.

Principle 6   Work principle. Material handling work should be minimized without sacrificing productivity or the level of service required for the operation. The measure of material handling work is flow rate (volume weight or count per unit of time) multiplied by distance moved.

Consider each pickup and set-down, or placing material in and out of storage, as distinct moves and components of the storage, as distinct moves and components of the distance moved.

Simplifying processes by reducing, combining, shortening or eliminating unnecessary moves will reduce work. Where possible gravity should be used to move materials or to assist in their movement while respecting consideration of safety and the potential for product damage.

Principle 7   Shortest test distance principle. Movement of materials should be over the shortest distance possible. This depends on the plant layout design. This principle is implemented best by appropriate layout planning, i.e., by locating the production equipment into a physical arrangement tends to minimize the distances that must be travelled by the materials being processed.

Principle 8   Straight-line flow principle. The material handling path should be in a straight line from the point of origination to the point of destination. This rule is consistent with the shortest distance principle.

Principle 9   Minimum terminal time principle. Movement of a unit load consists of the move time plus the time required for loading, unloading and other activities that do not involve actual transport of the materials. Minimize these non-move times.

Principle 10   Part orientation principle. In automated production systems, the orientation of the workpart should be established and maintained throughout the material handling process.

Principle 11   Carry loads both ways. The handling system should be designed and scheduled to the extent possible, to carry loads in both directions. Return trips with empty loads are wasteful.

21.19   CONVEYOR SYSTEMS

A conveyor system is used when materials must be saved in relatively large quantities between specific locations over a fixed path. Most conveyor systems are powered to move the loads along the pathways. Other conveyors use gravity to cause the load to travel from one elevation in the system to a lower elevation. Conveyors have the following attributes:

(a) They are generally mechanized and sometimes automated.

(b) They are fixed-in-position to establish the paths.

(c) They can be either floor mounted or overhead.

(d) They are almost always limited to one-directional flow of materials.

(e) They generally move discrete loads but certain types can be used to move bulk or continuous loads.

(f) They can be used for delivery-only or delivery-plus-storage of items.

Types of conveyors

Roller conveyors

Skate-wheel conveyors

Belt conveyors

Chain conveyors

Slat conveyors

Overhead trolley conveyors

In-floor towline conveyors

Cart-on-track conveyors

21.19.1 Roller Conveyors

This is a very common form of conveyor system. The pathway consists of series of tubes (rollers) that are perpendicular to the direction of travel. Figure 21.28 shows a roller conveyor. The rollers are contained in a fixed frame which elevates the pathway above floor level from several inches to several feet. Flat pallets or tote pans carrying unit loads are moved forward as the rollers rotate. Roller conveyors can be either powered or gravity types. The powered types are driven by any of several different mechanisms, belts and chains are common.

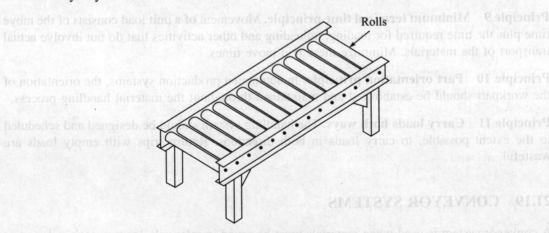

Rolls

Fig. 21.28 Roller conveyor.

The gravity types are arranged so that the pathway is along a downward slope sufficient to overcome rolling friction.

Roller conveyors can be used for delivering loads between manufacturing operations, delivery to and from storage and distribution applications.

Automated systems of conveyors are useful for merging and storing operations.

21.19.2 Skate-wheel Conveyors

These are similar in operation to the roller conveyors. Instead of rollers, skate wheels rotating on shafts connected to the frame are used to roll the pallet or tote pan or other container along the pathway. Figure 21.29 shows a skate-wheel conveyor.

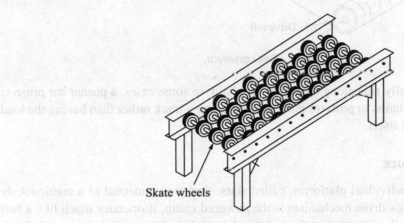

Skate wheels

Fig. 21.29 Skate-wheel conveyor.

The applications of skate-wheel conveyors are similar to those for roller conveyors, except that the loads must generally be lighter since the contact between the loads and the conveyor are much more concentrated.

21.19.3 Belt Conveyors

This type is available in two common forms: flat belts for pallets, parts or even certain types of bulk materials and troughed belts for bulk materials. Materials are placed on the belt surface and travel along the moving pathway. Figure 21.30 shows a Belt conveyor. The belt is made into a continuous loop so that half of its length can be used for delivering materials and the other half is the return run (usually empty).

The belt is supported by a frame that has rollers or other supports spaced every few feet. At the each end of the conveyor (where the belt loops back) are driver rolls (pulleys) that power the belt.

21.19.4 Chain Conveyors

Chain conveyors are made of loops of endless chain in an over-and-under configuration around powered sprockets at the ends of the pathway. There may be one or more chains operating in parallel to form the conveyor. The chains travel along channels that provide support for the flexible chain sections. Either the chains slide along the channel or they use rollers to ride in the

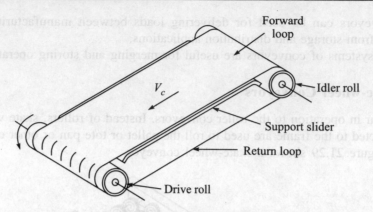

Fig. 21.30 Belt conveyor.

channel. The loads generally ride along the top of the chain; in some cases, a pusher bar projects up between two parallel chains to push (or pull) the load along a track rather than having the load ride directly on the chain itself.

21.19.5 Slat Conveyors

The slat conveyor uses individual platforms, called **slats**, that are connected to a continuously moving chain. Although its drive mechanism is the powered chain, it operates much like a belt conveyor. Loads are placed on the flat surface of the slats and are transported along with them.

Straight-line flows are common in slat conveyor systems. However, because of the chain drive and the capability to alter the chain direction using sprockets, the conveyor pathway can have turns in its continuous loop.

21.19.6 Overhead Trolley Conveyors

A trolley in material handling is a wheeled carriage running on an overhead rail from which loads can be suspended. A trolley conveyor consists of multiple trolleys, usually equally spaced along the rail system by means of an endless chain or cable. Figure 21.31 shows an overhead trolley conveyor. The chain or cable is attached to a drive wheel that supplies power to the system.

The path is determined by the configuration of the rail system. It has turns and changes in elevation to form an endless loop: suspended from the trolleys are hooks, baskets or other receptacles to carry the loads. Overhead trolley conveyors are often used in factories to move parts and assemblies between major production departments. They can be used both for delivery and storage purposes.

21.19.7 In-floor Towline Conveyors

These conveyors make use of wheeled carts powered by means of moving chains or cables located in trenches in the floor. The chain or cable is called the **towline**. Figure 21.32 shows an In-floor towline conveyor. The path ways of the conveyor system; switches between powered

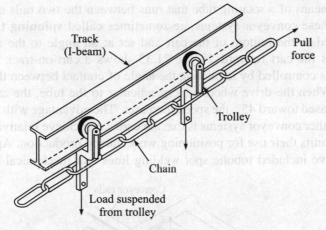

Fig. 21.31 Overhead trolley conveyor.

pathways are possible in the towline system to achieve some flexibility in the handling routes. The carts use hardened steel pins (dowels) that project below the floor surface into the trench to engage the chain for towing. The pin can be pulled out of the trench to disengage the cart for (loading, unloading, accumulation of parts, etc.) off the main pathway. Tow line systems are used for moving unit loads in plants and warehouses.

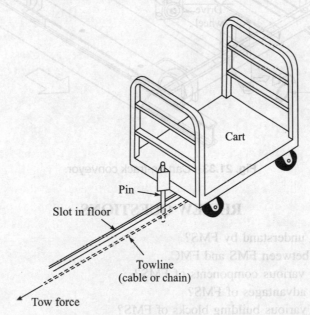

Fig. 21.32 In-floor towline conveyor.

21.19.8 Cart-on-Track Conveyors

These conveyor systems use individual carts riding on a two-railed track contained in a frame that places the track a few feet above floor level. The carts are not individually powered; instead, they are driven by means of a rotating tube that runs between the two rails owing to the use of the rotating tube. These conveyor systems are sometimes called **spinning tube conveyors**. A drive wheel, attached to the bottom of the cart and set at an angle to the rotating tube, rests against it and drives the cart forward. Figure 21.33 shows a cart-on-track conveyor.

The cart speed is controlled by regulating the angle of contact between the drive wheel and the spinning tube. When the drive wheel is perpendicular to the tube, the cart does not move. As the angle is increased toward 45°, the speed increases. The advantage with this type of system when compared to other conveyor systems is that the carts can achieve relatively high accuracies of position. This permits their use for positioning work during production. Applications of cart-on-track systems have included robotic spot welding lines and mechanical assembly systems.

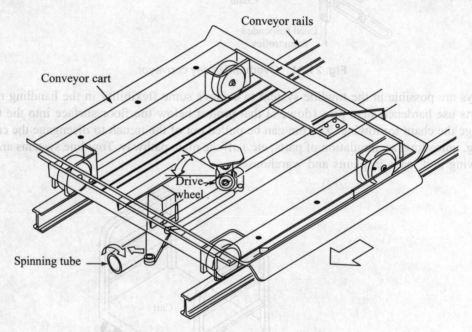

Fig. 21.33 Cart-on-track conveyor.

REVIEW QUESTIONS

21.1 What do you understand by FMS?

21.2 Differentiate between FMS and FMC.

21.3 What are the various components of FMS?

21.4 What are the advantages of FMS?

21.5 What are the various building blocks of FMS?

21.6 Discuss the various types of layouts used in FMS design. Explain briefly about their applications.

21.7 Discuss the machining systems of FMS.

21.8 What are the various methods available for tool monitoring in FMS?

21.9 Discuss the importance of material handling devices used in FMS.

21.10 Discribe the types of material handling devices used in FMS.

21.11 Discuss the tool management system of FMS.

21.12 What are the applications of AGVs in FMS?

21.13 Discuss the data files and system reports generated by the computer control system of FMS.

21.14 Compare FMS and transfer line automation for production rate and product variety.

21.15 Discuss the planning of FMS in an automated factory.

21.16 How is an FMS optimized?

21.17 Classify and discuss various flexible manufacturing systems.

21.18 Explain the term flexibility in FMS.

21.19 Describe the principal objectives of FMS.

21.20 What is the need of tool and work piece coding? Explain with an example.

21.21 Explain various principles of tool monitoring? How does it help the productivity in FMS?

21.22 Discuss with the help of an information flow in a tool management system of FMS.

21.23 Compare the advantages and suitability of various methods of work piece transport in FMS.

21.24 What are the limitations of FMS?

21.25 Discuss the following terms in FMS:

    (a) Machine flexibility

    (b) Routing flexibility

    (c) Production flexibility

    (d) Expansion flexibility

21.26 Discuss the applications of FMS.

21.27 Explain the different types of machines used in FMS workstations.

21.28 Compare FMS with other types of manufacturing approaches.

21.29 Define automated storage and retrieval system.

21.30 Explain the operation of an automated storage and retrieval system.

21.31 What are the major components present in an automated storage and retrieval system.

21.32 Explain briefly the advantages to be gained by the use of automated storage and retrieval system.

21.33 Write short notes on the automatic storage and retrieval systems and their application areas in FMS.

COMPUTER INTEGRATED MANUFACTURING

22

INTRODUCTION

Computers are controlling machines and robots are performing processes. Both are doing it far more efficiently than human operators. The high degree of automation that was recently reserved for mass production only, is applied now, with the aid of computers and robots also to small batches.

This requires a change from hard automation in the production line to flexible automation which can more readily be rearranged to meet new requirements of the customer. Flexible manufacturing systems (FMS) combine with automatic assembly and product inspection on one hand and CAD/CAM system on the other are the basic components of the futuristic factory. The overriding and encompassing world trend in the automation of manufacturing is towards computer integrated manufacturing (CIM). It is this technology which holds the capability to accomplish overall automation of the total system of manufacturing. Computer integrated manufacturing encompasses the entire range of product development and manufacturing activity with all the functions being carried out with the help of application software packages. All the manufacturing and related activities are carried out under computer control. CIM includes all of the engineering functions of CAD/CAM but it also includes the business functions of the firm as well. CIM reduces the human component of manufacturing and thereby relieves the process of its slow expensive and error-prone component.

CIM stands for a methodological approach to all the activities from the design of the product to customer support in an integrated way, using various methods, means and techniques in order to achieve production improvement, cost reduction, meeting scheduled delivery dates, quality improvement and total flexibility in the manufacturing system. CIM covers the whole lot of related technologies including total quality management, business process reengineering, concurrent engineering, work flow automation, enterprise resource planning and flexible manufacturing. CIM is the term used to describe the most modern approach to manufacturing. CIM makes use of the capabilities of the computer for manufacturing. Two of them are variable and programmable automation and real time optimization. The primary objective of such a computer-integrated manufacturing system is to deal effectively with the real-time analysis, planning and control of the manufacturing process.

The effectiveness of CIM greatly depends on the use of a large-scale integrated communications system, involving computers, machines and their controls. Furthermore, because CIM ideally should involve the total operation of a company, it requires an extensive database aspects of the operation. Consequently, if planned all at once, CIM can be prohibitively expensive, particularly for small and medium-size companies.

22.1 COMPONENTS OF CIM SYSTEM

The five main components of the CIM system are:

(a) Business planning and execution
(b) Product design
(c) Manufacturing process planning
(d) Process automation and control
(e) Factory-floor monitoring systems

Business planning and execution

Business planning functions include activities such as forecasting, scheduling, material requirement planning, invoicing and accounting. Business execution functions include production and process control, material handling, testing and inspection.

Product design

The design department of the company establishes the initial database for production of a proposed product. In a CIM system, this is accomplished through activities such as geometric modelling and computer aided design while considering the product requirements and concepts created by the creativity of a design engineer. The design process should be constrained by the costs that will be incurred in actual production and by the capabilities of the existing production equipment and processes.

Manufacturing process planning

The planning department takes the database established by the product design and enriches it with production data and information of the product. In a CIM system, this planning process should be constrained by the production costs and by the production equipment and process capability to generate an optimized plan.

Process automation and control

This requires further enrichment of the database with performance data and information about the production equipment and processes. In a CIM system, this requires activities like NC programming simulation and computer aided scheduling of the production activity. This should include process automation and control based on the real-time performance of the equipment and processes to assure continuous optimization of the production activity.

Factory-floor monitoring systems

Factory automation equipment further enriches the database with equipment and process data and information resident either in the operator or the equipment to carry out the production process. In CIM system this consists of computer controlled process machinery such as CNC machine tools, flexible manufacturing systems (FMS), computer controlled robots, material handling systems, computer controlled assembly systems, flexibly automated inspection systems and so on.

22.2   TYPES OF MANUFACTURING SYSTEMS

The middle range shown, in Fig. 22.1, covering the medium part variety and medium production volume, can be further divided into finer categories. These categories represent different levels of compromise between the objective of flexibility versus production capacity. Kearney and Trecker Corporation defines three types of manufacturing systems to satisfy the variety of processing needs within this middle range. They are:

(a)   Special manufacturing system

(b)   Manufacturing cell

(c)   Flexible manufacturing system (FMS)

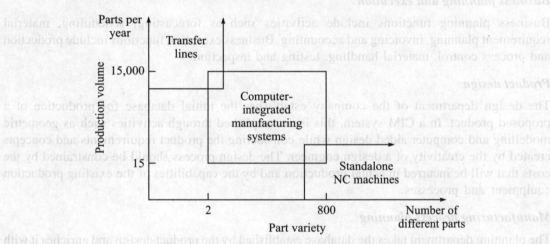

Fig. 22.1   General application guideline of each of the three types.

Figure 22.2 illustrated the general application guide lines of each of the three types. The special manufacturing system is the least flexible computer-integrated manufacturing system. It is designed to produce a very limited number of different parts (perhaps two to eight) in the same manufacturing family. The annual production rate per part would typically lie between 1500 and 15,000 pieces. The configuration of the special system would be similar to the high-production transfer line. The variety of processes would be limited, and specialized machine tools would not be uncommon.

At the opposite end of mid-volume range is the manufacturing cell. It is the most flexible but generally has the lowest production rate of the three types. The number of different parts manufactured in the cell might be between 40 and 800, and annual production levels for these parts would be between 15 and 500.

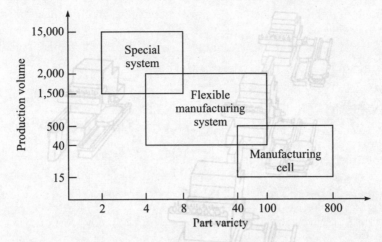

Fig. 22.2 Application guidelines for the types of computer-integrated manufacturing systems.

Figures 22.3 and 22.4 illustrate the special system and the manufacturing cell, respectively. The highly integrated and inline flow is evident in the workpart handling system. As pictured, the manufacturing cell might consist of several separate NC machines without an interconnecting materials handling system.

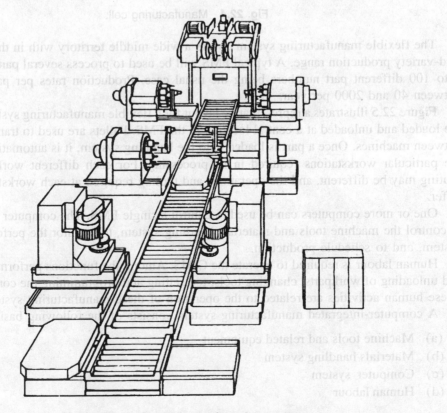

Fig. 22.3 Special manufacturing system.

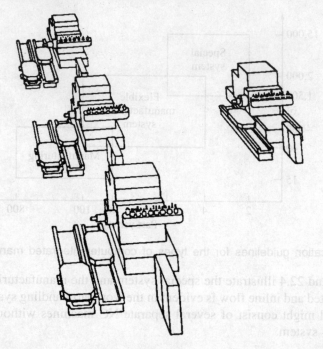

Fig. 22.4 Manufacturing cell.

The flexible manufacturing system covers a wide middle territory with in the mid-volume, mid-variety production range. A typical FMS will be used to process several part families, with 4 to 100 different part numbers being the usual case. Production rates per part would vary between 40 and 2000 per year.

Figure 22.5 illustrates a representative layout for a flexible manufacturing system. Workparts are loaded and unloaded at a central location in the FMS. Pallets are used to transfer workparts between machines. Once a part is loaded onto the handling system, it is automatically routed to the particular workstations required in its processing. For each different workpart type, the routing may be different, and the operations and tooling required at each workstation will also differ.

One or more computers can be used to control a single FMS. The computer system is used to control the machine tools and materials handling system, to monitor the performance of the system, and to schedule production.

Human labour is required to operate the CIMS. Among the functions performed are loading and unloading of workparts, changing tools, toolsetting and programming the computer system. These human activities are related to the operation of these manufacturing systems.

A computer-integrated manufacturing system consists of the following basic components:

(a) Machine tools and related equipment
(b) Materials handling system
(c) Computer system
(d) Human labour

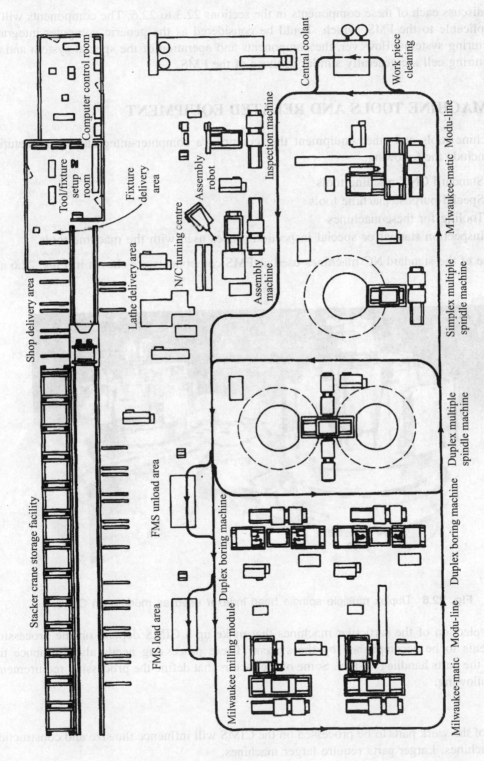

Fig. 22.5 Sample FMS layout.

We discuss each of these components in the sections 22.3 to 22.6. The components will be most applicable to the FMS, which should be considered as the generic computer-integrated manufacturing systems. However, the components and operation of the special system and the manufacturing cell are generally similar to those of the FMS.

22.3 MACHINE TOOLS AND RELATED EQUIPMENT

The machine tools and other equipment that comprise a computer-integrated manufacturing system include the following:

(a) Standard CNC machine tools
(b) Special-purpose machine tools
(c) Tooling for these machines
(d) Inspection stations or special inspection probes used with the machine tools

Some of the standard NC machines used as FMS components are shown in Figs. 22.6 and 22.7.

Fig. 22.6 Duplex multiple spindle head indexer used as module on CIMS.

The selection of the particular machines that make up a CIMS depend on the processing requirements to be accomplished by the system. These processing needs also influence the design of the parts handling system. Some of the factors that define the processing requirements are the following.

Part size

The size of the work parts to be processed on the CIMS will influence the size and construction of the machines. Larger parts require larger machines.

Fig. 22.7 Machining centre module used on CIMS.

Part shape

Machined workparts usually divide themselves naturally into two types accordingly to shape: round and prismatic. Round parts such as gears, disks, shafts require turning and boring operations. Prismatic workparts, which are cube-shaped and non-rotational, require milling and drilling operations.

Part variety

If the part variety is limited, the machine tools would be more specialized for higher production. The CIMS would be designed as a special system. If a wide variety of parts are to be processed, standard machine tools which are more versatile would be selected.

Product life cycle

The influence of product life cycle is similar to that of part variety. If the product life is relatively long, the CIMS can include more specialized and less flexible machine tools.

Definition of future parts

Another factor that affect the versatility of the CIMS is the level of knowledge about parts which are to be processed. We can distinguish two cases. The first case is where the manufacturing system is designed to process a family of parts that are completely known in advance. An example of this case is the dedicated manufacturing system for machining components of the XM-1 tank. The range of components is precisely known at the time of CIMS design and the system can be configured to meet these specific needs. The other case is where the future parts are not known in advance. New part designs must be accommodated by the system. Its machine tools must posses a significant degree of flexibility.

Operations other than machining

Most computer-integrated manufacturing systems are designed for machining exclusively. In some cases, the processing requirements include other operations, such as assembly or inspection.

22.4 MATERIAL HANDLING SYSTEM

The material handling system in a CIMS must be designed to serve two functions. The first function is to move workparts between machines. The second function is to orient and locate the workparts for processing at the machines. These two functions are often accomplished by means of two different but connected materials handling systems. We shall refer to them as the primary handling system and the secondary handling system.

The primary work handling system is used to move parts between machine tools in the CIMS. The requirements usually placed on the primary material handling system are:

(a) It must be compatible with computer control.
(b) It must provide random, independent movement of palletized workparts between machine tools in the system.
(c) It must permit temporary storage or banking of workparts.
(d) It should allow access to the machine tools for maintenance, tool changing and so on.
(e) It must interface with the secondary work handling system.

The term *random, independent movement of parts* means that the parts must be able to flow from any one station (machine tool) to any other station. The requirement is not always necessary in the case of the manufacturing system, which may involve an inline flow of parts based on a fixed processing sequence.

The secondary parts handling system must present parts to the individual machine tool in the CIMS. The secondary system generally consists of one transport mechanism for each machine. The specifications placed on the secondary materials handling system are:

(a) It must interface with the primary handling system. Parts must be transferred automatically between the primary system and secondary system.
(b) It must be compatible with computer control.
(c) It must permit temporary storage of workparts.
(d) It must provide for parts orientation and location at each workstation for processing.
(e) It should allow acess to the machine tool for maintenance, tool changing and so on.

An illustration of the primary and secondary work handling systems is found in Kearney and Tracker's FMS concept (see Fig. 22.8). The primary work handling system is an under-the-floor towline system which pulls a series of carts between the different workstations. The carts have four wheels which roll on the floor surface. Slots in the floor define the permissible pathways for the carts. These paths may branch and merge in a manner similar to the one depicted in

Fig. 22.8 Infloor towline cart system combined with pallet shuttle system on a flexible manufacturing system.

Fig. 22.5. The layout of the pathways depend on the design of the FMS. Guide pins at the front and back of the carts engage the slot in the floor to follow the correct path. The front guide pin engages a moving chain in the slot, which propels the cart.

The secondary work handling system is the shuttle system at each machine. The shuttles are designed to transfer palletized parts to and from the towline carts, and to move them onto and off the machine tool table for accurate registration with the cutting tools. If the processing cycle at the workstation is relatively short, the shuttle system may also be required to maintain an inventory of parts ready for machining. Both the towline cart and shuttle system are shown in Fig. 22.8.

The use of both a primary and secondary work handling system is not always required to satisfy the material handling objectives of a CIMS. In some cases, the objectives can be satisfied with one-work handling system. In the special system illustrated in Fig. 22.3, the fact that all parts are processed in the same sequence means that the same roller conveyor system can fully accomplish the inline flow of parts needed for this manufacturing system.

A relatively recent innovation in the design of flexible manufacturing systems involves the use of industrial robots to perform a portion of the parts handling chore. The geometry of the workparts to be processed on the CIMS has a big influence on the type of work handling system. Round workparts have been found to be ideal candidates for handling by industrial robots. The robot is physically located in the centre of a group of machine tools and transfers parts from one machine to the next in an independent, random fashion depending on processing requirements for each part. Figure 22.9 shows the layout for a robotic manufacturing cell.

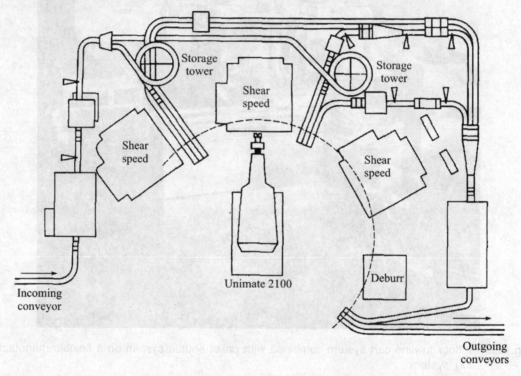

Fig. 22.9 Layout of manufacturing cell with robot used for parts transfer between machines.

It has been estimated that 75% of all rotational parts can be accommodated in a machining cell which uses a robot as the handling mechanism. Most round parts are within the lift capacities of the common commercially available robots. The design problem is to develop the gripper device that is suited to the variety of parts handled in the cell.

Prismatic parts which are processed by machining are usually heavier and larger. Robots cannot generally be used to transport these parts within the manufacturing system. The handling problem is typically solved by mounting the part on a pallet fixture and designing a materials handling system (primary system and secondary system) which will transport the standard-size pallet as required. Typical solutions include the following types of materials handling systems:

(a) Towline cart combined with shuttle system, in a layout similar to that of Fig. 22.5. Figure 22.3 illustrates this form of handling system.

(b) Guided vehicle combined with shuttle system. A wire-guided cart is illustrated in Fig. 22.10.

Fig. 22.10 Wire-guided cart used in variable mission manufacturing system. CNC machining centre in background.

A means of coordinating the activities of the materials handling system with those of the machine tools in the CIMS must be achieved. This is done by the computer control system.

22.5 COMPUTER CONTROL SYSTEM

This section describes how the digital computer system is used to manage the operation of a complex manufacturing system. We discuss the various functions performed by the computer, the data files needed to carry out these functions, and the various types of reports that the computer can be programmed to prepare.

22.5.1 Functions of the Computer in a CIMS

The functions accomplished by the computer control system can be divided into eight categories. The following descriptions apply best to the case of the flexible manufacturing system. To a slightly lesser extent, they also apply to the special system and the manufacturing cell.

Machine control

This is usually accomplished by computer numerical control (CNC). The advantage of CNC is that it can be conveniently interfaced with the other elements of the computer control system.

Direct numerical control (DNC)

Most computer-integrated manufacturing systems operate under DNC. In some of the special systems which are dedicated to a limited part variety, CNC may be a sufficient control method for the system. The purpose of direct numerical control is to perform the usual DNC functions, including NC part program storage, distribution of programs to the individual machines in the system, post-processing and so on.

Production control

This function includes decisions on part mix and rate of input of the various parts onto the system. These decisions are based on data entered into the computer, such as desired production rate per day for the various parts, numbers of raw workparts available pallets. The computer performs its production control function by routing a pallet to the load/unload area and providing instructions to the operator to load the desired raw part. A data entry unit (DEU) is located in the load/unload area for communication between the operators and the computer.

Traffic control

This term refers to the regulation of the primary work piece transport system which moves parts between workstations. This control can be accomplished by dividing the transport system (towline chain, conveyor, etc.) which is individually controlled by the computer. By allowing only one cart or pallet to be in a zone, the movement of each individual workpart is controlled. The traffic controller operates the switches at branches and merging points, stops workparts at machine tool loading points, and moves parts to operate load/unload stations.

Shuttle control

This is concerned with the regulation of the secondary part handling systems at each machine tool. Each shuttle system must be coordinated with the handling system, and it must also be synchronized with the operations of the machine tool it serves. In cases where there is only one part handling system (rather than a primary and secondary handling system), the functions of traffic control and shuttle control may be combined. This would be the case in some of the special systems and certain robotic work cells.

Work handling system monitoring

The computer must monitor the status of each cart and/or pallet in the primary and secondary handling systems as well as the status of each of the various workpart types in the system.

Tool control

Monitoring and control of cutting tool status is an important feature of the computer system. There are two aspects to tool control: accounting for the location of each tool in the CIMS and tool-life monitoring.

The first aspect of tool control involves keeping track of the each station in the system. If one or more tools required in the processing of a particular workpart are not present at the workstation specified in the part's routing, the computer control system will not deliver the part of the station. Instead, it will determine an alternative machine to which the part can be routed, or it will temporarily *float* the part in the handling system. In the second case, the operator is notified via the data entry unit what tools and notifies the computer accordingly. Any type of tool transaction (e.g., removal, replacement, addition) must be entered into the computer to maintain effective tool control.

The second aspect of tool control is tool-life monitoring. A tool life is specified to the computer for each cutting tool in the CIMS. Then a file is kept on the machining time usage of each tool. When the cumulative machining time reaches the life for a given tool, the operator is notified that the placement is required.

System performance monitoring and reporting

The system computer can be programmed to generate various reports desired by management on system performance. The types of reports are discussed in section 22.5.3.

These computer functions can be accomplished by any of several different computer configurations. One computer can be used for all components of the CIMS, or several different computers can be used. Upto three levels are practical in a given manufacturing system. CNC would be used for control of each individual machine tool. DNC would be appropriate for distribution of part programs from a central control room to the machines. A third level would concern itself with the production control, the operation of the work handling system, tool control and generation of managements reports.

22.5.2 CIMS Data Files

To control the operation of the manufacturing system, the computer relies on data contained in files. The principal data files required are of the following six types:

Part program file

The part program for each workpart processed on the system is maintained in this file. For any given workpart, a separate program is required for each station that performs operations on the part.

Routing file

This file contains the list of workstations through which each workpart must be processed. It also contains alternate routings for the parts. If a machine in the primary routing is down for repairs or there is a large backlog of work waiting for the machine, the computer will select an alternate routing for the part to follow.

In the special system, where the routings would be identical or similar and the parts processed through the system in an inline flow, the importance of the routing file is substantially reduced.

Part production file

A file of production parameters is maintained for each workpart. It contains data relative to production rates for the various machines in the routing, allowance for in-process inventory, inspections required and so on. These data are used for production control purposes.

Pallet reference file

A given pallet may be fixtured only for certain parts. The pallet reference file is used to maintain a record of the parts that each pallet can accept. Each pallet in the CIMS is uniquely identified and referenced in this file.

Station tool file

A file is kept for each workstation, identifying the codes of the cutting tools stored at that station. This file is used for tool control purposes.

Tool-life file

This data file keeps the tool-life value for each cutting tool in the system. The cumulative machining time of each tool is compared with its life value so that a replacement can be made before complete failure occurs.

22.5.3 System Reports

The data collected during monitoring can be summarized for preparation of performance reports. These reports are tailored to the particular needs and desires of management. The following categories are typical:

Utilization reports

These are reports that summarize the utilization of individual workstations as well as overall average utilization for the system.

Production reports

Management is interested in the daily and weekly quantities of parts produced from the CIMS. This information is provided in the form of production reports which list the required schedule together with actual production completions. One possible format for the production report is shown.

Status reports

Line supervision can call for a report on the current status of the system at any time. A status report can be considered as an instantaneous *snapshot* of the present condition of the CIMS. Of interest to supervision would be status data on workparts, machine utilization, pallets and other system operating parameters.

Tool reports

These reports relate to various aspects of tool control. Reported data might include a listing of missing tools at each workstation. Also, a tool-life status report can be prepared at the start of each shift.

22.6 HUMAN LABOUR IN THE MANUFACTURING SYSTEM

The computer-integrated manufactured system is a highly automated production facility. However, human resources are required to operate the system. In the majority of CIMS installations, the individual machines are operated under CNC or DNC control (or a combination of these). The machines are not manually operated except in certain special operations, such as assembly. Personnel are required principally to manage, and service the CIMS. Paprocki lists the following personnel requirements for a computer-integrated manufacturing system:

1. **System manager.** This person has overall responsibility for the operation of the CIMS. The functions include production planning, responding to deviations and exceptions to normal operations, and supervision of the other human resources which support the system.
2. **Electrical technician.** This person is often a member of the plant's electrical maintenance crew. Duties preformed include maintenance and repair services on the electrical components of the machine tools and materials and material handling system.
3. **Mechanical/hydraulic technician.** Again, this person is likely to be a regular member of the plant maintenance department. Technical services consist of maintenance and repair of the mechanical and hydraulic components of the CIMS.
4. **Tool setter.** The tool setter is responsible for the tooling inventory and making the tools ready for production.
5. **Fixture setup and lead man.** This person is responsible for setting up the fixtures, pallets and tools for the system.
6. **Load/unload man.** This person is responsible for loading raw workparts and unloading finished parts. This is typically done according to instructions and schedules generated by the computer. The load/unload area is at a convenient central location in the manufacturing system.
7. **Rover operator.** The duties of the rover operator include reacting to unscheduled machine stops, identifying broken tools or tools in need of immediate replacement, tool adjustments, and so forth. This person may also be responsible for certain manual production tasks or inspection operations.

Although each of these functions must be accomplished, this does not necessarily mean that seven people are required full time to operate the CIMS. The electrical technician and the mechanical/hydraulic technician are required on an on-call basis. In most companies, these workers would report to the plant maintenance department. In the startup of new manufacturing system they might be required full time to solve the many problems which are typically associated with a new installation of complex equipment. Under normal operation, these technicians would devote only part of their time to the CIMS.

The remaining five functions might be combined to some extent to reduce the actual number of people required. The duties of the tool setter, lead man, and rover operator might be shared by two persons rather than three. The amount of human resources needed to operate the manufacturing system will depend to a large degree on its size, number of processing machines, and level of sophistication and automation.

In addition to these seven operational functions, NC part programmers, computer programmers, and related support staff are required to prepare the programs by which the computer system will control production and monitor performance.

22.7 COMPUTER-INTEGRATED MANUFACTURING

Computer-integrated manufacturing (CIM) systems have emerged as a result of the developments manufacturing and computer technology. The computer plays an important role in integrating the following functional areas of a CIM system which is shown in Fig. 22.11.

(a) Part (component) and product design
(b) Tool and fixture design
(c) Process planning
(d) Programming of numerically controlled (NC) machines material handling systems (MHS), etc.
(e) Production planning
(f) Machining
(g) Assembly
(h) Maintenance
(i) Quality control
(j) Inspection
(k) Storage and retrieval

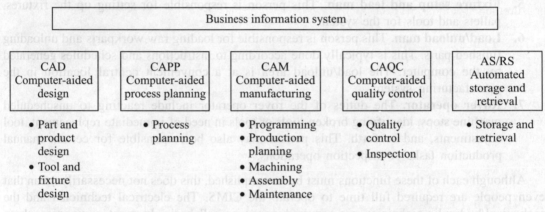

Fig. 22.11 Structure and functional areas of a CIM system.

To emphasize the computer aspect, the terms computer-aided design, computer-aided process planning, computer-aided manufacturing, computer-aided quality control and automated storage and retrieval storage have been used. Each term refers to one or more of the listed functional areas. CAD refers to part and product design and tool and fixture design, CAPP refers to process planning, CAM refers to programming of manufacturing hardware, production planning, machining, assembly and maintenance, CAQC refers to quality control and inspection and AS/RS involves storage and in-process inventory.

The computer plays a leading role in the automation and integration of the hardware components, i.e., machines, material handling carriers and the software components of the manufacturing system. In order to understand the CIM subsystems and the importance of integrating them, it is first necessary to explore the functional areas covered by these subsystems. The functional areas covered by CIM subsystems are discussed hereinafter.

22.7.1 Part and Product Design

A typical product consists of a number of parts (components). The design of a product involves the design of its parts, that is, determining their shape, dimensions, tolearences and the like. It is a complex task that requires the intuition and creativity of the design engineer. The design engineer must have a good working knowledge of the manufacturing processes, tools and fixtures in order to ensure that the part can be manufactured according to the specification, economically and efficiently.

22.7.2 Tool and Fixture Design

Based on the part data stored in the database, the tooling engineer designs the tools required to produce the part. The jigs and fixtures required to hold the part are also designed by the tool engineer. A CAD system allows reduction of the design time of tools and fixtures from 12 to 25% of the time required by the traditional design.

The geometrical representation of the part available in a CAD system is useful in designing fixtures for complicated components. The CAD system allows the user to rotate the part on any axis and measure its dimensions from any reference plane, line or point. Therefore, dimension measurements, which are necessary for tool design, may be easily accomplished.

22.7.3 Process Planning

Process planning for a part involves the preparation of a plan that outlines the routes, operations, machines and tools required to produce that component. Based on the design specifications provided by the design engineer, the process planner determines the process plan for a part that minimizes production cost, manufacturing time and machine idle time and maximizes rate of production and quality of the part. It is a tedious task and requires the service of an experienced process planner who has a good knowledge of production processes, machine capabilities and the like. Many attempts have been made to automate process planning.

22.7.4 Programming

There are two basic types of programming in computer-integrated manufacturing: programming numerically controlled (NC) machines and programming material handling systems (MHS). In NC machine programming, machining instructions for the part are prepared in computer-usable form and provided to the computer. The instructions include operation sequence, tool path, tool feed rate and cutting speed. The component geometry is also provided to the computer. The computer performs certain calculations based on the input data and ensures that the machines perform the necessary processes.

A computer numerically controlled (CNC) system has a central processing unit (CPU) which supervises logic control, geometric data processing and NC program execution. It is essentially a software control system and may be integrated with a direct numerically controlled (DNC) system. A DNC system schedules and supervises several NC machine tools directly with the aid of a computer.

By linking the controllers of the CNC system, manufacturing resource planning (MRP II) system, AS/RS system and so on, one can integrate part processing, material storage and material handling functions. Also, using a suitable data acquisition system, information on these functions can be easily obtained. This, of course, means that the databases corresponding to these functions must be networked.

Programming of a material handling system is concerned with the description of the travelling path of automated material handling carriers (e.g., for robots, automated guided vehicles).

22.7.5 Production Planning

Production planning involves establishing production levels for a known length of time. This forms basis for the following two functions:

(a) Material requirement planning (MRP)
(b) Machine loading and scheduling

Based on the information regarding the type and quantity of parts or products to be manufactured, the materials required to produce the parts or products must be ordered with an appropriate lead time to ensure their availability for production. This function constitutes material requirement planning. Based on the material required, capacity requirements—for example, machine hours and labour hours—can be calculated. Thus, the capacity planning, material requirement planning and master scheduling can be incorporated into a system. Based on the feedback generated, one can determine how far the actual results deviate from the original plans.

MRP II is a system in which the master scheduling material requirement planning and other functions are integrated with the company's business plan. MRP II requires massive data processing, as it uses shop floor data, inventory data, scheduling data, customer orders, the business plan and so on. Efficient data base management is necessary for an MRP II system to work effectively.

Machine loading involves assignment of parts and products to machines and machine cells in order to distribute the production loads. Scheduling determines the sequence in which machines perform operations. Operation processing times, machine capacity, due dates and number of other factors have to be considered while generating schedules. Machine scheduling is a very complex task, especially when the number of machines and jobs are large. In many instances, the schedule has to be recomputed frequently, especially when unexpected events such as machine breakdowns and rush jobs occur.

In order to perform scheduling in computer-integrated manufacturing (CIM) systems, interaction among various databases is necessary. As shown in Fig. 22.12, the scheduling database interacts with part, machine and tool, material handling system, pallet and fixture databases.

22.7.6 Machining

All activities that consist of metal-removal operations constitute machining. Examples are turning, drilling and face milling. The introduction of numerical control has led to the automation of these operations. Tool feed rate, speed and other production parameters are automatically calculated and the data are passed to a controller which, in turn, instructs the machines to perform the required machining operations.

22.7.7 Assembly

Assembly involves assembling the manufacture parts to form the required products. In order to perform this task efficiently, the parts have to be machined according to the required specifications and tolerances. Also the assembly function can be greatly simplified by appropriate design. For example, if the parts designed are not symmetrical in shape, then the assembly

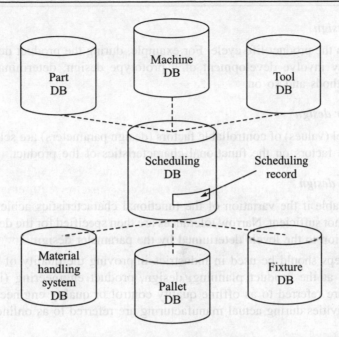

Fig. 22.12 Databases required for scheduling in a CIM System.

equipment has to orient the part in a specific manner before assembly. On the other hand, if the parts are symmetrical, then the assembly equipment may pick up the component in any manner for assembly. In recent years, robots have been increasingly used for assembly. By using barcodes on manufactured parts and scanners, which read the bar codes, it is now possible to instruct robots automatically about the sequence in which to pick up and assemble the different parts.

22.7.8 Maintenance

Maintenance involves diagnosing and correcting malfunctions that may arise in various elements of a manufacturing system. In traditional manufacturing systems, this function was typically performed by technicians with expertise mostly in mechanical engineering. In a CAM system, it is done by personnel with expertise in electronics, software and mechanical engineering. It appears that in future manufacturing systems the maintenance function will perhaps be performed by intelligent systems.

22.7.9 Quality Control

The purpose of the quality control system is to manufacture products and parts that are robust with respect to all factors causing deviation from the target values, frequently called the **noise factors**. In order to achieve robustness, quality control efforts must begin in the early product (part) design phase and be continued through process planning, machining and assembly. The following three steps should be followed:

Step 1 *System design*

This step relates to the product life cycle. For example, during the product development phase, system design may involve development of a prototype design, determination of materials manufacturing methods and so on.

Step 2 *Parameter design*

In this step the level (values) of controllable factors (design parameters) are selected to minimize the effect of noise factors on the functional characteristics of the product.

Step 3 *Tolerance design*

This step is applicable if the variation of the functional characteristics achieved by parameter design in Step 2 is not sufficient. Narrow tolerances are then specified for the deviations of design parameters in relation to the levels determined by the parameter design.

These three steps should be used in industries improving the quality of products. Quality contains activities at the product planning, design, product engineering (including process planning) phases are referred to as offline quality control or quality engineering, whereas the quality control activities during actual manufacturing are referred to as online quality control.

22.7.10 Inspection

Inspection of manufactured parts and products is done in order to determine defects in the components and inform the process department about the defects. With such information, the error in the process can be corrected and future production of defective components can be prevented. The level of automation of inspection in CIM systems is increasing.

22.7.11 Storage and Retrieval

Storage and retrieval involves storing and retrieving of items such as raw materials, in-process inventory, finished goods and tools as well as computer control of handling equipment like stacker cranes, carousels and palletizers.

22.8 BENEFITS OF CIM

CIM provides a means to manage and control the uncontrollable disturbances while meeting customer demands and requirements.

The following is the list of a few of the benefits that can be achieved by the use of CIM:

(a) CIM increases machines utilization by
   Eliminating or reducing machine setup.
   Utilizing automated features to replace manual intervention to the extent possible.
   Providing quick transfer device to keep the machines in the machining cycle.

(b) CIM reduces inventory by
   Reducing lot sizes.
   Improving inventory turnovers.
   Providing the planning tools for just-in-time manufacturing.

(c) CIM provides greater flexibility and improves the short run responsiveness consisting of:
Engineering changes
Processing changes
Machine down time
Operator unavailability
Cutting tool failure
Late material delivery

(d) CIM improves scheduling flexibility. Routing the work parts to proper workstations. Equipping the workstations in advance with tooling required to process the work parts.

(e) CIM reduces manufacturing lead time by providing tool setup for a workstation, which consists loading the preset tools required for the job into the tool drum at the station, and pallets which can be off the system being loaded or unloaded while other pallets are being used on the system.

(f) CIM improves long run accommodations through quicker and easier assimilation of changing product volumes, new product additions and different part mixes.

(g) CIM helps to increase in manufacturing productivity. Productivity of the manufacturing operations can be increased by 40 to 70%.

(h) CIM helps to improve product quality. The product quality can be increased dramatically as measured by the yield of acceptable product, offering 2 to 5 times the previous level.

22.9 LEAN MANUFACTURING

Lean manufacturing pioneered by Toyota, Japan has been adopted successfully by many manufacturing companies as a way to reduce costs, satisfy customers and increase profitability. Lean manufacturing is a way of thinking that develops a culture of eliminating non-value adding activities while responding to customer needs. Lean manufacturing reaches into every aspect of a company. The process of becoming lean may mean transforming the current company's existing style of operations to an entirely different one.

Lean manufacturing may involve process reengineering, adopting new technologies or adding new and different equipment. This may generally involve significant changes in human resources such as education, training, practices and policies.

Womack and Jones defined lean as doing more and more with less and less—less human efforts, less equipment, less time and less space—which coming closer and closer to providing customers with exactly what they want.

Lean manufacturing can be defined as an adoption of mass production in which workers and work cells are made more flexible and efficient by adopting methods that reduce waste in all forms.

Lean manufacturing is based on four principles:

(a) Minimize waste
(b) Perfect first-time quality
(c) Flexible production lines
(d) Continuous improvement

Minimize waste Taiichiohno's list of waste forms can be listed as follows:

    (a) Production of defective parts

    (b) Production of more than the number of items needed

    (c) Unnecessary inventories

    (d) Unnecessary processing steps

    (e) Unnecessary movement of people

    (f) Unnecessary transport of materials

    (g) Workers waiting

The various procedures used in the Toyota plants were developed to minimize these forms of waste.

Perfect first-time quality

In the area of quality, the comparison between mass production and lean production provides a sharp contrast. In mass production, quality control is defined in terms of an acceptable quality level or AQL. This means that a certain level of fraction defects is sufficient. In lean production by contrast, perfect quality is required. The just-in-time delivery discipline used in lean production necessitates a zero-defect level in parts quality, because if the part delivered to the downstream workstation is defective, production stops. There is minimum inventory in a lean system to act as a buffer. In mass production, inventory buffers are used just in case these quality problems occur. The defective work units are simply taken off the line and replaced with acceptable units. However, the problem is that such a policy tends to perpetuate the cause of the poor quality. Therefore, defective parts continue to be produced. In lean production, a single defect draws attention to the quality problem, forcing corrective action and a permanent solution. Workers inspect their own production minimizing the delivery of defects to the downstream production station.

Flexible production systems

Lean production makes use of worker teams to organize the tasks to be accomplished and worker involvement to solve technical problems. In lean production, procedures are designed to speed the changeover. Reduced setup times allow for smaller batch sizes, thus providing the production system with greater flexibility.

Flexible production systems were needed in Toyota's comeback period because of the much smaller car market in Japan and the need to be as efficient as possible.

Continuous improvement

Lean production supports the policy of continuous improvement called **Kaizen** by the Japanese. Continuous improvement means constantly searching for and implementing ways to reduce cost, improve quality and increase productivity. The scope of continuous improvement goes beyond factory operations and involves design improvements as well. Continuous improvement is carried out one project at a time. The projects may be concerned with any of the following problem areas: cost reduction quality improvement, productivity improvement, setup time reduction, cycle time reduction, manufacturing lead time and work-in-process inventory reduction, and improvement of product design to increase performance and customer appeal.

22.10   AGILE MANUFACTURING

Agile manufacturing can be defined as:

(a)   An enterprise level manufacturing strategy of introducing new products into rapidly changing markets.

(b)   An organizational ability to thrive in a competitive environment characterized by continuous and sometimes unforeseen change.

Four principles of agility were identified by the 1991 study. Manufacturing companies that are agile competitors tend to exhibit these principles. The four principles are:

Organize to master change

"An agile company is organized in a way that allows it to thrive on change and uncertainty."

In a company that is agile, the human and physical resources can be rapidly reconfigured to adapt to changing environment and market opportunities.

Leverage the impact of people and information

In an agile company knowledge is valued, innovation is rewarded, authority is distributed to the appropriate level of the organization. Management provides the resources that personnel needs. The organization is entrepreneurial in spirit. There is a "climate of mutual responsibility for joint success."

Cooperate to enhance competitiveness

"Cooperation—internally and with other companies—is an agile competitor's operational strategy of first choice". The objective is to bring products to market as rapidly as possible. The required resources and competencies are found and used, wherever they exist. This may involve partnering with other companies, possibly even competing companies to form virtual enterprises.

Enrich the customer

"An agile company is perceived by its customers as enriching them in a significant way, not only itself". The products of an agile company are perceived as solutions to customer's problems. Pricing of the product can be based on the value of the solution to the customer rather than on manufacturing cost.

22.11   REORGANIZING THE PRODUCTION SYSTEM FOR AGILITY

Companies seeking to be agile must organize their production operations differently than the traditional organization. The changes in three basic areas are:

Product design

Reorganizing production for agility includes issues related to product design. Decisions mode in product design determine approximately 70% of the manufacturing cost of a product. For a company to be more agile, the design engineer department must develop products that can be characterized as follows:

Customizable. Products can be customized for individual niche markets. In some cases, the product must be customizable for individual customers.

Upgradeable. It should be possible for customers who purchased the base model to subsequently buy additional options to upgrade the product.

Reconfigurable. Through modest changes in design, the product can be altered to provide it with unique features. A new model can be developed from the previous model without drastic and time-consuming redesign effort.

Design modularity. The product should be designed so that it consists of several modules (e.g, subassemblies) that can be readily assembled to create the finished item. In this way, if a module needs to be redesigned, the entire product does not require redesign. The other modules can remain the same frequent model changes within stable market families. Even for products that succeed in the market place, the company should nevertheless introduce new versions of the product to remain competitive.

Platforms for information and services. Depending on the type of product offering, it should include some aspect of information and service. Information and service might be in the form of an embedded microprocessor to carry out seemingly intelligent functions. For example, the capability of a VCR to display instructions on the TV screen to guide the viewer through a procedure or service by the company in the form of telephone number that can be called for an immediate response to an important issue troubling the customer.

Marketing

A company's design and marketing objectives must be closely linked. The best efforts of design may be lost if the marketing plan is flawed. Being an agile marketing company suggests the following objectives, several of which are related directly to the preceding product design attributes:

Aggressive and proactive product marketing. The sales and marketing functions of the firm should make change happen in the marketplace. The company should be the change agent that introduces the new models and products.

Cannibalize successful products. The company should introduce new models to replace and obsolete its most successful current models.

Frequent new product introductions. The company must maintain a high rate of new product introductions.

Life cycle product support. The company must provide support for the product throughout its life cycle.

Pricing by customer value. The price of the product should be established according to its value to the customer rather than according to its own cost.

Effective niche market competitor. Many companies have become successful by competing effectively in niche markets using the same basic product platform. The product has been reconfigured to provide offerings for different markets.

Production operations

A substantial impact on the agility of the production system can be achieved by reorganizing factory operations and the procedures and systems that support these operations.

Objectives in production operations and procedures that are consistent with an agility strategy are the following:

(a) Be a cost-effective low-volume producer
This is accomplished using flexible production systems and low setup times.

(b) Be able to produce to customer order
Producing to customer order reduces inventories of unsold finished goods.

(c) Master mass customization
The agile company is capable of economically producing a unique product for an individual customer.

(d) Use reconfigurable and reusable processes, tooling and resources. Examples include computer numerical control machine tools, parametic part programming, robots that are reprogrammed for different jobs, programmable logic controllers, mixed-model production lines and modular fixtures.

(e) Bring customers closer to the production process. Provide systems that enable customers to specify or even design their own unique products. As an example, it has become very common in the personal computer market for customers to be able to order exactly the PC configuration and software that they want.

(f) Integrate business procedures with production. The production system should include sales marketing, order entry, accounts receivable and other business functions. These functions are included in a computer integrated production planning and control system based on manufacturing resource planning (MRP-II).

(g) Treat production as a system that extends from suppliers through to customers. The company's own factory is a component in a larger production system that includes suppliers that deliver raw materials and parts to the factory.

TABLE 22.1 Four Principles of Lean and Agile Manufacturing

Lean manufacturing	Agile manufacturing
Enhancement of mass production	Break with mass production and emphasis on mass customization
Flexible production for product variety	Greater flexibility for customized products
Focus on factory operations	Scope is enterprise wide
Emphasis on supplier management	Formation of virtual enterprises
Emphasis on efficient use of resources	Emphasis on thriving in environment marked by continuous unpredictable change
Relies on smooth production schedule	Acknowledges and attempts to be responsive to change

TABLE 22.2 Comparison of Lean and Agile Manufacturing Attributes

Lean manufacturing	Agile manufacturing
Minimize waste	Enrich the customer
Perfect first-time quality	Cooperate to enhance competitiveness
Flexible production lines	Organize to master change
Continuous improvement	Leverage the impact of people and information

REVIEW QUESTIONS

22.1 Define computer integrated manufacturing.

22.2 Briefly explain the advantages that will be gained by the implementation of CIM.

22.3 Discuss the computerized elements of CIM systems.

22.4 Explain the SME manufacturing enterprise wheel.

22.5 Explain the aspects that one should consider in implementing CIM.

22.6 What do you understand by the term lean manufacturing?

22.7 Explain the steps used in implementing lean manufacturing.

22.8 What are the steps to be considered on a shop floor for implementing lean manufacturing principles?

22.9 What do you understand by the term agile manufacturing?

22.10 Differentiate between lean and agile manufacturing.

22.11 State and elaborate the advantages of CIM in a manufacturing unit.

22.12 What is the significance of quality control in CIM?

22.13 Define total quality management (TQM) and explain its relevance to CIM.

22.14 Explain the computerized business functions of CIM.

22.15 Discuss the role of computer networks in CIM.

22.16 Explain the applications of computer integrated manufacturing systems.

22.17 What are the functions of shop floor control?

22.18 Explain the three shop floor control systems.

Part VIII

Intelligent CAD and Manufacturing Systems

ARTIFICIAL INTELLIGENCE AND EXPERT SYSTEM

Chapter

23

INTRODUCTION

An expert system is a computer program that relies on knowledge and reasoning to perform a difficult task usually undertaken only by a human expert. The expert system technology has emerged from research in artificial intelligence (AI). The expert system approach is to take knowledge from human experts and represents it as a knowledge base, which can then be processed to solve difficult problems in the same way the expert would. A knowledge base is formulated and encoded in such a way that the system can readily explain why it arrived at its answer.

An expert system can free up valuable human expertise for difficult problems and for more creative activities such as research. It often expands the scope and flexibility of applying expertise. A computer system is much easier than a human to copy and transport to various locations where it is needed. An expert system successfully deals with problems for which clear algorithmic solutions do not exist.

There are a wide variety of ways that expert system technology can assist the CAD process. The routine type CAD systems provide means of representing the final form of the design, whereas designers also need a continued stream of advice and information to assist in decision making. Properties that can be modelled on a CAD system are: form, dimension, tolerance, material, surface condition, structure and function, but only the first two of these are covered extensively by CAD. The others are generally covered by annotation of a drawing or by attaching attributes to a 3D model. A system that captures a complete model of a produced will require notations for all of the properties of the design. In an expert system, the problem solving is done correctly using symbolic processing.

23.1 ARTIFICIAL INTELLIGENCE

Artificial intelligence (AI) is an area of computer science concerned with systems that exhibit some characteristics which are usually associated with intelligence in human behaviour, e.g. learning, reasoning, problem solving and the understanding of language. The goal of AI is to stimulate such human endeavours on the computer. The art of bringing relevant principles and tools of AI to bear on difficult application problems is known as knowledge engineering. Artificial intelligence is having a major impact on the design, the automation and the overall economics of manufacturing operations in large part because of advances in computer-memory expansion (VLSI chip design) and decreasing costs. Artificial intelligence packages costing on the order of a few thousand dollars have been developed. Many of which can now be run on personal computers. Thus, AI has become accessible to office desks and shop floors.

Artificial-intelligence applications in manufacturing generally encompass the following:

(a) Expert systems
(b) Natural language
(c) Machine vision
(d) Artificial neural networks
(e) Fuzzy logic

23.1.1 Expert Systems

Also called a **knowledge-based system**, an expert system is generally defined as an intelligent computer program that has the capability to solve difficult real-life problems, using knowledge base and inference procedures (see Fig. 23.1). The goal of an expert system is to develop the capability to conduct an intellectually demanding task in the way that a human expert would desire. The field of knowledge required to perform this task is called the **domain** of the expert system. Expert systems use a knowledge base containing facts, data, definitions and assumptions. They also have the capacity to follow a heuristic approach, that is, to make good judgements on the basis of discovery and revelation and to make high-probability guesses, just as a human expert would. The knowledge base is expressed in computer codes, usually in the form of if-then rules and can generate a series of questions. The mechanism for using these rules to solve problems is called an **inference engine**. Expert systems can also communicate with other computer software packages.

To construct expert systems for solving the complex design and manufacturing problems encountered, one needs

(a) a great deal of knowledge
(b) a mechanism for manipulating this knowledge to create solutions

Because of the difficulty involved in accurately modelling the many years of experience of an expert or a team of experts and the complex inductive reasoning and decision-making capabilities of humans, including the capacity to learn from mistakes, the development of knowledge-based systems requires much time and effort.

Expert system operates on a real time basis and their short reaction times provide rapid responses to problem. The programming languages most commonly used for this application are C++ , LISP and PROLOG, other languages can also called framework systems. These software packages are essentially expert-system outlines that allow a person to write specific applications to suit special needs. Writing these programs requires considerable experience and time.

Several expert systems have been developed and used for such specialized applications as:

(a) Problem diagnosis in machines and equipment and determination of corrective actions
(b) Modelling and simulation of production facilities

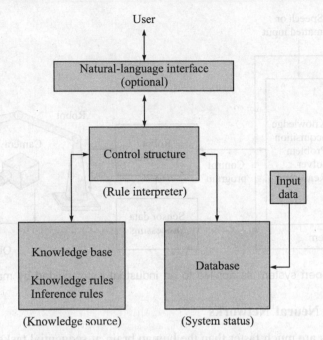

Fig. 23.1 The basic structure of an expert system.

(c) Computer-aided design, process planning and production scheduling
(d) Management of a company's manufacturing strategy

23.1.2 Natural-Language Processing

Traditionally, obtaining information from a database in computer memory has required the use of computer programmers to translate questions expressed in natural language into "queries" in some machine language. Natural-language interfaces with database systems are in various stages of development. These systems allow a user to obtain information by entering English-language commands in the form of simple, typed questions. Natural-language software shells are used in such applications as scheduling of material flow in manufacturing and analyzing information in databases. Major progress continue to be made in the development of computer software that will have speech-synthesis and voice recognition capabilities, thus eliminating the need to type commands on keyboards.

23.1.3 Machine Vision

The basic features of machine vision are described in Chapter 18. In systems that incorporate machine vision, computers and software implementing artificial intelligence are combined with cameras and other optical sensors. These machines then perform such operations as inspecting, identifying and sorting parts and guiding robots (see Fig. 23.2). In other words, operations that would otherwise require human involvement and intervention.

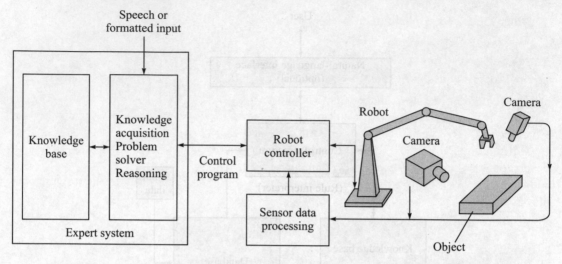

Fig. 23.2 Expert system as applied to an industrial robot guided by machine vision.

23.1.4 Artificial Neural Networks

Although computers are much faster than the human brain at sequential tasks, humans are much better at pattern-based tasks that can be dealt with parallel processing such as feature recognition (e.g., in faces and voices, even under noisy conditions) assessing situations quickly and adjusting to new and dynamic conditions. These advantages are also due partly to the ability of humans to use, in real times their senses (i.e., sight, hearing, smell, taste and touch) simultaneously, a process called **data fusion**. The branch of artificial intelligence, called **artificial neural networks (ANN)**, attempts to gain some of these human capabilities through computer imitation. The way that data are processed by the human brain. The human brain has about 100 billion linked neurons and more than a thousand times that many connections. Each neuron performs only one simple task: It receives input signals from a fixed set of neurons. It is now believed that human learning is accomplished by changes in the strength of these signal connections between neurons.

A fully developed feed-forward network is the most common type of ANN and it is built according to this principle from several layers of processing elements (simulating neurons). The elements in the first (input) layer are fed with input data, for example, forces, velocities or voltages. Each element sums up all its inputs—one per element in the input layer and many per element in succeeding layers. Each element in a layer then transfers the data (according to a transfer function) to all the elements in the next layer. Each element in that next layer, however, receives a different signal, because of the different connection weights between the elements.

The last layer is the output layer, within which each element is compared with the desired output to that of the process being simulated. The difference between the desired output and the calculated output (error) is fed back to the network by changing the weights of the connections in a way that reduce this error. After this procedure has peen repeated several times the network has been trained and it can thus be used on input data that have not previously been presented to the system.

Other kinds of artificial neural works are:

- Associative memories
- Self-organizing ANN
- Adaptive-resonance ANN

The feature common to these neural networks is that they must be trained with concrete examples. It is, therefore, very difficult to formulate input-output relations mathematically and to predict an ANN's behaviour with untrained inputs.

Artificial neural networks are used in such applications as noise reduction (in telephones), speech recognition and process control in manufacturing. For example, they can be used to predict the surface finish of a work piece obtained by milling on the basis of input parameters such as cutting force, torque, acoustic emission and spindle speed. Although this field is still controversial the opinion of many through advances are being made in ANN.

23.1.5 Fuzzy Logic

An element of artificial intelligence that has important applications in control systems and pattern recognition is fuzzy logic (fuzzy models). Introduced in 1965, and based on the observation that people can make good decisions on the basis of imprecise and non-numerical information, fuzzy models are mathematical means of representing vagueness and imprecise information (hence the term *fuzzy*). These models have the capability to recognize, represent, manipulate, interpret and use data and information that are vague or lack precision. Fuzzylogic methods deal with reasoning and decision making at a level higher than do neural networks. Typical linguistic examples of concepts used in fuzzy logic are the following words and terms: very, few, more or less, small, medium, extremely and almost all.

Fuzzylogic technologies and devices have been developed and successfully applied in areas such as robotics and motion control, image processing and machine vision, machine learning and the design of intelligent systems. Some applications of fuzzy logic include the automatic transmission of lexus automobiles, a washing machine that automatically adjusts the washing cycle for load size, fabric type and amount of dirt and a helicopter that obeys vocal commands to go forward, up, left and right, to hover and to land.

23.2 ARTIFICIAL INTELLIGENCE IN CAD

The application of AI to CAD systems is concerned with studying how designers apply human intelligence to design and with trying to make computer aids to design more knowledgeable. The main themes in the application of AI are to explore the formal representation of design knowledge and also to develop techniques for reasoning with or applying this knowledge. AI may allow the representation of heuristic (or rule of thumb) knowledge known as **expert systems** or **knowledge based expert systems (KBES)**. The large body of concepts and techniques developed in AI can be applied to CAD systems software so as to free the designer to concentrate on stages of the design process in which decisions have a greater impact on the utilitarian value of the final solution.

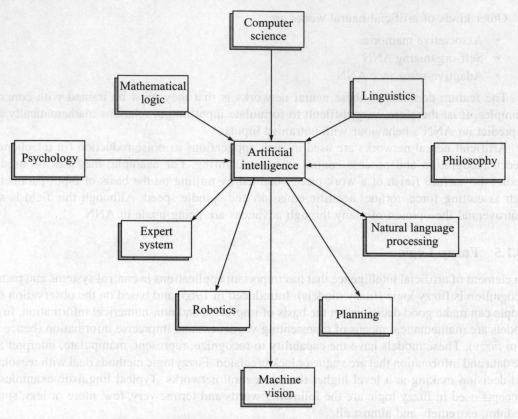

Fig. 23.3 Input/output model for artificial intelligence.

The three main topics in CAD systems software are design automation, man-machine interaction and database management. The effects of AI on this CAD systems software are:

• Design Automation

To augment CAD resource capabilities by making them aware of the goal of the design problems, learning techniques can be used to progressively improve the man-machine problem solving strategy. Heuristic search techniques may be more adequate for solving some problems than traditional algorithms.

• Man-Machine Interaction

Spoken natural language input could release the designer from tedious keyboard inputs. Several perception techniques could be combined, e.g. coupling hand-drawing input on a CRT with vocal input.

• Database Management

The computer may be provided with deductive mechanisms to display common-sense reasoning capabilities. The computer may maintain for each problem or sub-problem a 'context' to be used for restricting as much as possible data retrieval to current relevant information. The primary concern of AI is to write down descriptions of the world in such a way that an intelligent by formally manufacturing these descriptions. A characteristic of knowledge-based system is a formal and explicit representation stored in a knowledge base of the knowledge pertaining uses symbolic terms defined in a notation or language with semantics which are used to define their meaning and well defined syntax governing the form of statements. The symbols define both concepts with the domain and the relationship between them. An input-output model for artificial intelligence is shown in Fig. 23.3.

- **Machine vision** aims at recognizing patterns in much the same way as the human visual systems does.
- **Robotics** focuses on producing mechanical devices capable of controlled motion.
- **Speech processing** aims at recognizing and synthesizing spoken human speech.
- **Natural language** processing attempts to understand and produce written natural language.
- **Theorem proving** attempts to automatically prove theorems in mathematics and logic.
- **General problem solving** aims at solving general classes of problems expressed in a formal language.
- **Pattern recognition** focuses on the recognition and classification of patterns.
- **Machine learning** aims at producing machines that accumulate programs by observing examples.

23.3 APPLICATIONS OF AI IN DESIGN

AI techniques are particularly relevant in the selection of materials, manufacturing processes, analytical techniques diagnosis and assessment. There are several other classes of design problems for which AI techniques may be relevant. Some of these are discussed hereinafter.

Decomposition

Engineering products are often highly complex. A way of managing complexity is to subdivide or decompose the design knowledge and the design process into smaller elements. Once these smaller problems have been solved, it is then necessary to assemble the results into an overall solution. Decomposition may be achieved in a variety of ways.

(a) Dividing the problem according to the nature of the solution technique adopted
(b) Dividing the design itself into a series of elements

If the design process progresses from the upper levels to the lower ones, this is known as **top-down approach**. Conversely, if the detailed component parameters are established first and then the results assembled at a higher level, then this is known as **bottom-up approach**. In mechanical design, a hybrid approach is generally adopted.

Plan selection and refinement

In many design problems, the design approach can, atleast in part, be reduced to that of identifying a generic design type and then filling in details of dimensions, materials and component arrangement. In AI, this approach is called **plan refinement**.

Constraint-based reasoning

The idea of constraint-based reasoning is that design can be modelled in a network of design attributes and their associated constraints. A feasible design within the constraint space may be identified by chaining through the network.

Case-based reasoning

This technique seeks to provide computer-based design tools that make judgements on the basis of past experience, often by drawing connections between the problem to be solved and previous similar problems.

Consistency maintenance

A product design is normally developed over a period of time as the designer collects information and experiments with different design arrangements. During the progress of design process, assumptions will be made and values assigned to design attributes. The problem of maintaining these different contexts and the data dependencies that exist in such complex situations is known as **truth maintenance**.

23.4   EXPERT SYSTEMS

An expert system is a machine system which embodies useful human knowledge in machine memory in such a way that it can give intelligent advice and also can offer explanation and justifications of its decision or demand. Expert systems are problem solving computer programs.

An expert system may be defined as a computer program that relies on knowledge and reasoning to perform a difficult task usually performed only by a human expert. Expert systems are based on heuristics knowledge. A heuristic is a rule of thumb most often used to make problem solving or arrives at conclusions based on personal knowledge. An expert system reasons and arrives at conclusions based on the knowledge it possesses.

An expert system is so called because it contains the skill of a human expert in solving problems in specific area or domain and is defined as "A computer system emulating the skill of an expert in a narrow area using knowledge and reasoning to produce valuable advice justifying its conclusion, explaining its line of reasoning and possibly handling uncertainty. They attempt to encapsulate the skills of the expert and to dispense advice to less knowledgeable users or to assist people who do not have access to an expert. The generic applications tend to be in diagnosis, design and planning—all of which rely heavily upon human expertise and problem solving ability. It is a tool which has the capability to understand problem specific knowledge and use domain knowledge intelligently to suggest alternate path of action.

The advantages of an expert system are significant enough to justify a major effort to built them. Decisions can be obtained more reliably and consistently. A problem area can be standardized and formalized through the process of building an expert system for it. An expert system may be especially useful in consolation made on difficult cases, where humans may over-look obscure factors. An expert system can often serve as an example of good strategy in approaching a problem, which might be useful in training situations. Expert systems can be more easily expandable than conventional software, so that they can be gradually improved as their problem domain evolves. Expert systems are often implemented in an interactive decentralized environment, taking advantage of emerging, cost-effective personal computing resources. Ready availability of an expert consultant program can improve the turning environment in industrial settings.

23.5   STRUCTURE OF AN EXPERT SYSTEM

The structure of an expert system includes the following, as shown in Fig. 23.4.

(a)   Knowledge base
(b)   Working memory
(c)   Inference engine
(d)   User interface module

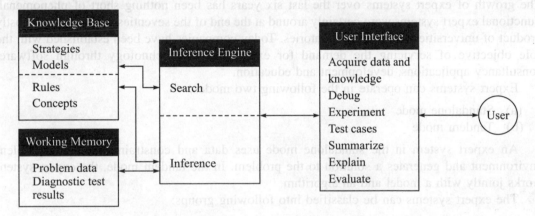

Fig. 23.4   Structure of an expert system.

The architecture of a knowledge based system is shown in Fig. 23.5. The heart of the expert system is the knowledge base. The part of an expert system that carries out the reasoning function is called the **inference engine**. The knowledge base is developed by a process of knowledge acquisition, which may be defined as the transfer and transformation of a potential problem solving-expertise from some knowledge source to a program. In addition to the knowledge base and inference engine, the expert system environment will include a number of tools for help in the various people who build or use the expert system. When building the expert system, the developer uses debug knowledge within the knowledge base tools for the developer include tools for acquiring knowledge, knowledge-base editors, debuggers, compilers and validation tools.

The user interface module is an essential part of an expert system. It is the task of the interface to handle all the communication between the user and the expert system. The way that information is presented to the user should conform to the user's model of the task and expectations.

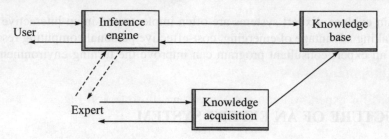

Fig. 23.5 Architecture of knowledge based system.

23.6 DEVELOPMENT OF AN EXPERT SYSTEM

23.6.1 Conceptual Framework

The growth of expert systems over the last six years has been nothing short of phenomenal. Functional expert systems were certainly around at the end of the seventies but they were mostly product of universities and R&D laboratories. Today companies have been established with the sole objective of servicing the demand for expert systems technology through software, consultancy applications, development and education.

Expert systems can operate in the following two modes:

(a) Standalone mode
(b) Tandem mode

An expert system in the standalone mode uses data and constraints from the problem environment and generates a solution to the problem. In the tandem mode, the expert system works jointly with a model and an algorithm.

The expert systems can be classified into following groups:

(a) Interpretation
(b) Prediction
(c) Diagnosis
(d) Debugging
(e) Design
(f) Planning
(g) Monitoring
(h) Instruction
(i) Control
(j) Repair

Although expert systems vary in structure, they have in general two distinct parts: a domain specific base which consists of factual knowledge and procedural knowledge, the latter being the production rules of the system. The inference engine contains the operating rules or the rules that control the production rules.

All expert systems depend on expert elicitation and two main approaches are used. Firstly, based on the assumption that experts can articulate their knowledge, a programme of interviewing is carried out. This approach is prone to error as experts are often better at doing than analyzing why. The second approach, knowledge induction, relies on second party analysis of case studies that the expert has worked on and is often preferred.

23.6.2 Development Steps

The major steps for expert system development are illustrated in Fig. 23.6.

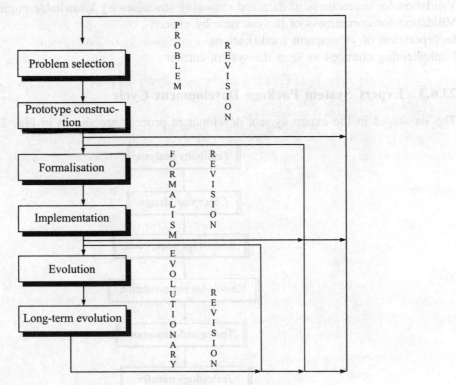

Fig. 23.6 Steps involved in development of an expert system.

Problem selection

Requirements for development
Characteristics required for deciding the appropriateness of the approach
Identifying beneficiaries and estimating time and cost
Identifying limitations

Prototype construction

Design, development, implementation and testing
Identifying potential for a full file system

Formalisation

Forced planning and record of decisions
Provide visibility and check points
Allow for participative involvement

Implementation

Prototype revision and system framework development

Provide training, support and documentation to the user

Evaluation

Validation for correctness of data and reasoning strategies by knowledge engineers

Validation for correctness of the outcome by experts

Incorporation of subsequent modifications

Implementing changes to keep the system current

23.6.3 Expert System Package Development Cycle

The six stages in the expert system development process are shown in Fig. 23.7.

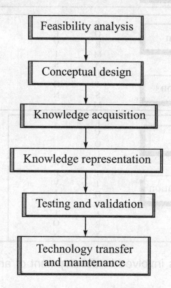

Fig. 23.7 Stages in expert system development.

Feasibility analysis

The domain in which the expert system is to operate and the task which will be performed by the expert system are studied and analyzed by the expert system builders. Identification of an appropriate task is a vital step in the development of any expert system. The decision about whether to build an expert system should be based on a specification of what tasks will be performed by the expert system and whether or not they meet appropriate needs.

Conceptual design

The conceptual structure of the system is defined along with a specification that describes the way in which the expert system will carry out the task. A good conceptual design tells the knowledge engineer what to look for and can be used to decide which issues are important and which are not. All hardware required for delivery and convention to the application is also identified.

Knowledge acquisition

The knowledge required for performing the task is acquired from a human expert. Case histories, reference sources, etc. are included in the expert system's knowledge base. Knowledge acquisition is a crucial part of the expert system construction process.

Knowledge representation

The knowledge is formalized represented within a symbolic program so that it is executable by the inference engine. Knowledge must be expressed in the knowledge representation method and the language of the expert system tools for building the expert system.

Testing and validation

The working expert system is moved to the plant floor environment for final testing. The expert system is made to run on a large number of representation samples of test cases. If bugs or problems are detected, the necessary corrections are made and the system is evaluated again.

Maintenance of the system

The expert system structure and use are gradually modified through maintenance. Maintenance tasks include: adapter user interfaces, porting the system to other hardware or software environments, reflecting new user requirements, fixing bugs and errors and enhancing the performance.

23.7 EXPERT SYSTEM SHELLS

Expert systems are generally rule-based and the programs are coded as a set of "condition-action pairs or a variation of this, i.e.

IF condition
THEN action

The rule which has its condition true will be executed and this cycle be repeated potentially forever. Expert system shells are commonly used to develop protypes of expert systems quickly. The structure of an expert system shell is shown in Fig. 23.8. They are thus non-specialized and then to have inadequate user-interfaces.

A shell is like an *empty* expert system, because it has predefined forms of knowledge representation and inference, but no knowledge about your application. It is like a software package.

23.8 KNOWLEDGE REPRESENTATION

Knowledge may be defined as the information about the world that allows an expert to make decisions. Knowledge is of two forms: qualitative and quantitative. The designer would generally

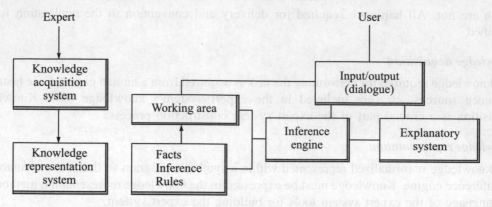

Fig. 23.8 The structure of an expert system shell.

use both forms in the design/manufacturing process. Knowledge structures are used to store knowledge and reason with it, just as data structures are used to store and deal with data. The process of represented at different levels, depending on the degree to which fundamental principles and casual relationships are taken into account. Shallow knowledge handles only surface level information that can be used to deal with specific situations. Deep knowledge, on the other hand, represents the interval and casual structure of a system and considers the interactions between its underlying components. The relationship between knowledge and intelligence is shown in Fig. 23.9.

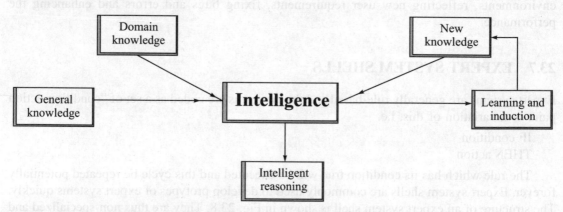

Fig. 23.9 Relationship between knowledge and intelligence.

The knowledge of an expert occupies three distinctive layers:

(a) Factual knowledge
(b) Heuristic knowledge
(c) Meta knowledge

Factual knowledge is accepted by all human experts. Heuristic knowledge is acquired by experience in the field. Meta knowledge helps in efficient utilization of heuristic and factual knowledge.

The knowledge that is used for solving engineering problems can be classified as:

(a) Compiled
(b) Qualitative
(c) Quantitative

Compiled knowledge consists of experience of the experts in the domain, knowledge gathered from sources like handbooks, old records, etc. The quantitative knowledge deals with numerical techniques, closed from solution, domain concepts etc. A design engineer would generally use all of these forms of knowledge in his design process.

The compiled as well as qualitative knowledge can be further classified into two broad categories:

(a) Declarative knowledge
(b) Procedural knowledge

Declarative knowledge primarily deals with the knowledge on the physical properties of the problem domain. The procedural knowledge deals with the problem solving techniques. The most common ways of representing such knowledge are:

(a) Production rules
(b) Decision tables
(c) Frames
(d) Semantic networks
(e) Predicate logic
(f) Conventional programs

23.8.1 Production Rules

In production systems, knowledge is represented as a set of production rules which may consist of both procedural and heuristic knowledge. Production rules are in effect, a subset of predicate calculus with an added prescriptive component indicating how the information in the rules is to be used during reasoning. Each rule in the knowledge base has a left hand side (antecedent) consisting of either one or more conditions and a right hand side (consequent) consisting of one or more actions. The actions specified in the right hand side are taken only if all the conditions in the left hand side are satisfied. The general form is:

IF (some combination of conditions is true)

THEN (draw some conclusion or take some action)

Consider, for example, the following production rule:

IF part P_i is to be dispatched to machine M_a that is occupied by another part P_j

THEN check availability of an alternative machine $M_b, c_f 100$

where f refers to the certainty factor of 100%.

A production rule states that a conclusion or an action or a consequent is sure to take place, if the situation on the IF part happens. If the situation in the IF part is satisfied, the rule is said to be triggered. If the action THEN part of the rule is carried out then the rule is said to have fired.

Each of the IF conditions are called **clauses**. A set of clauses joined by logical ANDs constitute HORN CLAUSES. The ACTION part of one production rule becomes the

CONDITION part of another production rule. This is called **production rule chain** or **networking of rules**.

The conditions of rules are:

Rules should not be inconsistent.

A rule should not cause a loop.

A rule can be changed without changing other rules.

23.8.2 Decision Tables

Decision tables are simple form of knowledge representation technique, which is capable of representing heuristic as well as intutive knowledge in a simple but efficient manner. The decision table approach can be used either only at the knowledge acquisition stage and/or as a formal technique for knowledge representation.

23.8.3 Frames

A frame provides a concise structural representation of useful relations. It can be viewed as a generalization of a property list that provides a structured representation of an object or a class of objects. Each individual object or class of objects may be represented by a frame. A frame is a unit of knowledge source described by a set of slots. There can be two types of slots in a frame: concrete slot which contains a specific value, and abstract slot which contains description that characterizes any possible value for the slot. The basic structure of a frame consists of name of the frame, parents of the frame, slots of the frame and their values, the attached predicates for each slot. The most important features of a frame are as follows:

(a) It provides a structured representation of objects and closes of objects in an application domain.

(b) It provides a mechanism (called **inheritance**) that guides description movement from class descriptions to individual descriptions.

(c) It allows the specification of procedures (called **demons**) for computing descriptions. This feature is known as **procedural attainment**.

(d) It allows one to determine descriptions in the absence of specific knowledge. This feature is referred to as the default feature. Frames can represent not only objects being reasoned about but rules as well. When each rule is represented as a frame, rules can be grouped into classes and the description of a rule can include arbitrary attributes of the rule.

In a manufacturing context, frames have been used in job shop scheduling and in project management. Frames have also been used in scheduling activities outside the manufacturing context.

23.8.4 Semantic Networks

Semantic network representation of knowledge is very useful in representing static or declarative type of knowledge. In its simplest form, a semantic network is a graph whose nodes represent individual objects, concepts or events and whose directed arcs (or links) represent binary relationships between nodes.

Figure 23.10 is an example of how a body of facts can be represented using a semantic network. It represents the following facts:

Machine M_1 is idle at time t

Machine M_2 is idle at time t

Machine M_1 can perform operation O_1

Machine M_1 can perform operation O_2

Machine M_2 can perform operation O_2

Machine M_2 can perform operation O_3

O_1 is the first operation that must be performed on part P_1

O_2 is the second operation that must be performed on part P_1

O_3 is the last operation that must be performed on part P_1

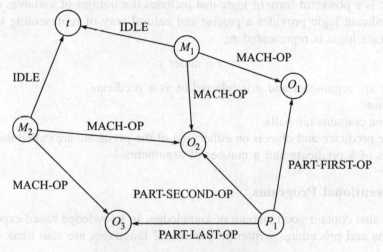

Fig. 23.10 An example of semantic network.

The semantic network representation of knowledge can be viewed as a graphical representation of the binary predicate version of predicate calculus, where an arc labelled R directed from node x to node y.

$$R$$
$$\text{o------------------------o}$$
$$x \qquad\qquad\qquad\qquad y$$

represents the assertion $R(x,y)$.

Hence the semantic network in Fig. 23.10 represents the following assertions:

IDLE (M_1, t)

IDLE (M_2, t)

MACH-OP (M_1, O_1)

MACH-OP (M_1, O_2)

MACH-OP (M_2, O_2)

MACH-OP (M_2, O_3)

PART-FIRST-OP (P_1, O_1)

PART-SECOND-OP (P_1, O_2)

PART-LAST-OP (P_1, O_3)

It may be noted that all nodes in Fig. 23.10 are labelled by constant symbols. In general, however, nodes can be constants, variables or terms constructed using function symbols. Arcs can represent conditions as well as conclusions and can be grouped into classes. The overall graph is then referred to as an extended semantic network.

23.8.5 Predicate Logic

Predicate logic is a powerful form of logic that includes the notions of variables quantifiers and predicates. Predicate logic provides a precise and natural way of representing knowledge.

The predicate logic is represented as:

$$x \ p \ \text{under} \ y$$

where x and y are arguments and p is referred to as a predicate.

In the expression:

Ball bearing one contains ten balls.

Contains is the predicate and objects on either side of the predicate are called the arguments. A clause consists of a predicate and a number of arguments.

23.8.6 Conventional Programs

The programs also contain good amount of knowledge. In knowledge based expert systems for CAD, functions and procedures written in procedural languages are also used.

23.9 INFERENCE ENGINE

The knowledge base could not be used unless it had a good inference mechanism enabling it to use the knowledge. Different types of inference engines are possible, depending on the type of knowledge representation scheme adopted. In a rule-based system, the inference engine (rule interpreter) examines facts and executes rules contained in the knowledge base according to the inference and control procedures selected.

The process of combining facts and rules is referred to as inference. Inference process is a kind of search technique, where a given pattern is matched against a set of stored patterns. Simple search techniques can be used for problem solving using knowledge stored in rules. The most commonly used inferencing models are forward chaining and backward chaining. The current state of patterns are stored in a working memory called **contex**.

23.9.1 Forward Chaining

Reasoning by the invoking of inference rules can proceed in different ways according to control procedures. One procedure is to start with a set of facts (or data) and to look for those rules in the knowledge base for which the IF portion matches the facts. When such rules are found, one of them is selected based upon an appropriate conflict resolution criterion and executed (fired). This generates new facts in the knowledge base, which in turn causes other rules to fire. The reasoning operation stops when no more new rules can be fired. This kind of reasoning is known as **forward chaining** or **data-driven reasoning** (inference). Forward-chaining inference mechanism is adopted in those cases where there are large number of goal states (or conclusions) with less initial states (or facts). An example of forward chaining is illustrated in Fig. 23.10.

23.9.2 Backward Chaining

In backward chaining, the system is to being with a specific clause called a **goal** to be proved and try to establish the facts needed to prove it by examining those rules which have the desired goal in the THEN portion. If such facts are not available in the knowledge base, they are set up as sub goals. The process continues until all the required facts are found, in which case the original goal is proved or the situation is reached when one of the subgoals cannot be satisfied, in which case the original goal is disproved. This method of reasoning is called **backward chaining** or **goal-directed reasoning** (inferencing). Backward chaining is favoured when the number of rules is large. An example of backward chaining is illustrated in Fig. 23.11.

23.9.3 Hybrid Chaining

Hybrid chaining mechanism first starts with forward chaining and whenever a data is required from the user, backtracks to the leaf node of the knowledge net and gets it to continue with the forward chaining mechanism. The hybrid process is found to be very suitable in such cases, where the knowledge net is very large and the use of the expert system does not know about what are the data to be supplied in the beginning.

23.10 ROLE OF ARTIFICIAL INTELLIGENCE IN MANUFACTURING

The generic applications of AI in manufacturing are:

   (a) error diagnosis and repair (machines and systems)
   (b) operation of machines and complex systems
   (c) design of systems, equipment and usage of intelligent design aids and invention of innovative systems
   (d) management of manufacturing planning activities, computer-aided scheduling and condition monitoring
   (e) industrial activities like factory management, computer-aided process planning, computer-aided inspection, use of AI in industrial robots, etc.
   (f) visual perception and guidance (inspection, identification verification, AGV guidance, screening)

Knowledge base
(Initial state)

Fact:

F_1 — L-001 is a lathe

Rules:

R_1 — IF X is a machine, THEN X has a motor
R_2 — IF X is a lathe, THEN X has a turret
R_3 — IF X is a lathe, THEN X is a machine

$\Leftrightarrow$ F_1 & R_2 match

Knowledge base
(Intermediate state)

Facts:

F_1 — L-001 is a lathe
F_2 — L-001 has a turret

Rules:

$\left.\begin{array}{l} R_1 \\ R_2 \\ R_3 \end{array}\right\}$ as before

$\Leftrightarrow$ F_1 & R_3 match

Knowledge base
(Intermediate state)

Facts:

$\left.\begin{array}{l} F_1 \\ F_2 \end{array}\right\}$ as before

F_3 — L-001 is a machine

Rules:

$\left.\begin{array}{l} R_1 \\ R_2 \\ R_3 \end{array}\right\}$ as before

$\Leftrightarrow$ F_3 & R_1 match

Knowledge base
(Final state)

Facts:

F_1 — L-001 is a lathe
F_2 — L-001 has a turret
F_3 — L-001 is a machine
F_4 — L-001 has a motor

Rules:

R_1 — IF X is a machine THEN X has a motor
R_2 — IF X is a lathe, THEN X has a turret
R_3 — IF X is a lathe, THEN X is a machine

Fig. 23.11 An example of forward chaining.

Knowledge base
(Initial state)

Fact:

F_1 — L-001 ia a lathe

Rules:

R_1 — IF X is a machine, THEN X has a motor
R_2 — IF X is a lathe, THEN X has a turret
R_3 — IF X is a lathe, THEN X has a machine

Goal stack

Goal:		Satisfied:
G_1 — L-001 has a motor		?

↶ G_1 & R_1 match

Knowledge base
(Intermediate state)

Fact:

F_1 — as before

Rules:

$\left.\begin{array}{l} R_1 \text{—} \\ R_2 \text{—} \\ R_3 \text{—} \end{array}\right\}$ as before

Goal stack

Goals:		Satisfied:
G_1 — as before		?
G_2 — L-001 is a machine		?

↶ G_2 & R_3 match

Knowledge base
(Intermediate state)

Fact:

F_1 — as before

Rules:

$\left.\begin{array}{l} R_1 \text{—} \\ R_2 \text{—} \\ R_3 \text{—} \end{array}\right\}$ as before

Goal stack

Goals:		Satisfied:
$\left.\begin{array}{l} G_1 \text{—} \\ G_2 \text{—} \end{array}\right\}$ as before		? ?
G_3 — L-001 is a lathe		?

↶ F_1 & G_3 match

Knowledge base
(Intermediate state)

Fact:

F_1 — as before

(Contd.)

Rules:

$$\left.\begin{array}{l} R_1 \text{—} \\ R_2 \text{—} \\ R_3 \text{—} \end{array}\right\} \text{ as before}$$

Goal stack

Goals:		Satisfied:
$\left.\begin{array}{l} G_1 \text{—} \\ G_2 \text{—} \end{array}\right\}$ as before		?
		?
G_3 — L-001 is a lathe		Yes

$\circ$ F_1 & R_3 match

Knowledge base
(Intermediate state)

Facts:

F_1 — as before
F_2 — L-001 is a machine

Rules:

$$\left.\begin{array}{l} R_1 \text{—} \\ R_2 \text{—} \\ R_3 \text{—} \end{array}\right\} \text{ as before}$$

Goal stack

Goals:		Satisfied:
G_1 — as before		?
G_2 — L-001 is a machine		Yes

$\circ$ F_2 & R_1 match

Knowledge base
(Final state)

Facts:

F_1 — L-001 is a lathe
F_2 — L-001 is a machine
F_3 — L-001 has a motor

Rules:

R_1 — IF X is a machine, THEN X has a motor
R_2 — IF X is a lathe, THEN X has a turret
R_3 — IF X is a lathe THEN X has a machine

Goal stack

Goal:		Satisfied:
G_1 — L-001 has a motor		Yes

Fig. 23.12 An example of backward chaining.

The current and future AI manufacturing applications are listed in Table 23.1.

TABLE 23.1 Applications of AI in Manufacturing

AI technology	Application
Expert systems	Design
	Maintenance
	Process control
	Monitoring
	Alarm analysis
	Equipment diagnosis
	Process planning
	Scheduling
	Inspection
Machine vision	Identification
	Measurement
	Welding
	Material handling
Robotics	Parts positioning
	Assembly
	Spray painting
Natural language understanding	Database information retrieval
	Data entry
	Inventory control
Voice recognition	Quality inspection
	NC programming
	Robotics
Speech synthesis	Control room alarms

23.11 CHARACTERISTICS OF AN EXPERT SYSTEM

The characteristics of an expert system are:

(a) Ability to solve complex programs in a domain as good as or better than human experts
(b) Consistent results can be obtained
(c) Reliable sources for prediction, interpretation, fault diagnosis, debugging, monitoring and control
(d) Complete goal can be achieved with the available rules and facts
(e) Ease of use of the system

23.12 LANGUAGES FOR EXPERT SYSTEMS

There are several approaches available for building an expert system. One can take a conventional programming approach, using one of the programming languages suitable for AI, LISP (list processing), PROLOG (logic programming), small talk or one of a variety of other specialized languages for system-building. It is often necessary to combine such techniques with a frame-based system, a language formalism for representing declarative facts. A number of tools have been developed that allow a higher-level approach to building expert system. A few provide an integrated knowledge engineering environment combining features of all of the above mentioned AI languages. A common term for these more powerful tools is expert system shell, referring to their origins as specialized expert systems whose knowledge base has been removed, leaving only a shell that can perform the functions of inference engine, user interface and knowledge storage medium.

23.13 EXAMPLE OF EXPERT SYSTEMS

ABSTRIPS	Robotics planning
CATS-1	Computer-assisted trouble shooting system
GUIDON	Instructions
IMS	Automated factory management
MECHO	Mechanical problems analysis
MYCIN	Medicine diagnosis
VM	Ventilator manager for monitoring and control
XCON	Computer system configuration design

23.14 BENEFITS OF EXPERT SYSTEM

Expert system provides companies with improved productivity levels and increased competitive advantages owing to the following potential benefits:

(a) Some expert systems have proved to do a better job than humans. They make few mistakes and are more consistent in their recommendations.

(b) Expert systems can be used as a training vehicle to train non-experts and even to improve the expertise of experts.

(c) Expert systems can free experts from time-consuming routine tasks, e.g. repetitive advising and training data search tests, etc.

(d) Expert systems are compatible with manager's decision styles and are based on judgement which is extensively used by managers. Human expertise is usually very expensive compared with a frequently used expert system.

(e) Expert systems can preserve scarce expertise.

(f) Expert systems enable operation in hazardous environment.

REVIEW QUESTIONS

23.1 Define artificial intelligence.

23.2 What are the effects of artificial intelligence on CAD systems software?

23.3 Define an expert system.

23.4 Discuss the structure of an expert system.

23.5 How an expert system can be built up?

23.6 List out the benefits of expert systems.

23.7 Discuss the languages used for expert systems.

23.8 What are the functions of the knowledge-base and inference engine in knowledge-based systems?

23.9 What is knowledge? Differentiate between shallow knowledge and deep knowledge.

23.10 Discuss the various techniques of representing knowledge.

23.11 Outline the principles of forward and backward chaining inference in a knowledge base comprising production rules.

23.12 Explain the various components of knowledge.

23.13 Define the following terms with respect to AI in design.

(a) Decomposition   (b) Plan section and refinement
(c) Constraint-based reasoning   (d) Case-based reasoning

23.14 Discuss the various strategies for knowledge acquisition.

23.15 Explain the following terms:

(a) Expert system shell   (b) Inference engine

23.16 Explain the expert system techniques for design.

23.17 Explain the term fuzzy logic and discuss its applications.

23.18 Describe the elements of artificial intelligence. Why is machine vision a part of it?

23.19 Give examples in manufacturing where artificial intelligence could be effective.

23.20 Give examples in manufacturing where artificial neural networks are particularly useful.

23.21 Differentiate between expert systems and artificial intelligence.

COMMUNICATION SYSTEMS IN MANUFACTURING

INTRODUCTION

The starting point of any economy is manufacturing the making of products. Look at the manufacturing systems today, most of them assume that the business of manufacturing takes place within the four walls of the factory, but manufacturing doesn't only occur within the four walls of the factory. In the real world the manufacturing business extends far beyond the factory floor. That is why the need for communication systems in manufacturing arises. Computer and manufacturing are like the bread of a sandwich. The meat is integrated without the integration we would not have achieved growth in communication. Just as computer technology is moving towards massively parallel architectures, network technology should be moving towards massively parallel networks. Network performance must be kept in balance with the computer and software performance. In this chapter, communication networks in manufacturing, hierarchy of computers in manufacturing, communications standards and manufacturing communication systems are discussed in detail.

24.1 COMMUNICATION NETWORKS IN MANUFACTURING

In order to maintain a high level of coordination and efficiency of operation in integrated manufacturing, an extensive, high-speed and interactive communications network is required. A major advance in communications technology is the local area network (LAN). In this hardware and software systems, logically related groups of manufacturing cell communicate with each other. The network links these groups to each other, bringing different phases of manufacturing into a unified operation. A local area network can be very large and complex, linking hundreds, or even thousands, of machines and devices in several buildings; various network layouts (see Fig. 24.1) of fibre-optic or copper cables are typically used, over distances ranging from a few metres to as much as 32 km. For longer distances, wide area networks (WANs) are used. Different types of networks can be linked or integrated through *gateways* and *bridges*.

Access control to the network is important. Otherwise, collisions can occur when several workstations are transmitting simultaneously. Thus, continuous scanning of the transmitting

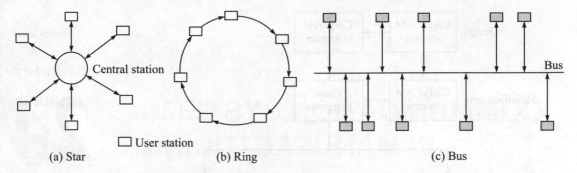

Fig. 24.1 Three basic types of topology for a local area network.

medium is essential. In the 1970s, a carrier sense multiple access with collision detection (CSMA/CD) system was developed and implemented in Ethernet. Other access-control methods are token ring and token bus, in which a *token* (special message) is passed from device to device. The device that has the token is allowed to transmit, while all other devices only receive.

A new trend is the use of wireless local area networks (WLAN). Conventional LANs require routing of wires, often through masonry walls or other permanent structures and require computers or machinery to remain stationary. WLANs allow equipment such as mobile test stands or data collection devices, such as bar code readers, to easily maintain a network connection. A communication standard (IEEE 802.11) currently defines frequencies and specifications of signals, and defines two radio frequency and one infrared methods. It should be noted that wireless networks are slower than desirable, especially for situations where slow tasks such as machine monitoring are the main application.

Personal area networks (PAN) are under development. PANs are based on communications standards such as bluetooth, IRDA and home RF, and are designed to allow data and voice communication over short distances. For example, a short range bluetooth device will allow communication over a 10 m distance. PANs are undergoing major changes and communications standards are continually being refined.

24.2 HIERARCHY OF COMPUTERS IN MANUFACTURING

Computers and computer-driven devices (e.g., CNC machine tools, robots) in a manufacturing firm tend to form a pyramidal control structure. The term sometimes used to describe this computer pyramid is *hierarchical computer system*. Its general configuration is shown in Fig. 24.2. The hierarchical arrangement depicted here is a management oriented structure indicating various levels of responsibility, some of which are subordinate to others. The configuration shown in the figure should not be interpreted as the physical arrangement of the communications lines between different computers in the system. Rather, it represents the command structure that exists between the computers and computer-driven devices in the factory.

The various computers in the hierarchy are tied together by communication links to form a distributed computer system. The individual links provide a computer-communications network to forward data and information up through the various levels from direction. Product designs, process plans, production schedules, machine commands and so on are passed down

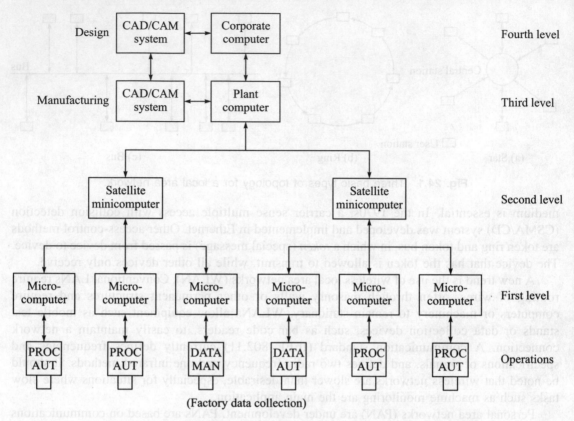

Fig. 24.2 Hierarchy of computers in a manufacturing organization.

to the individual production cells. The hierarchy allows for the design engineering and manufacturing engineering functions to be included within the computer network.

24.2.1 Levels in the Computer Hierarchy

There are several distinct levels in the hierarchical network, each with its own purpose and function. As indicated by Fig. 24.2, we are proposing that this hierarchical computer structure consists of four levels in the firm. This may be the appropriate number of levels for all organizations. We are also proposing a very ideal structure, one that is consistent with the realization of computer-integrated manufacturing.

At the lowest level within the hierarchy are the computers that are connected directly to the process, either to monitor or control it. The computers that make up this level are typically microcomputers, located in close proximity to the manufacturing operations they monitor and control. In many cases, the computer is an integral part of the production machine (e.g., a CNC machine tool or a robot). The computer is dedicated to the process and performs tasks that serve the process and facilitate communication with the second level in the computer hierarchy. This level also includes the devices used for factory data collection systems (e.g., bar code readers, manual data entry terminals, etc.)

The second level in the hierarchy consists typically of minicomputers (more recently, super microcomputers) that are located throughout the plant and report to the larger plant computer at the third level. The second-level minicomputers are sometimes referred to as *satellites* (or are *controllers*) in the plant. Their purpose is to serve in a supervisory capacity, coordinating and controlling the activities of the first-level computers in the various sections of the plant. Performance data are collected from the individual machine tools, production cells, data input terminals, and inspection stations; and operating commands are relayed back to the separate processes and workstations.

The third level in the hierarchical structure is the central plant computer. Operating data from the various satellite computers in the plant are collected and summarized to prepare periodic (e.g., daily, weekly, monthly) reports for plant management. Real-time information on the current status of plant operations (e.g., machine utilization, customer order progress, inventory levels, etc.) can be accessed from this level by plant personnel with a need to know. The manufacturing engineering CAD/CAM system interface would probably be at this third level in the organization, since most of its functions would be implemented in the plant. The computer at the plant level would be a larger, more complete data processing system, either a large minicomputer or a small mainframe. Its use must be shared between plant manufacturing operations and other business-related functions that are performed at the plant level. Payroll, cost accounting and personnel are examples of these other functions.

The fourth level is the corporate mainframe computer. Data are compiled from the various plants in the corporation at this level. Communication with the individual plant computers can be achieved by means of telephone long lines, satellites or other wide area communications technologies. The purpose of the corporate computer in the hierarchy is to summarize plant operations and performance for the entire corporation. The CAD/CAM system interface for design engineering is probably at this level although it may be at the plant level depending on corporate philosophy and organization. In addition, the corporate computer must be shared with other departments at the corporate level: sales, marketing, accounting and so on.

24.2.2 Benefits of the Hierarchical Structure

The hierarchical computer structure has become the most effective and efficient arrangement for implementing computer systems in manufacturing. At one time in the evolution of computer control of production, it seemed most feasible to use one large mainframe computer to handle all planning, monitoring and control functions. Indeed, this mode of control became typical of computer control in the process industries. Today, with the proliferation of small computers (e.g., personal computers), the advantage has gone to the hierarchical computer network. The following is a list of the important benefits of the hierarchical approach to CIM:

Gradual implementation

The hierarchical computer system can be installed gradually rather than all at once. Each individual computer project can be justified on its own merits. The company is not required to make an all-or-nothing ("you bet your company") commitment to install a single plant wide computer. The risk is therefore reduced, and the expense can be spread over a number of years, with paybacks realized for each step in the implementation of the hierarchical system.

Redundancy

The hierarchical structure contains redundancy. In the event of a computer breakdown, other computers in the system are programmed to assume the critical tasks of the broken-down computer.

Reduced software development problem

Software development, a very major portion of the total expense in most integrated computer systems, can be managed more easily in the hierarchical configuration. Since the computers are separated in the pyramidal arrangement, programming for each project can be handled separately. Once the project is installed, changes in software are more easily accomplished, with less chance of disrupting the system.

There are significant problems involved in connecting a large number of computers and computer-controlled programmable machines together into the hierarchical structure described here. The following two sections discuss some of these problems.

24.3 COMMUNICATIONS STANDARDS

Often, one manufacturing cell is built with machines and equipment purchased from one vendor, a third cell with machines purchased from yet a third vendor, and so on. As a result, a variety of programmable devices is involved. Driven by several computers and microprocessors purchased at various times from different vendors and having various capacities and levels of sophistications and proprietary standards and they cannot communicate beyond the cell with others unless equipped with custom-built interfaces. This situation creates islands of automation and in some cases, upto 50% of the cost of automation has been related to overcoming difficulties in the communications between individual manufacturing cells and other parts of the organization.

The existence of automated cells that could function only independently of each other, without a common base for information transfer, led to the need for a communications standard to improve communications and efficiency in computer integrated manufacturing. After considerable effort and on the basis of existing national and international standards, a set of communications standards known as **manufacturing automation protocol (MAP)** was developed. The capabilities and effectiveness of MAP were first demonstrated in 1984 with the successful interconnection of devices from a number of vendors.

The international organization for standardization open system interconnect (ISO/OSI) reference model accepted worldwide. This model has a hierarchical structure in which communication between two users is divided into seven layers. Figure 24.3 shows the ISO/OSI reference model for open communication. Each layer has a special task:

    (a)   Mechanical and electronic means of data transmission
    (b)   Error detection and correction
    (c)   Correct transmission of the message
    (d)   Control of the dialog between users
    (e)   Translation of the message into a common syntax
    (f)   Verification that the data transferred have been understood

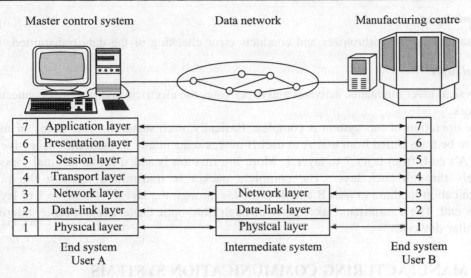

Fig. 24.3 The ISO/OSI reference model for open communication.

The OSI model divides the communication functions into seven layers application, presentation, session, transport, network, data-link and physical.

Application layer

The application layer manipulates information exchange and supports distributed processing between two parties in a communication network.

Presentation layer

The presentation layer ensures that the data transmitted in a network have format and syntax suitable for the receiving system.

Session layer

The session layer provides control functions that ensure that communication in a network is synchronized and managed properly. For example, it ensures that communication is not prematurely terminated. It also provides two-way simultaneous communication, i.e., the parties can receive and send messages simultaneously or two-way alternate communication, i.e., the parties alternate in communicating with each other.

Transport layer

The transport layer controls end-to-end data transmission and also ensures that the data are appropriately segmented before sending them to the upper layer.

Network layer

The network layer handles routing of the data in the network. It also determines the address of the receiving party.

Data-link layer

The data-link layer synchronizes and conducts error checking of the data transmitted.

Physical layer

The physical layer maintains, activates and deactivates the electrical and physical connections in a network.

The operation of this system is complex. Basically, each standard-sized chunk of message or data to be transmitted from user A to user B moves sequentially through the successive layers at user A's end, from layer 7 to layer 1. More information is added to the original message as it travels through each layer. The complete packet is transmitted through the physical communication medium to user B and then moves through the layers from layer 1 to layer 7 at user B's end. The transmission takes place through fibre-optic cable, coaxial cable, microwaves and similar devices.

24.4 MANUFACTURING COMMUNICATION SYSTEMS

Communication protocols have been extended office automation as well with the development of technical and office protocol (TOP) which is based on the ISO/OSI reference model. In this way, total communication (MAP/TOP) can be established across the factory floor and among offices at all levels of an organization. MAP is a factory communications standard developed by general motors and is specifically designed to provide communication in a CIM environment. Its objective is to make communication possible between mainframe systems, cell controllers, workstation terminals, programmable logic controllers, material handling systems, robots and other types of factory equipment. Implementation of the MAP standard results in a number of benefits. Some of which are:

(a) Equipment purchased from different vendors can be linked through a factory local area network (LAN), since each vendor adopts the MAP standard. As a result, the manufacturing system designers can select the most appropriate equipment from various vendors.

(b) The manufacturing system is reliable because the hardware and software parts are built in accordance with the MAP standard and the manufacturing system can be installed in relatively short time.

TOP is a communication standard developed by Boeing company for non-proprietary multiple-vendor data communication in technical and office environment. It shares a core of common protocols with MAP. Although the MAP and TOP specifications are derived from the seven-layer OSI reference model, the services provided at the application layer of these specifications are suitable for their corresponding environment. For example, a process controller requires real-time data transfer between two points, while office users require communication oriented file transfers.

If the benefits of MAP/TOP implementation are to be fully realized, one must determine to what extent the products offered by vendors conform to the MAP/TOP standards. This requires testing products to verify whether they conform to the desired specifications. Work on this area is currently under way in the United States. The National Institute of Standards and Technology provides software for testing conformance of products to various OSI protocols.

Another aspect that is of great importance is network security. Communication networks are to be adequately protected from unauthorized access and fraudulent misuse. Since data exchange between corporations, electronic funds transfer and so on, take place in a communication network, a network architecture that outlines a policy for network security is essential.

There are three issues in manufacturing common systems that are worth mentioning:

(a) Data redundancy
(b) Bottom-up versus top-down approach to design
(c) Volume of information

24.4.1 Data Redundancy

Data redundancy in CIM databases is undesirable but is frequently unavoidable given the limitations of existing database software technology. For example, the database of the part design module may have a three-dimensional model of a part that contains a complete representation of the part geometry and the database corresponding to the NC part programming module may also have the part geometry stored without the representation of the part geometry, unless instructions for creating it from the three-dimensional model are stored in the database. Such technology is not available in existing software and current research is proceeding in this direction.

24.4.2 Bottom-up versus Top-down Approach to Design

In the bottom-up design approach in computer-integrated manufacturing (CIM) systems, the computer is used mainly as a support system, i.e., to perform calculations, print data specified by the user and so forth. Skilled personnel are required to work with the computer, interpret the data and request the computer to perform calculations. Thus, in the bottom-up approach, starting from the initial data, a step-by-step procedure is followed to arrive at the goal. For example, in the case of a computer-aided process planning system, starting from initial data regarding parts, machines tools and fixtures, a procedure that involves generation of process routes and determination of production parameters (such as depth of cut, tool feed rate, speed and tolerances) is used to arrive at the goal, which is a process plan (general strategy) for the part.

In the top-down approach, the computer system initially considers various strategies to arrive at the goal. Then, based on the constraints of the problem it either accepts, rejects or modifies these strategies. In some cases, two or more strategies may be acceptable and the decision to accept one of these strategies may be a difficult task.

These two approaches have advantages as well as disadvantages. The bottom-up approach is simple, but it does consider alternative strategies in arriving at the goal. On the other hand, the top-down approach is rather difficult to implement but has a number of advantages. This is why some researchers feel that a combination of the top-down and bottom-up approaches is perhaps the most promising.

24.4.3 Volume of Information

The volume of information handled at various levels in a manufacturing organization depends on the structure of the organization. As shown in Fig. 24.4 the volume of information handled is least at the management level and greatest at the shop floor level.

Communication among the various levels shown in Fig. 24.4 is necessary because the information gathered at one level is used in another level. For example, at the machine cell level, information regarding such factors as machine status and production statistics is available. The machine cell level collects such information from the individual machine and other entities in the shop floor level. This information is used at the plant level in order to plan and control overall production. An efficient communication protocol will ensure that the information is used by the various levels in an effective manner. Thus, a particular level, for example, the cell level receives only the information that is required. Also it passes only the necessary information to another level.

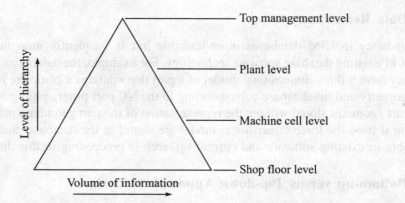

Fig. 24.4 Volume of information handled by various levels in a manufacturing organization.

24.5   INTERNET

Internet is the mother of all networks and is a gigantic network of networks that connects presently about 50 million people and more than 2.5 or 3.0 million computer sites. These computer sites are spread all over the world at various government locations, corporate offices, colleges and universities and research institutions.

There are literally hundreds of databases that let people access information ranging from medical transcription economic data to GIS and engineering tables. This superhighway of information makes available millions and millions of computer files filled with sound and video clips. Literally anybody can take these and people can exchange notes, reviews and views with each other.

Whenever one wants to send a message on the network, all we have to do was to get onto your computer that would put one's data in an envelope, called an **internet protocol (IP)** packet, "address" the data or packets correctly. The communicating computers were also given the responsibility to ensure that the communication was accomplished. The international standard organization was instrumental in the development of the internet protocol who spent years and years also designing, according to them the ultimate standard in computer networking. The manufacturers of computers realized that it is the logical method by which their computers could communicate to each other.

24.6 INTRANET

Intranet can be defined as a manufacturing organization's use of world wide web and related internet technology to accomplish its primary objective helping in production of superior quality goods with substantial decrease in operating costs or achieving total customer satisfaction, maintaining healthy employer-employee relationship.

An intranet is designed to distribute corporate information within an organization to organization's own people using internet, web and related technology can be put to maximum use to improve the purpose of the organization. People may include both employees and valuble customers of that organization. In the broadest sense, both internet and intranet use the same technology. However, intranet is an organizational network, whereas internet is a global one, which embraces thousands of servers providing wide range of information. Intranet has access to internet but not vice versa.

Intranets are internal networks that the employee's can access for information by simple browsing as on the world wide web. It acts as a reservoir of company information. It is a very effective platform for the employees to discuss projects and share ideas. Intranet enhances project management and reporting systems and improves work flow and increases productivity. Another very significant advantage of intranet is that it permits employees share information with colleagues spread over various locations simultaneously.

Following are the major applications of intranet mail:

(a) Publishing corporate documents
(b) Access to searchable directories
(c) Workgroup collaboration
(d) Software distribution
(e) User interface using HTML

A company which has manufacturing division spread geographically can make use of internet technology for effective work flow automation and collaborative product development. The company can set up intranet at the headquarters. Each division can have its own intranet which can be linked to the intranet at the headquarters through internet. The divisional intranets can be called **extranets**. Thus, the entire manufacturing organization can be linked into a single networked entity.

24.7 HARDWARE ELEMENTS OF A NETWORK

Introduction of computer-integrated manufacturing necessitates every manufacturing engineer to be aware of the basic hardware elements of a network. The components required for a small network, restricted to a computerized design or production planning department have been discussed earlier. This topic gives a brief description of these elements used in an entire manufacturing plant.

LAN adapters

These adapters are used for client workstations running DOS and windows or OS/2 or for servers with ISA buses. The LAN adapter for ethernet provides installation flexibility. It provides software configurability.

Ethernet switch

Ethernet switch is used to increase bandwidth 400 percent by using high-speed, parallel paths among connected ethernet LAN segments and devices. By using this switch, LAN congestion is reduced, so that the frequent LAN users get extra capacity without crowding out others on the same network. It connects to a variety of hubs and concentrators. It improves security by controlling traffic flow to specific addresses or within a defined group of virtual switch ports. It enables remote configuration and microcode download.

Ethernet workgroup hubs

Workgroup hubs allows us to add up to 8 or 16 stations to an existing ethernet network to prevent network congestion. They automatically partition any port connected to a device which generates repeated collisions and re-enable the port when the condition clears. They are also pathways to continued network growth. It indicates status of power, collisions and connections via LEDs.

Wireless LAN adapters

These adapters provide seamless LAN access to any unwired PC. We can even create a LAN among notebook PCs in conference rooms, airplanes and wherever required. It provides seamless bridging and roaming support for token-ring and ethernet networks. It extends existing LANs via infrared enabled bridges or routers. It supports different drivers. It works with industry-standard computers. It provides a 1 Mbps data transfer rate.

FDDI adapters

The fibre distributed data interface (FDDI) delivers the bandwidth we need for even the most demanding network applications. Microchannel, ISA and EISA systems alike benefit from 100 Mbps speed to the desktop—just the capacity, we require for today's graphics intensive applications. It provides synchronous frame transmission and bandwidth allocation for multimedia and time critical applications. It includes remote program update.

FDDI work group concentrator

The work group concentrator is a low cost hub solution. It works with existing hubs to serve as a high speed link or backbone.

Token-ring adapters

Token-ring adapters are available for ISA, EISA, PCMICA, microchannel and PCI buses. They offer low price easy installation, reduced network maintenance costs and enhanced functions. They are software configurable with graphical installation software to simplify setup. They are switchless.

Router

Many branch networks need a cost-effective, high solution that connects to the main network. A router serves this purpose.

24.8   NETWORKING IN A MANUFACTURING COMPANY

A manufacturing company carries out a number of operations using computers. These include:

(a)   Master production and material requirement planning
(b)   Pay roll and human resources management
(c)   Purchasing and receiving
(d)   Order entry and invoicing
(e)   Shop management
(f)   Warehouse management
(g)   Tool crib management
(h)   Time keeping
(i)   Quality control
(j)   Shipping
(k)   Materials handling
(l)   Inventory control
(m)   Processing operations
(n)   Marketing

There are software solutions to carry out the above operations and many other tasks. These are implemented in different types of computers. In addition, several equipments like bar code readers will be used for data collection and input. All these equipments are to be connected in a LAN for effective operation. Figure 24.5 shows a typical factory LAN and the various equipment connected to the LAN.

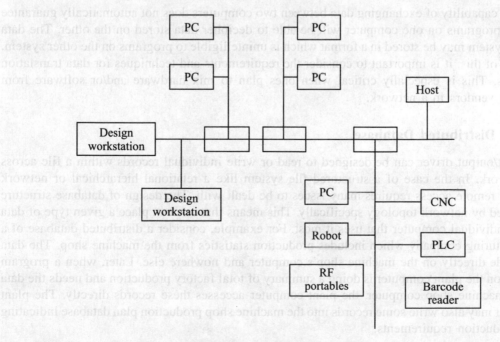

Fig. 24.5   A typical factory LAN.

24.9 ISSUES IN INTER-SYSTEM COMMUNICATION

There are several issues to be considered in network communication.

24.9.1 Terminal Networking

Any given network usually contains several different kinds of computers, with different kinds of data. Networking makes it possible for a given individual terminal or workstation to access different computers on the network. This is done by terminal networking. Terminal networking allows the user to enter some commands which instruct the appropriate network device to establish communication between the user's terminal and the desired computer. A lot of flexibility is provided by this arrangement.

24.9.2 File Sharing

When it is necessary to share and/or transfer files across a network from one network node to another, file sharing is used. The steps involved in file sharing are: Understand the service file requests coming from anywhere in the network, realize when the user needs to access a file which is stored at another node, then request the file from the input/output driver on the other system, via the network.

From the stand point of a given network node, this technique can be used to access and read a file from another node and also send a file to another node.

24.9.3 Data Translation

Physical capability of exchanging data between two computers does not automatically guarantee that the programs on one computer will be able to decipher data stored on the other. The data on one system may be stored in a format which is unintelligible to programs on the other system. Because of this, it is important to consider the requirements and techniques for data translation networks. This is especially critical when ones plan to mix hardware and/or software from different vendors in a network.

24.9.4 Distributed Database

An input/output driver can be designed to read or write individual records within a file across the network. In the case of a structured file system like a relational hierarchical or network database remote access requires many issues to be dealt with. The design of database structure is affected by network topology specifically. This means that one can place a given type of data on the individual computer that uses it most. For example, consider a distributed database of a manufacturing company which includes production statistics from the machine shop. The data can reside directly on the machine shop's computer and nowhere else. Later, when a program running on the plant computer is doing a summary of total factory production and needs the data on the machine shop computer, the plant computer accesses these records directly. The plant computer may also write some records into the machine shop production plan database indicating new production requirements.

24.10 FUNDAMENTALS OF COMPUTER COMMUNICATIONS

Some common terminologies relating to communications are:

(a) Data
(b) Information
(c) Signals
(d) Signalling
(e) Transmission

24.10.1 Data

Data is raw whereas information is processed data. Information displays some useful message whereas data merely contains it. In manufacturing, data and information are generated, collected, manipulated, transmitted, stored, retrieved, plotted and shared. Before data can be transferred, they must be represented in some suitable format. Representation of data is a function of the mode of its transfer. There are four different ways to transfer manufacturing data: spoken, written, analog or digital form. Since the industrial revolution, most manufacturing data have been transferred either orally or written on paper as memos, letters, reports, drawings and the like. These methods of information transfer are unsuitable for CIM. Data can be lost on the way. They could be misunderstood by the recipient or their storage may be cumbersome as with designer's drawings. CIM cannot afford these deficiencies. It requires more reliable and faster information transfer. The other two formats, analog and digital are based on whether digits are used to represent data. Historically, we have used analog data in manufacturing. An example of analog representation is the physical model of a component.

In such a case, the data is embodied in the model rather than described verbally or written (expressed through drawings). Other examples of analog data representation are forming tools, jigs, fixtures and prototypes. For CIM purposes, digital representation proves superior. Paper tapes, floppy diskettes, compact discs and direct communication between machines and computers are all based on digital data representation.

Data represented in digital form are most convenient for the needs of CIM. Since integration is achieved through computers, digital representation is, of necessity, binary rather than in the decimal system of the human world. Thus, data conversions cannot be avoided, since computer processing is all binary, whereas human processing is non-binary.

24.10.2 Coding

Whenever the prevalent practice of data representation is unsuitable, inefficient or both, some sort of coding is done. For example, the Greek letter Φ is used in drawings as a code for the diameter of a shaft or hole. This symbol saves the time and effort of expressing the term *diameter*. Since we use this term quite often in design and technical communications, coding could be justified. An acronym used to shorten a term, such as CIM for computer-integrated manufacturing, is another example of coding.

Communication requires two types of coding: data and control. While data coding translates the data to be transmitted, control coding is essential for controlling the communicating devices.

24.10.3 Transmission

Once manufacturing data have been coded, the next step in effective communication is to transmit the bits. The bits are transmitted as signals. The two bits, 0 and 1 are represented by two different signals, which flow through the transmitting medium. Each signal is at a certain level of voltage. In practice, bit 1 is represented by a negative voltage, usually between –5 and –25 volts, and bit 0 by a positive voltage between 5 and 25 volts.

Signals transmitted through a cable may become distorted. Distortion can take two forms:

Attenuation

The signal gets weaker due to power loss to the medium. Amplifiers are used along the transmission path to correct for this. Attenuation is not the same for each frequency; higher and lower frequencies lose more than mid frequencies.

Delay

The delay arises due to the fact that signals have different speeds of propagation at various frequencies. This effect, which is hardly noticeable in voice transmission except over long distances such as overseas calls, can give rise to errors in data transmission.

Equalizers are used to compensate for distortion, for example, frequencies losing power are boosted, those travelling faster are delayed and so on. Communication lines supported by equalizers are said to be "conditioned". The lines leased by telephone companies are usually conditioned.

24.10.4 Error Detection

Electrical noise can create errors in data communication. In a telephone conversation, the receiver can judge whether the message was error-free. In case of an error, the receiver responds: could you repeat that please? In computer communications with no human reasoning capability, the software is designed to detect errors.

These are two techniques for error detection: parity check and cyclic redundancy check (CRC). In the parity technique, a bit, called a **parity bit** is appended to the character bits at the left in the most significant position. The parity bit is either 0 (zero) or 1. Parity check may be odd or even. In the odd method, the parity bit renders the number of 1's in the character to be odd.

In the even parity method, the number of 1's in the eight-bit group is even parity method. The number of 1's in the eight-bit group is even.

Error detection based on parity is effective when one bit has suffered transmission error. If two bits are in error, the transmission error will go undetected. In fact the parity scheme is able to detect only when an odd number of errors have occurred. In general, this method works well, since the chance of two or more bits out of eight being in error simultaneously is extremely low.

The CRC technique detects errors by performing calculations on the bits. The sender appends the results of the calculation to the message. The receiver carries out the same calculation and compares it with the appended results to determine whether the transmission was error-free.

24.10.5 Modulation and Demodulation

Prior to transmission over telephone lines, digital signals called a carrier signal. The process of imprinting is termed as *modulation*. At the receiver end, a reverse process called *demodulation* takes place to recover the digital information. The device that implements these processes is a modem, an acronym for modulation and demodulation. Modulation may be of three types:

Amplitude modulation (AM)
Frequency modulation (FM)
Phase modulation (PM)

The binary bits ready for transmission enter a modulator that generates sine waves. The sine wave is modified depending on the bit pattern. In the AM method, the modulated signal looks like the one shown in Fig. 24.6(a), where bit 1 is represented by the high-amplitude and bit 0 by the low-amplitude portion of the signal.

A frequency modulated signal looks like that in Fig. 24.6(b) where the high-frequency portion represents bit 1 and the low-frequency portion bit 0.

In the FM technique, the phase of the sine wave is shifted by the bits so that the peaks do not occur where they should. This technique can result in dibits, two bits per signal state, using phase shifts of 90, 180, 270 and 360 degrees.

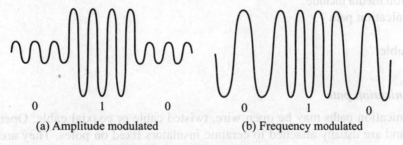

(a) Amplitude modulated   (b) Frequency modulated

Fig. 24.6 Modulated signals.

24.10.6 Simplex and Duplex

When communications is one-way, as with a doorbell, it is called **simplex**. The two-way mode, called **duplex**, may be half-duplex or full-duplex.

In half-duplex communications, only one device or person can use the line at a time whereas in full-duplex mode both can do it simultaneously. The citizen band radio is an example of half duplex with the word "over" used to indicate the end of transmission so that the other party may use the line. The common telephone is an example of full-duplex communications. Even when the medium is capable of full-duplexing, the protocol may limit the communication to half-duplex.

24.10.7 Line Speed

Communications speed is usually expressed in baud rate. The baud rate indicates the number of times the line condition changes state every second when the state can be either 1 or 0, as in AM or FM modulation. The baud rate equals line speed in bps. If there are four states, as in phase modulation, then the baud rate is twice the bps value. To avoid this confusion about line speed, it is better to use bps than baud rate.

Depending on the ratio between signal strength and background noise in the medium, there is an upper limit to the transmission rate. The theoretically possible maximum transmission rate (MTR) through a medium in bps is expressed by Shanon's law:

$$MTR = W \log 2\left(\frac{1+S}{N}\right)$$

where W is the band width and S/N is the signal-to-noise ratio.

The S/N ratio is sometimes expressed in db (decibel) which is given by

$$10 \times \log 10 \ (S/N)$$

24.10.8 Medium

A medium is that through which data is transmitted. The medium links the sender with the receiver. It provides for both offline and online communications. Paper tapes and magnetic disks or tapes are examples of offline media. Alternatively, a cable connecting the sender and receiver can be used for on-line transmission.

Communication media include:
Wire communication path
Microwave
Fibre-optic cable
Satellite

Wire communication path

Wire communication paths may be open wire, twisted cable or coaxial cable. Open wires have no covering and are usually attached to ceramic insulators fixed on poles. They are suitable for low traffic only. In twisted cables, copper wires insulated from each other are twisted in pairs. There is a risk of cross talk (undesirable signals from one wire to another) when multiple twisted pairs have been warped in the cable. Coaxial cables avoid this risk, since they have a grounded shield around the conductor. Coaxial cables are used in LANs since such cables can transmit at higher frequencies than twisted-pair cables.

Microwave

Microwave transmission uses very high-frequency radio waves in the range of 4.6 to 12.0 GHz. This broad-band facility provides line-of-site transmission capability. A repeater is used every 30 miles—provided the path is not obstructed—to transmit to the next receiving station.

Fibre-optic cable

In fibre-optic cable, a strand of glass with diameter as thin as that of human hair is used to transmit data. A fibre-optic cable can carry high bandwidth signals, since it is immune to noise and distortion. A light source—laser diode or LED (light emitting diode)—allows modulation of data at high transmission rates. Fibre-optic cables are impossible to tap which makes them a secure medium. However, this makes it difficult to use such cables in bus topology LANs. Figure 24.7 shows the cross-sectional view of typical fibre-optic cable.

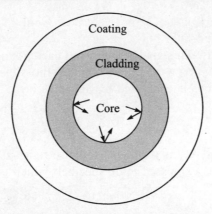

Fig. 24.7 Cross-sectional view a typical fibre-optic cable.

Satellite

Satellites work as relay stations. A station receives signals at one frequency and transmits them at another frequency to avoid interference with the incoming signals. A device called a **transponder** accomplishes this. Propagation delays are longer (about one-fourth of second) with this method as compared to data transmission through ground links (approximately six microseconds). Manufacturing companies engaged in worldwide operations use satellite communications to implement CIM.

REVIEW QUESTIONS

24.1 Explain the functions of a local area network.

24.2 What are the different components of LAN?

24.3 Differentiate between LAN, WAN and PAN.

24.4 Discuss the need and advantages of networking in a manufacturing shop.

24.5 Describe the three basic network topologies.

24.6 Describe the salient features of MAP and TOP.

24.7 What are the advantages of a communications standard?

24.8 Explain the functions of each layer in ISO-(OSI).

24.9 What are the differences between ring and star networks?

24.10 How will you use internet to advantage to implement CIM?

24.11 Discuss the limitations of parity check in detecting transmissions errors.

24.12 List out five major benefits of fibre-optic cabling.

24.13 What are the differences between serial and parallel transmission?

24.14 Briefly explain the serial transmission method used in CNC machine tools.

24.15 Explain why serial transmission is generally preferred in CNC machine tools.

BIBLIOGRAPHY

Alexander, G. Jr., *Effective Database Management*, Prentice-Hall, Englewood Cliffs, NJ, 1988.

Andrew, Kusiak, *Intelligent Manufacturing Systems*, Prentice-Hall International Edition, 2004.

Arora, J.S., *Introduction to Optimum Design*, McGraw-Hill, New York, 1989.

Bedworth, D.D., Henderson, M.R. and Wolfe, P.M., *Computer-Integrated Design and Manufacturing*, McGraw-Hill, Inc, 1991.

Besant, C.B. and Lui, C.W.K., *Computer-aided Design and Computer-aided Manufacturing*, 3rd ed., East-West Press, New Delhi, 1988.

Bethuno, J.D. and Kee, B.A., *Modern Drafting: An Introduction to CAD*, Prentice-Hall, Englewood Cliffs, NJ, 1989.

Boothroyd, G., Dewhurst, P, and Knight, *Product Design for Manufacturing and Assembly*, Marcel Dekker, Inc, 1994.

Boothroyd, Geoffrey, Corrado Poli, and Laurence E. Murch, *Automated Assembly*, Marcel Dekker, New York, 1982.

Bourne, Dravid, A. and Marks Fox, *Autonomous Manufacturing: Automating the Job Shop*, Computer, September, 1984.

Carl Machover, *The CAD/CAM Handbook*, McGraw-Hill, 1996.

Carrie, A.S., E., Adhami, A. Stephens, and I.C. Murdoch, *International Journal on Production Research*, March, 1985.

Chao-Hwa Chang, and Michel A. Melkanoff, *NC Machine Programming and Software Design*, Prentice-Hall International, NJ, 1989.

Charniak, Eugene and Drew V., *Introduction to Artificial Intelligence*, Addison-Wesley, Reading M.A, 1985.

Chevalier, Peter, W., "Group technology as a CAD/CAM integrator in batch manufacturing", *International Journal of Operations and Production Management*, Vol 4, No 3, 1984.

Childs, *Numerical Control Part Programming*, Industrial Press, New York, 1973.

Chis, McMohan, and Jimmie Browne, *CAD/CAM Principles: Practice and Manufacturing Management*, Addison-Wesley, Longman 2000.

Cook, R.D., *Concepts and Applications of Finite Element Analysis*, 2nd ed., John Wiley, New York, 1981.

David F. Rogers, and J. Alan Adams, *Mathematical Elements for Computer Graphics*, McGraw-Hill, 1990.

Davies, B.L., Robotham, A.J., and Yarwood, A. *Computer-aided Drawing and Design*, Chapman & Hall, 1991.

Donald Hearn and M. Pauline Baker, *Computer Graphics*, Prentice-Hall, NJ, 2005.

Earle, James H., *Engineering Design Graphics*, 5th ed., Addison-Wesley, 1987.

Encarnacao, J. and Schlechtendahl, E.G., *Computer-aided Design*, Springer-Verlag, Berlin, Germany, 1983.

Foley, J.D. and Van Dam, A., *Fundamentals of Interactive Computer Graphics*, Addison-Wesley Inc, Reading, Mass., 1982.

Gary Dunning, *Introduction to Programmable Logic Controllers*, Thomson & Delmar Learning, 2nd ed., 2005.

Gerald Farin, *Curves and Surfaces for Computer Aided Geometric Design*, 3rd ed., Academic Press, 1993.

Giloi, W.K., *Interactive Computer Graphics*, Prentice-Hall Inc., Englewood Cliffs, NJ, 1978.

Goetsch, David L., *Introduction to Computer-aided Drafting*, Prentice-Hall, Englewood Cliffs, New Jersey, 1983.

Goetsch, David, L., *Advanced Manufacturing Technology*, Delmar Publishers Inc., 1990.

Goetsch, David, L., *Fundamentals of CIM Technology*, Delmar Publishers Inc., 1988.

Groover, Mikell, P. and Emory W. Zimmers Jr., *CAD/CAM: Computer-Aided Design and Manufacturing*, Prentice-Hall, New Jersey, 1980.

Gunn, Thomas, G., *Computer Applications in Manufacturing*, Industrial Press, New York, 1981.

Haigh, M.J., *An Introduction to CAD and CAM*, Blackwell Scientific Publications, Oxford, 2000.

Ishrat Mirzana, *Computer Aided Design and Manufacturing*, Premier Publishing House, 2006.

Janakiraman, V.S., Sarukesi, K. and Gopalakrishnan, P., *Foundations of Artificial Intelligence and Expert systems*. Macmillan, New Delhi, 1993.

Koren, Y., *Computer Control of Manufacturing Systems*, McGraw-Hill, New York, 1983.

Korgaonker, M.G., *Just in Time Manufacturing"*, Macmillan, New Delhi, 1992.

Krar, S. and Gill, A., *CNC Technology and Programming*, McGraw-Hill, 1990.

Krouse, John, K., *Automation Revolutionizes Mechanical Design*, High Technology, March, 1984.

Larry Horath, *Computer Numerical Control Programming of Machines*, Macmillan, New York, 1993.

Luggen, W.W., *Flexible Manufacturing Cells and Systems*, Prentice-Hall, New Jersey, 1991.

Luggen, William, *CNC: A First Look Primer*, Delmar Publishers, New York, 1997.

Mair, G.M., *Industrial Robotics*, Prentice-Hall, NJ, 1988.

Malcolm, D.R. Jr, *Robotics: An Introduction*, 2nd ed., Delmar Publishers Inc., 1988.

Maleki, R.A., *Flexible Manufacturing Systems: The Technology and Management*", Prentice Hall, Englewood Cliffs, 1991.

Manfred, Weck, and H. Bibring, *Handbook of Machine Tools,* Vol. 3, *Automation and Controls*, John Wiley & Sons, 1984.

Marsland, D.W., "A review of computer-aided part programming system developments", *Computer-aided Engineering Journal*, 1984.

Mikell P. Groover, *Automation, Production Systems and Computer Integrated Manufacturing*, Prentice-Hall, 2005.

Mikell P. Groover, Mitchell Weiss, Roger N. Nagel and Nicholas G. Odrey, *Industrial Robotics Technology*, *Programming and Applications*, McGraw-Hill, New York, 1979.

Miller, R.K., "Automated guided vehicles and automated manufacturing, *Society of Manufacturing Engineers*, 1987.

Morris A. Cohen and Uday M. Apte, *Manufacturing Automation*, McGraw-Hill, New York, 1997.

Mortenson, M.E. and *Heinemann Newnes, Computer Graphics: An Introduction to the Mathematics and Geometry*.

Mortenson, M.E., *Geometric Modeling*, John Wiley & Sons, New York, 1985.

Newman, W.M. and Sproull, R.F., *Principles of Interactive Computer Graphics*, McGraw-Hill, New York, 1979.

Nyman, L.R., *Making Manufacturing Cells Work*, McGraw-Hill, New York, 1993.

Ohsuga, S., *Towards Integrated CAD Systems, Computer-aided Design*, 21(5) pp. 315–337.

Optiz, H., *A Classification System to Describe Workpieces*, Pergamon Press, Oxford, 1970.

Pahl, G. and Beitz, W., *Engineering Design*, Springer-Verlag, New York, 1984.

Pao, Y.C., *Elements of Computer Aided Design and Manufacturing*, John Wiley & Sons, New York, 1984.

Parrish, D., *Flexible Manufacturing*, Butterworths-Heinemann, Oxford, 1990.

Parsaye, K. and Chignell, M., *Expert Systems for Experts*, John Wiley & Sons, Inc., 1989.

Pimentel, J.R., *Communication Networks in Manufacturing*, Prentice-Hall, New Jersey, 1990.

Pressman, R.S. and Williams, J.E., *Numerical Control and Computer-aided Manufacturing*, John Wiley & Sons, Inc., New York, 1977.

Radhakrishnan P. and Subrahmanyam, *CAD/CAM/CIM*, New Age Publishers, New Delhi, 2005.

Ranky, P.G., *Flexible Manufacturing Cells and Systems in CIM*, CIMware, 1990.

Rao, C.S.P., *Computer Aided Design and Manufacturing*, Hi-Tech Publishers, 2005.

Rao, P.N., Tewari, N.K., and Kundra, T.K., *Computer-aided Manufacturing*, Tata McGraw-Hill, New Delhi, 1993.

Rao, S.S., *The Finite Element Method in Engineering*, 2nd ed., Pergamon Press, 1989.

Reddy, J.N., *An Introduction to the Finite Element Method*, McGraw-Hill, New York, 1984.

Rehg, J.A. and Kraebber, H.W., *Computer Integrated Manufacturing*, 2nd ed., Prentice-Hall, New Jersey, 2001.

Rehg, J.A., *Introduction to Robotics—A System Approach*, Prentice-Hall, Inc., Englewood Cliffs, New Jersey, 1985.

Rembold, U. and Dillmann, *Computer-aided Design and Manufacturing Methods and Tools*, Springer-Verlag, 1984.

Rembold, U. Bo Nanji and A. Storr, *Computer Integrated Manufacturing and Engineering*, Addison-Wesley, 1993.

Rogers, D.F. and Adams, J.A. *Mathematical Elements for Computer Graphics*, McGraw-Hill, New York, 1985.

Rooney, J. and Steadman, P., *Principles of Computer-aided Design*, Affiliated East-West Press, 1990.

Ryan, D.L., *Computer-aided Graphics and Design*, Marcel Dekker Inc., New York, 1979.

Rychener, M.D., *Expert Systems for Engineering Design*, Academic Press Inc., 1988.

Sadhu Singh, *Computer Aided Design and Manufacturing*, Khanna Publishers, New Delhi, 2003.

Scheer, A.W., *Computer Integrated Manufacturing—Towards the Factory of the Future*, Springer, Verlag, 1991.

Seames, W.S., *Computer Numerical Control*: *Concepts and Programming*, 2nd ed., Delmar Publishers Inc., 2005.

Serope Kalpakijian and Steven R. Schmid, *Manufacturing Processes for Engineering Materials*, Thomson Learning, 2005.

Shanmugam, N., *Mechatronics*, Anuradha Publishers, 2003.

Singh, N., *Systems Approach to Computer-Integrated Design and Manufacturing*, John Wiley, New York, 1996.

Smith, G.T., *CNC Machining Technology*, Vols 1, 2 & 3, Springer-Verlag, London, 1993.

Strasser, W., *Theory and Practice of Geometric Modeling*, Springer-Verlag, London, 1989.

Surender Kumar and A.K. Jha, *Technology of Computer Aided Design and Manufacturing*, Dhanpat Rai Publishers, New Delhi, 2002.

Talavage, J. and R.G. Hannam, *Flexible Manufacturing Systems: Applications, Design and Simulations*, Marcel Dekker, New York, 1988.

Teicholz, E.W., Stawood, and J. Tchijov, *Computer Integrated Manufacturing*", R.U. Ayres (Ed.), Chapman & Hall, 1992.

Tercholz, E., *CAD/CAM Handbook*, McGraw-Hill, New York, 1985.

Thyer, G.E. *Computer Numerical Control of Machine Tools*, Newnes, 1994.

Thyer, G.E., *Computer Numerical Control of Machine Tools*, Heinemann, Oxford, 1988.

Todd, D.J., *Fundamentals of Robot Technology*, Kogan Page, 1986.

Vail, P.S., *Computer Integrated Manufacturing*, PWS-KENT Publishing Company, Boston, 1988.

Vajpayee, S.K., *Computer Integrated Manufacturing*, Prentice-Hall, 2004.

Valliere, *Computer Aided Design in Manufacturing*, Prentice-Hall, New Jersey, 1990.

Viswanadham, N. and Y. Narahari, *Performance Modeling of Automated Manufacturing Systems*, Prentice-Hall, New Jersey, 1992.

Voisinet, D.D., *Introduction to Computer-aided Drafting*, McGraw-Hill, New York, 1986.

Wakil, E.L., *Processes and Design for Manufacturing*, Prentice-Hall, New Jersey, 1989.

Warnecke, H.J. and R.D. Schraft, *Industrial Robots: Applications Experience*, I.F.S. Publications, Bedford, 1982.

Warren S. Seames, *Computer Numerical Control—Concepts and Programming*, Pearson Publishers, 2005.

Weatherall, Alan, *Computer Integrated Manufacturing*, Butterworths, 1988.

William, N. Newman and Robert F. Sproll, *Principle of Interactive Computer Graphics*, McGraw-Hill, New York, 2000.

Womack, J.P. and Daniel T. Jones, *Lean Thinking: Banish Waste and Create in Your Corporation* (Revised edition), Free Press, New York, 2003.

Womack, J.P., Daniel T. Jones, and Daniel Roos, *The Machine Changed the World: The Story of Lean Production*, Harper Perennial, New York, 1991.

Yu-Cheng, Liu and Glenn A. Gibson, *Microcomputer Systems*, Prentice-Hall, New Jersey, 2005.

Zandi, M., *Computer-aided Design and Drafting*, Delmar Publishers Inc., 1985.

Zeid, Ibrahim, *CAD/CAM: Theory and Practice*, McGraw-Hill, New York, 1991.

Zhou, Mengchu, *Petrinets in Flexible and Agile Automation*, Kluwear Academic Publishers, London, 1995.

GLOSSARY

Absolute coordinates:	The location of a point in terms of distances and/or angles from a fixed origin.
Absolute system:	A numerical control system in which all the positional dimensions, both input and feedback are given with reference to a common datum point.
Access time:	The time it takes to transfer an instruction or unit of data from computer memory to the processing unit of a computer.
Accumulator:	A special temporary storage register in the CPU, used during arithmetic or logical operations. It handles the transfer of data by holding the data to be operated on by the arithmetic-logic unit until it is ready and by receiving the results when it is done.
Acoustic tablet:	This makes use of strip microphones located around the perimeter of the tablet, which creates noise or buzzing sound, which is then fed to the computer, in order to locate the position.
Adaptive control:	A type of computer control system that compensates for sources of variability to optimize performance.
Address:	The coded representation of a specific computer workstation or host computer used in transferring data through a network. A unique number is assigned to each computer.
AGV:	An acronym for automatic guided vehicle. This is a battery powered vehicle that can move and transfer materials by following prescribed paths around the manufacturing floor without being physically tied to the production operation or being driven by an operator.
AI:	An acronym for artificial intelligence. AI is a field of computer science that deals with computers performing human-like functions, such as reasoning and interpretation. AI normally

	takes the form of a set of software that permits a computer to deal with very high level languages, adapt to sensory inputs, interpret data, and "learn" from experience.
Algorithm:	A computational method for solving problems.
Aliasing:	Refers to the jagged lines or edges that can appear in computer drawn images. Aliasing occurs when smooth lines or edges in an image are drawn with pixels.
Alphanumeric:	A system of code that consists of the characters A–Z and numerals 0–9.
ALU:	Arithmetic and logic unit. This is the section of the CPU of a computer which performs the logical and computational functions.
Analog circuit:	A circuit in which the output varies as a continuous function of the input, as contrasted with digital circuit.
Analog:	Pertaining to a system that uses electrical voltage magnitudes or ratios to represent physical axis positions.
Annotation:	The process of interesting text or a special note, explanation or to provide relevant detail on a CAD/CAM drawing.
ANSI (American National Standards Institute):	The American National Standards Institute (ANSI) is a privately funded, non-profit organization, which coordinates the development of voluntary standards in the United States and is the agency that approves standards (as American National Standards).
Anti-aliasing:	A technique for smoothing out jaggies—the jagged edges on diagonal lines and curves on-screen. To compensate, graphics cards blur the edges by adding various shades of grey or colour to surrounding pixels (this is called dithering).
API (Application Programming Interface):	Provides standard documented access to software functions, allowing customers and third parties to develop and customize their own commercial applications.
APICS:	American Production and Inventory Control Society.
Application profile:	A number of application protocols required for a specified task or industry sector. Associated with STEP.
Application protocol (AP):	Defines the context for the use of product data and specifies the use of the standard in that context to satisfy an industrial need. Associated with STEP.
Application:	A computer software program that performs specific functions such as page layout, word processing, accounting, drawing and spreadsheet formation.
Application software:	It performs the data entry, design, analysis, drafting, and manufacturing functions.

APT programming:	Automatically Programmed Tools. A universal computer assisted programming system for multi-axis contouring programming.
ASCII:	American Standard Code for Information Interchange. A data transmission code, which has been established as an American Standard by the American Standards Association. It is a code in which 7 bits are used to represent each character.
Aspect ratio:	The shape of the display device on which an image will be viewed. The aspect ratio of a rendered image is expressed as the width of the image divided by its height.
Assembly:	A final gathering of piece-parts and subassemblies to make one unique assembled product.
Assembly drawing:	A drawing, which can be created on the CAD system to represent a major subdivision of the product or the complete product.
Assembly modelling:	The process by which individual solid models are brought together to form an assembly model.
AS/RS:	An acronym for automatic storage and retrieval system. It is an automated materials handling system for bulk or in-line storage of parts, materials or products. It includes several different types of equipment, such as stacker-cranes, narrow-aisle, and mini-load equipment.
Automatic dimensioning:	A CAD capability that will automatically compute and insert the dimensions of a design or drawing or designated section of it.
Automation:	The mechanization and control of the physical movement, fabrication or data handling operations in manufacturing.
Automated tool system:	It is composed of tool setup devices, central tool storage, tool management systems and tool transfer systems. They are used to prepare tools for the machining centres as well as transfer tools between machining centres and the central tool storage.
Auxiliary function:	A function of a CNC machine tool denoted by M-word address.
Auxiliary storage:	A computer storage device which provides high memory capacity for storing large amounts of data that is not currently being processed by the CPU. This is also known as **secondary storage.** Typical devices of this type include magnetic tape units and magnetic disk files.
Axis:	A principal direction along which the relative movements of a tool or workpiece occur. Three linear axes, occurring at 90 degree angles from each other, named X, Y and Z.
Axisymmetric:	A solid geometry entity that is symmetric and typically revolves about a common axis.

Backbone: A central high-speed network that connects smaller, independent networks.

Background: A feature which facilitates in the execution of low priority work when high priority work is not using the computer.

Backlash: A relative movement between interacting mechanical parts as a result of looseness.

Barcode: A standard format of bars and space that can be printed on labels to identify parts or products which can be read automatically by a machine for data collection.

Base: A number base is a quantity used implicitly to define some system of representing numbers by positional notation.

Batch: A number of items being dealt with as a group.

Batch processing: A manufacturing operation in which a specified quantity of material is subject to a series of treatment steps. Also, a mode of computer operations in which each program is completed before the next is started.

Batch manufacturing: The manufacture of parts or products in groups at each step of the process. It usually involves small quantities and a variety of types.

Baud: A unit of signalling speed equals to the number of discrete conditions or signal events per second, e.g. 1 bit per second in a train of binary signals.

BCL: Basic control language. ELA/ANSI 494.

Behind the tape reader: A means of inputting data directly into a machine tool control unit from an external source connected behind the tape reader.

Benchmark: A standard example against which measurements can be made.

Bezier curves: A quadratic (or greater) polynomial for describing complex curves and surfaces.

Bill of materials: A listing of all the parts that constitute an assembled product.

Binary: A numbering system based on 2. Only the digits 0 and 1 are used when written.

Binary code: Code based on binary numbers expressed as either 1 or 0, true or false, on or off.

Binary digit (BIT): A character used to represent one of the two digits in the binary number system, and the basic unit of information or data storage in a two-state device.

BIT: Binary digit. A binary digit has only two possible states.

Bitmap: Generally, a bitmap is associated with graphics objects. The bits are a direct representation of the picture image.

Blending: Blending is the combining of two or more objects by adding them on a pixel-by-pixel basis.

Block:	A set of words, characters, digits or other elements handled as a unit. In a CNC part program, it consists of one or more characters or rows across that collectively provide enough information for an operation.
Block delete:	A function that permits selected blocks of code to be ignored by the control system at the operator's discretion.
BLU (Basic length unit):	It is the smallest possible incremental linear distance in CNC machine.
Boolean algebra:	An algebra named after George Boole for logical operations.
Boundary:	A 3D outline of a data volume.
Boundary evaluation:	An operation that generates a B-Rep solid from a CSG solid.
Bounded objects:	In solid modelling, an object is considered bounded if it has a complete set of bounding surfaces and is restricted to occupying a finite volume.
B-Rep (Boundary representation):	A database method that defines and stores a solid as a set of vertices, edges and faces (points, lines, curves and surfaces) which encloses its volume completely.
B-spline (Bicubic spline):	A sequence of parametric polynomial curves (typically quadratic or cubic polynomials) forming a smooth fit between a sequence of points in 3D space. The piecewise defined curve maintains a level of mathematical continuity that is dependent upon the polynomial degree that is chosen. B-splines are used extensively in the mechanical design applications in the automotive and aerospace industries.
Buffer storage:	A place in which information in a control system or computer is stored for use at a later time.
BUS:	A conductor used for transmitting signals or power between elements.
BYTE:	A sequence of eight adjacent bits operated on as a unit.
Cache:	An area in the memory used for temporary storage. Cache is used to optimize operation and retrieval speeds.
CAD/CAM:	Computer aided design and computer aided manufacturing.
CAD:	Computer aided design. Software for geometric design of parts.
CADD:	Computer aided design and drafting.
CAE:	Computer aided engineering. The use of computer and digital technology to support basic error checking, analysis, optimization, manufacturability, etc., of a product design. Finite element analysis (FEA) is one example of CAE.
CAM:	Computer aided manufacturing. Software used for preparing the CNC part programs.

Canned cycle:	A preset sequence of events initiated by a single 'G' coded command.
CAPP:	An acronym for computer-assisted part programming or computer-automated process planning. This is an approach to using a computer to perform the computational work involved in programming tools to perform manufacturing tasks.
Capacity requirement planning (CRP):	It is a technique for determining what personnel and equipment capacities are needed to meet the production objective embodied in the master production schedule and the material requirement planning.
Cartesian coordinates:	A three-dimensional system where by the position of a point can be defined with reference to a set of axes at right angles to each other.
Cathode ray tube (CRT):	A display device in which controlled electron beams are used to present alphanumeric or graphical data on a luminescent screen.
Charged coupled device (CCD):	An electronic memory made of a metal-oxide semiconductor (MOS) transistor that can store patterns of charges sequentially.
Compact disc read only memory (CD-ROM):	The CD-ROM when used in computer can store text, sounds and images, as well as video information.
Concurrent engineering (CE):	A systematic approach to creating a product design that considers all elements of the product life cycle from conception of the design to disposal of the product.
Central processing unit (CPU):	The portion of a computer system consisting of the arithmetic and control units and the working memory.
Chip:	A single piece of silicon cut from a slice by scribing or breaking. A chip can contain one or more circuits.
Computer integrated manufacturing (CIM):	Automated use of computer and digital technology to completely integrated all manufacturing processes with engineering design.
Circular interpolation:	A mode of contouring control, which uses the information, contained in a single block to produce an arc of a circle.
CISC:	Complex instruction set computing.
CLDATA:	Processor output between the tool and the work piece when the change is made from rapid approach to feed movement to avoid tool breakage.
CLFILE:	Cutter location file. A file of data generated from a machine tool path created on a CAD/CAM system or on an APT processor. The CL file contains X, Y and Z coordinates and other NC information to be post-processed to program NC machines.

Client: A computer or computer program that is one side of the client-server communication.

Client-server architecture: An information-passing scheme that works as follows. A client program, such as Netscape, sends a request to a server. The server takes the request, processes it, and transfers the information to the client.

Clock: A device, which generates periodic synchronization signals.

Closed loop: A signal path in which outputs are fed back for comparison with desired values to regulate system behaviour.

Closed-loop MRP: An enhancement to materials requirements planning which adds capacity planning functions to adjust schedules or manufacturing resources in response to over or under scheduling conditions.

Coordinate measuring machine (CMM): Computer controlled equipment used to inspect part dimensions.

CNC computer (computerized) numerical control: A numerical control system wherein a dedicated, stored program computer is used to perform some or all of the basic numerical control functions.

Coincidence: Refers to geometry that occupies the same spatial location. For example, coincident vertices are points that occupy the same x, y and z coordinates. Coincident lines can have differing lengths while one occupies the same locations the other.

Command: An operative order, which initiates a movement or a function.

Compile: To generate a machine language program from a computer program written in a high-level source code.

Compliance: A spring-mounted fixture on the end of a robot arm that allows parts with chamfered edges to be grasped quickly and easily even when they are misaligned by several millimetres.

Concentric: Having a common centre or origin point with varying radii.

Configuration: A particular combination of a computer, software and hardware modules, and peripherals at a single installation and interconnected in such a way as to support certain requirements.

Conformity: The degree to which physical and performance characteristics of a product match pre-established standards.

Conic section: Curve formed by the intersection of a plane with a cone.

Console: The part of a computer system used by the operator for communication with the computer system.

Contouring: A numerical control method where the control program generates a contour by keeping the cutting tool in constant contact with the work piece.

Controller: Device attached to a machine tool, which reads and stores machine data file and passes movement information to the machine tool. Also controls other machine activities such as turning coolant on or off.

Control system: A system of hardware and software which controls the operation of a machine, such as a robot or NC machine tool. For motion control, it may use either non-servo techniques, which control end points only, or a servo control of the path and speed.

Control chart: A typical control chart is a graphical display of a quality characteristic that has been measured or computed from a simple versus the sample number or time.

Conveyor: A materials transport mechanism that provides a moving link between delivery points in a process. There are several types of conveyors, including continuous belt, roller, track, overhead trolley and air tables.

Coons patch: Surface defined by four boundary curves, tangents, corner twists and blending functions. Curvature and continuity can be maintained between patches.

Coordinate system: Geometric relation used to denote the location of points in 3D space. The most common is the rectangular coordinate system, where by points are located by traversing the X, Y and Z-axis of 3D space.

Coordinates: Ordered set of absolute or relative data values that specify a location in a coordinate system.

Coplanar: Refers to two or more entities that lie on the same plane.

C-Rep: A CSG model is based on the topological notion that a physical project object can be divided into a set of primitives that can be combined in a certain order following a set of rules to form the object.

Constructive solids geometry (CSG): A scheme for representing solid objects. It is a tree representing instances of solids and Boolean operations (union, intersection, difference).

Curve smoothing: The process of curve or surface approximation using path may be altered to compensate for cutter diameter differences.

Cutter offset: The distance from the part surface to the axial centre of a cutter (the radius of the cutter).

Cutter diameter compensation: A feature of a control system in which the inputted cutter diameter and part profile data are used to automatically place the tool on the part boundary. This feature is most desirable when compensating for tool wear.

Cutter path: The path defined by the centre of the cutter.

Cutting plane:	A tool that slices through the 3D data exhibiting characteristics of a plane in the data, including scalar and vector measurements.
Cycle:	A sequence of operations that is repeated. The time the repetition requires is cycle time.
Cylinder:	A solid primitive defined as a right-circular cylinder. The ends are circular and of equal radius. The axis is normal to the ends.
Data exchange (in CAD/CAM):	Data exchange covers all the issues of exchanging models between different CAD systems. Standard like IGES and STEP are parts of this issue.
Data management:	Coordinated management of all aspects of electronic manipulation of machine readable data in a computer environment; includes data capture, communication, storage, retrieval and associated processing.
Database:	A collection of data. The collection is organized in a logical structure for the primary purpose of automating information.
Datum:	Point or plane of reference.
DBMS (Database management system):	Software designed to manipulate the information in a database. It can create, sort, display selected information, search for specific information, and perform many other tasks of a database.
DDA:	Digital differential analyzer
Debug:	To troubleshoot, detect, locate and remove mistakes from a program.
Decompression:	Process of returning a compressed file to its full bitmap.
Decision support system (DSS):	A form of artificial intelligence which uses knowledge representation to model the process and provide a rule-base expert system predict or react to the performance of a production operations.
Dedicated:	Process or system that is available for only one function or use.
Default:	An automatic decision that is made by computer software and hardware programs. The decision will automatically be carried out unless the user changes the default settings.
Degrees of freedom (DOF):	The number of axes or independent types of movement a robot manipulator can make. It takes three coordinates to locate the centre of gravity of an object and three more to determine its orientation.
Delta dimensioning:	A system of specifying part dimensions whereby each new dimension is taken relative to the last dimension given. This is also known as **incremental dimensioning.**

Design for manufacturing: Design for manufacturing (DFM)—DFM seeks to minimize manufacturing information content of a product design to the fullest extent possible within constraints imposed by functionality and performance.

Detail drawing: The drawing of a single object design including graphics, dimensions, and annotations complete enough for manufacturing.

Digit: A single character in any numbering system.

Digitizers: Digitizers board is an electromagnetic graphic input device that resembles a drafting board.

DNC: Direct numerical control. A system in which machine coded programs are introduced into the CNC controller from a remote computer.

Documentation: A generic term for a wide variety of hard-copy or online reports, drawings, and lists to be used by various departments involved in any aspect of design to fabrication of a part.

DOS: Disk operating system.

Dot pitch: Dot pitch is the space between pixels. The smaller the number, the sharper the image will appear.

Download: To transfer to your computer a copy of a file that resides on another computer.

Downstream process: All subsequent operations or processes performed on or with a solid model. Download processes include (but are not limited to) analysis, NC code generation and rapid prototyping.

Downtime: Time during which equipment is inoperable because of faults.

Dots per inch (DPI): A measurement of linear resolution for a printer or scanner. For example, a resolution of 600 dpi means that there are 600 dots across and 600 dots down. A higher number of dots create a finer resolution.

Dragging: Dynamically moving image of graphical entity across the display screen to a new location using a puck, mouse, or stylus.

DRAM: Dynamic random access memory is the memory at any location in a computer that can be accessed immediately for reading and writing operations.

Drawing: An engineering document or digital data file(s) that discloses (directly or by references), by means of graphic or textual presentations, or a combination of both, the physical and functional requirements of an item.

Drawing conversion: The process of moving engineering data from hardcopy or raster format to CAD format.

Drive: An internal or external assembly that can read and/or write electronic data using disk-storage media.

Dwell time: A timed delay of programmed or established duration used in specific machining operations.

Data exchange format (DXF): A translation format developed by auto CAD to transfer geometry data to and from AutoCAD.

EBCDIC: Extended binary coded decimal interchange code.

Edge: A bounded line or curve that forms the intersection of two faces on the surface of an object.

EIA: A conventional code that is used for systems that execute straight-cut and contouring operations. EIA code is used in eight-track one-inch punched tape.

Electrostatic printer: An electrostatic printer is a raster scan device. Basically it operates by deposing small particles of toner onto electrostatically charged areas of a special paper.

Electrostatic printing: Printing large-format prints in a process similar to, but not the same as, colour photocopiers.

Emulate: The use of software to allow one device to imitate another.

Encoder: A transducer that produces digital pulses based on mechanical displacement.

Encryption: A method of ensuring data secrecy. The message is coded using a key available only to the sender and the receiver. The coded message is sent to the receiver and decoded upon receipt.

End of program: A miscellaneous function (M02) that represents the completion of a programmed cycle.

End-of-block character: A character representing the end of a programmed block of information.

Engineering analysis: Engineering analysis may take the form of stress-strain calculations, heat transfer analysis, or dynamic simulation.

Entity: Any information that is being displayed on an interactive CRT, which can be identified. Entities consist of geometry (points, lines, circular arcs, conics, splines, surfaces, solids, etc.), or as text items (notes, dimensions, lists, tables, etc.), or as "information-adding" things (coordinate systems, surface normal vectors, etc.).

ETHERNET: Xerox trademark name for a network cable system that allows communication between the workstations and servers connected to the network.

Executive program: A set of programming instructions that allows a CNC lathe to have the capability to perform lathe functions or a CNC mill to perform milling functions. A set of instructions designed to output specific functions.

Expert system: A form of AI which a computer system provides decision-making capabilities for specific applications that require expert knowledge. Fault tolerance is the ability accomplished by redundancy in the hardware or error-recovery software. Feature matching an image analysis technique used in vision systems that involves gray scale interpretation to match objects to design rules and shape characteristics.

Export: Sending a model to a file (IGES, DXF, STEP, TIFF, etc.) so that it can be read or imported into another program.

Extrusion: A process used in geometric modelling to convert 2D shapes into 3D shapes. A 3D object is created by displacing a copy of the 2D shape, than linking the copy to the original to form a closed, solid object.

Face: A type of element used in geometric modelling. A face can be a flat, planner polygon or a curved, bounded surface. Some systems define a face as the bounded portion of an infinite surface region.

Family of parts: A collection of parts with similar shape, but differing in physical measurement.

Finite element analysis (FEA): A method used in CAD for determining the structural integrity of a mechanical part or physical construction under design by mathematical simulation of the part and its loading conditions.

Feature: Software abstraction of the mechanical concept of hole, rib, slot and pocket used to describe the model in a more functional way.

Featured-based modelling: Performs functions that were previously performed using primitive Boolean operations. Example, the through-hole feature understands the rule that it must pass completely through the part and will do so no matter how the part changes.

Feed: A programmed or manually established rate of movement of the cutting tool into the work piece for the required machining operation.

Feedback: In a closed loop system this represents the return signal or response of the system to an inputted instruction.

Feedback override: A manual switch that enables the operator to alter the programmed feed rate during a cutting operation.

FEM: Finite element modelling. A type of software made to compute and simulate the reaction of CAD models in respect of stress, magnetism, plastic injection, etc.

File format: The specific organization of data within a file. There are multiple raster and vector file formats, including JPEG, STEP, IGES and DXF.

File server:	The local area networks allow users to share the peripherals (printers, modems, scanners) and thereby conserve their software cost. The file server is a device on the LAN where shared software is stored.
File:	An organized collection of relevant, orderly data.
Fillet surface:	The transition surface that blends together two surfaces.
Firmware:	Programs or controlled instructions that are not changeable by the user, and that are often held in ROM, read only memory.
Fixed cycle:	A preset sequence of events initiated by a single G coded command.
Flexible automation:	The mechanization and control of a sequence of operations which can be reprogrammed or changed.
Flowchart:	A flowchart is defined as a pictorial representation describing a process being studied or even used to plan stages of a project.
Floating zero:	A characteristic of a machine control unit that allows the zero reference point to be established at any point of travel along an axis.
FMS:	An acronym for flexible manufacturing system. A computer control system that ties together multiple manufacturing operations into an integrated production process which fabricates a finished product. A formal definition of an industry group (The Machine Tool Task Force) is: "A series of automatic machine tools or items of fabrication equipment linked together with an automatic materials handling system, a common hierarchical digital pre-programmed computer control, and providing for randomly fabricating parts or assemblies that fall within predetermined families".
FMM:	FMM consists of a standalone NC-lathe or machining centre equipped with a transport function such as a robot, automatic pallet changer (APC).
Form features:	In solid modelling, parts of solid objects that can be specified in familiar engineering terms (e.g., fillets, slots and through-holes).
Format:	Arrangement of data
FORTRAN:	An acronym for FORmula TRANslation. A high-level mathematical source language developed for scientific and engineering applications.
Free-form surface	Surfaces that are not limited to mathematically simple linear or quadric surfaces.
FTP:	File transfer protocol is a protocol that allows the transfer of files from one computer to another.

Full duplex:	Allows simultaneous transmission of information in both directions.
Functionality:	Refers to a set of system capabilities in terms of the functions they provide.
G code:	Preparatory function. An NC word addressed by the letter G followed by a numeric value.
Gateway:	A device used to connect networks with radically different communications architectures.
Gauge height:	A predetermined Z-axis clear plane retraction point along the Z-axis to which the cutter retreats allowing safe *XY* axis travel.
Generative process planning:	Computer-based process planning method whereby new process plans are created based on part and product information, as well as manufacturing information.
Geometric modelling:	Methods used to create the exact geometric model of a part in the computer system.
Geometry:	Elements that make up a model, such as points, lines, surfaces, solids, etc.
Geometric transformation:	The process in a computer graphics system which converts data from a geometric model of an object into an image which can be displayed on a graphics screen.
Gauge:	Damaging the part due to tool motion entering the bounds of the desired finished part.
Gouraud shading:	Gouraud shading is a process by which colour information is interpolated across the face of the polygon to determine the colours at each pixel.
Grey code:	Binary code in which the successive values differ by only one bit.
Gripper:	A mechanism on the end of a robot arm which permits it to grasp and pick up objects. It may be a simple clamp or a complex device with sensors and multiple "fingers" like a human hand.
Group:	A collection of elements.
Group technology:	Is a manufacturing philosophy in whom similar parts are identified and grouped together to take advantage of their similarities in design and production.
Hard copy:	A readable output of data on paper.
Hardware:	The CAD hardware includes the computer and input-output devices. It introduces the requirements and specifications necessary for CAD equipment.
Hardwired:	Having logic circuits interconnected on a back plane to give a fixed pattern of events.

Hertz:	Cycles per second or some repeated action per second (Hz).
Hidden line:	A wire frame display option that displays only the lines that should be visible from the current view.
Hidden surface removal:	Hidden surface removal or visible surface determination entails displaying only those surfaces that are visible to a viewer because objects are a collection of surfaces or solids.
Hierarchical data model:	In this type of approach, the data is represented by a tree structure. The top of the tree is usually known as the "root" and the hierarchy, of the tree levels relative to each other descends from the root down.
High-level language:	A problem-oriented programming language using words, symbols and command statements, which closely resemble english language statements.
Home position:	A fixed location in the basic coordinate axis of the machine tool.
Homogeneous transformation system:	Homogeneous transformation is based on mapping an N-dimensional space into $(N + 1)$ dimensional space.
Host computer:	A large computer in a system of computers which stores the common database. Human factors considerations in the design of a workstation or piece of equipment that deals with the effects on humans. This includes operator safety, convenience and efficiency. It may involve making the operator more comfortable, making the task easier or preventing hazards.
Hybrid solid modeller:	A solid modelling database that actively maintains two or more substantially different representations of solid objects such as CSG and B-Rep.
IC:	Integrated circuit.
IEEE:	Institute of electrical and electronics engineers.
Initial graphics exchange specification (IGES):	A standard for exchanging mechanical design data between CAD systems.
Image:	Electronic representation of a document stored and displayed as a bitmap.
Implicitly defined:	Information that is defined by a situation rather than by explicit definition. In solid modelling, an edge defined as the intersection of two surfaces is implicitly defined.
Incremental dimensioning:	A method of expressing a dimension with respect to previous point.
Independent demand:	Demand for an item that occurs separately.
Inkjet printer:	A type of printer that sprays tiny streams of quick-drying ink onto paper. An inkjet printer produces high quality printing like that of a laser printer.

Intelligent robot: A type of computer-controlled robot that is not restricted in its actions to only the preprogrammed instructions. Intelligent robots are automated manufacturing systems which integrate manipulation, sensing, computation, and control functions to perform complex tasks.

Intelligent workstations: A workstation in a system that can perform certain data processing functions in a standalone mode, independent of another computer.

Interactive computer graphics: The computer technology which permits graphics displays to be created and modified interactively with the user.

Interactive input: Techniques used in computer graphics which permit the user to create and change images frequently. These include the use of high-level commands and interactive devices, such as light pens and function keys.

Interface: The communication that takes place between various elements in a system.

Interference checking: A CAD/CAM capability that enables mechanical designers to automatically examine intersection of objects within a 3D model.

Interpolation: A function of a control whereby data points are generated between given coordinate positions.

Inventory: Stores or goods including raw materials, WIP, finished products, etc.

IP: The abbreviation for internet protocol. IP refers to the set of communication standards that control communications activity on the internet.

ISDN: Integrated services digital network: a high-speed digital phone system.

ISO 9000: A series of international standards that provides quality management guidance and identifies quality system elements that are necessary for quality assurance.

ISO: International organization of standardization.

Jig: A fixturing device used most often for drilling operations.

Jog: A control function that momentarily operates a drive into the machine.

Just-in-time system (JIT): JIT is a management philosophy that strives to eliminate sources of manufacturing waste by producing the right part in the right place at the right time. Waste results from any activity that adds cost without adding value, such as moving and storing. JIT (also known as **lean production** or **stockless production**) should improve profits and return on investment by reducing inventory levels (increasing the inventory turnover

Kaizen: rate), improving product quality, reducing production and delivery, lead times, and reducing other costs (such as those associated with machine setup and equipment breakdown).

Kaizen: Japanese word for continuous improvement.

Local area network (LAN):   A network designed to connect devices over short distances. A data communications system that offers high-speed communication channels optimized for connecting information processing equipment over short distances. Consists of protocols and software to drive networks.

Layer: A logical separation of data to be viewed individually or in combination.

Lead time: The time between ordering and receiving goods.

Lead-through programming: A technique for programming robots which uses a control panel, called a **teach pendant**, with buttons or switches that control the motion of the robot through a cable connected to the control system. The operator or programmer can lead the manipulator through a task, one step at a time, recording each incremental move along the way.

Leading zeros: Redundant zeros to the left of a number.

Light emitting diode: A light-emitting diode (LED) is a semiconductor device that emits visible light when an electric current pass is through it.

Linear interpolation: A control function whereby data points are generated between given coordinate positions to allow simultaneous movement one, two or more axes of motion in a linear path.

Loop: Repetitive operations can be programmed in a continuous mode until the desired functions have been completed.

LSI: Large scale integration used in connection with the integrated circuits.

Machining centre: Machine tools, normally numerically controlled, capable of automatically repeating many operations such as drilling, reaming, tapping, milling and boring multiple faces on a work piece.

Machine cell: A set of complementary, computer-controlled machine tools that together perform a series of machining operations on a work piece.

Machine language: The lowest level programming language. It uses a binary code which directly relates to the on/off switching operations of the computer logic circuits.

Machine flexibility: Machine flexibility refers to the capability of a machine to perform a variety of operations on a variety of part types and sizes.

Macro:	A group of instructions that can be stored and recalled to solve a recurring problem.
Manual data input:	A mode of control that allows the operator to input data into the control system, the data input is identical to the data that can be input by other means such as tape or DNC.
Manual part programming:	The preparation of a manuscript in machine control language and format to define a sequence of commands for processing by a CNC machine.
Manipulator:	The base and arm assembly of a robot which performs the movements and tasks.
Manufacturing control system:	A computer system used to handle the technical data for process or tool control applications in manufacturing.
Manufacturing Resource Planning (MRP II):	A method for the effective planning of all the resources of a manufacturing company. Ideally it addresses operational planning in units, financial planning in money, and has a simulation capability to answer what if questions. It is made up of a variety of functions, each linked together: business planning, master (or production) planning, master production scheduling, material requirements planning, capacity requirements planning and the execution systems for capacity and priority.
MAP:	An acronym for manufacturing automation protocol. A token ring-based data communication protocol compatible with the international standards organization model of standard system architecture.
Markov models:	In this approach, the given system is modelled as a discrete parameter or continuous parameter markov chain or as a semi-markov process with discrete state space, and it assumes existence of a steady state and the performance estimates are obtained as functions of the steady state probabilities.
Master production schedule:	It is developed from customer orders and forecasts of future demand. The master schedule represents the plan of production for the firm, which serves as an input to the material requirements planning function.
Material requirement planning (MRP):	"It is a technique for determining the quantity and timing for the acquisition of dependent demand items needed to satisfy master production schedule requirements".
MCU:	MCU has to read and decode the part program to provide the decoded instructions to the control loops of the machine axes of motion, and to control the machine tool operation.
MDI:	A mode of control that allows the operator to input data into the control system, the data input is identical to the data that can be input by other means such as tape or DNC.

Memory:	An organized collection of storage elements into which a unit of information consisting of binary digits can be stored and later be retrieved.
Meshing:	Action of computing a set of simple elements giving a good approximation of the designed part. A good meshing must be precise where the computation must be precise but with a minimum number of elements.
Mil:	One thousandth of an inch.
MIPS:	Million instructions per second.
Miniload system:	A type of automatic materials storage system which is relatively small and used to store and retrieve materials or small parts within a production line.
Mirror image:	The reversal of plus and minus values along an axis. Mirror imaging is used to make a left-handed part from a right-handed tool path.
Miscellaneous functions:	Machine on/off control functions. They are designated as M codes in word address format. Thus, M03 signals for spindle on clockwise.
Modal:	Pertaining to information that is retained by the system until new information that replaces it is obtained.
Model (in CAD):	Precise term for the data managed by a CAD system to represent the parts. It is the virtual model of the designed part.
Model space:	The geometric space in terms of three-dimensional coordinates where 3D modelling takes place.
Model (or world, database) Coordinate System (MCS):	The reference space of the model with respect to which all of the geometrical data is stored.
Modem:	Modem stand for modulator/demodulator. It is a device that encodes data for transmission over a particular medium, such as telephone lines.
Monocode or hierarchical code:	The structure of these code is like a tree in which each symbol amplifies the information provided in the previous digit.
Material requirements planning (MRP):	The computerized method for planning the utilization of a company's resources in manufacturing, including scheduling, vendor selection, material alternatives.
MSI:	Medium scale integration used in connection with the integrated circuits.
Multimedia:	The discipline of integrating audio and pictorial data, in information technology, often for education and training applications.
Multipoint network:	A type of computer network in which a number of terminals or satellite computers share one line to the host computer.

Multiprocessing: A type of computer system architecture in which more than one processor is used simultaneously to increase the speed of computation for a particular task.

Multiprogramming: A mode of operation of a computer system in which the operating system must control the execution of multiple programs at the same time. This technique was developed to increase the efficiency and utilization of the CPU.

NC: Numerical control. A technique of operating machine tools by software commands.

Network: A system of computers, terminals and databases connected by communications lines, which allows the exchange of information and files.

Network architecture: The organizational concept enabling communications between data processing equipment at multiple locations. The network architecture specifies the processors and terminals and defines the protocols and software that must be used to accomplish accurate data communications.

Network data model: This approach permits modelling of many-to-many correspondence more directly than the hierarchical approaches.

Node: A generic term for any device attached to a network. A node uses the network as a means of communication and has an address on the network.

Non-uniform rational B-splines (NURBS): A mathematical description of a surface created by two (or more) B-splines.

Offset: A displacement in the axial direction of the tool equal to the difference between the actual tool length and the programmed tool length.

Offline programming: A technique whereby a part program is developed away from a particular CNC machine. After it is written, manually or with the aid of a computer, the program can be transferred to the CNC machine.

Open loop system: A control system that has no means of comparing the output with the input for control purposes. This means feedback is absent.

Open order: A customer order that has been launched into production and is in process.

Operating system: A set of special programs which manage all of the hardware resources of a computer system during its operation and provide the primary interface to the user.

Optional stop: A miscellaneous function similar to program stop (M01) except that the control ignores the command unless the operator has previously pushed a button to validate the command.

Overshoot:	The amount by which axis motion exceeds the target value.
PAN:	The process of moving the display window to view different areas of a drawing. This is done most commonly by depressing the pan button and moving the mouse.
Parity check:	An extra hole punched in one of the track columns or channels of a tape to make an even number of existing holes odd as required by the RS-244A format. The extra hole would be punched in RS-358B tape to make the existing odd number even.
Part programs:	A complete set of instructions written by a part programmer in a programming language (word, address, etc.). The MCU follows the part program's instructions in manufacturing the part.
Part programming:	The process of planning and specifying every step and movement of an NC machine into a complete process sequence in the form of an NC program.
Part families:	A part family is a collection of parts that are similar either because of geometric shape and size or because similar processing steps are required in their manufacture.
PCB:	Printed circuit board.
PDES:	Acronym for product data exchange using STEP.
PDES/STEP:	A set of standards under development for communicating a complete product model with sufficient information content that advanced CAD/CAM applications can interpret.
PDM:	Product data management. A complete set of software to manage all data and files related to a product.
Perforated tape:	Punched tape in which the hole pattern corresponds to the instructions of a part program.
Peripheral equipment:	This term refers to external input or export devices that are physically not part of a computer's housing. Examples include printers, scanners, external drives, modems, monitors, etc.
PHIGS:	Programmer's hierarchical interface graphics standard.
Phong shading:	Phong shading is a sophisticated smooth shading method, and is best known for its ability to render precise and realistic highlights.
Piezoelectric:	An inkjet printing technology that uses a mechanical-electric charge instead of heat to drive microdroplets through the nozzle.
Pixel:	The smallest unit of data in a digital image.
Planned order:	A customer order that is planned for production but not yet released.
Platform:	A combination of computer hardware and an operating system.

Plotter: A term that refers to the CAD origins of wide-format printers. A printer that graphs computer output.

Point-to-point network: A type of computer system network in which each terminal or satellite computer is connected directly to the host computer on its own communication line.

Point-to-point programming: A type of NC programming in which the tool is moved to a predetermined position with no control over the tool's speed or path.

Polycode: This is also known by many other names such as chain code, discrete code and fixed digit code. In polycode, the code symbols are independent of each other. Each digit in a specific location of the code describes a unique property of the work piece.

Port: An outlet or connection location on a computer, which allows a peripheral device to operate. A communications port (COM port) allows the modem to operate, and a local port (LPT) enables the printer to operate.

Post-processor: A computer procedure, that takes CL file information and translates into NC machine specific programming terms.

Preparatory functions: The preparatory function or cycle code refers to some mode of operation of the machine tool or NC system.

Primitives: A solid or surface that is not derived from other elements. A valid volume defined by simple standard geometrical shapes, such as a box, cone and cylinder.

Process control: The monitoring, comparing and controlling operations of a manufacturing process to maintain its operation.

Process planning: The set of instructions for product manufacturing.

Product cycle: The total of all steps leading from concept of a product to its manufacture.

Product data: All engineering data, in processable form, necessary to define the geometry, the function and the behaviour of an item over its entire life span.

Production flow analysis (PFA): PFA is a method for identifying part families and associated machine groupings that uses the information contained on production route sheets rather than on part drawings.

Product flexibility: Product flexibility, also known as **mix-change flexibility**, refers to the ability to change over to a new set of products economically and quickly in response to markets or engineering changes or even to operate on a make-to-order basis.

Production management: Production management is defined as the "planning and control of all activities necessary to produce a set of products".

Product model: A data model that contains the functions and physical characteristics of each unit of a product throughout its complete life cycle.

Product structure: The definition and organization of the objects that is appropriate for the design.

Productivity: Efficiency as a ratio of output divided by input.

Program stop: A miscellaneous (M00) function entered in a program to stop the CNC machine. Coolant, spindle and feed activity is halted after the completion of a tool movement. The operator can restart the program again by pushing a button.

Projection: The process of reducing three dimensions to two dimensions for display is called **projection**.

Protocol: A defined communication format that contains the control procedures required to facilitate data transfer across the link interfaces, end to end from the user's application programs.

Quality: Quality confirms to requirements.

Quality assurance: All the planned and systematic activities, implemented within the organization for quality management, to provide adequate confidence that a product or service will satisfy given requirements for quality.

Quality control: Activity and procedures designed to assure that the manufacturing process does not produce defective products. It is usually performed by a quality control organization whose job is to acquire and analyze data which will determine whether the process is under control and the product being manufactured will meet its specifications.

RAM: Random access memory needed for running programs such as Netscape.

Raster: A two-dimensional array of elements, called **pixels** or **picture elements,** which when displayed on a screen or paper, form an image or representation of an original document.

Raster display: In raster displays, the display screen area is divided horizontally and vertically into a matrix of small elements called **picture elements** or pixels.

Rasterization: Translating an image into pixels.

Real-time: The description for an operating system that responds to an external event within a short and predictable time frame.

Reboot: The process of turning a computer system or printer off and then back on again, to reload the software.

Rendering: Process of adding shading, colours, reflectivity, textures, etc. to a model to make it appear realistic.

Repeatability: Closeness of agreement of tool movement positions from one part to another when cutting several copies of the same part.

Reset:	Returning a storage location in the MCU to zero or to a specified initial value.
Retract:	Machining move which removes the tool from the cut.
RGB (red, green, blue):	RGB is an additive colour model used in colour monitors, conventional photo film and paper to create full colour.
RISC:	Reduced instruction set computing.
Robot:	A reprogrammable, multifunctional manipulator designed to move material, parts, tools or specialized devices, through a variety of tasks. A technology comprised a mechanical manipulator, sensors, software, and computer control which provides programmable automation capabilities.
Routing:	The processing steps needed to create a product.
Ruled surface:	A surface generated by linear interpolation between two lines or curves, or a point and a line or curve.
Screen coordinate system:	Phosphor grains coated on the screen posses longer persistence than the phosphor crystals of CRT-screen.
Sculptured surface:	A free-form surface that is curved in more than one direction, typically by NURBS, Bezier or other mathematical definitions.
SDRAM:	Synchronous DRAM is a type of DRAM to which reads or writes can be performed synchronously with the memory clock.
SERVER:	A computer dedicated to a single purpose for multiple users on a network. An example is the print server, which is dedicated to the handling of print requests.
SGRAM:	Synchronous graphics random access memory (SGRAM) is a type of memory that is optimized for graphics use.
Shop floor control system:	Control of the flow of the product and materials on a factory floor involving the quantities, types of parts, schedule dates, priorities, and the status of jobs and orders.
Simulation:	Use of a model of an operation to evaluate its behavior under varied conditions. Software computer programs that have been developed to accomplish specific tasks.
Software:	Programs, data files, procedures, rules and any associated documentation pertaining to the operation of a computer system or of a computer application.
Solid modelling:	Software capability of representing the sense of material with its familiar operations like drilling a hole or adding a slot. Solid modellers will be able to produce automatically cross-sections and display with hidden lines removed.
Spline:	Mathematical interpolation routine for describing curves or surfaces.

Standard for the exchange of product model data (STEP): ISO standard 10303. An international standard under development, which will be used to describe a product in a neutral format over its complete life cycle in a hardware independent way.

Surface: A boundary defining an exterior face of a solid model.

Surface modelling: Geometric modelling method that describes a part by its surfaces.

System: Data processing or computer system comprising all the hardware, software, and data tied together to perform a particular set of tasks. A basic computer system is made up of a CPU, main memory, input, and output peripheral equipment, with operating system software and applications programs.

Tables: A representation of data in a relational database. Information is arranged in columns and rows.

Tactile sensor: Type of contact sensor which detects objects by touch. It may use a microswitch or pressure-sensitive elements.

Task: The smallest group of work that can be assigned to a workstation.

TCP-IP: The protocols controlling applications on the internet.

Teach pendent: Control panel which can be held by an operator or programmer of a robot which is used to program the motion of the robot. It has buttons or switches which are connected through a cable to the control system. The manipulator is led through a task, one step at a time, recording each incremental move along the way. Technical data system: a computer system which controls the technical information about the design and manufacture of a product. It drives the tools and physical manufacturing activities on the factory floor.

Thermal-transfer printer: A machine that digitally prints by transferring inks from a foil ribbon onto media such as paper or vinyl.

Thumb wheels: Two wheels are provided one for x-movement and the other for y-movement of the cursor.

Throughput: The total items going through a conversion process.

Token ring: A network topology originated and promoted by IBM Corporation. At the conceptual level, a token is created for the transfer of data along the network. The token is passed from one networked device to another until the matching device is located. The data is then delivered.

Tolerance stack-up: The accumulative tolerance of mating parts in an assembly.

Tool changer: Mechanisms which automatically change tools on a machine under program control. It may take the form of a carousel with a variety of cutting tools for an NC machining centre or a special device at the end of a robot arm which provides for quick changes of the end-effectors or tool.

Tool control:	The control of the operation of a piece of manufacturing equipment or tool by a predetermined program of instructions.
Tool function:	A command that identifies a tool and calls for its selection. In word address, this is programmed as the letter T followed by the tool register number. Thus, T01 signals for tool 1 to be selected for use.
Tool length offset:	The distance between the bottom of the fully retracted tool and the part Z_0.
Tool offset:	A correction entered for a tool's position parallel to a tool movement axis. This feature allows for compensation to be applied due to tool wear and for executing finish cuts.
Tool path:	In numerical control, the path of a cutting tool as it passes over stock of material to produce desired shape.
Topological data:	Data that includes the connectively relationships among geometric components.
TORUS:	A solid primitive defined by the revolution of a circle about an axis in the plane of the circle. The axis must not pass through the centre of the circle, and must lie outside the circle.
Transformation:	Change of coordinates; a series of mathematical operations that act on output primitives and geometric attributes to convert them from modelling coordinates to device coordinates.
Trimmed surface:	A surface that has a distinct boundary. The boundary of a trimmed surface is typically where the surface intersects other surfaces.
User interface:	User interface describes how to interact with CADD by various means of data entry.
UNIX:	An open, multiple-user operating system supported on a wide range of hardware and software. It is an operating system that supports multi-users and multi-tasking.
Variational geometry:	A method of representing a solid model as a set of interrelated equations defining its shape and dimensions.
Variant process planning:	A variant process-planning system uses the similarity among components to retrieve existing process plans.
Vector:	An image plotted by lines on an XY axis. This image is different from a bitmap, which is composed of dots.
Vertex:	An X-Y-Z location (a point in space) used to define an element. All types of elements consists of one or more vertices.
Viewing coordinate system (VCS):	A 3D cartesian coordinate system (right hand of left hand) in which a projection of the modelled object is formed.
VLSI:	Very large scale integration.
Walk-through programming:	A method used to program robots which involves counter balancing the manipulator arm so that it can be moved manually

	through the intended motions while its path is being recorded by the control system. It is also known as the **guiding** or **playback method**.
WAN:	Wide area network. Private network facilities, usually offered by public telephone companies that link business network nodes.
Work-in-process (WIP):	Raw material undergoing change in the manufacturing process before it becomes finished inventory.
Wireframe modelling:	A method of modelling geometry by using "wire". The geometry is described with lines, arcs, splines, etc. in 3D space.
Word address:	Often called **NC words** such as X, Y, Z, F, G, M.
Work centre:	A facility or a set of machines where a service is offered on a job.
Working coordinate system (WCS):	An auxiliary coordinate system for conveniently entering graphical data into a CAD system.
Work envelope:	The maximum reach or range of movement of a robot arm. It will vary in shape and size depending on the configuration and size of the manipulator.
Work flow:	The sequential management of document images through work queues and various application processes.
Workstation:	A computer that can serve only one operator at a time that commonly uses specialized software designed for engineering or other scientific applications.
X-charts:	The X bar chart is developed from the average of each subgroup data.
Z-buffering:	A process of removing hidden surfaces using the depth value stored in the Z-buffer.
Zoom:	The process of magnifying the display of an image in a window to more closely inspect areas of a drawing.
Z-sorting:	A process of removing hidden surfaces by sorting polygons in back-to-front order prior to rendering.
Zero offset:	A feature of an MCU that allows the programmer to shift the zero or starting point for movements to a new position over a specified range. The system can be switched back to its old permanent origin if desired.
Zero shift:	Operates in a manner similar to zero offset except that the system cannot be switched back to the origin set prior to the shift.
Zero suppression:	The capability of eliminating zeros either before or after the significant digits entered in an instruction.

QUESTION BANK
COMPUTER AIDED DESIGN

1. Explain the differences in scope between automation and CAD/CAM through the mathematical model of the product life cycle.

2. What are the various hardware components of a standalone CAD system? Explain the various types of input devices.

3. What is the most commonly used graphics terminal? Explain its working.

4. (a) Explain the terms analytical and synthetic curves and their characteristics.
   (b) With schematic sketches explain the zero-order, first-order and second-order continuity of curves.

5. Develop the equation of a Bezier curve, find the points on the curve for $t = 0$, 1/4, 1/2, 3/4 and 1.

   The coordinates of the four control points are given by

   $$V_0 = [0,0,0] \qquad V_1 = [0,2,0]$$
   $$V_2 = [4,2,0] \qquad V_3 = [4,0,0]$$

6. Find the reflected coordinates of a triangle ABC about the line $y = 3x + 4$. The original coordinates of vertices of triangle are A(5,10), B(10,10) and C(5,15).

7. Write briefly about the following:
   (a) Homogeneous transformation   (b) Advantages of CAD/CAM

8. (a) Explain the design-related tasks which are performed by modern computer aided design system.
   (b) List out the various benefits of implementing computer aided design.

9. (a) Define the terms: ROM, RAM, Secondary storage, Binary coded decimal system
   (b) Explain the different types of operator output CAG devices.

10. (a) Explain solid modelling.
    (b) A line is defined in two-dimensional space by its end points (1,2) and (6,4). Express this in matrix notation and perform the following transformations in succession on this line:
    (i) Rotate the line by 90° about the origin.
    (ii) Scale the line by a factor of 0.5.

11. Explain the Cront's procedure listing out the various steps involved.

12. Find out the inverse of the square matrix given below by the method of partitioning:

$$[k] = \begin{bmatrix} 12 & -6 & -6 & -15 \\ -6 & 4 & 3 & 0.5 \\ -6 & 3 & 6 & 15 \\ -15 & 0.5 & 15 & 1 \end{bmatrix}$$

13. Explain any one method of iteration schemes giving examples.

14. Write short notes on:
    (a) Graphics workstation
    (b) Relative merits and demerits of FEM and FDM
    (c) Jacobic transformation

15. (a) Explain the concept of design for manufacturability? How CAD is useful in this context?
    (b) Discuss the integrated product design concept in the concurrent engineering.

16. (a) A line is represented by the matrix of its end points as follows:

$$L = \begin{bmatrix} 1, & 2 \\ 3, & 4 \end{bmatrix}$$

    Find the resultant matrix if it is scaled by a factor of 2 in X direction and 3 in Y direction.

17. What is order of continuity in the case of
    (i) Bezier curve   (ii) B-spline curve

18. (a) List 3 modes of providing the menu facility in CAD.
    (b) Distinguish between a compiler and assembler.

19. (a) Explain the terms analytical and synthetic curves and their characteristics.
    (b) With schematic sketches explain the zero-order, first-order and second-order continuity of curves.

20. (a) Define the cubic spline and Bezier curves. Which of them is more popular in CAD?
    (b) Explain the blending procedure for cubic spline segments.

21. (a) Discuss the relative features of wireframe and solid modelling.
    (b) Explain the concept of finite element analysis with a simple example of one dimensional element.

22. (a) Briefly describe the hardware components associated with CAD.
    (b) Explain how the Boolean operations can be used to construct geometric models.

23. (a) Explain the term transformations of graphic elements. Illustrate the rotation transformation with the help of a simple two-dimensional example.
    (b) How are the NC systems classified? Briefly expain the associated features.

24. Name any three of the graphic input devices.
25. What is hardware integration?
26. Differentiate between fillets and chamfers.
27. Define curve fitting.
28. What are Boolean operations?
29. (a) Discuss the features of any three of the graphic output devices?
    (b) Explain the various sub-systems those constitute for the CAD system configurations.
30. (a) Explain the concept of O.S. Kernel and its utilities.
    (b) Perform a 45° degree rotation of a triangle A(0,0), B(1,1) and C(5,2).
31. (a) Discuss the importance of interpolation and approximation Φ curves in 2D graphics with their importance in CAD.
    (a) Discuss about B-rep and C-rep approaches in solid modelling with advantages and limitations.
32. What are the geometric entities for geometric modelling?
33. (a) Explain boundary representation technique of geometric modelling.
    (b) How CAD differs from conventional design? List the advantages and applications of CAD.
    (c) What is CAD system software? Discuss any three softwares and their functions.
    (d) Compare the spline for the same control points created by B-spline, Bezier spline and Hermite cubic spline techniques.
34. (a) A pyramid is defined by the coordinates $X(0,0,0)$, $Y(1,0,0)$, $Z(0,1,0)$ and $W(0,0,1)$ is rotated through 60° about a line 1. It has a vector $(i = j + k)$ and passing through the point $(0,1,0)$. Find the coordinates of the rotated figure.
    (b) Differentiate between NC, CNC and DNC.
    (c) Discuss the techniques for surface modelling and database structure for graphic modelling.
35. Name any four reasons for implementing CAD in design.
36. Mention the CAD tools required to support the various phases of the design process.
37. What is PDES?
38. Write the equation of an ellipses in parametric form.
39. What are the different types of entities that a user finds in a typical CAD package for creating a drawing.
40. List out at least five types of algorithms used for hidden line removal.
41. What are the various features that a design database must have?
42. Briefly explain the term constructive solid geometry (CSG).
43. What is post-processing in finite element modelling.
44. Mention the various types of software modules used in CAD.
45. Explain the general procedure of the finite element analysis with help of an example.

46. (a) (i) What is product cycle and explain the various steps involved in the product cycle.

(ii) With the help of a neat sketch explain the implementation of a typical CAD process on a CAD/CAM system.

(b) (i) Briefly explain the benefits of computer aided design.

(ii) What is meant by computer aided engineering? Discuss how CAE process will be helpful to optimize the product design.

47. (a) (i) List the various output devices for graphics and state their functions.

(ii) Discuss the special techniques used for inputting graphic information?

(b) (i) Classify different logical input devices and mention their physical equivalents, i.e., devices.

(ii) Mention the different types of data structures. Discuss about network data structure.

48. (a) (i) What are the different types of geometric modelling. Compare 2D and 3D wire frame models.

(ii) Briefly explain the features of a drafting package.

(b) (i) What is a DXF file. Explain the general DXF file structure.

(ii) What are the various schemes for representing solid objects. Discuss boundary representation (B-rep) technique.

(b) (i) Explain the various stages in the design process.

(ii) What do you understand by design for manufacture and assembly.

49. What is the use of computer in the design of a product?

50. What is phal and Bezier model of the design process?

51. What is meant by system layout models?

52. Name any four applications of CAD.

53. What is IGES?

54. Write the equations of a circle in parametric form.

55. What is meant by satellite station in a typical computer network?

56. Define the term computer interface.

57. Mention any four elements used in finite element techniques.

58. What is preprocessing in finite element modelling?

59. Describe the direct methods of formulating characteristic matrices and vectors for the FEM.

60. (a) (i) Define CAE. List main applications of computer aided engineering.

(ii) Differentiate between form and structure models. Discuss their applications.

(b) (i) Describe the six steps proposed by Earle for the design process.

(ii) Write short notes on construction models.

61. (a) What is geometric modelling? Explain the geometric models, bringing out their limitations and applications.

    (b) What are geometric entities for geometric modelling? How will you generate the geometric entity for (i) straight line and (ii) complete circle.

62. (a) (i) List the various input devices for graphics and state their functions.

        (ii) Write short notes on mathematical formulation for graphics.

    (b) What are the 2D display control facilities? Explain with sketches.

    (c) (i) What do you understand by design for manufacture and assembly?

        (ii) What are the various steps involved in the product cycle? Explain briefly.

63. (a) What are the processes involved in the general design process? Explain each process.

    (b) Explain the function areas of a basic CAD system and their application in design process.

64. (a) Distinguish between hardware and software. What are the advantages of CAD workstation? Discuss briefly the working of workstation.

    (b) Discuss the classification of digital computers.

65. (a) Explain the geometry and geometric transformations.

    (b) A triangle lamina has corners P, Q, R. The coordinates of the points are (20, 20), (40, 25), (30, 40) respectively. The lamina is rotated about P through 30° in clock-wise direction:

        (i) Obtain the transformation matrix.

        (ii) Calculate the new coordinates of the triangle.

66. (a) What results do you expect by introducing CAD and CAM in an industry?

    (b) Discuss the computer requirements of process control.

67. Describe how surface composition is made through

    (a) Ruled surface   (b) Surface of revolution

    (c) Composite surface   (d) Cylindrical surface

68. Give the general configuration of a CAD computer systems.

69. With suitable figures, explain the difficulty in displaying holes and curved ends in wire frame modeling.

70. Describe the various methods and operation required in each approach for the following objects

    (a) Connecting rod   (b) Journal bearing

71. Bring out the importance of studying geometric modeling in CAD.

72. What is meant by a concatenation matrix?

Demonstrate how translation, scaling and rotation operations can be performed simultaneously on a graphic element using concatenation matrix.

73. Explain the model structure used in database organization?

74. Explain how the curves are represented in

    (a) Generic form   (b) Parametric form

75. Distinguish between interpolation and approximation approaches used in design of curves.

76. With a suitable example illustrate the 'Duct surface Modeling'. How does it differ from wire frame modeling?

77. Develop an algorithm that can enable the user to create and manipulate boundary model by using set operations.

78. What do you understand by
    (a) Analysis of Mass properties   (b) Finite element analysis
    What is their importance in engineering analysis in CAD systems?

79. Explain how CAD software fits into the hierarchical structure of software system.

80. How is a host satellite CAD system superior to a conventional centralized configuration using "dumb terminals"?

81. (a) (i) What are the main differences between sequential and concurrent engineering.
    (ii) Define computer aided manufacturing. Explain the implementation of a typical CAM process on a CAD/CAM system.
    (b) (i) Explain the various stages in the design process.
    (ii) What do you understand by design for manufacturing and assembly.
    (c) What are the various output devices used in conjunction with CAD system? Explain?
    (d) What are the currently available capabilities of CAD/CAM?

82. (a) Explain geometric modelling and wireframe modelling computer graphic software.
    (b) Distinguish between retrieval type process planning systems and generative process planning systems.

83. Explain various stages of a product life cycle with CAD/CAM.

84. Define CAD/CAM tools and explain application of automation and CAD/CAM technologies to the industries using a mathematical model.

85. With the help of a neat sketch, explain the working of an electrostatic colour plotter.

86. What is PLC? Explain its architecture, functions and applications.

87. State the important functions of a graphics package and discuss the software configuration of a graphic system?

88. What is FMS? Discuss important elements of FMS and their functions.

89. Explain function of computers in a CIM system and discuss CIMS data files.

90. Write short notes on the following:
    (a) Graphics standards
    (b) Ladder diagrams

91. Write on the introduction of CAD. Present schematic of the design and manufacturing process. Give an example.

92. Explain the following with respect to computers:
    (a) CPU   (b) I/O device.
    (c) Operating systems   (d) Storing devices

93. Explain the following with examples:
    (a) Software in CAD
    (b) Three-dimensional transformations

94. Explain the following giving examples:
    (a) Graphics system
    (b) Display windowing and display files for 3D data

95. Present the computer aided design and drafting of a simple gear. Draw the flowchart and algorithm.

96. Present the computer aided design of a simple CAM. Draw its flowchart and algorithm.

97. What are the steps of finite element method? Explain with a small example.

98. What are the present software available for FEM? Give their features. What is automatic mesh generation?

99. Write notes on:
    (a) Basic concepts of Auto CAD
    (b) Numerical solutions of heat exchangers

100. Write notes on:
    (a) Successive linear interpolation
    (b) "Nastron" CAD package

101. (a) Describe the role of CAD/CAM in the product development cycle.
    (b) Distinguish between automation and CAD/CAM.

102. (a) What is a programmable controller? What are its functions and advantages?
    (b) What are the various design-related tasks performed by a modern CAD systems?

103. (a) What are the functions of a graphic package?
    (b) What are the common editing functions available in a drafting package?

104. (a) What is raster display?
    (b) What is DDA?
    (c) Give three different display files.
    (d) State the function of input device stick.
    (e) What do you understand by aliasing?
    (f) What is event-hanging?
    (g) What is flood-filling?
    (h) What is a pixel?
    (i) What is aspect ratio?
    (j) Define the term window.
    (k) Name two input pointing devices.
    (l) Define interaction.
    (m) Define clipping.

105. (a) Distinguish between a line drawing graphics system and a point plotting graphics system.
    (b) Write the simple DDA algorithm for drawing a line from (2, 3) to (6, 8).
    (c) Describe the working of raster refresh graphic display.
    (d) Trace the Bresenham's circle drawing algorithm with centre at (0, 0) and radius 6.

106. (a) Explain scan-line algorithm to fill the polygon.

(b) Write any two methods to test whether the point is inside the polygon or not.

(c) Explain the need and use of homogeneous coordinate system.

(d) Describe the visual effect on an image of each of the following transformations:

(i) $\begin{vmatrix} 0 & 2 \\ 0 & 0 \end{vmatrix}$ (ii) $\begin{vmatrix} 0 & 2 \\ 2 & 0 \end{vmatrix}$ (iii) $\begin{vmatrix} 0 & 2 \\ 0 & 2 \end{vmatrix}$ (iv) $\begin{vmatrix} 2 & 2 \\ 2 & 2 \end{vmatrix}$

107. (a) Write pseudocode for creation and closing operations on segments.

    (b) Explain windowing and clipping process.

    (c) Define window and view port, and find viewing transformation.

    (d) Explain Cohen-Sutherland outcode algorithm for line clipping.

108. (a) Explain the use of polling and interrupts in user interaction.

    (b) Explain the working of joystick and light pen.

    (c) Explain Bezier's technique to generate surface.

    (d) Write Pointer's algorithm for hidden surface removal.

109. Discuss CAD/CAM overlaid product cycle and elaborate on the scope of application of automation and CAD/CAM technologies.

110. Explain the scope of knowledge for the development of CAD, CAM and CAD/CAM tools and discuss typical utilization of CAD/CAM system in an industrial environment.

111. Discuss the features of CAD/CAM workstation. Compare it with personal computers.

112. With the help of a neat sketch explain the colour raster display with 8 planes.

113. How can you draw a 600 pixel wide square on a 1280 horizontal × 1024 vertical screen where aspect ratio is 5:4.

114. Explain the differences in scope between automation and CAD/CAM through a mathematical model of product life cycle.

115. (a) Explain with diagrams, the main components of an engineering CAD workstations.

    (b) Give certain examples of application of computer aided design in Indian industry.

116. Briefly discuss the function of each of the following computer hardware elements:

    (a) CPU   (b) Numeric coprocessor   (c) RAM

    (d) Ports   (e) Monitors   (f) Bus expansion slots

Also write the general configuration of a design work station.

117. What is PLC? Explain its programming.

118. (a) With the help of a neat sketch. Explain working of a electrostatic plotter.

    (b) Explain the principle of raster scanning.

119. (a) Explain the importance of a 3D geometry.

    (b) Explain three different types of geometric modelling techniques available.

120. (a) Explain the role of ICG in CAD system.

    (b) Explain in detail about geometric modelling, a functional area of the modern CAD system.

    (c) What are the various functions that could be performed by a graphics package? Give an elaborate discussion.

121. (a) Give representative configuration of various register that constitute several functional areas of CPU and explain in detail the function of each register.

122. (a) Explain the term turnkey system as applicable to CAD/CAM.

   (b) Enumerate the selection criteria of hardware and software alternatives available in CAD/CAM systems.

   (c) Give an account of CAD/CAM benefits/cost ratio in the evaluation of alternative systems.

123. (a) Delineate the basic functions that could be performed by a programmable controller.

   (b) Describe the various types of computer input/output devices and discuss their relative merits.

   (c) What is the need to have different levels of computer programming languages? Explain in detail the three levels of computer programming languages.

124. (a) What do you understand by the term product cycle?

   (b) Discuss the common application areas in CAD with practical examples.

125. (a) List the different components and configuration of a typical CAD system for automobile industry.

   (b) State and explain the six phases involved in the design process.

   (c) Explain the design process. How is it modified with application of computers.

   (d) What are the advantage of CAD.

126. Write a short note on any three of the following:

   (a) Wire-frame vs. solid modelling

   (b) Benefits of CAPP

   (c) Combined CNC/DNC systems

   (d) Input/Output devices of graphic workstation

127. What is design workstation? Discuss briefly the features of a design workstation.

128. With the help of a neat sketch, explain principle of working of electrostatic plotter.

129. An eight-plane raster display has a resolution of 1280×1024 and a refresh rate of 60 MHz non-interlaced. Find (a) RAM size of the bit map, (b) the time required for displaying a scan line and pixel, (c) the active display area of screen if the resolution is 78 pixels per inch, (d) the optimal design if the bit map size is to be reduced by half.

130. What is the programmable controller? Explain its architecture, applications and programming.

131. Explain the software configuration of a graphics package and its functions in detail.

132. A point in two dimensions is located at $(3,4)$. It is desired to relocate the point by means of rotation and scaling transformations only to a new position defined by $(0,8)$.

   (a) Describe the sequence of transformations required to accomplish the movement of the line as specified.

   (b) Write the transformation matrix for each step in the sequence.

133. What is data structure? Discuss various data structures.

134. Write short notes on the following:

   (a) CAD/CAM database　　　　　　　　(b) Solid model vs. wireframe model

135. Explain life cycle of a product and how CAD/CAM technologies influence product life cycle?

136. (a) (i) Distinguish between internal and external storage?

 (ii) How many storage positions does a 16 KCPU contain?

 (iii) Distinguish between multiprogramming and multiprocessing.

 (iv) What are the advantages and diadvantages of the three media: Cre, Cips and bubble memory?

 (b) (i) What are flexible cells? What are the desirable attributes of an FMS materials handling systems?

 (ii) Explain the several levels in a hierarchical system of manufacturing organization.

137. (a) Describe the various types of computer display devices and discuss their relative merits.

 (b) List the three types of graphics terminals which are used in commercially available CAD systems. Explain in detail about each of them.

138. (a) State any five significant points to be considered in designing graphics software.

 (b) Explain the following functions of the graphics package:

 (i) Segmenting function

 (ii) Windowing

 (iii) Transformations

139. (a) With the aid of a block diagram explain the concept of integrated CAD and CAM and discuss its benefit.

 (b) "The impact of CAD/CAM is manifest in all of the various activities in the product cycle". Discuss.

140. (a) Explain how a computer can be put to use effectively in the design-related tasks.

 (b) Explain schematically data transfer between CAD/CAM systems.

QUESTION BANK
COMPUTER AIDED MANUFACTURING

1. With CNC machines, both productivity and accuracy can be increased. Defend this statement explicitly.

2. Explain the functions of machine control unit of a NC machine and how the axes of motion are determined.

3. Discuss the following canned cycle operations:

   (i)   G81   (ii) G82
   (iii) G84   (iv) G80

   and using these codes write a manual part program to machine holes shown in Fig. 1.

4. Write a manual part-program to machine the part shown in Fig. 2.

5. Write an APT program to cut a profile shown in Fig. 3.

6. Write four different APT statements to define the following:

   (i)   Points   (ii) Lines
   (iii) Circles   (iv) Patterns

7. With the help of neat sketches explain the different configurations of industrial robots.

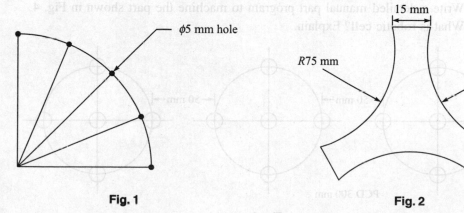

Fig. 1

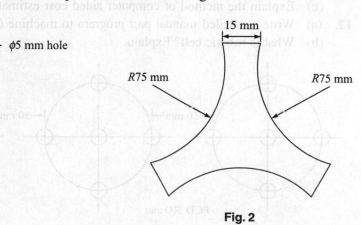

Fig. 2

ϕ5 mm hole

15 mm

R75 mm   R75 mm

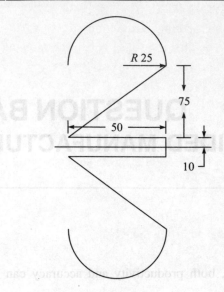

Fig. 3

8. Explain the characteristics of robots used for the following applications:
   (i) Spray painting   (ii) Arc welding
   (iii) Assembly.
9. (a) What is CIM cycle? With a neat sketch explain various parts of CIM cycle.
   (b) Does CIM required for Indian industry? Discuss various issues of implementation, challenges and trends in CIM.
10. (a) What is CAD/CAM database? Discuss the need of such database in a CIM environment.
    (b) Discuss the various types of network architectures available and write about MAP/TOP systems.
11. (a) Compare MRP-II with MRP system.
    (b) What is advanced planning and scheduling (APS) system?
    (c) Explain the method of computer aided cost estimation.
12. (a) Write a detailed manual part program to machine the part shown in Fig. 4.
    (b) What is robotic cell? Explain.

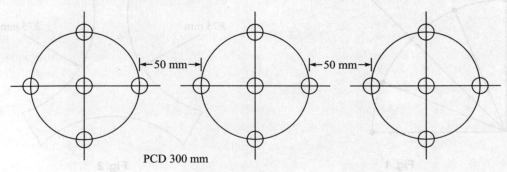

PCD 300 mm

Fig. 4

13. (a) What is reverse engineering? Explain the use of CMM in reverse engineering.

    (b) What is statistical quality control? Explain the tools of SQL.

14. Explain the following terms:

    (i) Throughput analysis

    (ii) Manufacturing lead time

    (iii) WIP and TIP ratios

    (iv) Plant capacity and machine utilization

15. (a) Explain the functions of computers in the control of FMS.

    (b) Discuss planning of FMS.

16. Write notes on the following:

    (a) Design for manufacturability

    (b) Deadlocks in manufacturing

    (c) Advanced CAD system

17. (a) What is computer integrated manufacturing? What is components? Discuss each of these components.

    (b) Explain step-wise implementation of CIM in a discrete parts of manufacturing system.

18. (a) What is enterprise wide integration? Explain various network topologies and architectures required for such integration.

    (b) Discuss the features of MAP/TOP technologies.

19. (a) Discuss the role of PPC in a CIM environment.

    (b) Briefly discuss the kind of research work done and developments made in the area of CAPP.

    (c) Enumerate the features of MRP-II and explain the role of MRP-II in the present context of CRM.

20. (a) Explain the important features of CNC machines compared to conventional machines.

    (b) Discuss the developments made in the programming of NC machines in the last two decades with suitable examples.

    (c) Explain the role of conveyors, robots, AGVs in a CIM system.

21. (a) What is process capability? Explain one method of estimation of process capability of a machine tool.

    (b) Discuss the various types of CMMs, their features and applications in the industry.

22. (a) List various measures of performance of an automated manufacturing system (AMS). Develop and discuss simple mathematical models of performance measures.

    (b) What is meant by flexibility? Discuss the various flexibilities that an AMS have.

23. (a) What is deadlock? How to avoid deadlocks in AMS?

    (b) How do you analyze the flexible automated assembly as a game of chance?

24. (a) A stepper motor of 300 steps/rev is mounted on the lead screw of a drilling machine. If the pitch of the lead screw = 0.3 mm. Find

      (i) BLU

      (ii) Number of pulses required for a movement of 10 mm

      (iii) Spindle speed if the motor reserves 2000 pulses per sec

  (b) In a NC papertape there are 8 tracks. The sprocket holes are provided non-symmetrically after third track rather than symmetrically after fourth track. What can be the reason?

25. (a) What do the following abbrevations stand for:

      EIA, APT, TAB

  (b) What is the meaning of the following codes for spindle speed and feedrate?

26. (a) What is the difference in hierarchical and chain type structure of GT coding?

  (b) For robot systems and NC machine tool systems which are the common classification types?

27. (a) Explain the concept of open loop and closed loop control.

  (b) Briefly describe the relative features of point to point, paraxial and continuous path systems of NC machines.

28. (a) Write in word address format a manual part program for drilling a hole of 20 mm diacentrally in a block of 100 × 100 × 10 mm dimensions. Cutting speed = 50 m/min, feed rate = 0.2 mm/rev.

  (b) Define the terms CNC, DNC and FMS.

29. (a) For the component shown in Fig. 5, write the geometry statements and motion statements in APT language. Take origin as target.

  (b) Explain the ACO and ACC techniques of adaptive control.

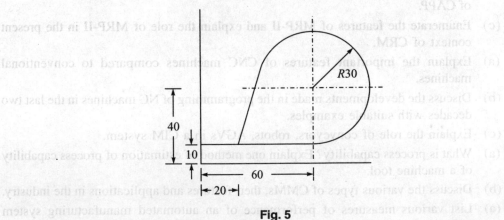

Fig. 5

30. (a) With neat schematic sketches describe the configuration and relative features of cartesian and articulated robot arms.

  (b) Briefly explain the classification of CAPP systems and compare the characteristics.

31. (a) Explain the walk through and lead through methods of robot programming.

  (b) Define and explain the concept of group technology. Why it is not popular for practical implementation?

32. What is part classification?
33. Out of *X, Y* and *Z* axes which is first identified? How is it done?
34. What for the fifth and eigth tracks on NC tape meant for?
35. Expand the abbreviations ACO and ACC.
36. What is a MACRO statement in NC programming?
37. (a) Discuss the terms: interpolation, post-processor and 3D axes.
    (b) Write down the APT Geometry and motion statements for machininig the work part shown in Fig. 6.

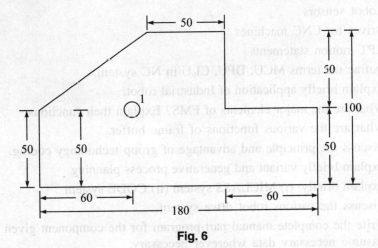

Fig. 6

38. (a) Explain the features of word address and tab-sequential formats of tape preparation.
    (b) What do you understand by G and M functions?
    (d) Explain the concept of tool length and cutter radius compensation.
39. (a) Discuss the importance of NC, DNC and CNC systems in automated machining. What is control loop?
    (a) Name the different types of robot programming methods. Explain the manual lead through programming of Robots.
40. (a) Briefly describe any four types of anatomy of robots.
    (b) Explain the working of adaptive control with its advantages and its limitations.
41. (a) What is coding systems? Explain the part family concept associated with GT.
    (b) Discuss the relative features of retrieval and generative types of CAPP.
42. (a) Explain the roles of AI and ES in manufacturing.
    (b) Define the FMS and discuss the features of a typical FMS with a schematic diagram
43. (a) Discuss the salient features of modern computer technology required for CAD.
    (b) Define group technology. List the advantages.
44. Classify and explain the advantages of DNC system.
45. Differentiate between generative and retrieval type process planning and explain retrieval type process planning system.

46. (a) Differentiate between ACC and ACO type adaptive controllers.

    (b) Explain the advantages of CNC.

47. Explain the methods of NC part programming.

48. Explain optiz classification and coding system and explain the generation of form code taking a suitable example.

49. (a) Explain the robot configurations in view of its applications with neat sketch.

    (b) Discuss teach pendant method of robot programming.

50. Write short notes:

    (a) Robot sensors

    (b) Drives for CNC machines

    (c) APT-motion statements

51. (a) Define the terms MCU, DPU, CLU in NC system.

    (b) Explain briefly application of Industrial robot.

52. (a) What are the major elements of FMS? Explain their functions.

    (b) What are the various functions of frame buffer.

53. (a) Discuss the principle and advantage of group technology coding.

    (b) Explain briefly variant and generative process planning.

54. (a) Explain briefly: (i) MICLASS system (ii) CODE system

    (b) Discuss the various robot drive system.

55. (a) Write the complete manual part-program for the component given in Fig. 7. Assume necessary data wherever necessary.

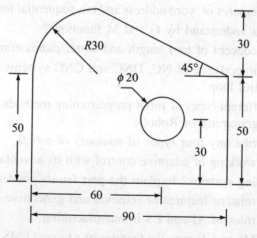

Fig. 7

    (b) Write APT language for the component given in Fig. 7.

56. (a) Explain various principles of tool monitoring in FMS? How does it help the productivity in FMS?

57. Write short notes on the following:
    (a) Robotic sensors
    (b) Difference between absolute and incremental dimensions in NC systems
    (c) Elements of CAM system
58. (a) (i) What are the main differences between sequential and concurrent engineering.
       (ii) Define computer aided manufacturing. Explain the implementation of a typical CAM process on a CAD/CAM system.
59. (a) (i) Define computer aided manufacturing. What are the broad applications of CAM?
       (ii) Describe various steps in simultaneous engineering. What are its applications?
60. Differentiate clearly in all aspects, between CNC and DNC.
61. (a) Explain the following:
    (i) Product cycle   (ii) CAM hierarchy
    (iii) CAM database
    (b) Explain the procedure for part programming an arc and a circle.
62. (a) List and explain the major components of FMS.
    (b) Explain the term computer integrated manufacturing. How does it differ from FMS?
63. Discuss briefly on any three of the following:
    (a) CAD input devices
    (b) Components of an NC system
    (c) Composite part concept in group technology
    (d) Material requirement planning
64. Classify machine tools by degree of automation criterion.
65. Give the various parameters for the working motions of machine tools.
66. List out the various factors affecting stiffness of machine tool structure.
67. What are LM guideways?
68. What are the various design requirements of CNC machine tools?
69. Why a gear box is used in a spindle drive of a CNC lathe? Explain.
70. What is tool length compensation?
71. What are canned cycles?
72. What is multiaxis machining?
73. Give the various work holding devices used in CNC.
74. (i) Explain the basic design procedure of machine tool structure based on strength stiffness
    (ii) Explain the functioning of CNC control system with help of a block diagram. Briefly describe the various parts.
75. (a) (i) List the various techniques that are used to reduce the frictional resistance on CNC machine tool slides. Explain any three.
       (ii) Explain the following:

        (a) Tool nose radius compensation

        (b) Block tooling systems.

   (b) (i) Explain the constructional features and applications of incremental and absolute encoders.

     (ii) Explain the following:

        (a) Qualified tools

        (b) Conversational programming

76. (a) Write a CNC part program to machine the part shown in Fig. 8 from a bar of diameter 70 mm with a length of 140 mm:

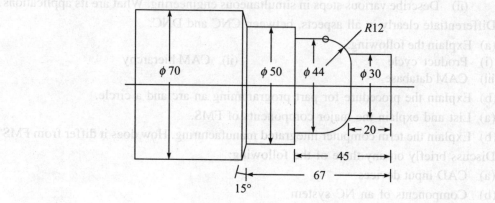

Fig. 8

   (b) Write a APT program to machine the following part shown in Fig. 9.

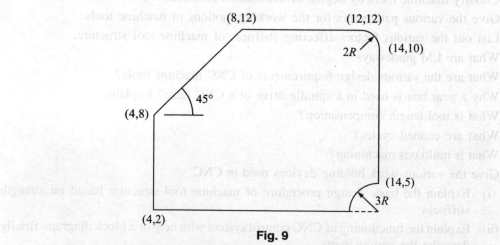

Fig. 9

77. What is computer aided manufacturing? Write on the advantages of it.

78. (i) Classify computer aided processing planning methods. Explain in detail.

(ii) Explain the following:
(a) CAM packages
(b) Tool presting

79. (a) Distinguish between mass production and batch production.
(b) What are the several fundamental reasons for implementing a computer aided design system.

80. (a) Distinguish between point to point, straight and contouring motion control systems.
(b) What are the steps performed in computer assisted part programming.

81. (a) What are the advantages of group technology.
(b) Explain with sketch an FMS system.

82. (a) What are the benefits of CAD carry over into manufacturing.
(b) Describe the principle of working of a CNC machine.

83. (a) What is the need for CAD based process planning.
(b) Explain with sketch the articulated arm configuration of a robot.

84. (a) What are the different types of control system used in robots.
(b) Describe any one computer aided process planning software.

85. (a) Briefly explain significant application areas of robot.
(b) Differentiate between NC, CNC and DNC system.

86. (a) Discuss the tool control system in FMS. How tool coding is done in FMS.
(b) How is productivity improve in a CNC turning centre.

87. With a neat sketch, explain the functioning of a NC machine. State two important differences between NC and CNC.

88. Explain the point to point and contouring NC systems and state the various features of a machining centre.

89. Write a manual part program to cut letter 'Q' on a plate 100 mm × 100 mm size using a 3-axes CNC machine (Assume suitable dimensions).

90. Write an APT program to cut the following profile shown in Fig. 10 on a Triac-Denford control milling machine (Assume suitable machining data).

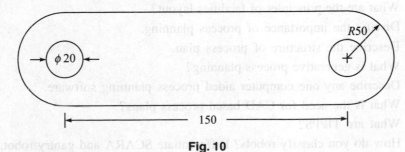

Fig. 10

91. Answer the following:
(a) Explain G81, G82, G84 codes.
(b) The macro statement in APT.

92. Explain the various design related tasks which are performed by modern computer aided design.

93. What is design workstation? Explain its features.

94. (a) Explain the term robot. What characteristics of robot made them to use in the industry.

95. (b) How do you specify a robot?

96. (a) Discuss the teaching methodologies of a robot.

    (b) Explain various grippers used for industrial robots.

97. (a) Explain the benefits of group technology.

    (b) Discuss the coding structures used for GT.

98. Define CAPP. State the various approaches to CAPP and explain one approach where GT is used.

99. Define adaptive control. Explain the adaptive control optimization system for a milling machine.

100. (a) What is direct numerical control? Explain the classification and information flow in the DNC.

     (b) What is distributed numerical control?

101. Answer the following:

     (a) ACC system for a lathe machine.

     (b) MI Class.

     (c) Advantages of adaptive control.

102. Explain the following terms:

     (a) CNC

     (b) CAM

     (c) Adaptive control

     (d) NC part program

103. (a) What is group technology? What are its merits and demerits?

     (b) What is classification and coding system? Give examples.

     (c) What are the principles of facilities layout?

104. (a) Discuss the importance of process planning.

     (b) Describe the structure of process plan.

     (c) What is generative process planning?

105. (a) Describe any one computer aided process planning software.

     (b) What is the need for CAD based process plans?

     (c) What are TIPPS?

106. (a) How do you classify robots? Differentiate SCARA and gantry robot.

     (b) Explain (i) pick and place and (ii) point-to-point operations of a robot.

     (c) Define and specify a robot. Give an example.

107. (a) What are AGVS? What are advantages of using them in FMS?
    (b) Explain gantry setup and RGV and their applications.
    (c) What is MRP? Explain the various aspects of MRP.
108. (a) What are the objectives of computer aided quality control?
    (b) Explain the method of part inspection using a CMM.
109. (a) Discuss the integration of CAD database and CMM operation.
    (b) Describe the features of a flexible inspection system.
110. (a) What is CIM? How can it be implemented in an Indian company?
    (b) Comment on the use of:
      (i) CNC machine tools
      (ii) Robots in manufacture
111. Write notes on:
    (a) Management information systems   (b) TQM
112. (a) Describe the basic components of a CNC system.
    (b) Discuss the major applications of numerical control.
113. (a) Describe briefly the following types of statements used in APT:
      Geometry statements
      Motion statements
      Post-processor statements
    (b) What is macrostatement in APT?
114. (a) Describe the procedure adopted to create a NC program using CAD/CAM technique.
    (b) What are the advantages of CAD/CAM in NC programming?
115. (a) Describe the basic physical configurations of an industrial robot.
    (b) Discuss the application of robots to welding.
116. (a) Describe the cycle of activities in a computer integrated production management system.
117. (b) Describe the application of computers in inspection.
118. (a) Discuss the basic structure of any one type of part classification and coding system.
    (b) What are the benefits of group technology?
119. What is group technology (GT)? Explain the same with reference to a layout problem and also mention the advantages of GT.
120. Write short notes on the following:
    (a) Optiz classification—coding system
    (b) Input devices
121. Explain the following machining operations with neat sketches:
    (a) Straddle milling   (b) Thread cutting
    (c) Slotting   (d) Counter-sinking
    (e) End milling

122. (a) Explain the K23 kinematic structure with a neat sketch.

(b) Explain pitch error correction mechanism.

123. Write an APT program to machine the part shown in Fig. 11 on a 2 ½ axis CNC milling machine.

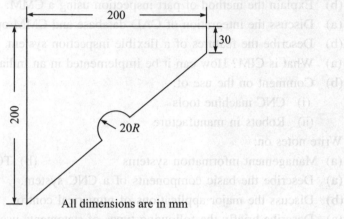

All dimensions are in mm

Fig. 11

124. (a) Explain briefly about the different approaches used in CAPP.

(b) Bring out the various advantages of CAPP.

125. (a) What are the various basic components of an NC system? How are NC machines classified?

(b) Explain the differences between CNC and DNC systems.

126. (a) Explain the physical configuration of a robot. What are the basic motions of a robot?

(b) Give certain practical examples of the use of robots in industrial situations.

127. Explain the term group technology. List the various benefits of group technology.

128. Write short notes on:

(a) CIM   (b) FAPT language

(c) programmable controllers

129. (a) What is BLU? What is its importance in a NC system?

(b) Define industrial robot.

(c) Define accuracy and resolution with respect to an NC system. Give the relation between them.

(d) What is design review and evaluation?

(e) What is the main function of an ICG system?

(f) What does the term 16-bit or 32-bit machine mean?

(g) Write APT statement for a line passing through a point P_1 and tangential to a circle C_1.

(h) Name different coding structures followed in GT coding and classification system.

(i) Write three differences between CMM and CNC machine.

(j) Define flexible manufacturing system.

130. (a) Explain the functions of computer numerical control.
    (b) Write a manual part program for the part shown in Fig. 12.

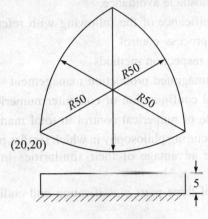

(20,20)

5

Fig. 12

131. (a) What are the advantages of computer assisted part programming over manual part programming?
    (b) Write complete APT code for the part shown in Fig. 13.

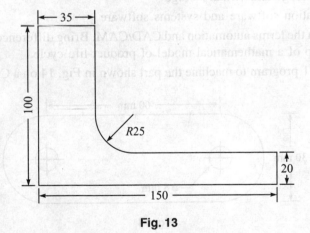

Fig. 13

132. (a) With the help of a neat sketch, explain all the motions of a robot.
    (b) Explain walk through method of programming a robot.
133. (a) Enumerate the various functions of material handling in FMS.
    (b) Explain the concept of adaptive control of NC machines.
134. (a) What are the major benefits of group technology?
    (b) List 3 classification and coding systems and write short notes on one of them.
135. (a) What are the benefits of CAPP?
    (b) Write a short note on MRP.

136. (a) Give the classification of robotics.

   (b) Discuss several range sensing techniques that address various problems like robot navigation and obstacle avoidance.

   (c) Discuss the significance of the following with reference to CIM system:
      (i) Computer process control
      (ii) Automated inspection methods
      (iii) Computer integrated production management

137. (a) Give the general configuration of computer numerical control system.

   (b) Describe the role of numerical control in total manufacturing system.

138. (a) "GT is a manufacturing philosophy in which similar parts are identified and grouped together to take advantage of their similarities in manufacturing and design". Discuss.

   (b) Give a review of three parts classification and coding systems which are widely recognized.

139. Differentiate between:

   (a) Absolute and incremental positioning system
   (b) CNC and DNC
   (c) Fixed and floating point zero methods for specifying zero points
   (d) Main storage and virtual storage
   (e) Application software and systems software

140. (a) Explain the terms automation and CAD/CAM. Bring differences between them with the help of a mathematical model of product-life cycle.

141. Write an APT program to machine the part shown in Fig. 14 on a CNC milling machine.

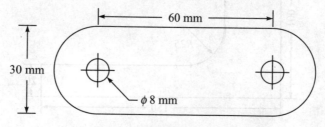

Fig. 14

142. Classify different NC systems and briefly explain them.

143. What are the general characteristics of an industrial situation where installation of robots is practicable? Give certain practical examples for the use of robots.

144. (a) Explain the term group technology.

   (b) What are the various factors to be considered in parts coding and classification system.

   (c) List out the various benefits of the group technology.

145. Explain the term CIM. What are the advantages of CIM. What are the steps involved in the selection of a turnkey CAD/CAM systems.

146. Sketch the following robot configurations:
     (a) TRR:TL
     (b) ROL:TR
     (c) TLR:OO
     (d) LRT:LR
     (e) VOL:LR

147. Develop a program in VALLII to command a robot to unload a cylindrical part of 15 mm diameter from Machine 1 positioned at point P_1 and load the part on Machine 2 positioned at P_2. The operational speed of robot is 2 m/sec. The operational speed of the robot must be kept at 0.5 m/sec.

148. What do you understand by the term Group technology. Discuss the uses of GT in various functional departments of an organization.

149. Based on the Optiz coding system, develop a form code for the part shown in Fig. 15.

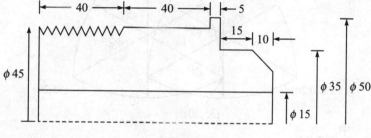

Fig. 15

150. What is FMS? Discuss basic components of an FMS.

151. (a) Discuss the variant CAPP system.
     (b) Write differences between variant and generation process planning systems.

152. Answer the following:
     (a) TIPPS process planning software
     (b) MIPLAN process planning software

153. Answer the following :
     (a) Robotic grippers
     (b) Rank order clustering algorithm

154. "With CNC Machines, both productivity and accuracy can be increased". Define this statement explicitly.

155. Write a manual part-program to machine the part shown in Fig. 16.

156. Write an APT program to cut a profile shown in Fig. 17.

157. Write short notes on the following:
     (a) Machining centres
     (b) Canned cycles

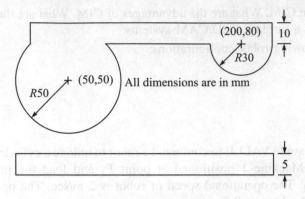

All dimensions are in mm

Fig. 16

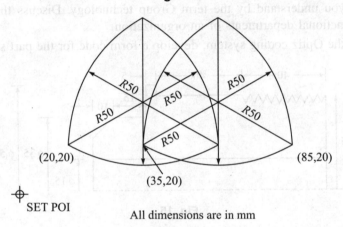

SET POI

All dimensions are in mm

Fig. 17

158. (a) Identify the basic characteristics that distinguish different types of robots.
    (b) Distinguish clearly between resolution, accuracy and repeatability as applicable to a robotic arm.
    (c) Discuss the terms: robotics and automation, bringing out clearly the distinguishing features and scope with illustrative examples.

159. (a) What are the four types of statements in APT language?
    (b) What do you understand by the term canned cycles?

160. The rectangular component, shown in Fig. 18, is to be manufactured by NC.
    (a) List out the steps to be carried out in machining the component using NC.
    (b) Write down the manual part programme for the same.

161. What are the different types of GT machine cells? Give the benefits of GT.
    (a) Discuss how part classification is done in the context of GT. What are the essential attributes such a coding system should take care of?

162. Explain what do you understand by flexibility and reconfigurability. What are hybrid systems and give their application areas?

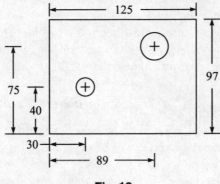

Fig. 18

(a) Write short notes on:
   (i) CAPP and its benefits
   (ii) Machining centres and CNC

163. Write short notes on the following:
  (i) Turnkey CAD/CAM systems
  (ii) Adaptive control
  (iii) Generative process planning
  (iv) Advantages of CIM

OBJECTIVE TYPE QUESTIONS

1. CAD/CAM is the relationship between
   (a) Science and engineering
   (b) Manufacturing and marketing
   (c) Design and manufacturing
   (d) Design and marketing.

2. The different phases that a product undergoes from the conceptualization of the product until the end product reaches the customer is known as:
   (a) Manufacturing cycle
   (b) Product life cycle
   (c) Production process
   (d) None of these.

3. The use of computer systems to assist in the creation modification, analysis or optimization of a design is known as:
   (a) CAD
   (b) CAM
   (c) CADD
   (d) All of the above.

4. The final stage in the implementation of CAD in a CAD/CAM system is:
   (a) Geometric modelling
   (b) Drafting and detailing
   (c) Documentation
   (d) Design analysis.

5. The use of computer systems to plan, manage and control the operations of manufacturing plant through either direct or indirect computer interface with the plant's resources is known as:
   (a) CAD
   (b) CAM
   (c) CADD
   (d) All of the above.

6. Following is one of the important stage in the implementation of CAM process on a CAD/CAM system:
   (a) Design analysis
   (b) Documentation
   (c) Inspection
   (d) Drafting.

7. Following is not the important step in design process:
   (a) Problem identification
   (b) Preliminary process
   (c) Implementation
   (d) Forecasting process.

669

8. The process in which the detailed specifications materials, dimensions, tolerances and surface rough is made is known as:
   (a) Decision process
   (b) Analysis process
   (c) Implementation process
   (d) Refinement process.

9. Which two disciplines are tied by a common database?
   (a) Documentation and geometric modelling
   (b) CAD and CAM
   (c) Drafting and documentation.

10. The term that is used for geometric modelling like solid modelling, wire frame modelling and drafting is known as:
    (a) Software package
    (b) Operating system
    (c) Application software
    (d) None of these.

11. The system environment in a mainframe computer consists of
    (a) Central processing
    (b) Storage devices
    (c) Printers and Plotters
    (d) Both (a) and (b).

12. The nerve centre or brain of any computer system is known as:
    (a) CPU
    (b) Storage devices
    (c) ALU
    (d) Monitor.

13. Locating devices are classified as:
    (a) Text input device
    (b) Graphic device
    (c) Both (a) and (b)
    (d) None of these.

14. A potentiometric device that contains sets of variable registers which feed signals that indicate the device-position to the computer is known as:
    (a) Track ball
    (b) Mouse
    (c) Joystick
    (d) All of the above.

15. Which of the following devices do not produce a hard copy?
    (a) Impact printers
    (b) Plotters
    (c) CRT-terminals
    (d) Non-impact printers.

16. The software that is used to control the computer's work flow, organize its data and perform house keeping functions is known as:
    (a) Operating software
    (b) Graphics software
    (c) Application software
    (d) Programming software.

17. The software that is used to provide the users with various functions to perform geometric modelling and construction is known as:
    (a) Operating software
    (b) Graphics software
    (c) Applications software
    (d) Programming software.

18. The software that performs the data entry, design, analysis, drafting and manufacturing functions is known as:
    (a) Operating software
    (b) Graphics software
    (c) Application software
    (d) Programming software.

19. The software that enables the user to implement custom applications or modify the system for specialized needs is known as:
   - (a) Operating software
   - (b) Graphics software
   - (c) Application software
   - (d) Programming software.

20. Following is not a operating system software:
   - (a) Windows
   - (b) UNIX
   - (c) VAX/VMS
   - (d) IDEAS.

21. The basic geometric building blocks provided in a CAD/CAM package are:
   - (a) Points
   - (b) Lines
   - (c) Circles
   - (d) All of the above.

22. The references space of the model with respect to which all of the geometric data is stored is known as:
   - (a) Model co-ordinate system
   - (b) Working co-ordinate system
   - (c) Screen co-ordinate system
   - (d) None of these.

23. The transformation based on mapping an N-dimensional space into (N+1) dimensional space is known as:
   - (a) Homogeneous
   - (b) Non-homogeneous
   - (c) Heterogeneous
   - (d) None of these.

24. The transformation matrix for reflection through the xy plane in a 3-dimensional rotation is:
   - (a) 1 0 0 0
         0 1 0 0
         0 0 1 0
   - (b) 1 0 0 0
         0 1 0 0
         0 0 1 0
   - (c) Both (a) and (b)
   - (d) None of these.

25. Best quality graphics is produced by
   - (a) Line printer
   - (b) Plotters
   - (c) Daisy wheel printers
   - (d) Dot-matrix printers.

26. Applications software
   - (a) Is used to control the operating system
   - (b) Is designed to help programmers
   - (c) Performs specific task for computer users
   - (d) All of the above.

27. The interaction of user and computer for generating figures is called
   - (a) ICG
   - (b) Computer graphics
   - (c) CAD
   - (d) CAPP.

28. Which key hardware item ties a CAD/CAM system together?
   - (a) Keyboard
   - (b) Graphics work station
   - (c) Digitizer
   - (d) Plotter.

29. The key element for a graphics work station is:
   - (a) Plotter
   - (b) Keyboard
   - (c) CRT
   - (d) Digitizer.

30. A raster CRT eliminates
    (a) Flicker and slow update
    (b) Flicker only
    (c) Slow update only
    (d) Has no effect.

31. The following is not a method of input control in a CAD system:
    (a) Function keyboard
    (b) Joystick
    (c) Plotters
    (d) Touch terminal.

32. Digitizer can be converted from "dumb" to "smart" through the addition of
    (a) Microprocessor
    (b) Electromagnetic coupling
    (c) Mouse
    (d) Scanner.

33. Interaction between user and CRT is accomplished via
    (a) Vector scanning
    (b) Digitizing tablet
    (c) Stylus
    (d) Touch terminal.

34. A tablet cursor is:
    (a) Digitizer
    (b) Light pen
    (c) Joystick
    (d) Programmable dials.

35. Plotter accuracy is measured in terms of
    (a) Vertical dimension
    (b) Intelligence
    (c) Resolution
    (d) Buffer size.

36. Flood guns bombard the phosphor surface of the CRT in
    (a) DVST
    (b) RVT
    (c) RST
    (d) Stroke CRT.

37. Skipping every other line and picking up the skipped lines on the second pass is called
    (a) Interlacing
    (b) Raster-scan
    (c) Refresh vector
    (d) Stroke writing.

38. The form of inkjet printing is:
    (a) Impact
    (b) Dot matrix
    (c) Thermal
    (d) Serial impact.

39. A rectangular boundary around a specific area on the display is called
    (a) Window
    (b) View port
    (c) Conic section
    (d) View plane.

40. A display of volume model that shows all of edges is called a
    (a) 3D model
    (b) Surface model
    (c) Wire frame
    (d) Solid model.

41. The representation of complex 3D geometry of the outer shell of a product is called
    (a) 2D modelling
    (b) Solid modelling
    (c) 3D modelling
    (d) Surface modelling.

42. A technique used to mould and shape using a series of lines is called
    (a) Solid modelling
    (b) Surface modelling
    (c) Wire-frame modelling
    (d) FEM.

43. The method of solid modelling that defines the topology of faces, edges and vertices is called
    - (a) Layering
    - (b) CSG
    - (c) b-rep
    - (d) Isometric.

44. A compiler translates a computer program from
    - (a) High-level language to low-level language
    - (b) High-level language to assembly language
    - (c) Low-level language to high-level language
    - (d) Low-level language to assemble language.

45. A register is used for
    - (a) Storing only a single bit of information
    - (b) Providing random access data memory
    - (c) Storing a group of related data memory
    - (d) All of the above.

46. A computer program written in a high-level language is called a
    - (a) Source program
    - (b) Object program
    - (c) Machine language program
    - (d) Application program.

47. Pixels are associated with
    - (a) Data processing
    - (b) Sound
    - (c) Graphics
    - (d) Bytes.

48. A place through which data can enter or leave a computer is called a
    - (a) Bus
    - (b) Dock
    - (c) Mouse
    - (d) Port.

49. The storage places that hold data and instruction temporarily within the CPU are called
    - (a) Accumulators
    - (b) Buses
    - (c) Registers
    - (d) Addresses.

50. Volatile memory means that the contents are:
    - (a) Lost when power is put on
    - (b) Lost when power is put off
    - (c) Written at two places in the memory
    - (d) Permanently put into the memory.

51. To ensure that data is accessed by authorized users only, computers make use of what are known as
    - (a) Cipher codes
    - (b) Identification numbers
    - (c) Secret numbers
    - (d) Passwords.

52. The main use of secondary storage device is:
    - (a) To hold programs during execution
    - (b) To hold data which can be retrieved during processing
    - (c) As back up for floppy disks
    - (d) As an alternative to primary storage device.

53. The CRT screen is coated with
    - (a) Silicon
    - (b) Radium
    - (c) Uranium
    - (d) Phosphor.

54. A list of displayed items from which a user can select is called a
    - (a) Program
    - (b) Menu
    - (c) Cursor
    - (d) Windows.

55. A rectangular specific area on a video display is a
    - (a) View port
    - (b) I/O port
    - (c) Window
    - (d) View point.

56. The rate at which scanning is repeated in a CRT is called
    - (a) Resolution
    - (b) Pitch
    - (c) Band width
    - (d) Refresh.

57. Utility programs are:
    - (a) Special programs to do common repetitive jobs
    - (b) Programs to supervise computer actions
    - (c) Used to translate high level language programs to machine language programs
    - (d) Used to block operation of unwanted programs.

58. Which of the following printer can produce carbon copies?
    - (a) Laser printer
    - (b) Thermal printer
    - (c) Ink-jet printer
    - (d) Line printer.

59. The most popular device for online data entry is:
    - (a) Floppy disk
    - (b) Hard disk
    - (c) Keyboard
    - (d) Printer.

60. Which of the following is a data-recording media?
    - (a) Plotter
    - (b) Digitizer
    - (c) Mouse
    - (d) Joystick.

61. A data arranged in intelligible form is called
    - (a) Software
    - (b) Program
    - (c) Information
    - (d) Processed data.

62. The binary information contained in a computer is called
    - (a) Memory
    - (b) Program
    - (c) Instruction
    - (d) Address.

63. Which of the following is not a hardware?
    - (a) Tap
    - (b) Printer
    - (c) CRT
    - (d) Assembler.

64. Voice data entry terminals
    - (a) Are speaker dependent
    - (b) Are speaker independent
    - (c) Have limited vocabulary
    - (d) Convert human speech into audible signals.

65. The linking of computers with a communication system is called
    (a) Networking
    (b) Pairing
    (c) Interlocking
    (d) Assembling.

66. The process of putting data into a storage location is called
    (a) Reading
    (b) Writing
    (c) Controlling
    (d) Hand shaking.

67. The process of copying data from a memory location is called
    (a) Reading
    (b) Writing
    (c) Controlling
    (d) Hand shaking.

68. Which of the following computers is the least powerful?
    (a) Micro computer
    (h) Mini computer
    (c) Mainframe computer
    (d) Supercomputer.

69. The hardware unit which is used to monitor computer processing is the
    (a) Console
    (b) Printer
    (c) Mouse
    (d) ROM.

70. Which of the following is the internal memory of the computer?
    (a) CPU register
    (b) Cache
    (c) Main memory
    (d) All of the above.

71. The program that controls the execution of application programs and acts as an interface between user of a computer and computer hardware is known as:
    (a) Linker
    (b) Text editor
    (c) Compiler
    (d) Operating system.

72. The memory which is programmed at the time of manufacture is called
    (a) ROM
    (b) RAM
    (c) PROM
    (d) EPROM.

73. A storage device used to compensate the differences in flow of data is known as:
    (a) Main storage
    (b) Auxiliary storage
    (c) Buffer
    (d) Core memory.

74. Which memory is volatile?
    (a) RAM
    (b) ROM
    (c) PROM
    (d) EPROM.

75. Which of the following storage is volatile?
    (a) Floppy disk
    (b) Bubble memory
    (c) Semiconductor memory
    (d) Core memory.

76. Icons are:
    (a) Typed commands
    (b) Verbal commands
    (c) Picture commands
    (d) Imaginary commands.

77. The ALU of a computer normally contains a number of high speed storage elements called

(a) Semiconductor memory   (b) Registers

(c) Hard drums   (d) Magnetic disk.

78. Which of the following is not used as secondary storage?

(a) Semiconductor memory   (b) Magnetic disks

(c) Magnetic drums   (d) Magnetic tapes.

79. Utility program include

(a) Editors   (b) Spread sheets

(c) Operating systems   (d) All of the above.

80. Formatting the disk involves

(a) Copying the contents of one disk to another disk

(b) Preventing a user from copying the disk

(c) Erasing the disk

(d) None of the above.

81. Booting the computer means

(a) Logging in   (b) Loading the operating system

(c) Turning the computer on   (d) Both (a) and (b).

82. A 2 ½ D CAD model will have

(a) Zero thickness   (b) Constant thickness

(c) Variable thickness   (d) None.

83. The functions of geometric modelling does not include

(a) Manufacturing   (b) Inspector of quality

(c) Marketing   (d) Design.

84. The core product model required for any CAD/CAM system

(a) Data model   (b) Solid model

(c) Prototype   (d) Geometric model.

85. The database memory is small to store the following geometric model:

(a) Wire frame model   (b) Solid

(c) Surface   (d) Product shape.

86. Ambiguous representation of real objects is seen in

(a) Wire frame model   (b) Surface model

(c) Solid model   (d) Product model.

87. Number of lines required to represent a cube in wire frame model are:

(a) 8   (b) 12

(c) 16   (d) 24.

88. Which is not an analytical entity in the following?

(a) Line   (b) Circle

(c) Fillet   (d) Hyperbola.

89. One of the following does not belong to family of conics:

(a) Ellipse   (b) Parabola

(c) Spline   (d) Hyperbola.

90. One of the following is not a synthetic entity:
    (a) B-spline
    (b) Bezier curve
    (c) Cubic spline
    (d) Hyperbola.

91. Number of tangents required to describe cubic splines are:
    (a) 1
    (b) 2
    (c) 3
    (d) 4.

92. C^0 continuity of curves refer to
    (a) Common tangent
    (b) Common point
    (c) Common normal
    (d) Common curvature.

93. C^1 continuity refers to
    (a) Common tangent
    (b) Common point
    (c) Common normal
    (d) Common curvature.

94. C^2 continuity of curves refer to
    (a) Common tangent
    (b) Common point
    (c) Common normal
    (d) Common curvature.

95. The minimum order polynomial that has C^0, C^1 and C^2 continuity
    (a) B-spline
    (b) Bezier
    (c) Hermite cubic spline
    (d) Quadratic polynomial.

96. Two end points and two tangent vectors are required to define
    (a) B-spline
    (b) Cubic spline
    (c) Bezier curve
    (d) None.

97. The order of the cubic spline is always
    (a) 2nd order
    (b) 3rd order
    (c) Any order
    (d) No order.

98. Furguson curves are also called as:
    (a) B-splines
    (b) Cubic spline
    (c) Rational B-spline
    (d) Bezier curve.

99. The shape of the Bezier curve is controlled by
    (a) Control points
    (b) Spline points
    (c) Nodal points
    (d) End points.

100. The curve constructed using Einstein polynomial
    (a) Bezier
    (b) B-spline
    (c) Cubic spline
    (d) NURBS.

101. The curve that follows a convex hull property
    (a) Bezier curve
    (b) Cubic spline
    (c) B-spline
    (d) Both (a) and (b).

102. The degree of the Bezier with n control point is:
    (a) $n + 1$
    (b) n
    (c) $n - 1$
    (d) 3 always.

103. The degree of the B-spline with varying k not vector is:
    (a) Increases
    (b) Decreases
    (c) Same
    (d) None.

104. Piecewise polynomial functions are seen in
    (a) Bezier curves
    (b) B-splines
    (c) Cubic splines
    (d) Conics.

105. The accuracy is very high in one of the following solid modelling schemes:
    (a) B-Rep
    (b) C.S.G
    (c) Sweep
    (d) A.S.M.

106. The accuracy of the solid model is low in
    (a) B-Rep
    (b) C.S.G
    (c) Sweep
    (d) Cell decomposition.

107. In a B-spline surface the control mesh is:
    (a) An array of nm control points
    (b) An array of m control points
    (c) An array of n control points
    (d) An array of mn control points.

108. The number of control points required for Bi-quadratic patch
    (a) 12
    (b) 16
    (c) 24
    (d) 32.

109. The convex hull property is satisfied by the following surface:
    (a) Bezier surface
    (b) B-spline
    (c) NURBS
    (d) All of the above.

110. Solid models contain the following information:
    (a) Geometrical
    (b) Topological
    (c) Both (a) and (b)
    (d) None.

111. Mass properties calculation can be done by the following modelling techniques:
    (a) Wire frame
    (b) Surface
    (c) Solid
    (d) Surface manipulation.

112. Polyhedral objects that do not have holes and face is bounded by a single set of connected edges is called
    (a) Simple polyhedron
    (b) Polyhedron through holes
    (c) Complex polyhedron
    (d) None of the above.

113. The minimum body of a shell is:
    (a) Point
    (b) Circle
    (c) Cuboid
    (d) Cube.

114. The topological name for the number of handles or through holes in an object is:
    (a) Shell
    (b) Loop
    (c) Genus
    (d) Polyhedra.

115. The Euler question for satisfying polyhedra is a
    (a) $F - E + V = 2$
    (b) $F + E - V = 2$
    (c) $F + E - V = 12$
    (d) $F - E - V = 2$.

116. Numerical control
    (a) Applies only to milling machines
    (b) Is a method for producing exact number of parts per hour
    (c) Is a method for controlling by means of set of instructions
    (d) All of the above.

117. Computer will perform the data processing functions in
    (a) NC
    (b) CNC
    (c) DNC
    (d) Both (a) and (b).

118. Control loop unit of M.C.U is always
    (a) A hardware unit
    (b) A software unit
    (c) A control unit
    (d) None of the above.

119. B.L.U of an NC system is 0.01 mm. How many pulses are required to move the table of NC to a distance of 10 mm?
    (a) 100
    (b) 1000
    (c) 10
    (d) 10,000.

120. The positional command to move a distance of 20 mm is 20,000. What is B.L.U of the system?
    (a) 0.01
    (b) 0.1
    (c) 1
    (d) 0.001.

121. The step angle of a motor is 2°, then the number of steps per revolution occur
    (a) 150°
    (b) 180°
    (c) 300°
    (d) 360°

122. The first commercial CNC machine was developed in the year
    (a) 1952
    (b) 1970
    (c) 1976
    (d) 1980.

123. The first NC machine was developed in the year
    (a) 1948
    (b) 1950
    (c) 1952
    (d) 1954.

124. The person who is named after the development NC machine
    (a) John parson
    (b) John mark
    (c) Forneaur
    (d) None of the above.

125. The repeatability of NC machine depends on
    (a) Control loop errors
    (b) Mechanical errors
    (c) Electrical errors
    (d) None of the above.

126. Rotation about Z-axis is called
    (a) A-axis
    (b) B-axis
    (c) C-axis
    (d) None of the above.

127. Rotation of spindle is designated by one of the following axis:
    (a) A-axis
    (b) B-axis
    (c) C-axis
    (d) None of the above.

128. In a vertical lathe the axis of rotation of work piece is considered to be
    (a) A-axis
    (b) B-axis
    (c) C-axis
    (d) None of the above.

129. The secondary linear axis along Z-axis is termed as:
    (a) U-axis
    (b) V-axis
    (c) W-axis
    (d) None of the above.

130. The secondary linear movement along X-axis is considered to be
    (a) U-axis
    (b) V-axis
    (c) W-axis
    (d) None of the above.

131. CNC drilling machine is considered to be
    (a) P.T.P controlled machine
    (b) Continuous path controlled machine
    (c) Servo controlled machine
    (d) Adaptive controlled machine.

132. In word address format 'I' indicates
    (a) Simultaneous motion of X and Y axis
    (b) Tool changing function
    (c) Arc centre coordinate parallel to X-axis
    (d) Arc centre coordinate parallel to Y-axis.

133. EIA standard punch tape has
    (a) Eight regular column of holes
    (b) Seven regular column of holes
    (c) Nine regular column of holes
    (d) Ten regular column of holes.

134. In EIA punch type sprocket holes are provided between
    (a) Fourth and fifth columns
    (b) Third and fourth columns
    (c) First and second columns
    (d) None of the above.

135. In standard EIA type, parity check is provided on
    (a) Fifth column
    (b) Sixth column
    (c) First column
    (d) None of the above.

136. In the block No 10 G41 Dd, the value of Dd indicates
    (a) Cutter offset compensation
    (b) Cutter diameter compensation
    (c) Gauge height compensation
    (d) Tool number.

137. The canned cycle G85 stands for
    (a) Boring operation
    (b) Drilling operation
    (c) Tapping operation
    (d) Thread cutting operation.

138. The canned cycle G83 stands for
    (a) Boring operation
    (b) Peck drilling operation
    (c) Milling operation
    (d) None of the above.

139. The fixed cycle G84 stands for
    (a) Turning operation
    (b) Boring operation
    (c) Tapping operation
    (d) None of the above.

140. In the manual part program, the block numbers
    (a) Increase in the steps of 1
    (b) Increase in the steps of 5
    (c) Increase in the steps of 10
    (d) Are in increasing order.

141. A point is the intersection of two lines. The geometric statement in APT is described as:
    (a) POINT/INT OF, L_1, L_2
    (b) POINT/L_1, INTOF, L_2
    (c) POINT/L_1, L_2, INTOF
    (d) POINT/L, INTERSECT, L_2.

142. A plane is defined by $aX + bY + cZ = d$ equation, the appropriate command in APT is:
    (a) PLAN/a, b, c, Equal
    (b) PLANE/a, b, c, EQ.d
    (c) PLAN/a, b, c, d
    (d) PLANE/a, b, c, d.

143. A circle centre is p and radius is r, the appropriate APT statement to define circle is:
    (a) CIRCLE/p, r, CENTER, RADIUS
    (b) CIRCLE/CENTER, p, RADIUS, r
    (c) CIRCLE/RADIUS, p, CENTER, r
    (d) CIRCLE/CENTER, RADIUS, p, r.

144. A circle passing through 3-points, the appropriate APT code is:
    (a) CIRCLE/THRU, P_1, P_2, P_3
    (b) CIRCLE/P_1, THRU, P_2, P_3
    (c) C=CIRCLE/P_1, P_2, P_3
    (d) CIRCLE/P_1, P_2, P_3.

145. Which one of the following is a point-to-point statement?
    (a) GO HOME
    (b) GO/TO
    (c) GO DLTA/
    (d) FROM/.

146. The vocabulary word TLRGT is equivalent to one of the following G-codes:
    (a) G 40
    (b) G 41
    (c) G 42
    (d) G 80.

147. The vocabulary word TLLFT is equivalent to one of the following G-codes:
    (a) G 40
    (b) G 41
    (c) G 42
    (d) G 43.

148. In contouring motion statements, the tool always moves along
    (a) Part surface
    (b) Check surface
    (c) Drive surface
    (d) None of these.

149. In contouring motion statements, the tool will change direction with respect to
    (a) Part surface
    (b) Check surface
    (c) Drive surface
    (d) Change surface.

150. When the drive and check surfaces are tangent to each other, the positional relationship is specified by one of the following words:

(a) To   (b) PAST

(c) PASTAN   (d) TANTO.

151. In an APT statement feed rate is specified as FEED/1000. What are the units of 1000?

(a) Inches per minute   (b) mm/minute

(c) Inches per rev   (d) mm/rev.

152. The G code equivalent to LOAD TL/Ln statement is:

(a) G 06   (b) G 18

(c) M 06   (d) M 09.

153. The functions of a post processor is:

(a) To convert CL DATA file into part program

(b) To convert CL DATA into machine specific part program

(c) To convert CL DATA into APT program

(d) To convert APT program into part program.

154. In an APT program, the CL DATA file contains

(a) Coordinate information of cutter   (b) Machine limits

(c) Machine coordinates   (d) Cutter specifications.

155. Which is the last statement of APT program?

(a) CLOSE   (b) HOME

(c) FINI   (d) FINISH.

156. Which is the first statement of APT program?

(a) MACHIN/   (b) G POST/

(c) CL PRINT/   (d) FROM/.

157. Part surface defined in APT is always

(a) Parallel to cutter axis   (b) Parallel to check surface

(c) Perpendicular to cutter axis   (d) None of these.

158. The units of cutter speed mentioned in the statement SPEED/1200

(a) 1200 rpm   (b) 1200 sfm

(c) 1200 m/min   (d) 1200 m/hr.

159. The following are defined in auxiliary statement:

(a) Geometry of part   (b) Cutter motion

(c) Speed, feed and coolant   (d) Machine, tools, etc.

160. MACRO definition in APT is equivalent to

(a) Sub program   (b) Loop program

(c) Parametric program   (d) Point to point.

161. Which is not the component of a DNC system?

(a) Bulk memory   (b) Central computer

(c) Tape reader   (d) None of the above.

162. Applications of DNC system

(a) Greater security in data transmission

(b) Continuous monitoring of the machining process

(c) Simplified program data management

(d) All the above.

163. The functions of BTR in DNC system is:

(a) When DNC fails, the machine tool can be operated independently

(b) It eliminates the use of tape-reader

(c) It improves programme accuracy

(d) None of the above.

164. Implementation of adaptive control is difficult because of

(a) Complexity of the system

(b) Identification of objective function

(c) Lack of sensor technology

(d) All of the above.

165. The accuracy of interpolation is more in the case of

(a) BTR configuration unit configuration

(b) Special machine control of CNC

(c) Both (a) and (b)

(d) None of these.

166. Which of the following coding structure carries more information?

(a) Mono code   (b) Poly code

(c) Mixed code   (d) None of these.

167. Cellular manufacturing uses the information of

(a) Part-machine incidence matrix   (b) Classification and coding

(c) Coding structure   (d) None of these.

168. OPITZ code consists of

(a) 8 digits   (b) 9 digits

(c) 10 digits   (d) 12 digits.

169. In OPITZ code, from code is an example of

(a) Mixed code   (b) Poly code

(c) Mono code   (d) None of these.

170. In OPITZ code part geometry is represented in

(a) Supplementary code   (b) Secondary code

(c) Primary code   (d) Form code.

171. In OPITZ code, from code consists of

(a) Digits   (b) 5 digits

(c) 6 digits   (d) 9 digits.

172. In OPITZ code, supplementary code consists of

(a) 4 digits   (b) 5 digits

(c) 9 digits   (d) Depends on job complexity.

173. The form code OPITZ coding system consists of

(a) Design attributes   (b) Manufacturing attributes

(c) Both (a) and (b)   (d) Part process sequence.

174. In OPITZ code, type of material is specified in
    (a) Form code
    (b) Supply code
    (c) Design code
    (d) User code.

175. Advantages of cellular manufacturing system
    (a) Reduced step up time
    (b) Reduced MCT
    (c) Reduced WIP
    (d) All of the above.

176. Implementation of group technology can be done by
    (a) Visual inspection
    (b) Classification and coding system
    (c) Production flow analysis
    (d) All of the above.

177. MULTI CLASS coding system is based on
    (a) Hierarchical structure
    (b) Mixed structure
    (c) Poly code structure
    (d) None of these.

178. MULTICLASS system consists of
    (a) 17 digits
    (b) 18 digits
    (c) 19 digits
    (d) 20 digits.

179. A monocode consists of 3 digits and in that each of the digits, the numbers 0 to 9 are used. How many mutually exclusive characters can be stored in the code?
    (a) 10
    (b) 1000
    (c) 110
    (d) 1110.

180. In OPITZ code, plain surfaces are represented in
    (a) 1st digit
    (b) 2nd digit
    (c) 3rd digit
    (d) 4th digit.

181. Cellular manufacturing system is designed on the basis of
    (a) JIT
    (b) MRP
    (c) GT
    (d) Layout.

182. Advantages of GT are:
    (a) Reduced material handling system
    (b) Improved quality
    (c) Employee satisfaction
    (d) All of the above.

183. GT is suitable for production of
    (a) Medium range variety and high range quantity
    (b) High range variety and high range quantity
    (c) Medium range variety and medium range quantity
    (d) Low range variety and low range quantity.

184. KK-3 system consists of
    (a) 21 digits
    (b) 22 digits
    (c) 23 digits
    (d) 25 digits.

185. MULTI CLASS code consists of
    (a) Mono code system
    (b) Polycode system
    (c) Hybrid system
    (d) None of these.

186. For development of a process plan for new parts, the best is:
    (a) Variant process planing
    (b) Generative process planing
    (c) GT planing
    (d) None of these.

187. For development of a process plan for existing parts, the best process plan is:
    (a) Generative process planing
    (b) Variant process planing
    (c) APT
    (d) None of these.

188. When the number of part families are more, then the most suitable process plan is:
    (a) Variant process planing
    (b) Generative process planing
    (c) Manual process planing
    (d) None of these.

189. Generative process plans are generated by means of
    (a) Decision logic
    (b) Geometry based data
    (c) Technology algorithms
    (d) All of the above.

190. Generative process planing is used for
    (a) Production of consistency process plans
    (b) Production of new components
    (c) Integration of automated machining facilities
    (d) All of the above.

191. GARI is based on
    (a) Generative process planing
    (b) Variant process planing
    (c) Manual process planing
    (d) None of these.

192. Retrieval type process planing works on the principle of
    (a) Group technology
    (b) Knowledge base
    (c) Database
    (d) Networking.

193. When variant process planing method is useful?
    (a) Part families are more
    (b) Part families are less
    (b) Batch size is more
    (d) Batch size is less.

194. Improper gauge reading values can be best plotted by
    (a) Bi-modal
    (b) Cliff-like
    (c) Saw toothed
    (d) None of these.

195. C-charts are the best example of
    (a) Poisson distribution
    (b) Binomial distribution
    (c) Normal distribution
    (d) None of these.

196. In CMM, the accuracy depends on
    (a) Probe
    (b) Software
    (c) Machine structure
    (d) All of the above.

197. MRP input requires
    (a) BOM
    (b) Inventory file
    (c) MPS
    (d) All of the above.

198. The basic difference between MRP and MRP-II is:
    (a) Inventory
    (b) BOM
    (c) Finance
    (d) Capacity planning.

199. Inventory record file gives the following information:
    (a) Lot size
    (b) Machine details
    (c) Customer name
    (d) None of these.

200. Bill of material structure is used to
    (a) Calculate net requirements
    (b) Calculate due dates
    (c) Calculate man power requirements
    (d) All of the above.

201. Just in time manufacturing philosphy emphasizes on
    (a) Man power
    (b) Manufacturing
    (c) Profit
    (d) Inventory.

202. Forecasting is used for
    (a) Dependent demand items
    (b) Independent demand items
    (c) Both (a) and (b)
    (d) None of the above.

203. CRP takes material requirements from MRP and converts to
    (a) Standard hours of man power
    (b) Standard hours of machine
    (c) Standard hours of load
    (d) All of the above.

204. Capacity planning is concerned with
    (a) How many machines required
    (b) How much labour required
    (c) Both (a) and (b)
    (d) None of the above.

205. MRP-II system is called as closed loop system because it considers
    (a) Inventory
    (b) Finance
    (c) Man power
    (d) None of the above.

206. Wrist motion of y involves
    (a) Right-to-left rotation of the object
    (b) Up and down rotation of the object
    (c) Twisting of the object about the arm axis
    (d) None of these.

207. Robots are specified by
    (a) Pay load
    (b) Dimension of work envelope
    (c) Degree of freedom
    (d) All of the above.

208. Hydraulic drives are used for a robot when
    (a) High torque is required
    (b) High power is required
    (c) Rapid motion of robot arm
    (d) All of the above.

209. The following type of robot is most suitable for pick and place operations:
    (a) Rectangular
    (b) Cylindrical
    (c) Spherical
    (d) Jointed arm type.

210. Which type of actuator is used for nuclear reactor applications?
   (a) Hydraulic actuator system
   (b) Pneumatic actuator system
   (c) Electric actuator system
   (d) None of the above.

211. No programme code is required in the case of
   (a) Teach pendant method
   (b) Lead through method
   (c) Walk through method
   (d) Off-line method.

212. Which of the following is not a Robot language?
   (a) VAL
   (b) AML
   (c) AS
   (d) GOAL.

213. The following robot operation is used for placing the IC's on a printed circuit board:
   (a) Point-to-point
   (b) Pick and place
   (c) Continuous path
   (d) Controlled path.

214. A large amount of memory and high-speed sampling is needed in the case of.
   (a) Pick and place
   (b) Point to point
   (c) Continuous path
   (d) None of these.

215. To find the actual distance of the object, the following sensors are used:
   (a) Tactile sensors
   (b) Ranger sensors
   (c) Proximity sensors
   (d) Touch sensors.

216. A robot is designated as TLR: TR. It means
   (a) Body of robot having 2 degrees of freedom
   (b) Arm of robot having 3 degrees of freedom
   (c) Arm of robot having 2 degrees of freedom
   (d) None of the above.

217. The work volume is partial sphere in the case of
   (a) Polar configuration
   (b) Cylindrical configuration
   (b) Cartesian configuration
   (d) Jointed-arm type.

218. Flexible manufacturing module consists of
   (a) One NC machine
   (b) Two NC machines
   (c) Three NC machines
   (d) None of these.

219. Transfer lines are used for
   (a) Large variety of parts
   (b) Small variety of parts
   (c) Medium variety of parts
   (d) None of these.

220. AGV uses the following guidence technologies:
   (a) Wire guided system
   (b) Inertial guidence system
   (c) Infrared guidence system
   (d) All of the above.

221. Towing vehicles are used for
   (a) Transportation of bulk quantity items
   (b) Transportation of small quantity items
   (c) Transportation of medium quantity items
   (d) None of the above.

222. In FMS, the tools are identified by means of
    (a) Color code
    (b) Bar code
    (c) PLC
    (d) Digital code.

223. FMS hardware consists of
    (a) Traffic management
    (b) Tooling information
    (c) Work-order files
    (d) Tooling system.

224. Routing flexibility is used primarily to
    (a) Perform variety of operations
    (b) Manage internal changes when equipment break downs
    (c) Absorb changes in the product mix
    (d) None of the above.

225. The hierarchy of FMS control system are:
    (a) FMF, FMM, FMS, FMC
    (b) FMF, FMS, FMC, FMM
    (c) FMS, FMC, FMM, FMF
    (d) None of the above.

226. Arrange the following manufacturing system in the descending order of production rate:
    (a) Transfer machine(TM), Special purpose machines (SPM), Flexible manufacturing system (FMS)
    (b) FMS, SPM, TM
    (c) TM, FMS, SPM
    (d) None of the above.

227. Which one of the following is not the principal objective of FMS?
    (a) Reduced set up time
    (b) Reduced WIP
    (c) Reduced machine up time
    (d) Reduced tool change time.

228. Which of the following is the synchronized material handling system?
    (a) AGV
    (b) Conveyor
    (c) Fixtures
    (d) Trucks.

229. SIMULA programming language is used in
    (a) CAPP
    (b) Simultion
    (c) NC program
    (d) Tool programming.

230. In FMS, manufacturing lead time reduces due to
    (a) Reduced WIP
    (b) Reduced tool change time
    (c) Reduced part quality
    (d) Reduced production rate.

231. Line layout configuration is best suited for
    (a) No back flow of parts
    (b) Back flow of parts
    (c) Parts require high process time
    (d) None of these.

232. Gravity principle of material handling system is used for the transportation of
    (a) Fragile items
    (b) Explosive items
    (c) Brittle items
    (d) Sturdy items.

233. Conveyors are used for transportation of goods between
    (a) Variable path
    (b) Fixed path
    (c) Both (a) and (b)
    (d) None of these.

234. Which of the following is not the principle of material handling system?
    (a) Unit load principle
    (b) Gravity principle
    (c) Max. terminal time principle
    (d) Part orientation principle.

235. For heavy pay load on large distance, the type of AGV used is:
    (a) AGV pallet trucks
    (b) Driver less trains
    (c) AGV unit load carriers
    (d) None of the above.

236. Top-down design does not require
    (a) Step-down refinement
    (b) Loop invariants
    (c) Flow charting
    (d) Modularity.

237. An example of an expert system is:
    (a) A medical diagnosis program
    (b) Finite element analysis program
    (c) Stress analysis program
    (d) Structural analysis program.

238. An expert system
    (a) Simulates the reasoning of a human expert in a particular subject
    (b) Is an application of artificial intelligence research
    (c) Both (a) and (b)
    (d) None of the above.

239. Artificial intelligence is the main characteristic of
    (a) Second generation computers
    (b) Third generation computers
    (c) Fourth generation computers
    (d) Fifth generation computers.

240. Two major processing languages for artificial intelligence are:
    (a) COBOL and LISP
    (b) COBOL and FORTRAN
    (c) LISP and PROLOG
    (d) PROLOG and BASIC.

241. Flexible manufacturing allows for
    (a) Factory management
    (b) Automated design
    (c) Tool design
    (d) Quick and inexpensive product change.

242. A FMS may be
    (a) An automated assembly line
    (b) Expensive to alter
    (c) Difficult to change for new products
    (d) All of the above.

243. Robot is specified by
    (a) Control system
    (b) Axis of movement
    (c) Pay load
    (d) All of the above.

244. Which of the following power source(s) is used for the robot manipulator?
    (a) Hydraulic
    (b) Pneumatic
    (c) Electric
    (d) All of the above.

245. Robots are presently used for which of the following jobs:
    (a) System analysis and design
    (b) CAD/CAM
    (c) Welding, grinding and printing, etc.
    (d) Cutting operations.

246. What is not true about robots?
    (a) They are just like computers
    (b) They are used in oil and mineral exploration
    (c) They can replace human beings in many routine activities
    (d) They can respond to unforeseen circumstances and changed environments.

247. Various computer devices in an online can be connected by the use of
    (a) LAN   (b) WAN
    (c) Satellite   (d) MODEM.

248. A device that converts computer output for transmission over telephone lines is called a/an
    (a) Interface   (b) Interpreter
    (c) MODEM   (d) Port.

249. The physical arrangement of devices in a network is termed as:
    (a) Ergonomics   (b) Topology
    (c) Networking   (d) Pairing.

250. A collection of wires that connect several devices is called
    (a) Link   (b) Bus
    (c) Bidirectional wires   (d) Cable.

251. Which is used for modulation and demodulation?
    (a) Modem   (b) Protocols
    (c) Gateway   (d) Multiplexer.

252. LAN can transmit
    (a) Faster than telecommunications over public telephone lines
    (b) Slower than telecommunications over public telephone lines
    (c) Using coaxial cables
    (d) Both (a) and (b).

253. Devices on one network can communicate with devices on another network via a
    (a) File server   (b) Utility server
    (c) Printer server   (d) Gateway.

254. Artificial intelligence used communications between people and computers is:
    (a) Natural language processing   (b) Decision support
    (c) Symbolic processing   (d) Robotics.

255. A key element of artificial intelligence is:
    (a) Heuristic   (b) Cognition
    (c) Algorithm   (d) Digiton.

256. Artificial intelligence solves problems by processing
    (a) Numbers   (b) Symbols
    (c) Actions   (d) Algorithms.

257. An intelligent robot
    (a) Responds to changes in its environment

    (b) Follows instruction mindlessly

    (c) Possesses no more intelligence than a dish washer

    (d) Does all of the above.

258. A computer program that contains expertise in a particular domain is called an

    (a) Automatic processor   (b) Intelligent planner

    (c) Expert system processor   (d) Operations symbolizer.

259. The knowledge base of an expert system includes both facts and

    (a) Theories   (b) Heuristics

    (c) Algorithms   (d) Analysis.

260. A robot's arm is also known as its

    (a) Actuator   (b) End effector

    (c) Manipulator   (d) Servo mechanism.

261. Which type of actuator generates a good deal of power but tends to be messy?

    (a) Electric   (b) Hydraulic

    (c) Pneumatic   (d) Both (b) and (c).

262. If a robot can alter its own trajectory in response to external conditions, it is considered to be

    (a) Intelligent   (b) Mobile

    (c) Non-servo   (d) Open loop.

263. Programming a robot by physically moving it through the trajectory you want it to follow is called

    (a) Contact sensing control   (b) Continuous-path control

    (c) Pick and place control   (d) Robot vision control.

264. Reasoning from a goal state towards an initial state is called

    (a) Backward-chaining   (b) Bidirectional

    (c) Breadth-first   (d) Heuristic.

265. Which device is mostly associated with automation?

    (a) Flexible manufacturing   (b) Robots

    (c) Computer graphics workstation   (d) NC machine.

266. Choose the basic element for an automated machine tool

    (a) Logic   (b) NC tape programming

    (c) Software   (d) Workstation.

267. Choose the robot component from the following:

    (a) Micro computer   (b) Coaxial cable

    (c) Arm   (d) Software

268. A configuration for a robot is:

    (a) Octagonal   (b) Oblong

    (c) Square   (d) Spherical

269. Flexible manufacturing allows for

    (a) Automated design

(b) Factory management
(c) Tool design and production
(d) Quick and inexpensive product changes.

270. DFMA stands for
(a) Data for manual assembly
(b) Data for manufacturing analysis
(c) Design for manual assembly
(d) Design for manufacture and assembly.

Key to Objective Type Questions

1. (c)	2. (b)	3. (a)	4. (c)	5. (b)	6. (c)	7. (d)	8. (c)
9. (b)	10. (a)	11. (d)	12. (a)	13. (b)	14. (c)	15. (c)	16. (a)
17. (b)	18. (c)	19. (d)	20. (d)	21. (d)	22. (b)	23. (a)	24. (a)
25. (b)	26. (c)	27. (a)	28. (b)	29. (c)	30. (a)	31. (c)	32. (a)
33. (d)	34. (a)	35. (c)	36. (a)	37. (a)	38. (b)	39. (a)	40. (c)
41. (d)	42. (a)	43. (c)	44. (a)	45. (c)	46. (a)	47. (d)	48. (a)
49. (c)	50. (b)	51. (b)	52. (c)	53. (d)	54. (b)	55. (c)	56. (d)
57. (a)	58. (d)	59. (c)	60. (a)	61. (c)	62. (b)	63. (b)	64. (a) and (c)
65. (a)	66. (b)	67. (a)	68. (a)	69. (a)	70. (d)	71. (d)	72. (a)
73. (c)	74. (a)	75. (c)	76. (c)	77. (b)	78. (a)	79. (d)	80. (c)
81. (b)	82. (b)	83. (c)	84. (d)	85. (a)	86. (a)	87. (b)	88. (c)
89. (b)	90. (d)	91. (b)	92. (b)	93. (a)	94. (d)	95. (b)	96. (b)
97. (b)	98. (b)	99. (a)	100. (a)	101. (d)	102. (c)	103. (c)	104. (b)
105. (c)	106. (d)	107. (a)	108. (b)	109. (d)	110. (c)	111. (c)	112. (a)
113. (a)	114. (b)	115. (a)	116. (c)	117. (b)	118. (a)	119. (b)	120. (d)
121. (b)	122. (c)	123. (c)	124. (a)	125. (b)	126. (c)	127. (d)	128. (c)
129. (c)	130. (a)	131. (a)	132. (c)	133. (a)	134. (b)	135. (a)	136. (b)
137. (a)	138. (b)	139. (b)	140. (d)	141. (a)	142. (d)	143. (b)	144. (c)
145. (c)	146. (c)	147. (b)	148. (c)	149. (b)	150. (d)	151. (b)	152. (c)
153. (b)	154. (a)	155. (c)	156. (a)	157. (c)	158. (a)	159. (c)	160. (d)
161. (d)	162. (d)	163. (a)	164. (d)	165. (a)	166. (a)	167. (a)	168. (b)
169. (a)	170. (d)	171. (b)	172. (a)	173. (a)	174. (b)	175. (d)	176. (d)
177. (a)	178. (b)	179. (d)	180. (d)	181. (c)	182. (b)	183. (c)	184. (a)
185. (a)	186. (b)	187. (b)	188. (b)	189. (a)	190. (d)	191. (a)	192. (a)
193. (b)	194. (c)	195. (b)	196. (d)	197. (d)	198. (c)	199. (a)	200. (a)
201. (d)	202. (b)	203. (d)	204. (c)	205. (b)	206. (a)	207. (d)	208. (d)
209. (a)	210. (d)	211. (c)	212. (d)	213. (b)	214. (c)	215. (b)	216. (c)
217. (a)	218. (a)	219. (b)	220. (d)	221. (a)	222. (b)	223. (d)	224. (b)
225. (b)	226. (a)	227. (c)	228. (b)	229. (b)	230. (b)	231. (a)	232. (d)
233. (b)	234. (c)	235. (b)	236. (b)	237. (a)	238. (c)	239. (d)	240. (c)
241. (d)	242. (a)	243. (d)	244. (d)	245. (c)	246. (d)	247. (a)	248. (c)
249. (b)	250. (b)	251. (a)	252. (b)	253. (d)	254. (a)	255. (a)	256. (b)
257. (a)	258. (c)	259. (b)	260. (c)	261. (b)	262. (a)	263. (b)	264. (a)
265. (b)	266. (a)	267. (c)	268. (d)	269. (d)	270. (d)		

INDEX